CHARGED AND REACTIVE POLYMERS

VOLUME I

POLYELECTROLYTES

CHARGED AND REACTIVE POLYMERS

A SERIES EDITED BY

ERIC SÉLÉGNY

VOLUME I

POLYELECTROLYTES

PAPERS INITIATED BY A NATO ADVANCED STUDY INSTITUTE
ON CHARGED AND REACTIVE POLYMERS
HELD IN FRANCE, JUNE 1972

Edited by

ERIC SÉLÉGNY
Université de Rouen, France

Co-edited by

MICHEL MANDEL
University of Leyden, The Netherlands

and

ULRICH P. STRAUSS
Rutgers University, U.S.A.

D. REIDEL PUBLISHING COMPANY
DORDRECHT-HOLLAND / BOSTON-U.S.A.

Library of Congress Catalog Card Number 73–91435

ISBN-13: 978-94-010-2187-6 e-ISBN-13: 978-94-010-2185-2
DOI: 10.1007/978-94-010-2185-2

Published by D. Reidel Publishing Company,
P.O. Box 17, Dordrecht, Holland

Sold and distributed in the U.S.A., Canada, and Mexico
by D. Reidel Publishing Company, Inc.
306 Dartmouth Street, Boston,
Mass. 02116, U.S.A.

TABLE OF CONTENTS

AHARON KATCHALSKY

(1914–1972)

This book is dedicated to the memory of the late Professor Aharon Katchalsky, whose untimely death deprived the scientific community of one of its major representatives only a few days before the opening of the Institute.

INTRODUCTION

This book contains a number of articles inspired by the NATO Advanced Study Institute on 'Charged and Reactive Polymers I' held in France in June 1972. This general title indicates simply the intention of a series. The meeting dealt mainly with the fundamental problems of the physical chemistry of polyelectrolytes in solution. Some of the articles reproduce the lectures exactly as they were delivered. Some others have been modified to a greater or lesser extent, and this as a result of improvements or new inspiration arising from comments and discussions.

In previous larger conferences on macromolecules, polyelectrolytes constituted only a marginal problem and few were the individual communications or short was the time allotted to this subject. In other meetings of a biophysical character the uses of the techniques of charged macromolecules have been exposed with less attention given to the theories or to the creation or interpretation of these techniques. All of us felt that the time had come to enumerate and to evaluate this increasing science of polyelectrolytes which has become of major interest. During the whole period of the Institute physical chemists discussed their mutual problems for more than a week, and often far into the night!

One of the advantages of such an Institute is to enable the Directors and the members of the Scientific Committee to establish a logical order in the lectures; this order has been respected in the present edition.

Articles written in a more general theoretical vein are followed by contributions dealing with equilibrium properties in which the results found by different experimental techniques are confronted with each other as well as with the predictions of existing theories.

The multiple aspects of transport phenomena or non-equilibrium phenomena succeed and are widely illustrated and discussed.

Spectroscopic methods of absorption or of resonance are well represented and certain articles show clearly the future developments that are so necessary for a better understanding of polyelectrolytes at the molecular level.

Some works not strictly dealing with polyelectrolytes in solution but with membranes, for example, have been inserted in the written version close to those dealing with the respective experimental techniques.

According to the now well-established tradition of such meetings, the last few days' lectures of the Institute were devoted to illustrating the applications of charged polymers to other fundamental disciplines (enzymology, photo-chemistry) or their methodology in connected fields. Some of these contributions will be found at the end of the book; their aim is to draw the attention of specialists in one or another of these fields to the interest which might lie in future collaborations or in the borrowing and exchange of theoretical and practical methods.

A few interesting communications presented at the Institute constituted unfortunately only an oral contribution because their authors, absorbed by other pressing professional activities, were unable to submit their manuscripts in time for a not too delayed publication.

The Scientific Committee of the Institute encouraged the presentation of a small number of specialized short communications. Two further afternoons were devoted to the lectures of the annual meeting of the French group of concerted research on polyelectrolytes ('R.C.P. 194') which owes its being to the Centre National de la Recherche Scientifique. These communications have been published elsewhere and consequently are not reproduced here.

During the discussions the need to standardize some basic symbols and definitions (degree of neutralisation, ion-binding, hydrophobic interactions, etc....) became evident. Because of the difficulties encountered and of the necessity of continuity it was decided that for the present time it was preferable to let the authors continue to use the usual symbols of their own schools. However the influence of some of these discussions is reflected in the final versions of several papers. It should be the task of a future meeting or of a specialized commission to undertake and progressively complete this standardization.

Professor Champetier, member of the French Academy of Science, honoured us by delivering the opening lecture of the Institute. The full text of his speech is to be found at the beginning of the book in its original language as it was felt that translation might not do full justice to its high quality. This opening lecture took place in the presence of a number of personalities from the French scientific, academic, political and administrative worlds.

The field of polyelectrolytes can hardly be dissociated from the name of Professor Aharon Katchalsky whose tragic death unfortunately prevented his attending the Institute. He was a member of the Scientific Committee and he would have delivered one of his high-level and most brilliant lectures the secret of which was known only to him. His death is felt by all those connected with our discipline as an almost personal loss. By general and spontaneous accord this book is dedicated to his memory.

The omission of his lecture would have left too large a gulf. Responding to the personal invitation of the editors, Professor Silberberg, who was one of those who knew him the longest and the best, accepted the difficult task of regrouping the essence of the scientific work of Professor Katchalsky on polyelectrolytes.

We should like to acknowledge the financial help and the sponsorship of the Centre National de la Recherche Scientifique, the Délégation Générale à la Recherche Scientifique et Technique, the Scientific Affairs Division of NATO and the University of Rouen (Faculty of Science and ISHN Research Institute) and the National Science Foundation of the United States of America.

It is a pleasure too to acknowledge the help of all those who, either members of the Scientific Committee or not, have contributed to the organization of the Institute or to the material realization of this book: Professor Benoit, Dr Varoqui, the members of the 'R.C.P. 194', Mrs Mallet, Dr Meffroy-Biget, Dr Fenyo, Dr Merle, Mr Demarty, Mr Labbé.

The ladies' programme was conducted by Mrs Sélégny and Mrs Mallet fulfilled the secretarial and on-the-spot translation tasks.

It is now hoped that as a continuation of the Institute this book will prolong and widen its informative and activating action in the field of polyelectrolytes.

M. Moan, Labo. Prof. Wolff, Faculté des Sciences, Labo. Hydrodynamique moléculaire, Avenue Legorgeu, 29 N – Brest
M. Morcellet, Labo. Prof. Loucheux, Université des Sciences & Techniques, B.P. 36, 59 – Villeneuve d'Asq
G. Muller*, Labo. Prof. Sélégny, Faculté des Sciences, Chimie Macromoléculaire, 76 – Mont Sait Aignan
J. Néel, Faculté des Sciences, Labo. Chimie Organique Industrielle, 1, rue de Grainville, 54 – Nancy
M. Plaisance, Labo. Mme Ter-Minassian-Saraga, Faculté de Médecine, 45, rue des Saints Pères, 75 – Paris (6ème)
J. R. Puig, Chef CAPRI, Centre d'Étude Nucléaires Saclay, B.P. 2, 91 – Gif sur Yvette
M. Rinaudo*, Domaine Universitaire, CNRS CERMAV, Cédex 53, 38 – Grenoble
C. Ripoll, Labo. Prof. Sélégny, Faculté des Sciences, Chimie Macromoléculaire, 76 – Mont Saint Aignan
M. A. Rix Montel, Labo. Prof. Vasilescu, Biophysique, Université de Nice, Faculté des Sciences, Parc Valrose, 06 – Nice
B. Roux, Labo. Prof. Hanss, UER Expérimentale de Médecine & Biologie, 1, rue Marcel Cachin, 93 – Bobigny
A. Schmitt, C.R.M. 6, rue Boussingault, 67 – Strasbourg
E. and Mrs. Sélégny*, Labo. Chimie Macromoléculaire, Faculté des Sciences, 76 – Mont Saint Aignan
G. Spach, C.B.M., 45 – La Source Orleans
L. Ter-Minassian-Saraga*, Maître de Recherches, Faculté de Médecine, 45, rue des Saints Pères, 75 – Paris (6ème)
C. Thomas, Labo. Mme Ter-Minassian, 45, rue des Saints Pères, 75 – Paris (6ème)
R. Varoqui*, Maitre de Recherches, C.R.M., 6, rue Boussingault, 67 – Strasbourg
M. Vert*, Labo. Prof. Sélégny, Chimie Macromoléculaire, 76 – Mont Saint Aignan
J. Wallach, Labo. Prof. Hanss, UER Expérimentale de Médecine & Biologie, 1, rue Marcel Cachin, 93 – Bobigny
G. Weill*, C.R.M., 6, rue Boussingault, 67 – Strasbourg
C. Wolff, Faculté des Sciences, Labo. Hydrodynamique moléculaire, Avenue Legorgeu, 29 N – Brest
R. Zana*, maître de Recherches, C.R.M., 6, rue Boussingault, 67 – Strasbourg

Israel

J. Schechter, The Negev Institute, P.O. Box 1025, Beersheva

Italy

G. Barone, Universita di Napoli, Istituto Chimico, via Mezzocannone, 80134, Naples
A. Cesaro, Universita degli studi di Trieste, Istituto di Chimica, Chimica delle macromolecule, Trieste

Comme le souligne très justement la note d'information, la documentation sur les poly-électrolytes et leurs applications est très dispersée. Elle intéresse à la fois la physico-chimie, la biologie, les utilisations techniques, la catalyse chimique ou enzymatique. Bien que les polyélectrolytes avec les protéines aient été parmi les premiers composés macromoléculaires reconnus, leur étude s'est heurtée à de très nombreuses difficultés théoriques et expérimentales. Aussi, pendant plusieurs décennies, les recherches à leur sujet ont-elles été un peu délaissées, sauf dans le domaine structural et conformationnel. Grâce au développement des techniques instrumentales et par suite de l'intérêt des polymères réactifs ayant des actions catalytiques dont l'importance n'est pas à rappeler, les recherches sur les poly-électrolytes connaissent un regain d'activité. La Délégation Générale à la Recherche Scientifique a été conduite à créer un Comité spécial d'actions concertées que préside mon ancien collaborateur le Professeur J. Néel, devenu lui aussi mon collègue et ami. Des résultats particulièrement importants ont été obtenus au cours de ces dernières années. Le moment était venu de les confronter et d'en tirer des conclusions qui seront des promesses pour de futurs nouveaux développements scientifiques et techniques. Je souhaite que dans tous les domaines abordés puissent s'établir de fructueuses discussions et se nouer de nombreuses relations personnelles.

Je tiens à remercier, au nom des organisateurs, tous les participants à ces journées dont certains ont accepté de longs déplacements malgré les lourdes charges de leurs fonctions.

Le plaisir de nous retrouver est cependant tristement assombri par l'absence d'un collègue dont nous admirions la haute réputation scientifique et qui était un animateur écouté et apprécié des réunions comme celle-ci: Aharon Katchalsky, victime de l'inconcevable tuerie de l'aérodrome de Tel Aviv ne sera pas parmi nous. Sa tragique disparition, abominable exemple de l'aveuglement de certains partisans qui croient servir une cause en se livrant à des actes d'une barbarie que notre raison ne peut pas comprendre, prive la science d'un savant dont les apports ont été d'une importance capitale. Il comptait parmi les spécialistes les plus réputés des poly-électrolytes. Il faisait partie de notre petite équipe de la première heure qui avait été séduite par les concepts nouveaux de la chimie macromoléculaire et qui lui avait donné sa signification. Nous ressentirons lourdement son absence. Mais la mort absurde d'Aharon Katchalsky nous prive aussi d'un homme dont nous connaissions les qualités de coeur et de courage. Permettez-moi, avant d'ouvrir cette réunion, de vous demander de vous recueillir quelques instants pour honorer sa mémoire.

Merci Mesdames et Messieurs. Je souhaite bonne chance et travail fructueux aux Journées de l'Institut d'Études Avancées sur les Polymères chargés et réactifs.

behavior of polyelectrolytes, notably with polarizable interfaces. Together with Miller [16] electrochemical studies, both experimental and theoretical, were undertaken concerning the surface accommodation of charged macromolecules of various chemical constitution.

Interestingly enough, the conformational changes which give rise to the polyelectrolyte effect on the viscosity of the system are among the least well described by theory since the strongly non-Markoff character of the chains prevents a direct approach to the problem. On the other hand, the possibility of effecting mechanical work by changes of pH had continued to fascinate Katchalsky [17, 18] even though his original idea with Kuhn [18] that muscle action depended on this effect turned out to be wrong. The whole field of mechanochemistry [19] and the systematic investigation of polyelectrolyte gels and fibers [20] resulted from these early interests.

Many electrostatically induced conformational changes in polyelectrolyte systems involve site-site interactions of other character in addition. Under such circumstances it is a frequent occurrence that the changes are accompanied by a conformational hysteresis which effectively preserves the state until the system can be triggered out of it by a major environmental change. It is likely that memory processes share this character and for this reason Aharon Katchalsky, with Oplatka [21], Cox [22] and Neumann [23], has paid much attention to questions of molecular hysteresis, particularly in nucleic acid systems, in recent years.

These and other polyelectrolyte systems are increasingly being studied these days, when it is realized to what extent many biologically important processes are influenced by charge-charge interactions. Aharon Katchalsky's contribution to the field, his involvement in practically all the aspects where it is now finding its most intensive application, and in particular his fundamental work classifying the essentials of polyelectrolyte behavior, have made his name almost synonymous with the field itself. His untimely and tragic death leaves a gap which will not easily be filled.

References

1. Katchalsky, A. and Spitnik, P.: *J. Polymer Sci.* **2**, 432–446, 487. (1947).
2. Katchalsky, A. and Alexandrowicz, Z.: *J. Polymer Sci.* **A-1**, 2093. (1963a).
   Alexandrowicz, Z. and Katchalsky, A.: *J. Polymer Sci.* **A-1**, 3231 (1963b).
3. Kuhn, W., Künzle, O. and Katchalsky, A.: *Bull. Soc. Chim. Belg.* **57**, 421 (1948a)
   Kuhn, W., Künzle, O., and Katchalsky, A.: *Helv. Chim. Acta* **31**, 1994 (1948b).
   Katchalsky, A., Künzle, O., and Kuhn, W.: *J. Polymer Sci.* **5**, 283 (1950).
4. Katchalsky, A. and Lifson, S.: *J. Polymer Sci.* **11**, 409 (1953).
5. Fuoss, R. M., Katchalsky, A., and Lifson, S.: *Proc. Natl. Acad. Sci.* **37**, 579 (1951).
   Lifson, S. and Katchalsky, A.: *J. Polymer Sci.* **13**, 43 (1954).
6. Katchalsky, A., Alexandrowicz, Z., and Kedem, O.: in B. E. Conway and R. G. Barradas (eds.), *Chemical Physics of Ionic Solutions*, Wiley, New York, 1966, p. 295.
7. Katchalsky, A.: *Pure Appl. Chem.* **26**, 327 (1971).
8. Eisenberg, H.: *J. Polymer Sci.* **30**, 47 (1958).
   Sachs, S. B., Raziel, A., Eisenberg, H., and Katchalsky, A.: *Trans. Faraday Soc.* **65**, 77 (1969).
9. Katchalsky, A., Shavit, N., and Eisenberg, H.: *J. Polymer Sci.* **13**, 69 (1954).
10. Katchalsky, A., Mazur, J., and Spitnik, P.: *J. Polymer Sci.* **23**, 513 (1957).
11. Mazur, J., Silberberg, A., and Katchalsky, A.: *J. Polymer Sci.* **35**, 43 (1959).

12. Katchalsky, A. and Miller, I. R.: *J. Polymer Sci.* **13**, 57 (1954).
13. Katchalsky, A. and Zwick, M.: *J. Polymer Sci.* **16**, 221 (1955).
14. Michaeli, I.: *J. Polymer Sci.* **48**, 291 (1960).
15. Katchalsky, A. and Spitnik, P.: Interactions entre polyélectrolytes, Colloques Internationaux du Centre National de la Recherche Scientifique, No. LVII, *Quelques aspects généraux de la science des macromolécules*, Strasbourg, Editions du Centre National de la Recherche Scientifique, Paris 1955, p. 103.
16. Katchalsky, A. and Miller, I. R.: *J. Phys. Coll. Chem.* **55**, 1182 (1951).
Miller, I. R. and Katchalsky, A.: in Proc. 2nd Intern. Congr. of Surface Activity, *Gas/Liquid & Liquid/Liquid Interface*, Butterworths, London, 1957, p. 159.
Miller, I. R. and Katchalsky, A.: *Differential Capacity. Bull. Res. Counc. Israel* **7A**, 225 (1958).
Miller, I. R. and Katchalsky, A.: in J. Th. G. Overbeek (ed.), *Physics & Physical Chemistry of Surface Active Substances*, Gordon & Breach, New York, 1967, p. 275.
17. Katchalsky, A.: *Experientia* **5**, 319 (1949).
Katchalsky, A. and Eisenberg, H.: *Nature* **166**, 267 (1950).
18. Kuhn, W., Hargitay, B., Katchalsky, A. and Eisenberg, H.: *Nature* **165**, 514 (1950)
19. Katchalsky, A.: *J. Polymer Sci.* **7**, 393 (1951).
Katchalsky, A., Lifson, S., Michaeli, I., and Zwick, M.: in A. Wassermann (ed.), *Contractile Polymers*, Pergamon Press, London, 1960, p. 1.
Katchalsky, A. and Oplatka, A.: in *Handbook of Sensory Physiology*, Vol. 1: *Principles of Receptor Physiology* (ed. by W. R. Loewenstein), Springer-Verlag, Berlin 1971, p. 1.
20. Katchalsky, A.: in *Prog. Biophys. Biophysical Chem.* Vol. 4, Pergamon Press, London, 1954, p. 1.
Katchalsky, A. and Zwick, M.: *J. Polymer Sci.* **16**, 221 (1955).
Yonath, J., Oplatka, A., and Katchalsky, A.: in *Structure and Function of Connective and Skeletal Tissue*, Butterworths, London, 1965, p. 381.
Steinberg, I. Z., Oplatka, A., and Katchalsky, A.: *Nature* **210**, 568 (1966).
Katchalsky, A. and Michaeli, I.: *J. Polymer Sci.* **15**, 69 (1955).
21. Katchalsky, A. and Oplatka, A.: *Israel J. Medical Sci.* **2**, 4 (1966).
22. Cox, R. A. and Katchalsky, A.: *Biochem. J.* **126**, 1039 (1972).
23. Neumann, E., and Katchalsky, A.: *Ber. Bunsengesellsch. Physikalische Chemie* **74**, 868 (1970).
Neumann, E. and Katchalsky, A.: *Proc. 1st Europ. Biophys. Congr.* **6**, 91. (1971).
Katchalsky, A. and Neumann, E.: *Int. J. Neurosci.* **3**, 175 (1972).
Neumann, E. and Katchalsky, A.: *Proc. Natl. Acad. Sci. U.S.A.*, **69**, 993 (1972).
Revzin, A., Neumann, E., and Katchalsky, A.: *J. Mol. Biol.* **79**, 95 (1973).
Revzin, A. Neumann, E., and Katchalsky, A.: 'Metastable Secondary Structures in Ribosomal RNA. II: A New Method for Analyzing the Titration Behavior of rRNA, *Biopolymers*, in press.

LIMITING LAWS FOR EQUILIBRIUM AND TRANSPORT PROPERTIES OF POLYELECTROLYTE SOLUTIONS

GERALD S. MANNING

School of Chemistry, Rutgers University, New Brunswick, N.J., U.S.A.

It is proposed here to discuss the interactions of point ions with infinitely long line charges against the background of a dielectric continuum [1–8]. It is expected, both on a priori and empirical grounds, that this model is a good representation of a polyelectrolyte solution at low concentrations. Further discussion of the model is deferred to the end of the article. Fairly extensive comparions of published data with the predictions of the model are given in References [1] and [7]. Further comparisons are given below, but emphasis is primarily directed at data which serve to illustrate specific points.

A rigorous analysis of the model indicates that two modes of interaction must be considered: the Debye-Hückel 'ion atmosphere' (adapted to cylindrical geometry). and the condensation of counterions on the line charge.

1. The Debye-Hückel Free Energy

We begin by a calculation of the (Helmholtz) free energy in the Debye-Hückel approximation. The Debye screening parameter κ is defined by

$$\kappa^2 = (4\pi e^2/\varepsilon kT) \sum_{i=1}^{\nu} n_i z_i^2, \tag{1}$$

where e is the charge on a proton, ε the dielectric constant, k Boltzmann's constant, T the absolute temperature, n_i the concentration in (molecule cm^{-3}) of the ith species of small ion, z_i its valence, and ν the number of such species (the polyion is not included in the sum). Denote the linear charge density of the line charge by β. Then the potential $\psi(\varrho)$ at a distance ϱ from the line charge (taken along the x-axis) is given by a superposition of screened Coulomb (Debye-Hückel) potentials from infinitesimal segments of length dx:

$$\begin{aligned}\psi(\varrho) &= \frac{\beta}{\varepsilon} \int_{-\infty}^{\infty} \left(\frac{\exp[-\kappa(\varrho^2 + x^2)^{1/2}]}{(\varrho^2 + x^2)^{1/2}} \right) \mathrm{d}x \\ &= (2\beta/\varepsilon)\, K_0(\kappa\varrho),\end{aligned} \tag{2}$$

where K_0 is the zeroth-order modified Bessel function of the second kind [1, 9]. This function has the asymptotic behavior,

$$K_0(\kappa\varrho) \sim -\ln \kappa\varrho, \; \kappa\varrho \to 0, \tag{3}$$

Eric Sélégny (ed.), Polyelectrolytes, 9–37. All Rights Reserved.

unity to be consistent with the approximation mentioned above that electrostatic interactions among small ions are neglected except insofar as they influence their interaction with the polyion.

The procedure is first to eliminate n_{Cl^-} in Equations (13) and (14) by using Equation (12). Then, Equations (13) and (14) are differentiated with respect to n_e, noting that the right sides of these equations are independent of n_e. The resulting equations are evaluated at $n_e=0$, noting that in this limit $n_{\mathrm{Mg}^{++}}$ and n_{Na^+} reduce to $n'_{\mathrm{Mg}^{++}}$ and n'_{Na^+}, respectively. The derivatives $\partial\gamma_{\mathrm{Mg}^{++}}\gamma^2_{\mathrm{Cl}^-}/\partial n_e$ and $\partial\gamma_{\mathrm{Na}^+}\gamma_{\mathrm{Cl}^-}/\partial n_e$ are calculated from Equation (9) and evaluated at $n_e=0$. With the definitions,

$$x = (\partial n_{\mathrm{Mg}^{++}}/\partial n_e)_{n_e=0}, \tag{15}$$

$$y = (\partial n_{\mathrm{Na}^+}/\partial n_e)_{n_e=0}, \tag{16}$$

and

$$\delta = n'_{\mathrm{Mg}^{++}}/n'_{\mathrm{Cl}^-}, \tag{17}$$

the following two simultaneous equations are then arrived at:

$$(1+4\delta)\,x + 2\delta y = (2+\tfrac{3}{2}\xi)\,\delta \tag{18}$$

$$(1-2\delta)\,x + (1-\delta)\,y = \tfrac{1}{2}(1+\tfrac{1}{2}\xi) - (1+\tfrac{3}{4}\xi)\,\delta. \tag{19}$$

In obtaining these equations, δ was treated as a small quantity.

For $\delta=0$, i. e., for the case of no divalent ions, the equations reduce to $x=0$, $y=\frac{1}{2}(1+\frac{1}{2}\xi)$. Therefore, to first order, x may be set equal to $A\delta$ and y to $[\frac{1}{2}(1+\frac{1}{2}\xi)+B\delta]$, whereupon, again to first order, Equations (18) and (19) become,

$$A\delta + (1+\tfrac{1}{2}\xi)\,\delta = (2+\tfrac{3}{2}\xi)\,\delta \tag{20}$$

$$A\delta + B\delta = -\tfrac{1}{2}(1+\xi)\,\delta. \tag{21}$$

The factor δ cancels in each of the equations, and A and B are easily determined. Then,

$$x = (1+\xi)\,\delta, \tag{22}$$

$$y = \tfrac{1}{2}(1+\tfrac{1}{2}\xi) - \tfrac{3}{2}(1+\xi)\,\delta. \tag{23}$$

With the definitions of x, y, and δ provided by Equations (15)–(17), the ionic distributions are obtained as,

$$\lim_{n_e\to 0}\frac{n_{\mathrm{Mg}^{++}} - n'_{\mathrm{Mg}^{++}}}{n_e} = (1+\xi)\,(n'_{\mathrm{Mg}^{++}}/n'_{\mathrm{NaCl}}) \tag{24}$$

$$\lim_{n_e\to 0}\frac{n_{\mathrm{Na}^+} - n'_{\mathrm{Na}^+}}{n_e} = \tfrac{1}{2}(1+\tfrac{1}{2}\xi) - \tfrac{3}{2}(1+\xi)\,(n'_{\mathrm{Mg}^{++}}/n'_{\mathrm{NaCl}}). \tag{25}$$

These distributions are valid to first order in δ, i.e. excess univalent salt.

As with Equation (10) there might be a strong temptation to consider the left side of Equations (24) and (25) as representing, respectively, the number of Mg^{++} and

Na^+ ions bound to the polyion per charged group. One could even get carried away and interpret the dependence of Equation (25) on $n'_{Mg^{++}}$ as 'a decrease in the number of bound Na^+ due to the competition of the Mg^{++} ions with the Na^+ for sites on the polyion.' Such an interpretation, however, is completely wrong; the effects are due to the diffuse Debye-Hückel atmosphere. Actually, in this case, the effect persists even if $\xi = 0$, that is, even when all activity coefficients are unity; the electroneutrality constraint given by Equation (12) is sufficient to cause an asymmetric distribution of small ions.

2. Counterion Condensation

Consider now the behavior of a point ion of valence z in the immediate vicinity of a line charge. If ϱ is its distance from the line charge (of linear charge density β), then its electrostatic energy u is

$$u = -ze(2\beta/\varepsilon)\ln\varrho. \tag{26}$$

The screening effect of other ions is neglected, as ϱ is considered to be small. Let I be that part of the statistical mechanical phase integral which corresponds to the region close to the line charge:

$$I = \int_0^{0+} \exp(-u/kT)\, 2\pi\varrho\, d\varrho. \tag{27}$$

Substitution of Equation (26) together with Equation (8) yields,

$$I = 2\pi \int_0^{0+} \varrho^{(1 \pm 2z\xi)}\, d\varrho, \tag{28}$$

where the upper sign applies to a cationic polyion ($\beta > 0$), the lower to a polyanion ($\beta < 0$).

For definiteness, consider a polycation, so that the exponent in the integrand is $1 + 2z\xi$. Suppose first that $z > 0$, i.e., the point ion is a co-ion. Then, as the upper limit tends to zero, I also tends to zero; the behavior is regular. Now suppose that the point ion is a counterion, $z < 0$. If $\xi < |z|^{-1}$, the exponent $1 + 2z\xi$ falls between -1 and 1, and the behavior at $\varrho = 0$ is still regular. But if

$$\xi \geqslant |z|^{-1} \tag{29}$$

then the exponent equals -1 or less, and I diverges to $+\infty$ at the lower limit. The same conclusion is reached for a polyanion: for a co-ion, $z < 0$, the behavior of I is regular; for a counterion, $z > 0$, I diverges to $+\infty$ if the condition (29) is satisfied. The physical interpretation of the divergence is that the probability (proportional to the phase integral) of a counterion of valence z being located on the line charge is unity if $\xi \geqslant |z|^{-1}$; the counterion 'condenses' on the polyion.

Because of the condensation phenomenon one must carefully distinguish between the stoichiometric value of ξ, that is, the value computed from Equation (8) with only the charged groups on the polyion contributing to β, and the lower *net* value of ξ computed by considering the charge of the condensed counterions as part of the polyion charge density. Counterions of valence z will condense on the line charge only until the *net* value of ξ is lowered to the value $|z|^{-1}$; counterions in excess of the number thus required will remain in solution.

Of great interest is the situation for which two species of counterions of different valence are present. For definiteness, consider again Mg^{++} and Na^{+} counterions. If $\xi > 1$, the above criterion for condensation of either species is satisfied. However, it is evident from a consideration of Equation (28) that the counterion with larger valence causes the stronger divergence of the phase integral; the probability that Mg^{++} will condense is infinitely greater than the corresponding probability for Na^{+}. Therefore if there are sufficiently many Mg^{++} counterions, they will condense until the net value of ξ is lowered to 1/2; excess Mg^{++} as well as all Na^{+} will remain in solution. If there are only as many Mg^{++} ions as are required to lower ξ_{net} to a value between 1/2 and 1, then all the Mg^{++} will condense, whereas all the Na^{+} will remain in solution (since $\xi_{net} < 1$). Finally, if condensation of all the Mg^{++} results in a value of ξ_{net} which is still greater than unity, then sufficiently many Na^{+} counterions will condense in addition to all the Mg^{++} to lower the value of ξ_{net} to unity.

Attempts at spectroscopic confirmation of the condensation of counterions have not yet been undertaken*. Experimental studies have been made, however, which strongly suggest that condensation does indeed occur in typical polyelectrolyte systems. Ikegami [11] studied changes in the refractive index when NaCl was added to solutions of polyacrylic acid partially neutralized by tetrabutylammonium hydroxide. His data are consistent with the hypothesis that no Na^{+} ions are bound when the degree of neutralization α is less than 0.35, the value which for polyacrylic acid corresponds to $\xi = 1$, the critical charge density for univalent counterions. However, when α exceeds 0.35, the data suggest that Na^{+} ions are bound and that the concentration of bound Na^{+} is $(\alpha - 0.35)$ Cp, where Cp is the monomeric concentration of the polyacrylic acid, i.e., a concentration just sufficient to lower the net value of ξ to unity. In a similar experiment Zana *et al.* [12] demonstrated the onset of ultrasonic absorption increments at $\xi = 1$ for a series of carboxymethyl celluloses of varying degrees of substitution. Armstrong *et al.* [13] found that the electrophoretic mobility of DNA does not change when acridine dye cations are incorporated into the helix, a result which they noted is consistent with the requirement that condensed small counterions are released as dye cations are bound to maintain the net charge density at a constant value.

The concept of counterion condensation was discovered independently by a number of workers, in particular, Oosawa's group [14], Jackson and Coriell [15], and MacGillivray [16]. The author's indebtedness to Onsager in this connection has been acknowledged [1]. The author's particular contribution was to combine the conden-

* While this article was in press, van der Klink, Zuiderweg, and Leyte (to be published) have apparently found such direct confirmation by means of nuclear magnetic relaxation.

sation concept with the Debye-Hückel effect into a rigorous and comprehensive theory of the behavior of polyelectrolyte solutions. This theory will now be described.

3. Synthesis of the Two Effects

The Debye-Hückel theory for simple ionic solutions is known to provide 'limiting laws' for thermodynamic properties; that is, the expressions derived from the Debye-Hückel free energy are asymptotically valid as the concentration tends to zero. Accordingly, the Debye-Hückel free energy in Equation (6) also should be asymptotically accurate at zero concentration – unless some feature arises for line charges which is absent for simple ionic solutions. We have just seen that when the charge density parameter ξ is less than the reciprocal of the magnitude of the highest valence among the counterions present, the behavior of the interactions is regular. We may then conclude that Equation (6) is a limiting law under this condition.

When ξ is larger than the critical value, however, a new feature does arise; counterions of valence z condense on the line charge until the net value of ξ is lowered to the critical value or until the supply of such counterions from the solution is totally depleted. After the condensation is taken into account, the interaction of the line charge with the counterions which remain in solution becomes regular, so that at infinite dilution Equation (6) is a limiting law for the free energy of interaction between small ions in solution on the one hand and the entity consisting of line charge plus condensed counterions on the other. In fact, since in the consideration of the phase integral defined by Equation (27), the potential u given in Equation (26) was taken as unaffected by the presence of other small ions, an educated intuition would indicate that the condensation itself is a limiting law, valid only at infinite dilution [3]. Thus, all aspects underlying the theoretical expressions to be derived below are 'limiting laws'.

As important examples let us return to the case represented by Equations (10), (24), and (25). These equations are now understood to be valid at sufficiently low ionic strengths provided that $\xi<\frac{1}{2}$. For the case $\xi>\frac{1}{2}$, which in practice is the more relevant, modifications must be made. First suppose that $n_{\mathrm{Mg}^{++}}\leqslant\frac{1}{2}(1-\frac{1}{2}\xi^{-1})\,n_e$; that is, $n_{\mathrm{Mg}^{++}}$ does not exceed the concentration which would result in $\xi_{\mathrm{net}}=\frac{1}{2}$ ($n_{\mathrm{Mg}^{++}}$ is the total, i.e. stoichiometric, Mg^{++} concentration). For this condition, all the Mg^{++} ions are condensed on the polyions, so the Mg^{++} activity (and hence activity coefficient) is zero:

$$(n_{\mathrm{Mg}^{++}}-\gamma_{\mathrm{Mg}^{++}}\,n_{\mathrm{Mg}^{++}})/n_e=n_{\mathrm{Mg}^{++}}/n_e,\ [\xi>\tfrac{1}{2},\ n_{\mathrm{Mg}^{++}}/n_e\leqslant\tfrac{1}{2}(1-\tfrac{1}{2}\xi^{-1})]\,. \tag{30}$$

For the case $n_{\mathrm{Mg}^{++}}/n_e>\frac{1}{2}(1-\frac{1}{2}\xi^{-1})$, the value of ξ_{net} is $\frac{1}{2}$, and the concentration of uncondensed Mg^{++}, $n^{(u)}_{\mathrm{Mg}^{++}}$, is

$$n^{(u)}_{\mathrm{Mg}^{++}}=n_{\mathrm{Mg}^{++}}-\tfrac{1}{2}(1-\tfrac{1}{2}\xi^{-1})\,n_e\,. \tag{31}$$

Species 2 of concentration n_2 and valence z_2. It is assumed that $|z_2| > |z_1|$. The electroneutrality condition is expressed by

$$n_e = n_1 |z_1| + n_2 |z_2| . \tag{38}$$

Equivalent counterion mole fractions are defined,

$$x_i = n_i |z_i| / n_e, \quad i = 1, 2, \tag{39}$$

with

$$x_1 + x_2 = 1 . \tag{40}$$

As will be seen, the less general but more common system of only one counterion species may be obtained from the general results by setting x_1 or x_2 equal to zero. The following analysis, although not difficult, is intricate and must proceed by consideration of separate cases.

Case 1, $\xi < |z_2|^{-1}$. For this case there are no condensed counterions. From Equations (1) and (6) the excess free energy is the Debye-Hückel expression,

$$F^{\mathrm{ex}} = - kT\xi N_e \ln \kappa , \tag{41}$$

where N_e is the total number of charged polyion groups, and

$$\kappa^2 = (4\pi e^2/\varepsilon kT)(n_1 z_1^2 + n_2 z_2^2). \tag{42}$$

The activity coefficients of the counterion species are given by direct application of Equation (9) with use of Equation (39),

$$\ln \gamma_i = - \tfrac{1}{2}\xi z_i^2 (x_1 |z_1| + x_2 |z_2|)^{-1}, \quad i = 1, 2. \tag{43}$$

The excess osmotic pressure is

$$\pi^{\mathrm{ex}} = - (\partial F^{\mathrm{ex}}/\partial V)_{T, N_1, N_2, N_e} . \tag{44}$$

Then, from Equation (6),

$$\pi^{\mathrm{ex}} = - \tfrac{1}{2}kT\xi n_e . \tag{45}$$

For practical purposes, since the concentration of polyelectrolytes (as opposed to the equivalent concentration n_e) is negligible compared to the counterion concentration, the osmotic coefficient ϕ may be defined by

$$\pi = kT\phi (n_1 + n_2), \tag{46}$$

where π is the osmotic pressure. Since π^{ex} is defined by

$$\pi = \pi^{\mathrm{ideal}} + \pi^{\mathrm{ex}}, \tag{47}$$

where

$$\pi^{\mathrm{ideal}} = kT(n_1 + n_2), \tag{48}$$

it follows that

$$\pi^{\mathrm{ex}} = kT(\phi - 1)(n_1 + n_2). \tag{49}$$

Comparison of Equations (45) and (49) yields an expression for the osmotic coefficient,

$$\phi = 1 - \tfrac{1}{2}\xi\,(x_1\,|z_1|^{-1} + x_2\,|z_2|^{-1})^{-1}. \tag{50}$$

Case 2, $|z_2|^{-1} \leqslant \xi < |z_1|^{-1}$. For this case no counterions of Species 1 are condensed. There are two subcases:

Subcase 2a, $0 \leqslant x_1 \leqslant |z_2|^{-1}\xi^{-1}$. For this subcase, $n_1|z_1| \leqslant |z_2|^{-1}\xi^{-1}n_e$, so that there are sufficiently many condensed counterions of Species 2 to lower the net value of ξ to the critical value $|z_2|^{-1}$:

$$\xi_{\text{net}} = |z_2|^{-1}. \tag{51}$$

The concentration of polyion charged groups *not* compensated by condensed counterions is

$$n_e^{(u)} = |z_2|^{-1}\xi^{-1}n_e, \tag{52}$$

and the concentration of uncondensed counterions of Species 2 is

$$n_2^{(u)}\,|z_2| = n_e^{(u)} - n_1\,|z_1|. \tag{53}$$

The activity coefficient of Species 1 may be obtained from Equation (9) by replacing ξ by ξ_{net}, n_e by $n_e^{(u)}$, and n_2 by $n_2^{(u)}$:

$$\ln\gamma_1 = -\tfrac{1}{2}(z_1/z_2)^2\,[1 - \xi x_1\,(|z_2| - |z_1|)]^{-1}. \tag{54}$$

The starting point for the calculation of the activity coefficient of Species 2 is the statement that the activity of Species 2 equals the activity of the uncondensed portion of Species 2:

$$\gamma_2 n_2 = \gamma_2^{(u)} n_2^{(u)}. \tag{55}$$

The activity coefficient $\gamma_2^{(u)}$ of the uncondensed counterions of Species 2 is obtained from Equation (9) in the same way as γ_1. Then

$$\ln\gamma_2 = -\tfrac{1}{2}[1 - \xi x_1\,(|z_2| - |z_1|)]^{-1} + \ln[(|z_2|^{-1}\xi^{-1} - x_1)/(1 - x_1)] \tag{56}$$

The osmotic coefficient may be calculated by similar reasoning. The osmotic pressure is

$$\pi = kT\phi^{(u)}\,(n_1 + n_2^{(u)}), \tag{57}$$

where $\phi^{(u)}$ is the osmotic coefficient for a system with $\xi = \xi_{\text{net}}$, $n_e = n_e^{(u)}$, $n_2 = n_2^{(u)}$, and $n_1 = n_1$. It may be calculated from Equation (50) by replacing ξ by $|z_2|^{-1}$, x_1 by $n_1|z_1|/n_e^{(u)}$, and x_2 by $n_2^{(n)}|z_2|/n_e^{(u)}$. But the osmotic pressure may also be expressed in terms of the observable osmotic coefficient ϕ,

$$\pi = kT\phi\,(n_1 + n_2). \tag{58}$$

Comparison of Equations (57) and (58) then gives

$$\phi = [1 - \tfrac{1}{2}(1 + x_1\xi\,|z_2|\,(z_2 z_1{}^{-1} - 1))^{-1}] \times \\ \times [(|z_2|^{-1}\xi^{-1} + x_1\,(z_2 z_1{}^{-1} - 1))/(1 + x_1\,(z_2 z_1{}^{-1} - 1))] \tag{59}$$

Subcase 2b, $|z_2|^{-1}\xi^{-1} \leqslant x_1 \leqslant 1$. For this subcase there are insufficiently many counterions of Species 2 to lower the net value of ξ to $|z_2|^{-1}$, even after condensation of all such counterions. Since all counterions of Species 2 are condensed, the activity of that species is zero. But the activity equals $\gamma_2 n_2$. Hence,

$$\gamma_2 = 0. \tag{60}$$

For this subcase, since all of Species 2 is condensed and all of Species 1 is uncondensed,

$$n_e^{(u)} = n_e - n_2 |z_2| = x_1 n_e. \tag{61}$$

The ratio ξ_{net}/ξ equals $n_e^{(u)}/n_e$, so

$$\xi_{net} = x_1 \xi. \tag{62}$$

The activity coefficient of Species 1 is therefore obtained from Equation (9) by replacing ξ with ξ_{net}, n_e by $n_e^{(u)}$, and n_2 by zero:

$$\ln \gamma_1 = -\tfrac{1}{2} |z_1| \xi x_1. \tag{63}$$

The osmotic pressure π is the same as for a system with $n_2 = 0$, $\xi = \xi_{net}$, $n_e = n_e^{(u)}$, and $n_1 = n_1$. The excess osmotic pressure for such a system is given by Equation (44) with F^{ex} calculated from Equation (41) by making the appropriate replacements. The ideal contribution is kTn_1. Then,

$$\pi = kTn_1 (1 - \tfrac{1}{2} x_1 \xi |z_1|). \tag{64}$$

But, in terms of the osmotic coefficient ϕ, π is still given by Equation (46). Hence,

$$\phi = x_1 |z_1|^{-1} (1 - \tfrac{1}{2} x_1 \xi |z_1|) (x_1 |z_1|^{-1} + x_2 |z_2|^{-1})^{-1}. \tag{65}$$

Case 3, $\xi \geqslant |z_1|^{-1}$. There are three subcases.

Subcase 3(a), $0 \leqslant x_1 \leqslant |z_2|^{-1}\xi^{-1}$. This subcase is treated exactly like Subcase 2(a), since $\xi_{net} = |z_2|^{-}$ and no counterions of Species 1 are condensed. The activity coefficients are given by Equations (54) and (56), the osmotic coefficient by Equation (59).

Subcase 3(b), $|z_2|^{-1}\xi^{-1} \leqslant x_1 \leqslant |z_1|^{-1}\xi^{-1}$. This subcase is treated exactly like Subcase 2(b), since no counterions of Species 1 are condensed, while all counterions of Species 2 are condensed. The activity coefficients are given by Equation (60) and (63), the osmotic coefficient by Equation (65).

Subcase 3(c), $|z_1|^{-1}\xi^{-1} \leqslant x_1 \leqslant 1$. For this subcase, all counterions of Species 2 are condensed, but the net value of ξ due to this condensation is still greater than the critical value $|z_1|^{-1}$.

Hence, sufficiently many counterions of Species 1 also condense to lower the net value of ξ to $|z_1|^{-1}$,

$$\xi_{net} = |z_1|^{-1}. \tag{66}$$

Since all counterions of Species 2 are condensed,

$$\gamma_2 = 0. \tag{67}$$

The equivalent polyion concentration not compensated by condensed counterions is

$$n_e^{(u)} = |z_1|^{-1}\xi^{-1}n_e; \tag{68}$$

the equivalent concentration of uncondensed counterions of Species 1 is

$$n_1^{(u)}|z_1| = n_e^{(u)} = |z_1|^{-1}\xi^{-1}n_e. \tag{69}$$

The activity of Species 1 equals the activity of uncondensed counterions of Species 1,

$$\gamma_1 n_1 = \gamma_1^{(u)} n_1^{(u)}, \tag{70}$$

where $\gamma_1^{(u)}$ is the activity coefficient of uncondensed counterions of Species 1 and is obtained from Equation (9) by the replacements $\xi \to \xi_{\text{net}}$, $n_e \to n_e^{(u)}$, $n_1 \to n_1^{(u)}$, $n_2 \to 0$. Hence

$$\ln \gamma_1 = -\tfrac{1}{2} - \ln(\xi |z_1| x_1). \tag{71}$$

The osmotic pressure is,

$$\pi = kT\phi^{(u)} n_1^{(u)}, \tag{72}$$

where $\phi^{(u)}$ may be calculated from Equation (65) by replacing x_1 with $n_1^{(u)}|z_1|/n_e^{(u)} = 1$, x_2 with zero, and ξ with $|z_1|^{-1}$:

$$\phi^{(u)} = 1/2. \tag{73}$$

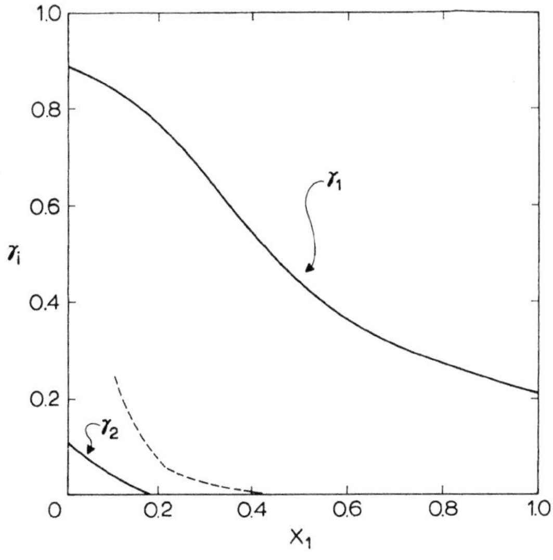

Fig. 2. Limiting laws for the counter-ion activity coefficients in polyelectrolyte solutions with uni- and divalent counterions; no added simple salt. The equivalent mole fraction of univalent counterion is x_1; γ_1 is the activity coefficient of the univalent counterions for $\xi = 2.83$, γ_2 of the divalent counter-ions for the same value of ξ. The dashed curve represents the values of γ_2 obtained in Table I from the data of Dolar and Juznic (Reference [19]).

But Equation (46) also holds for π; hence, the osmotic coefficient is,

$$\phi = \tfrac{1}{2}|z_1|^{-2}\xi^{-1}(x_1|z_1|^{-1} + x_2|z_2|^{-1})^{-1}. \tag{74}$$

It is to be noted that, with $x_2 = 1 - x_1$, all functions thus far derived in this section depend only on ξ and x_1.

Examples. In Figure 2 are plotted γ_1 and γ_2 as functions of x_1 for the case $z_2 = 2$, $z_1 = 1$. The value of ξ is 2.83, typical of highly charged vinylic polyelectrolytes and, in particular, the appropriate value for the system studied by Dolar and coworkers [19, 20]. Since $\xi > |z_1|^{-1} = 1$, the theoretical expressions used are those derived for Case 3 above. Dolar and Juznic [19], working with a lead electrode coupled to a hydrogen glass electrode, were able to measure the ratio $\gamma_{H^+}^2/\gamma_{Pb^{++}}$ as a function of x_{H^+} in a mixed solution of polystyrenesulfonic acid ($\xi = 2.83$) and its lead salt. Extremely high values of the ratio were found, although it was always finite; thus $\gamma_{Pb^{2+}}$ was never strictly zero, as predicted by Equations (60) and (67) for $x_{H^+} \geqslant 0.18$. However, a revealing treatment of the data is given in Table I. The last column consists of values

TABLE I

Correlation of activity coefficient data for a mixed polyelectrolyte system

x_1	γ_1(th)	$\log(\gamma_1^2/\gamma_2)$ measured	γ_2
0.1	0.84	0.45	0.25
0.2	0.75	0.96	6×10^{-2}
0.4	0.54	1.74	5×10^{-3}
0.6	0.36	2.25	7×10^{-4}
0.8	0.27	2.50	2×10^{-4}

of $\gamma_{Pb^{2+}}$ calculated from the *measured* values of $\gamma_{H^+}^2/\gamma_{Pb^{2+}}$ and the *theoretical* values of γ_{H^+} [Equations (54), (63), and (71)] from Figure 2. The values of $\gamma_{Pb^{2+}}$ thus obtained are also given by the dashed line in Figure 2. It can be seen that although the statement that *all* divalent counterions are condensed for $x_1 \geqslant \frac{1}{2}\xi^{-1}$ (Subcases 3(b) and 3(c)) is not strictly accurate, it is a close approximation to the real situation (and perhaps would be still closer at lower concentrations).

In Figure 3 the theoretical plot for the osmotic coefficient is given for $\xi = 2.83$, $z_1 = 1$, $z_2 = 2$ (Equations (59), (65), and (74)]. The maximum in this curve is in the region described by Subcase 3(b), and it is easily showed that, for $z_2 = 2$ and $z_1 = 1$,

$$x_1^{\max} = (1 + 2\xi^{-1})^{1/2} - 1. \tag{75}$$

The data points are taken from the report of Dolar and Kozak [20], and it is gratifying to observe that they pass through a maximum, even though agreement with Equation (75) is lacking.

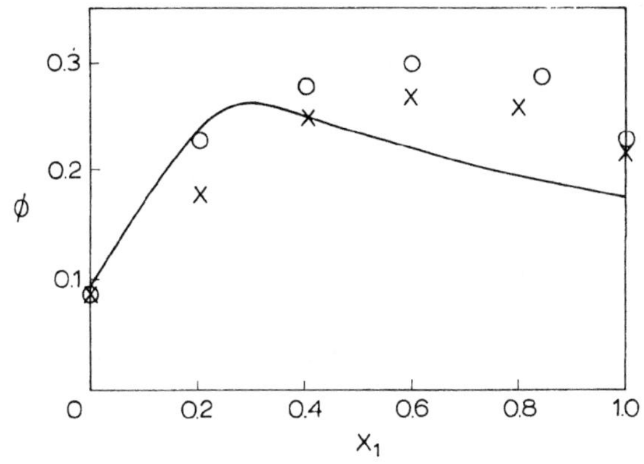

Fig. 3. The osmotic coefficient ϕ of a polyelectrolyte solution with a mixture of uni- and divalent counter-ions; no added simple salt. The solid curve is the limiting law for $\xi = 2.83$, the appropriate value for the polystyrenesulfonate used by Dolar and Kozak to obtain the portrayed data (Reference [20]). ○ – Species 1 is Na^{+}, Species 2 is Ca^{2+}; × – Species 1 is H^{+}, Species 2 is Ca^{2+}. The equivalent mole fraction of univalent counterion is x_1.

5. The Case of Only One Counter-ion Species

Suppose that only one counterion species, say Species 1, is present. The activity coefficient γ_1 and the osmotic coefficient are obtained from the general case treated above simply by setting $x_2 = 0$:

$$\ln \gamma_1 = \begin{cases} -\frac{1}{2}|z_1|\,\xi, & \xi \leqslant |z_1|^{-1} \\ -\frac{1}{2} - \ln |z_1|\,\xi, & \xi \geqslant |z_1|^{-1} \end{cases} \tag{76}$$

$$\phi = \begin{cases} 1 - \frac{1}{2}|z_1|\,\xi, & \xi \leqslant |z_1|^{-1} \\ \frac{1}{2}|z_1|^{-1}\,\xi^{-1}. & \xi \geqslant |z_1|^{-1} \end{cases} \tag{77}$$

For conceptual purposes it is very important to note the following about γ_1 and ϕ: *these quantities do not represent a 'free' fraction of counterions*; *nor do the quantities* $1-\gamma_1$ *and* $1-\phi$ *represent a 'bound' fraction of counterions*. In the first place, $\gamma_1 \neq \phi$. Secondly, even though at $\xi = |z_1|^{-1}$, for example, no counterions are condensed, the values of γ_1 and ϕ are 0.61 and 0.50, respectively; in other words, the Debye-Hückel effect is quite large in this case. The only meaningful statement about binding is that for $\xi \geqslant |z_1|^{-1}$, the fraction $1 - |z_1|^{-1}\xi^{-1}$ of the counterions is bound (condensed).

A most interesting quantity is the activity coefficient γ_p of the polyion, which we calculate now for the case of only one counterion species. If $\xi \leqslant |z_1|^{-1}$, Equations (6) and (7) yield,

$$F^{ex}/VkT = -\tfrac{1}{2}\xi P n_p \ln \kappa^2,\ \xi \leqslant |z_1|^{-1}, \tag{78}$$

where n_p is the concentration N_p/V of polyions, P is the number of charged groups per polyion, and

$$\kappa^2 = (4\pi e^2/\varepsilon kT)\, n_1 z_1^2\,, \quad \xi \leqslant |z_1|^{-1}\,. \tag{79}$$

Therefore,

$$\ln \gamma_p = [\partial (F^{ex}/VkT)/\partial n_p]_{T,V,n_1} = -\tfrac{1}{2}\xi P \ln \kappa^2, \quad \xi \leqslant |z_1|^{-1}\,. \tag{80}$$

In contrast to the osmotic coefficient and the activity coefficient of the counterion, γ_p depends on the concentration.

If $\xi \geqslant |z_1|^{-1}$, F^{ex} is calculated from Equations (78) and (79) by replacing ξ with $\xi_{net} = |z_1|^{-1}$, P with $|z_1|^{-1}\xi^{-1}P$ (the number of charged groups uncompensated by condensed counterions), and n_1 with $|z_1|^{-1}\xi^{-}n_1$ (the concentration of uncondensed counterions). Then,

$$\ln \gamma_p = -\tfrac{1}{2}|z_1|^{-2}\xi^{-1}P \ln \kappa^2, \quad \xi \geqslant |z_1|^{-1}, \tag{81}$$

where

$$\kappa^2 = (4\pi e^2/\varepsilon kT)\, \xi^{-1} n_1 |z_1|, \quad \xi \geqslant |z_1|^{-1}\,. \tag{82}$$

The mean activity coefficient $\gamma_\pm$ of the polyelectrolyte salt may also be calculated. If $\xi \leqslant |z_1|^{-1}$, the salt 'unit' consists of one P-valent polyion and $|z_1|^{-1}P$ counterions of valence z_1. The standard definition of $\gamma_\pm$ is, then,

$$(1 + |z_1|^{-1}P) \ln \gamma_\pm = \ln \gamma_p + |z_1|^{-1}P \ln \gamma_1\,. \tag{83}$$

If unity is neglected compared to $|z_1|^{-1}P$, and if Equations (76), (79), and (80) are used, it may be seen that

$$\ln \gamma_\pm = -\tfrac{1}{2}|z_1|\xi \ln n_e + \text{const.}, \qquad \xi \leqslant |z_1|^{-1}, \tag{84}$$

where the constant is independent of concentration.

If $\xi \geqslant |z_1|^{-1}$, the derivation must be done more carefully. The concentration-dependent part $\mu_s^{(c)}$ of the chemical potential of the polyelectrolyte salt may be expressed in terms of the stoichiometric mean activity $a_\pm$:

$$(kT)^{-1}\mu_s^{(c)} = (1 + |z_1|^{-1}P) \ln a_\pm \approx |z_1|^{-1}P \ln a_\pm\,. \tag{85}$$

But since a salt 'unit' in this case consists of one polyion and $|z_1|^{-2}\xi^{-1}P$ counterions (the uncondensed ones),

$$(kT)^{-1}\mu_s^{(c)} = \ln a_p + |z_1|^{-2}\xi^{-1}P \ln a_1\,, \tag{86}$$

with a_p the polyion activity and a_1 the counterion activity. The r.h.s. of Equations (85) and (86) are equated, and the relation $a_i = \gamma_i n_i (i = 1, p)$, is used:

$$\ln a_\pm = |z_1| P^{-1} \ln n_p + |z_1|^{-1}\xi^{-1} \ln n_1 + |z_1| P^{-1} \ln \gamma_p + |z_1|^{-1}\xi^{-1} \ln \gamma_1\,. \tag{87}$$

But

$$\ln a_\pm = \ln \gamma_\pm + \ln n_\pm\,, \tag{88}$$

where

$$(1 + |z_1|^{-1}P) \ln n_\pm = \ln n_p + |z_1|^{-1}P \ln n_1\,. \tag{89}$$

Combination of Equations (87)–(89), with $1 \ll |z_1|^{-1}P$, and also with Equations (76), (81), and (82), gives

$$\ln \gamma_\pm = -\left(1 - \tfrac{1}{2}|z_1|^{-1}\xi^{-1}\right)\ln n_e + \text{const.}, \qquad \xi \geqslant |z_1|^{-1} \tag{90}$$

the constant being independent of concentration.

It may be noted from Equation (77) that both Equation (84) and Equation (90) may be written

$$\ln \gamma_\pm = -\left(1 - \phi\right)\ln n_e + \text{const.}, \tag{91}$$

and, in fact, Equation (91) may be directly obtained [21] from an integration of the Gibbs-Duhem equation with the assumption that ϕ is independent of concentration.

Examples. Of particular interest are recent results for polyelectrolyte salts with divalent counterions which conform to the limiting laws. Although these laws depend only on the charge density of the polyion and the valence of the counterion, the impression has been widespread that thermodynamic properties for systems involving divalent counterions are necessarily dominated by specific effects. Examples involving a mixture of uni- and divalent counterions have been given above. In addition, for M^{2+}-polystyrenesulphonate, Kozak and Dolar [22] have measured ϕ for Mg^{2+} and Ca^{2+} and pointed out that the results are in close agreement at low concentrations with the prediction of Equation (77). Similar conclusions were reached by Reddy and Marinsky [23] for their measurements of ϕ using the same polyion with Ca^{2+}, Sr^{2+}, Zn^{2+}, and Cd^{2+}; for Co^{2+} and Ni^{2+}, their measured values of ϕ were somewhat higher than that predicted by the limiting law.

Measurements of γ_1 for M^{2+}-polymethylstyrenesulphonate conducted by Oman and Dolar [24] may also be mentioned. For Zn^{2+} and Cd^{2+} those authors found values over a wide concentration range which agree closely with the predicted value from Equation (76), $\gamma_1 = 0.11$. The value they found for Pb^{2+}, 0.05–0.07, is somewhat lower.

The mean activity coeffient of a polyelectrolyte salt was measured for the first time by Ise and Okubo [25–27], using a concentration cell with transference (actually, what is determined is the ratio of mean activity coefficients at two different concentrations). If their data for Na polyacrylate and Na polyglutamate are replotted as $\log\gamma_\pm$ *vs* $\log n_e$, straight lines with negative slopes are obtained, as predicted by Equations (84) and (90), but the magnitudes of the slopes are considerably less than predicted. Dolar and Leskovsek [28] measured $\gamma_\pm$ for polystyrenesulphic acid; if their data are replotted as $\log\gamma_\pm$ *vs* $-\log n_e$, a straight line is obtained with slope 0.67 as compared to the value 0.81 obtained from Equation (90). More recently, Dolar (private communication) has measured $\gamma_\pm$ for cadmium polystyrenesulphonate and obtained a linear lot with slope 0.89, which is very close to the predicted value 0.91 from Equation (90).

It is this author's opinion that the most important reason to study polyelectrolyte systems is to gain knowledge which bears on the properties of biological polyelectrolytes. In Table II a comparison is given of published data [29–37] for various biopolymers with the predictions of Equations (76) and (77). A word of caution is necessary. Although the observed concentration dependence of ϕ and γ_1 is usually very slight, there are examples [31, 32, 35, 38] for which the values are quite flat as a function of decreasing concentration only to increase or decrease sharply as the concentration is lowered still further. In these cases experimental values in Table II are those characteristic of the flat region; a severe concentration dependence is attributable either to experimental artifact or to a concentration region for which the model employed in the limiting laws breaks down (for unclear reasons). It is therefore impor

TABLE II

Comparison of theory with measured colligative properties of salt-free solutions of biopolyelectrolytes

System	Reference	ξ	γ_1(exp)	γ_1(th)	ϕ(exp)	ϕ(th)
Mg DNA	[29]	4.2	0.07	0.07	–	–
Na DNA	[30]	4.2	–	–	0.15–017	0.12
Na Cellulose sulfate	[31]	2.37	0.32	0.26	–	–
Na Cellulose sulfate	[32]	3.30	0.17	0.18	–	–
Na Carboxymethyl cellulose	[32]	1.05	0.58	0.58	–	–
Na Carboxymethyl cellulose	[31]	1.34	0.43	0.45	–	–
Na Carboxymethyl cellulose	[31]	1.00	0.65	0.61	–	–
Na Carboxymethyl cellulose	[33]	0.88	0.58	0.64	–	–
Na Carboxymethyl cellulose	[33]	1.38	0.41	0.44	–	–
Na Chondroitin sulfate	[33]	1.40	0.48	0.43	–	–
Na Hyaluronate	[33]	0.60	0.75	0.74	–	–
Na Hyaluronate	[34]	0.60	0.72–0.77	0.74	–	–
(Na, K)κ-carrageenan	[35]	1.05	0.44–0.56	0.58	–	–
(Na, K)λ-carrageenan	[35]	1.50	0.40–0.44	0.40	–	–
(Na, K)alginate	[36]	1.43	0.43–0.50	0.43	–	–
(Na, K, NH_4)alginate	[37]	1.43	–	–	0.30 ± 0.05	0.35
Mg alginate	[37]	1.43	–	–	0.16 ± 0.02	0.17

tant that measurements not be restricted to a narrow concentration range. In Table II the first and last entries involve Mg^{2+}; the agreement of theory and experiment is exceedingly close.

Finally, it may be noted that at low concentrations the heats of dilution of several polyelectrolyte salt solutions have been found [39–42] to be linear functions of the logarithm of the concentration. A limiting law for the heat of dilution may be easily obtained from the appropriate temperature derivative of F^{ex}, and good agreement has been found with the experimental results [41, 42].

6. Thermodynamic Properties for Systems with Added Simple Salt

The discussion in this section is restricted to the case for which one of the constituent ionic species of the added simple salt is identical to the counterion species (Species 1) of the polyelectrolyte salt. The results are presented without proof; for further details, Reference 1 should be consulted, although no new principles are involved. For

$\xi \leqslant |z_1|^{-1}$, all counterions, from both polyelectrolyte and added salt, as well as all co-ions (Species 2) from the added salt, are treated according to the Debye-Hückel law; for $\xi \geqslant |z_1|^{-1}$, the fraction $(1-|z_1|^{-1}\xi^{-1})$ of the equivalent polyion concentration is compensated by condensed counterions, while all uncondensed counterions and all co-ions are again in a Debye-Hückel atmosphere about the polyion.

For a uni-univalent salt of concentration n_s, define the mole fractions

$$x_e = n_e/(n_e + n_s), \quad x_s = n_s/(n_e + n_s), \quad x_e + x_s = 1\,. \tag{92}$$

Then, in a notation slightly different from that in Reference [1], the results for added uni-univalent salt are as follows. For the osmotic coefficient,

$$\phi = \begin{cases} [2x_s + (1-\tfrac{1}{2}\xi)\, x_e]/(2x_s + x_e), & \xi \leqslant 1 \\ (2x_s + \tfrac{1}{2}\xi^{-1}x_e)/(2x_s + x_e), & \xi \geqslant 1\,. \end{cases} \tag{93}$$

The activity coefficient of the counterion is,

$$\gamma_1 = \begin{cases} \exp[-\tfrac{1}{2}\xi x_e(2x_s + x_e)^{-1}], & \xi \leqslant 1 \\ (x_s + \xi^{-1}x_e)\exp[-\tfrac{1}{2}\xi^{-1}x_e(2x_s + \xi^{-1}x_e)^{-1}], & \xi \geqslant 1\,. \end{cases} \tag{94}$$

and the activity coefficient of the co-ion is given by,

$$\gamma_2 = \begin{cases} \exp[-\tfrac{1}{2}\xi x_e(2x_s + x_e)^{-1}], & \xi \leqslant 1 \\ \exp[-\tfrac{1}{2}\xi^{-1}x_e(2x_s + \xi^{-1}x_e)^{-1}], & \xi \geqslant 1\,. \end{cases} \tag{95}$$

The mean activity coefficient of the added salt, denoted by $\gamma_{\pm,\,s}$, is

$$\gamma_{\pm,s}^2 = \gamma_1\gamma_2\,, \tag{96}$$

γ_1 and γ_2 being given by Equations (94) and (95). Note that the above colligative properties are functions only of ξ and x_e. Note also that $\gamma_1 = \gamma_2$ when $\xi \leqslant 1$, but $\gamma_1 < \gamma_2$ when $\xi > 1$.

These limiting laws were found to be in close agreement with measurements [1]. A recent successful comparison of the law for $\gamma_{\pm,s}$ has been given for several polysaccharides by Preston and Snowden [33]. In comparison with data a limitation on Equations (93)–(96) is that ϕ, γ_1, γ_2, and $\gamma_{\pm,s}$ are all predicted to tend to unity as x_e tends to zero. Although that is perfectly consistent with the nature of these expressions as limiting laws which are strictly valid only as the ionic strength approaches zero, the experiments done at fixed values of n_s result. of course, in measured values of the concerned colligative property which, as n_e approaches zero, tend toward the value in pure simple salt solution of concentration n_s. Wells [43] has noted the extremely close agreement of the limiting laws for small x_e with data points which are 'reduced' by dividing each by the corresponding value for a pure simple salt solution of concentration equal to the concentration n_s of simple salt added to the polyelectrolyte solution; treated in this way, the data points also tend to unity as n_e tends to zero*.

* For an alternate treatment of such data see J. C. T. Kwak, *J. Phys. Chem.* **77**, 2790 (1973).

Polyelectrolyte systems with added salt have often been correlated in terms of so-called 'additivity rules'. With the notation introduced by Equation (92), these rules are as follows:

$$\phi = (2x_s + \phi_e x_e)(2x_s + x_e)^{-1} \tag{97a}$$

$$\gamma_1 = x_s + \gamma_1^e x_e \tag{97b}$$

$$\gamma_2 = 1 \tag{97c}$$

$$\gamma_{\pm,s}^2 \equiv \gamma_1 \gamma_2 = \gamma_1 \tag{97d}$$

In these equations ϕ_e and γ_1^e are the osmotic coefficient and counterion activity coefficient, respectively, in the absence of added salt (measured at the same equivalent polyelectrolyte concentration n_e).

Considered purely as empirical statements, Equations (97a) and (97b) may be looked at as two-point interpolation formulas. That is, each of these equations obviously is forced to agree with experiment at $x_e = 0$ and $x_e = 1$. (At $x_e = 0$, for example, Equation (97a) says that $\phi = 1$, the correct value for an infinitely dilute solution of simple salt in the absence of polyelectrolyte; for purposes of comparison with the limiting laws, the 'additivity rules' given above have been written as they apply to very dilute solutions.) Equation (97c) for the co-ion activity coefficient may not be given such an empirical rationale; Equation (97d) is merely the definition of the mean activity coefficient of a uni-univalent salt combined with Equation (97c).

In fact, these 'rules' are rarely viewed as merely empirical. They are usually considered as consequences of an assertion about the structure of a polyelectrolyte solution with added simple salt: the salt does not interact with the polyions. Although Equation (97) do indeed follow from this assertion (Equation (97b), for example, states that the total counterion activity is composed of independent contributions from the 'salt counterions' and the 'polyelectrolyte counterions.' while Equation (97c) clearly means that the co-ion does not interact with the polyion), the assertion itself is incorrect. It could be true only if *all* charged groups on the polyion were neutralized by bound counterions, thereby forming an electrically neutral complex which would not interact with the ionic species from the salt. But the very statement of the 'additivity rules' assumes that the activity of the 'polyelectrolyte counterions' is non-zero (being equal to $\gamma_1^e n_e$, which has never been measured as zero). The polyion, therefore, bears a net charge which *must* interact on an equal basis and to a finite extent with *all* unbound small ions of the same species, whether these ions originate from the polyelectrolyte salt or the simple salt. The structure which underlies the limiting laws, on the other hand, is fully consistent with this requirement: whether ξ is less than the critical value or not, all uncondensed small ions are treated on the same basis as part of a Debye-Hückel ionic atmosphere about the polyion.

A comparison of the 'additivity rule' with the limiting laws is instructive. The limiting law for the osmotic coefficient in the absence of added salt is given by Equation (77). If that result, with $|z_1| = 1$, is used in Equation (93), it is seen that the limiting law for ϕ in the presence of added salt is identical to the 'additivity rule', Equation

(97a); thus, it is proved with absolute rigor that the 'additivity rule' does not imply independent contributions to the osmotic pressure from the polyelectrolyte and added salt components.

The limiting laws for γ_1, γ_2, $\gamma^2_{\pm,s}$ are not identical to the 'additivity rules.' In Figure 4 the limiting laws for these quantities, Equations (94)–(96), are plotted for $\xi = 1.85$ (solid curves). They are compared there (dashed curves) with the 'additivity rules', Equations (97b–d): in the latter, Equation (76) is used for γ_1^e (with $|z_1| = 1$).

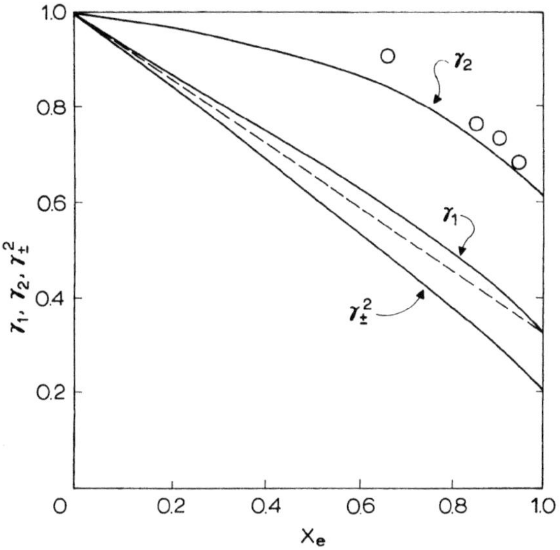

Fig. 4. Activity coefficients of univalent small ions in the presence of added simple salt. The solid lines are the limiting laws for (top to bottom) the co-ion activity coefficient γ_2, the counterion activity coefficient γ_1, and the square of the mean activity coefficient $\gamma_1\gamma_2$. The curves are drawn for $\xi = 1.85$. The dashed line is the 'additivity rule' for γ_1 and $\gamma_1\gamma_2$; the 'additivity rule' for γ_2 coincides with the top baseline $\gamma_2 \equiv 1$. The data points are measurements of γ_{Cl^-} (Reference [44]) for a polyvinylalcohol sulfate sample of degree of substitution corresponding to $\xi = 1.85$ and NaCl concentration 1×10^{-3} M.

From a coarse point of view the single sharpest distinction between the limiting laws and the 'additivity rules' is that the former predicts a sharp dependence of the co-ion activity coefficient γ_2 on x_e, which completely contradicts the 'additivity rule' $\gamma_2 \equiv 1$. The data points represented in Figure 4 are measurements of γ_{Cl^-} in a NaCl–Na (polyvinylalcohol sulfate) solution made by Nagaswa *et al.* [44]. The conclusion to be drawn is clear. It is especially important to realize that the departure of γ_2 from unity is due entirely to the Debye-Hückel mode of interaction, which, therefore, is again seen to be quite strong.

In systems with added salt the Donnan equilibrium is an important experimental tool. Let primed quantities refer to the compartment which contains only the permeable simple salt, unprimed quantities to the compartment which contains both the polyelectrolyte and salt. The Donnan exclusion parameter Γ may be defined as [45]

$$\Gamma = \lim_{n_e \to 0} (n_s' - n_s)/n_e , \tag{98}$$

where n_s' and n_s are equivalent concentrations of the simple salt. For a 1:1 electrolyte it may be proved [1] that the limiting law is

$$\Gamma = \begin{cases} \frac{1}{2}(1 - \frac{1}{2}\xi), & \xi \leqslant 1 \\ \frac{1}{4}\xi, & \xi \geqslant 1 . \end{cases} \tag{99}$$

This relation has been found [1] to be in excellent agreement with published data at low values of n_s'. Not to be confused with Γ is the ratio $(n_s' - n_s)/n_e$ before taking the limit $n_e \to 0$. Devore [46] has shown that the appropriate limiting law (low values of n_s') for a 1:1 electrolyte is

$$\begin{aligned}(n_s' - n_s)/n_e = 0.25\xi^{-1} &- 0.094\xi^{-2}(n_e/n_s') \\ &+ 0.052\xi^{-3}(n_e/n_s')^2 - 0.023\xi^{-4}(n_e/n_s')^3 \\ &+ 0.0031\xi^{-5}(n_e/n_s')^4 + \cdots, \qquad \xi \geqslant 1 .\end{aligned} \tag{100}$$

For $\xi < 1$, the expression is cumbersome and will not be given here. Note that Equation (100) is compatible with Equation (99). Again for a 1:1 electrolyte, the Donnan osmotic pressure π_D may be shown [46] to be closely related to the difference $n_s - n_s'$:

$$\begin{aligned}\pi_D/kT = n_p &+ 0.188\xi^{-2}(n_s')^{-1} n_e^2 - 0.104\xi^{-3}(n_s')^{-2} n_e^3 + \\ &+ 0.046\xi^{-4}(n_s')^{-3} n_e^4 - 0.0062\xi^{-5}(n_s')^{-4} n_e^5 + \\ &+ \cdots, \qquad \xi \geqslant 1 .\end{aligned} \tag{101}$$

Note that the coefficient of n_e^j in Equation (101), i.e., the jth virial coefficient, is exactly equal to minus two times the coefficient of n_e^{j-1} in Equation (100). Thus, the virial coefficients may be found either directly from π_D or from the salt exclusion.

7. Tracer Diffusion and Conductance in Salt-Free Polyelectrolyte Solutions

This section will briefly describe the influence of the polyion on the motion of its counterions. Although the detailed theory [2, 4] is more difficult than the considerations which led to the limiting laws for thermodynamic properties, the concepts involved are just as simple.

The condensed counterions are assumed to have negligible mobility along the length of the polyion. There is no implication here that the condensed ions are bound for long times to discrete sites or otherwise rigidly fixed to the polyion, only that their mobility is small compared to that of the uncondensed counterions. Justification for this assumption is its plausibility [47] and the close agreement with measurement of subsequently predicted transport properties.

The treatment of the uncondensed counterions bears the same logical relation to the equilibrium theory as does the Onsager conductance theory [48] to Debye-Hückel theory in simple salt solutions. The calculation shows that the net effect of the electrostatic potential set up by the polyion is to slow down the Brownian movement of the counterion and provides a formula for the effect which depends only on the charge density parameter ξ.

The results for both tracer diffusion and conductance with univalent counterions may be expressed in terms of a quantity f,.

$$f = \begin{cases} 1 - [0.55\xi^2/(\xi + 3.14)] & \xi \leqslant 1 \\ 0.87\xi^{-1} & \xi \geqslant 1 . \end{cases} \tag{102}$$

Then, if D_1 is the tracer diffusion coefficient of the counterion in the given salt-free polyelectrolyte solution and D_1° is its value in pure solvent (for practical purposes, in a very dilute solution of simple salt),

$$D_1/D_1^\circ = f. \tag{103}$$

The conductance equation is

$$\Lambda = f(\lambda_1^\circ + \lambda_p), \tag{104}$$

where Λ is the equivalent conductance of the salt-free polyelectrolyte solution, λ_1° the equivalent conductance of the counterion in pure solvent, and λ_p the equivalent conductance of the polyion.

It must be stressed, just as was done for the counterion activity coefficient γ_1 and the osmotic coefficient ϕ [Equations (76) and (77)], that, even though Equations (103) and (104) are identical to expressions based on the assumption that a fraction f of the counterions are free of the influence of the polyion while the fraction $(1-f)$ is bound to the polyion and moves with it [4, 49, 50] the quantity f of Equation (102) is *not* a fraction of 'free' counterions. When $\xi = 1$, so that no counterions are condensed, $f = 0.87$, the departure from unity being due entirely to the Debye-Hückel interaction and not to 'ion binding.' Comparison with the discussion following Equations (76) and (77) indicates that the Debye-Hückel effect on transport is smaller than on colligative properties. This feature is also characteristic of the Onsager conductance theory for simple salts; the percentage decrease of the effective field strength due to the 'relaxation effect' is less than the amount by which the activity is lowered.

Comparison of Equations (76), (77) and (102) for $|z_1| = 1$ shows that for $\xi \geqslant 1$,

$$f = 1.44\,\gamma_1 = 1.74\,\phi . \tag{105}$$

Alexandrowicz and Daniel [51] have plotted measured values of f against measured values of ϕ for various systems (Figure 3 of Reference [51] and it may be verified that the straight line $f = 1.74\,\phi$ provides a nice correlation of this data.

In Table III a comparison is presented of Equation (102) with values of f measured from tracer diffusion data, Equation (103), or conductance and transference data (the transference measurement being needed to obtain the value of λ_p in Equation (104)). Some details of the table to be noted are that the data for Na polyacrylate show a very close agreement of experimental values of f from tracer diffusion and from conduc-

TABLE III

Comparison of theory with measured values of the transport parameter f in salt-free polyelectrolyte solutions

System	Reference	ξ	f(exp)	f(th)
Na Polyacrylate	[52]	1.77	0.51[a]	0.49
Na Polyacrylate	[49]	1.77	0.45[b]	0.49
Na Polyacrylate	[52]	2.32	0.39[a]	0.37
Na Polyacrylate	[49]	2.32	0.38[b]	0.37
Na Polyacrylate	[52]	2.80	0.38[a]	0.31
Na Polyacrylate	[49]	2.80	0.38[b]	0.31
Na Polyacrylate	[53]	2.56	0.35[b]	0.34
Na Polyphosphate	[55]	2.85	0.30[b]	0.30
Polyvinylpyridinium Br	[56]	2.85	0.30[b]	0.30
Polyvinylbenzyltrimethylammonium Cl	[56]	2.85	0.28[b]	0.30
Polyvinylpyridinium Br[c]	[56]	9.20[c]	0.07[b]	0.09
Polystyrenesulfonic acid	[57]	3.14	0.30[b]	0.28
Na Polystyrenesulfonate	[57]	3.14	0.32[b]	0.28
Na (Denatured DNA)	[58]	1.05	0.67[b]	0.83
Na Alginate	[59]	1.43[d]	0.60[b]	0.61

[a] From tracer diffusion
[b] From conductance and transference
[c] The solvent is ethanol
[d] Structural value given in Reference [36].

tance (at fixed ξ); the agreement with theory of the measurement taken in ethanol provides impressive evidence of the correct dependence of the limiting law on the dielectric constant; and that the very close agreement between theory and experiment for alginate, found also in Table II for γ_1 and ϕ, indicates that this biopolymer is worth more extensive attention from polyelectrolyte chemists (Podlas and Ander [36] have also found good agreement with theory for activity coefficients in the presence of added salt).

Values of f not presented in Table III include those of Pefferkorn and Varoqui [59] from tracer diffusion for a maleic acid copolymer and of Bourdais [60] from conductance and electrophoresis (the latter to get λ_p) for polyacrylate; both sets of data agree well with the limiting law. In contrast to the entries in Table III from Reference [56], Dolar *et al.* [61] and Baumgartner *et al.* [62] found values of f for polystyrenesulfonate which are significantly higher than those predicted by the limiting law.

One of the most interesting applications of the limiting law thus far made is to the data of Sugai and Nitta [63] for polyglutamic acid, Figure 5. This synthetic polypeptide is known to undergo a helix-coil transition as its degree of neutralization with base increases; yet the data points alone in Figure 5 do not seem to reflect the transition. When coupled with the limiting laws for the helical structure and for the random coil structure, however, the data are clearly seen to do so.

Finally it may be mentioned that limiting laws are available for the tracer diffusion coefficients of counterion and co-ion in polyelectrolyte systems with added uni-uni-

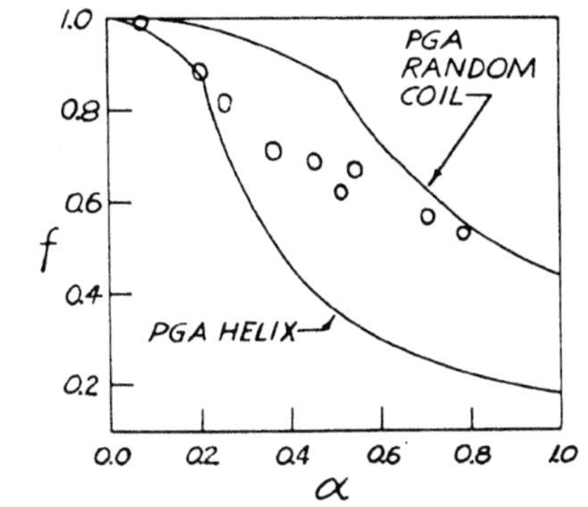

Fig. 5. The helix-coil transition of poly(D-glutamic acid) as revealed by conductance-transference. Solid lines are the limiting laws for f, Equation (102), for the helical structure, $\xi = 4.75\ \alpha$, and the coil structure, $\xi = 1.98\ \alpha$, α being the degree of neutralization. Conductance-transference data (open circles) from Reference [63].

valent salt [2], and that extensions to higher valences and to conductance in the presence of salt are being made.*

8. Properties of Titration Curves for Weak Polyacids

In this section the theory [8] of the titration curve of weak polyacids like polyacrylic acid or carboxymethylcellulose is discussed. A limiting law is derived for the concentration dependence of the titration curve, but the titration curve itself (pH *vs* degree of neutralization) is predicted to depend strongly on the hydration state of the polyelectrolyte.

Consider the equilibrium process,

$$P^{-(\nu-1)} \rightarrow P^{-\nu} + H^{+}, \tag{106}$$

where $P^{-\nu}$ represents a weak polyacid molecule with ν ionized groups. With μ_i the chemical potential (per molecule) of Species i, the condition for equilibrium of reaction (106) is,

$$\mu_{\nu} - \mu_{\nu-1} + \mu_{H^+} = 0; \tag{107}$$

for greater readability, the subscript ν has been used instead of $P^{-\nu}$. Since μ_i may be written in the form,

$$\mu_i = \mu_i^* (T, p) + kT \ln a_i, \tag{108}$$

* While this article was in press, Dixler and Ander published data on the counterion tracer diffusion coefficient for the system NaCl–Na polyacrylate which agree closely with the values predicted by the limiting laws [*J. Phys. Chem.* **77**, 2684 (1973)]. Ander (private communication) has also confirmed the limiting law for the tracer diffusion coefficient of the co-ion in this system.

where a_i is the activity of Species i, and since

$$a_i = \gamma_i m_i, \tag{109}$$

where γ_i is the activity coefficient and m_i the molality, it follows from Equation (4) and the standard definition of pH that,

$$\text{pH} = \log(\gamma_\nu/\gamma_{\nu-1}) + 0.434\,(\Delta\mu_\nu^* + \mu_{\text{H}^+}^*)/kT, \tag{110}$$

where

$$\Delta\mu_\nu^* \equiv \mu_\nu^* - \mu_{\nu-1}^*. \tag{111}$$

Moreover, in writing Equation (110), the equality $m_\nu = m_{\nu-1}$ has been used; that is, ν has been chosen as the number which maximizes the distribution of polyions according to the number of charged groups they bear.

Since the first term on the r.h.s. of Equation (110) depends on a ratio of polyion activity coefficients it may be shown from Equation (80) that

$$\log(\gamma_\nu/\gamma_{\nu-1}) = \begin{cases} -(\alpha/\alpha_1)\log\kappa^2, & \alpha \leqslant \alpha_1 \\ -\log\kappa^2, & \alpha \geqslant \alpha_1. \end{cases} \tag{112}$$

Here, the degree of neutralization α has been related to the charge density parameter ξ according to

$$\xi = \alpha/\alpha_1, \tag{113}$$

where α_1 is the value of α which corresponds to the critical value $\xi = 1$ (assuming all counterions are monovalent). Since μ_i^* is independent of concentration, the entire concentration dependence of pH is given by Equation (112): at fixed α,

$$\text{pH} = \begin{cases} -(\alpha/\alpha_1)\log\kappa^2 + \text{const}, & \alpha \leqslant \alpha_1 \\ -\log\kappa^2 + \text{const}, & \alpha \geqslant \alpha_1. \end{cases} \tag{114}$$

For example, in excess added salt, κ^2 is proportional to the added salt concentration n_s; Equation (114) then states that pH is a linear function of $\log n_s$ at fixed α and provides a simple expression for the slope. Comparison of Equation (114) with published data is given in Reference [8].

Equation (114) has the same status as a limiting law as the other expressions for colligative and transport properties presented in this article; it depends on no theoretical features other than the Debye-Hückel effect and counterion condensation. However, it is clear from Equation (110) that the dependence of pH on α, i.e., the titration curve itself, involves the term $\Delta\mu_\nu^*$, which describes the change in the self-energy of the polyion when one group is ionized at the degree of neutralization α. An appropriate theory for this term would have to involve a description of molecular forces operative in the immediate vicinity of the polyion. The details of the hydration layer become important as does the fact that the polyion charge is not characterized by cylindrical symmetry when viewed from a distance of only several angstroms. For these reasons no attempt is made at a calculation of $\Delta\mu_\nu^*$, and it is further asserted that no previous theory of the titration curve has seriously confronted the difficulties involved.

9. Further Discussion of the Model

All expressions given above are absolutely rigorous [3] in the limit of infinite dilution for a system consisting of infinitely long line charges and point ions immersed in a dielectric continuum and subject to the electroneutrality condition (except that interactions between line charges have not been considered). Until this point no indication has been offered that the model should be relevant to real polyelectrolyte solutions except for the excellent agreement with data of the theoretical expressions derived from the model. Indeed, the relevance of the model has been amply discussed [21, 64], and only several points will be stressed here.

Both the Debye-Hückel effect and the forces causing condensation of the counterions are long-range effects; they do not depend on the local structure of the polyion. Whether the charged groups project from the backbone of the chain or are on the backbone makes no difference; indeed, that the charged groups are discrete is actually irrelevant in the limit of infinite dilution and a continuous linear charge may be substituted for them. In addition, use of the bulk dielectric constant is appropriate [3]. These remarks are true only for the long-range effects considered in this article; they cease to be true when, as for the α-dependence of the titration curve, short-range effects influence the measurement.

Neglect of interactions between line charges makes the model more relevant to polyelectrolyte solutions rather than less. The rationale for replacing a real polyelectrolyte chain with an infinitely long line is that in the real solution at low but finite concentrations distant segments on the same chain are electrostatically screened from each other while nearby segments are in a more or less fully stretched, hence rod-like, configuration. To be consistent with the neglect of interactions between distant segments on the same chain it is obvious that interactions between separate chains must also be neglected. *It is inconsistent modeling to describe interactions between polyions as interactions between infinitely long lines (cylinders).* Moreover, it has not been demonstrated that such interactions contribute significantly to observable properties in dilute solution. It is much more likely that concentration effects not described by the limiting laws are due to departures from the Debye-Hückel interaction and incomplete condensation as the concentration increases.

Finally, it is important to discuss the empirical significance of the charge density parameter ξ, defined by Equation (8). All the numerical values of ξ used above for specific polyelectrolytes were calculated from known structural parameters by defining the charge density β as

$$|\beta| = e/b, \tag{115}$$

where, if L is the contour length of the real polyelectrolyte chain and P is the number of (monovalent) charged groups on the chain,

$$b = L/P. \tag{116}$$

For example, if an atactic vinylic chain has one charged group on each monomer,

$b = 2.5$ Å; if charged groups reside on the fraction α of the monomers, then $b = 2.5\alpha^{-1}$ Å. No other procedure for obtaining numerical values of ξ is consistent with the model. The actual distance between nearest-neighbor charged groups, which may be considerably larger than b as defined by Equation (116), is not a parameter of the theory. At infinite dilution the interaction of a small ion with the charged polyion chain is determined by charged groups far from the position of the small ion; it is then the linear charge density which is relevant.

Only under special circumstances should ξ be used as an adjustable parameter. An obvious case would arise if the structure of the chain is uncertain. A value of b slightly smaller than that given by Equation (116) may signify some sort of local folding which does not destroy the overall cylindrical geometry (perhaps the atactic vinylic chain contains small amounts of isotactic helicity). For a molecule such as double-stranded DNA, which can undergo gross conformational changes, great care must be exercised in the interpretation of data which do not conform to predictions of the limiting law based on the value $b = 1.7$ Å for the perfect helical conformation [5, 6]. Single-stranded denatured DNA may retain some local structure which results in a value of b somewhat smaller than the 6.9 Å which characterizes the fully extended strand.

If data for a particular measurement on a particular polyelectrolyte do not agree with the corresponding limiting law when Equation (116) is used for b, consideration should be given to the following possibilities. The data may be inaccurate, especially if they involve single-ion activities; measurements of other properties may be appropriate. The polyelectrolyte may not conform to the model; perhaps it is branched or tightly coiled. The same measurement on other polyelectrolyte systems with the same value of b may be revealing. The concentration range chosen may either be too high or too low for the applicability of the limiting laws. That the limiting laws become less accurate as the concentration increases is clear (for one thing, the detailed structure of the chain becomes important); that they may not apply if the concentration is too *low* is because the value of κ^{-1} may be too large to screen out interactions between distant segments on the same chain but not on the same cylindrical axis.

References

1. Manning, G. S.: *J. Chem. Phys.* **51**, 924 (1969).
2. Manning, G. S.: *J. Chem. Phys.* **51**, 934 (1969).
3. Manning, G. S.: *J. Chem. Phys.* **51**, 3249 (1969).
4. Manning, G. S.: *Biopolymers* **9**, 1543 (1970).
5. Manning, G. S.: *Biopolymers* **11**, 937 (1972).
6. Manning, G. S.: *Biopolymers* **11**, 951 (1972).
7. Manning, G. S.: *Ann. Rev. Phys. Chem.* **23**, 117 (1972).
8. Manning, G. S. and Holtzer, A.: *J. Chem. Phys.* **77**, 2206 (1973).
9. Abramowitz, M. and Stegun, I. A.: *Handbook of Mathematical Functions*, Dover, New York, 1965.
10. Davidson, N.: *Statistical Mechanics*, McGraw-Hill, New York, 1962.
11. Ikegami, A.: *J. Polymer Sci.* **A2**, 907 (1964).
12. Zana, R., Tondre, C., Rinaudo, M., and Milas, M.: *J. Chim. Phys. Physicochim. Biol.* **68**, 1258 (1971).
13. Armstrong, R. W., Kurucsev, T., and Strauss, U. P.: *J. Am. Chem. Soc.* **92**, 3174 (1970).

14. Oosawa, F.: *Polyelectrolytes*, Marcel Dekker, New York, 1971.
15. Jackson, J. L. and Coriell, S. R.: *J. Chem. Phys.* **40**, 1460 (1964).
16. MacGillivray, A. D. and Winkleman, Jr., J. J.: *J. Chem. Phys.* **45**, 2184 (1966).
17. Krakauer, H.: *Biopolymers* **10**, 2459 (1971).
18. Archer, B. G., Craney, C. L., and Krakauer, H.: *Biopolymers* **11**, 781 (1972).
19. Dolar, D. and Juznic, K.: Presented at Int. Conf. Calor. Thermodyn., 1st, Warsaw, August 31–September 4, 1969.
20. Dolar, D. and Kozak, D.: Presented at IUPAC Int. Symp. Macromol., Leiden, 1970.
21. Katchalsky, A.: *Pure and Appl. Chem.* **26**, 327 (1971).
22. Kozak, D. and Dolar, D.: *Z. Phys. Chem., Frankfurt* **76**, 93 (1971).
23. Reddy, M. and Marinsky, J. A.: *J. Phys. Chem.* **74**, 3891 (1970).
24. Oman, S. and Dolar, D.: *Z. Phys. Chem., Frankfurt* **56**, 13 (1967).
25. Ise, N.: *Advan. Polymer Sci.* **7**, 536 (1971).
26. Ise, N. and Okubo, T.: *J. Phys. Chem.* **69**, 4102 (1965).
27. Ise, N. and Okubo, T.: *Macromolecules* **2**, 401 (1969).
28. Dolar, D. and Leskovsek, H.: *Makromol. Chem.* **118**, 60 (1968).
29. Lyons, J. W. and Kotin, L.: *J. Am. Chem. Soc.* **87**, 1670 (1965).
30. Auer, H. E. and Alexandrowicz, Z.: *Biopolymers* **8**, 1 (1969).
31. Liquori, A. M. Ascoli, F., Botre, C., Crescenzi, V. and Mele, A.: *J. Polymer Sci.* **40**, 169 (1959).
32. Nagasawa, M. and Kagawa, I.: *J. Polymer Sci.* **25**, 61 (1957); Errata, *J. Polymer Sci.* **31**, 256 (1958).
33. Preston, B. N., Snowden, J. McK. and Houghton, K. T.: *Biopolymers* **11**, 1645 (1972).
34. Cleland, R. L.: *Biopolymers* **6**, 1519 (1968).
35. Podlas, T. J. and Ander, P.: *Macromolecules* **2**, 432 (1969).
36. Podlas, T. J. and Ander, P.: *Macromolecules* **3**, 154 (1970).
37. Katchalsky, A., Cooper, R. E., Upadhyay, J. and Wasserman, A.: *J. Chem. Soc. London* 5198 (1961).
38. Takahashi, A., Kato, T. and Nagasawa, M.: *J. Phys. Chem.* **74**, 944 (1970).
39. Skerjanc, J., Dolar, D. and Leskovsek, D.: *Z. Phys. Chem., Frankfurt* **56**, 207, 218 (1967).
40. Skerjanc, J., Dolar, D. and Leskovsek, D.: 1967, *Z. Phys. Chem., Frankfurt* **70**, 31 (1970).
41. Dolar, D: (private communication).
42. Ise, N.: (private communication).
43. Wells, J. D.: *Biopolymers* **12**, 223 (1973); *Proc. Roy. Soc. London* **B183**, 399 (1973).
44. Nagasawa, M., Izumi, M. and Kagawa, I.: *J. Polymer Sci.* **37**, 375 (1959).
45. Gross, L. M. and Strauss, U. P.: in B. E. Conway, R. G. Barrades (eds.), *Chemical Physics of Ionic Solutions*, Wiley, New York, 1966, pp. 361–89.
46. DeVore, D. I. and Manning, G. S. (to be published).
47. Ikegami, A.: *Biopolymers* **6**, 431 (1968).
48. Robinson, R. A. and Stokes, R. H.: *Electrolyte Solutions*, Academic Press, New York, 1955.
49. Huizenga, J. R., Griegor, P. F. and Wall, F. T.: *J. Am. Chem. Soc.* **72**, 2636 (1950).
50. Kurucsev, T. and Steel, B. J.: *Rev. Pure Appl. Chem.* **17**, 149 (1967).
51. Alexandrowicz, Z. and Daniel, E.: *Biopolymers* **1**, 447 (1963).
52. Huizenga, J. R., Griegor, P. F. and Wall, F. T.: *J. Am. Chem. Soc.*, **72**, 4228 (1950).
53. Wall, F. T. and Hill, W. B.: *J. Am. Chem. Soc.* **82**, 5599 (1960).
54. Wall, F. T. and Doremus, R. H.: *J. Am. Chem. Soc.* **76**, 868 (1954).
55. Darskus, R. L., Jordon, D. O. and Kurucsev, T.: *Trans. Faraday Soc.* **62**, 2876 (1966).
56. Jordon, D. O., Kurucsev, T., T. and Martin, M. L.: *Trans. Faraday Soc.* **65**, 606 (1969).
57. Okubo, T. and Ise, N.; *Macromolecules* **2**, 407 (1969).
58. Buchner, P., Cooper, R. E., and Wasserman, A.: *J. Chem. Soc. London*, 3974 (1961).
59. Pefferkorn, E. and Varoqui, R.: *Eur. Polymer J.* **6**, 663 (1970).
60. Bourdais, J.: *J. Chim. Phys. Physicochim. Biol.* **56**, 194 (1959).
61. Dolar, D., Span, J. and Pretnar, A.: *J. Polymer Sci. C.* **16**, 3557 (1968).
62. Baumgartner, E., Liberman, S. and Lagos, A. E.: *Z. Phys. Chem. Frankfurt* **61**, 211 (1968).
63. Sugai, S. and Nitta, K.: *Biopolymers* **7**, 495 (1969).
64. Armstrong, R. W. and Strauss, U. P.: 'Polyelectrolytes', in H. F. Mark, N. G. Gaylord, and N. M. Bikales (eds.), *Encyclopedia of Polymer Science and Technology*, Vol. 10, Wiley-Interscience, New York, 1969.

STATISTICAL THERMODYNAMICS OF POLYELECTROLYTE SOLUTIONS

M. MANDEL
Gorlaeus Laboratoria, Afdeling Fysische Chemie III, Rijksuniversiteit, Leiden, The Netherlands

1. Introduction

Since long it has been realized that the electrostatic interactions between charged macromolecules and small ions and eventually the interactions between the charges on a same macromolecule play an important role in the equilibrium properties of polyelectrolyte solutions. It is therefore not surprising that most efforts in the field of the theoretical description of these solutions have been focused on the solution of the Poisson-Boltzmann equation for the electrostatic potential around a polyion, usually in a given fixed conformation, and the subsequent calculation of the electrostatic free energy and its bearing on the thermodynamics of the system. Pioneering work was done by Hermans and Overbeek [1] and Katchalsky and coworkers [2]; many others have followed some of which have been listed in the references [3]. Less numerous papers have been devoted to a more general statistical thermodynamical approach. Some attempts, starting from a Kuhn-like chain of 'statistical elements' were made by Katchalsky [4] but these treatments were criticized by Rice and Harris [5] because of *inter alia* the use of undiminished statistical weight for all conformations of the polyion with the same end-to-end distance. The work of the latter [5, 6] however is also based on an analogous model and includes many specified and unspecified assumptions.

Marcus [7] has presented a calculation of the thermodynamic properties without the use of any particular model and based on a certain number of clearly stated assumptions but others which are not explicitly introduced and are more difficult to grasp.

It is the purpose of this paper to establish a statistical thermodynamical formalism for dilute polyelectrolyte solutions which may serve to discuss the assumptions which are usually introduced in the theoretical treatment of such systems. Use is made of a model which is kept as general as possible, and certainly is more realistic than a Kuhn-like chain, without being too complicated to be handled by ordinary statistical procedures. Canonical ensemble statistics are used, the external thermodynamic variables being the volume V, temperature T and composition of the system. Of course, for the problem thus outlined no exact solution of practical nature is presented which is in the present stage still beyond reach. It is hoped that such a formal treatment may help a better understanding of the problems underlying the theoretical approach to polyelectrolyte systems. It will also help to discuss the generality of theoretical description as presented by Marcus [7].

Eric Sélégny (ed.), Polyelectrolytes, 39–55.

2. General Formalism

Our treatment is based on the already old observation that a charged macromolecule differs from an equivalent collection of small ions only by the fact that for the latter the charges can, in principle, diffuse freely one with respect to another through the total volume whereas in the former the charges are more or less held together by chemical forces so that they must move in some coordinated way.

The system consists of N_1 molecules of solvent (usually water), N_2 molecules of the macromolecular component which for the sake of fixing ideas will be assumed to derive from a weak polycarboxylic acid (such as poly-acrylic acid) and N_3 molecules of salt in a fixed volume V and constant temperature T. Only solutions very diluted with respect to the polyelectrolyte will be considered in the limit of infinite dilution $N_p \to 1$. The salt will be assumed to be totally dissociated into $N_c = N_3$ monovalent cations and $N_a = N_3$ monovalent anions; the dissociation of the water molecules is neglected and the macromolecular component is dissociated into $N_p = N_2$ polyions with Z elementary negative charges each and $N_{H^+} = ZN_p$ hydrogen ions. Each macromolecule consists of n identical units, Z of which are dissociated (A^-) and carry an elementary negative charge and $(n-Z)$ are undissociated and uncharged (AH); n is directly related to the molecular weight of the macromolecule which is thus assumed to be uniform for all polyions.

The canonical ensemble statistics will be used to calculate the thermodynamic properties of the system which is characterized by the external variable V, T, N_1, N_2, N_3 and Z. Note that these state variables are really external variables in the sense that they can be changed, within reasonable limits, experimentally. Often another canonical ensemble is used in the description of a polymer (e.g. Hill [8]), the *isolated* macromolecule being treated itself as a macroscopic system with external variables T, n, Z and l, where l is the end-to-end distance of the polymer or an analogous conformational variable. Although formally this is allowed it must be realized that generally speaking not only is l not a true external variable, as it cannot be changed at will experimentally, but also that extension of such a treatment to the case of a non-isolated macromolecule is far from being straightforward.

In order to specify the *configuration of the system* solvent molecules and small ions will be treated as point masses the configuration of each being fixed by its position, $\mathbf{r}_i$ for a molecule of solvent and $\mathbf{r}_j$ for an ion, with respect to a given fixed reference coordinate system. The macromolecular conformation is represented by the location of n point masses (numbered in an ordered way from 1 to n) which have to be inter-related in a prescribed way (e.g. point mass k is always at a given constant distance from point mass $k-1$ and $k+1$). Note that this representation *does not imply* that the actual macromolecule behaves as a collection of interconnected point-masses; only its configuration is specified by the location of the center of mass of the n units. In order to define the macromolecular configuration entirely it is also necessary to define the distribution of the dissociated and undissociated groups along the chain of the n units. This can be done by introducing a parameter ζ_k for the kth unit which can have two values: $\zeta_k=0$ if this unit can be represented by AH and $\zeta_k=1$ for A^-. The configuration of the entire system is thus specified by the set of n values for $\mathbf{r}_k$ and ζ_k, $\{\mathbf{r}_k, \zeta_k\}$, defining the polyion (in the case of an infinitely diluted solution), the set of N_1 values of $\mathbf{r}_i$, $\{\mathbf{r}_i\}$, for the solvent and the set of $N_a+N_c+N_{\mathrm{H}}+$values $\mathbf{r}_j$, $\{\mathbf{r}_j\}$, for the small ions.

The center of mass of the macromolecule r_0 is defined by the following equation

$$\sum_{(k)} m_k \mathbf{r}_k = \mathbf{r}_0 \sum_{(k)} \mathbf{m}_k . \tag{1}$$

For systems, infinitely diluted with respect to macromolecules, $N_{\mathrm{p}}=1$, the origin of the coordinate system may be taken in the center of mass as long as the translational motion of the latter is not to be taken into account. New variables with respect to this origin may be defined as $\mathbf{R}_i$, $\mathbf{R}_j$ and $\mathbf{R}_k$.

$$\mathbf{R}_u = \mathbf{r}_u - \mathbf{r}_0 . \tag{2}$$

It should be observed that neither $\mathbf{r}_k$ nor $\mathbf{R}_k$ are continuously varying variables as is the case for $\mathbf{r}_i$, $\mathbf{R}_i$, $\mathbf{r}_j$ and $\mathbf{R}_j$. Only those sets $\{\mathbf{r}_k\}$ or $\{\mathbf{R}_k\}$ compatible with the possible conformations of the real chain have to be taken into account. The orientation of the coordinate system defined with its origin at the center of mass (1) will be arbitrarily fixed with respect to the orientation of the end-to-end vector of the macromolecule. $(\mathbf{R}_{k=n}-\mathbf{R}_{k=1})$. This means that in what follows only a given fixed orientation of the macromolecule will be considered. As, in the absence of an external field, all orientations are equally probable, the generality of the argument will not be harmed by this procedure. The term 'configuration of the system' also will refer to the configuration at fixed overall orientation of the polyion. Finally it should be observed that the total charge of the polyelectrolyte is related to the n parameters ζ_k

$$Z = \sum_{k=1}^{n} \zeta_k . \tag{3}$$

3. The Statistical Thermodynamics of a Polyelectrolyte Solution

As usual it will be assumed that the canonical partition function Q of the system may be written as the product of an 'internal' partition function $Q^{(0)}$ which does not depend on the configurational coordinates (but includes the contributions of the internal degrees of freedom and the kinetic energy of rotation and translation) and a configurational partition function $Q_t^{(c)}$.

$$Q = Q^{(0)} Q_t^{(c)} = Q^{(0)} Q^{(c)} V . \tag{4}$$

For a solution infinitely diluted in macromolecular component ($N_{\mathrm{p}}=1$), $Q^{(0)}$ can be defined as follows.

$$Q^{(0)} = \frac{(q_1^0)^{N_1}}{N_1!} \prod_{(r)} \frac{(q_r^0)^{N_r}}{N_r!} \left[(q_{AH}^0)^{(n-Z)} (q_{A^-}^0)^Z\right], \tag{5}$$

where q_1^0 and q_r^0 stands for the contribution of a solvent or ion of species r (with r equal to H^+, a or c) respectively. The factorials take into account that solvent molecules and small ions can be treated as non-localized particles. The contribution of the macromolecule is split up into a separate contribution of the AH and A^- units defined in such a way that the electronic energy of binding these units into a single chain has been taken into account. No factorials appear for these units as they may be considered localized in the macromolecule through their numbering.

Within the model described in Section 2 the configurational partition function can be put into the following form.

$$Q_t^{(c)} = \int d\mathbf{r}_0 \left[\sum_{(\varrho)} \int \cdots \int e^{-v/kT} \{d\mathbf{R}_i\} \{d\mathbf{R}_j\}\right] =$$
$$= V \sum_{(\beta)} \int \cdots \int e^{-v/kT} \{d\mathbf{R}_i\} \{d\mathbf{R}_j\}. \tag{6}$$

Here $v = v(\{\mathbf{R}_k, \zeta_k\}, \{\mathbf{R}_i\}, \{\mathbf{R}_j\})$ is the potential energy of interaction for the system in a given configuration. The integrations over all possible configurations of the solvent molecules and small ions have to be performed over the total volume V of the system at constant polymer configuration. The summation stands for the sum over all possible configurations $\{\mathbf{R}_k, \zeta_k\}_\beta$ of the macromolecule at fixed position of its center of mass $\mathbf{r}_0$. As all positions of the center of mass may be assumed to be equivalent in the system (neglecting surface effects) integration over $\mathbf{r}_0$ just yields V. Thus $Q^{(c)}$ is the configuration partition function of the system for an arbitrary position of $\mathbf{r}_0$. Equations (4), (5) and (6) are only valid if the interactions do not modify the internal energy levels of the particles and macromolecular units.

From (4) and (5) the following expression for the free energy may be obtained.

$$(F)_{N_p=1} = -kT \ln Q = N_1 f_1^0 + Z f_{A^-}^0 + (n-Z) f_{AH}^0 + \sum_{(r)} N_r f_r +$$
$$+ kT \left[N_1 \ln (N_1/e) + \sum_{(r)} N_r \ln (N_r/e)\right] - kT \ln V - kT \ln Q^{(c)}. \tag{7}$$

Here $f_u = -kT \ln q_u^0$ stands for the individual contribution to F of each particle (or unit) and Stirling's approximation $|x| = (x/e)^x$ has been used. It is, of course, the evaluation of $Q^{(c)}$ that makes an exact statistical treatment of a polyelectrolyte solution so difficult.

Formally it is possible to split up the free energy of configuration into an energy and an entropy contribution.

$$F^{(c)} = -kT \ln Q^{(c)} = U^{(c)} - TS^{(c)}, \tag{8}$$

$$U^{(c)} = \sum_{(\beta)} \int \cdots \int \frac{v}{Q^{(c)}} e^{-v/kT} \{d\mathbf{R}_i\} \{d\mathbf{R}_j\}, \tag{9}$$

$$S^{(c)} = -k \sum_{(\beta)} \int \cdots \int h \ln h \{d\mathbf{R}_i\} \{d\mathbf{R}_j\}. \tag{10}$$

Here $h = h(\{\mathbf{R}_k, \zeta_k\} \{\mathbf{R}_i\} \{\mathbf{R}_j\})$ stands for the distribution function in the configurational phase space (at constant $\mathbf{r}_0$).

$$h = \frac{1}{Q^{(c)}} e^{-v/kT}. \tag{11}$$

These thermodynamic quantities add up with the internal energy and entropy from $Q^{(0)}$. Note that it is possible to define the normalized probability of finding the macromolecule in a given configuration regardless the distribution of the small particles by the following equation.

$$P_\beta = \frac{1}{Q^{(c)}} \int \ldots \int e^{-v/kT} \{d\mathbf{R}_i\} \{d\mathbf{R}_j\} = \int \ldots \int h \{d\mathbf{R}_i\} \{d\mathbf{R}_j\}. \tag{12}$$

The difficulties involved in evaluating $Q^{(c)}$, $F^{(c)}$, $U^{(c)}$ and $S^{(c)}$ make a less general approach imperative. The usual procedure consists in assuming that, *in the absence of charge interactions* the system would behave ideally in as far as the mixing of small particles is concerned. In fact it is always possible to split up the free energy rigourously into two contributions the first of which F_0 represents the total free energy of the system if no charge interactions would occur, the other F_e that part which is related to these interactions.

$$F = F_0 + F_e, \tag{13}$$

$$F_0 = -kT \ln Q^{(0)} - kT \ln V - kT \ln Q_0^{(c)}, \tag{14}$$

$$F_e = -kT \ln \frac{Q^{(c)}}{Q_0^{(c)}}. \tag{15}$$

Here $Q_0^{(c)}$ stands for the configurational partition function in the absence of charge interactions. If v_0 represents the total potential energy of the system under this condition $Q_0^{(c)}$ is then defined by

$$Q_0^{(c)} = \sum_{(\beta)} \int \ldots \int e^{-v_0/kT} \{d\mathbf{R}_i\} \{d\mathbf{R}_j\}. \tag{16}$$

An expression for F_e may always be derived by an imaginary charging process. The contribution of $-kT \ln Q_0^{(c)}$ to F_0 can again be split up into an energy $U_0^{(c)}$ and an entropy $S_0^{(c)}$ contribution which are defined analogously to (9) and (10) with v_0 replacing v.

The free energy F_0 may be put into the following explicit form

$$F_0 = N_1 f_1^0 + \sum_{(r)} N_r f_r^0 + [Z f_A^0 + (n - Z) f_{AH}^0] - TS_{s,0} + F_{p,0}, \tag{17}$$

where $S_{s,0}$ represents the *ideal* entropy of mixing of the small particles and the macromolecule

$$S_{s,0} = -k \left[N_1 \ln \frac{N_1 e}{V} + \sum_{(r)} N_r \ln \frac{N_r e}{V} \right] + k \ln V \tag{18}$$

and $F_{p,0}$ is defined by

$$F_{p,0} = -kT \ln \frac{Q_0^{(c)}}{V^{N_1} V^{\Sigma(r) N_r}}. \tag{19}$$

In order to evaluate $F_{p,0}$ two different assumptions may be used:

(a) the small particles do not contribute to v_0, i.e. $v_{0,\beta}$ represents the potential energy of an isolated polyelectrolyte molecule in the absence of charge interactions;

(b) some interactions between the small particles and the macromolecule are taken into account which may have no bearing on the ideality of mixing of the small particles in the absence of charge interactions. The first assumption a) would lead to an expression

$$Q_0^{(c)} = V^{N_1} V^{\Sigma(r) N_r} \sum_{(\beta)} e^{-v_{0,\beta}/kT}$$

with a corresponding simple expression for $F_{p,0}$. This assumption is however unrealistic as it is known that for an ordinary uncharged polymer the conformational probabilities depend on the nature of the solvent (e.g. from the average dimensions of the polymer). For the assumption (b) we may introduce the potential of average force $W_{0,\beta}$ acting on the macromolecule in configuration $\{\mathbf{R}_k, \zeta_k\}_\beta$ after averaging over all configurations of the small particles, defined by

$$\int \cdots \int e^{-v_0/kT} \{d\mathbf{R}_i\} \{d\mathbf{R}_j\} = e^{-W_{0,\beta}/kT} V^{N_1} V^{\Sigma(r) N_r} \tag{20}$$

and which determines also the probability of finding the macromolecule in a given configuration β in the absence of charge interactions.

$$P_{0,\beta} = \frac{e^{-W_{0,\beta}/kT}}{\sum_{(\beta)} e^{-W_{0,\beta}/kT}}. \tag{21}$$

With the help of this potential of average force the following expressions may be derived for $U_{p,0}$ and $S_{p,0}$

$$U_{p,0} = \frac{\sum_{(\beta)} W_{0,\beta} e^{-W_{0,\beta}/kT}}{\sum_{(\beta)} e^{-W_{0,\beta}/kT}}, \tag{22a}$$

$$S_{p,0} = -k \sum_{(\beta)} P_{0,\beta} \ln P_{0,\beta}. \tag{22b}$$

Assumption (b) consists in admitting that $F_{p,0} = (U_{p,0} - TS_{p,0})$ does not contribute to the chemical potentials $\mu_{1,0}$ and $\mu_{r,0}$ which are obtained by taking the derivative of F_0 with respect to N_1 and N_r respectively, at constant V, T, $N_p = 1$ and constant N_r or N_1 and $N_{r' \neq r}$. Under these conditions the chemical potential of the polymer, in the absence of charge interactions, can formally be defined by

$$\mu_{p,0} = n[\alpha f_{A^-}^0 + (1-\alpha) f_{AH}^0] + k \ln V + U_{p,0} - TS_{p,0}, \tag{23}$$

where $\alpha = Z/n$ has been introduced. Here $F_{p,0}$ may be interpreted as the free energy arising, in the absence of charge interactions, from the configurational statistics of the macromolecule immersed in some kind of a continuum consisting of solvent and small chargeless ions. (Note that a third, rather trivial, assumption besides (a) and (b) could have been used in which $v_0 = 0$; this would have meant that the uncharged macromolecule can be represented as an ideal chain without interaction!)

Another assumption which is commonly introduced is that $W_{0,\beta}$ does not depend on the distribution of the AH and A^- groups along the macromolecular chain. If γ represents a given *conformation* of the chain $\{\mathbf{R}_k\}_\gamma$ irrespective this distribution, then $U_{p,0}$ and $S_{p,0}$ simplify according to this assumption into the following expressions [9]

$$U_{p,0} \sim U'_{p,0} = \frac{\sum\limits_{(\gamma)} W_{0,\gamma} e^{-W_{0,\gamma}/kT}}{\sum\limits_{(\gamma)} e^{-W_{0,\gamma}/kT}}, \tag{24}$$

$$\begin{aligned} S_{p,0} \sim S'_{p,0} &= k \ln \frac{n!}{(n-Z)!\,Z!} - k \sum_{(\gamma)} P_{0,\gamma} \ln P_{0,\gamma} = \\ &= k \ln \frac{n!}{(n-Z)!\,Z!} - \tilde{S}'_{p,0}. \end{aligned} \tag{25}$$

Here $P_{0,\gamma}$ represents the probability of finding the uncharged macromolecule in a given conformation γ regardless the distribution of AH and A^- groups. This distribution can be realized in $\chi = n!/(n-Z)!\,Z!$ different ways each of which is assumed to be equally probable.

$$P_{0,\gamma} = \sum_{\{\zeta_k\}} P_{0,\beta}(\{\mathbf{R}_k, \zeta_k\}_\beta) = \chi P_{0,\beta} = \frac{e^{-W_{0,\gamma}/kT}}{\sum\limits_{(\gamma)} e^{-W_{0,\gamma}/kT}}. \tag{26}$$

In this assumption $U_{p,0}$, $S_{p,0}$ and $P_{0,\gamma}$ are independent of Z. Under these circumstances $\mu^0_{p,0}$ can also be put into the following form

$$\begin{aligned} \mu_{p,0} = n\left[\alpha\mu^0_{A^-} + (1-\alpha)\,\mu^0_{AH}\right] + \\ + n\alpha kT \ln\alpha + n(1-\alpha)\,kT\ln(1-\alpha) + kT\ln V, \end{aligned} \tag{27}$$

where the standard chemical potentials μ^0_{AH} and $\mu^0_{A^-}$ of the AH and A^- units respectively have been introduced, which are defined as the free energy contribution per unit to $\mu_{p,0}$ for the macromolecule in either the $(AH)_n$ or $(A)_n$ state (the latter however without charge interactions).

$$\mu^0_{AH} = \frac{1}{n}(\mu_{p,0})_{\alpha=0} = f^0_{AH} + \frac{1}{n}\left[U'_{p,0} - T\tilde{S}'_{p,0}\right], \tag{28a}$$

$$\mu^0_{A^-} = \frac{1}{n}(\mu_{p,0})_{\alpha=1} = f^0_{A^-} + \frac{1}{n}\left[U'_{p,0} - T\tilde{S}'_{p,0}\right]. \tag{28b}$$

This expression justifies the use of formally defined standard chemical potentials

μ^0_{AH} and $\mu^0_{A^-}$. It should be pointed out that this formalism can easily be extended to the case where the polyelectrolyte exhibits a conformational transition on changing α.

It may be questioned how good approximations (24) and (25) really are. Their validity may be doubtful particularly in the case where the AH group along the chain may interact through hydrogen bonds. Although the formal definitions of μ^0_{AH} and $\mu^0_{A^-}$ may be maintained as referring eventually to the hypothetical case of a polyelectrolyte chain of which the properties in the absence of the charge interactions do not depend on the distribution of the units, the total expression for F_0 should be changed by adding a contribution $\Delta F_{p,0}$ defined by

$$\Delta F_{p,0} = (U_{p,0} - U'_{p,0}) - T(S_{p,0} - \tilde{S}'_{p,0}) \tag{29}$$

and which depends on Z or α. Therefore the total expression for F_0 can be written quite naturally as follows.

$$F_0 = N_1 f_1^0 + \sum N_r f_r^0 + n[\alpha\mu^0_{A^-} + (1-\alpha)\,\mu^0_{AH}] + n\alpha kT \ln\alpha + \\ + (1-\alpha)\,kT\ln(1-\alpha) + kT\ln V - TS_{s,0} + \Delta F_{p,0}\,. \tag{30}$$

Here in principle $\mu^0_{A^-}$, μ^0_{AH} and $F_{p,0}$ depend on the nature of solvent and salt as well as on the concentration of the latter.

An expression for F_e, as defined by (15), can formally be derived by an imaginary charging process in which the charge of all the particles is increased progressively from zero to its actual value. In any stage of this process the charge on the small ions and the polyion can be represented by a fraction λ of its final value. The potential energy of interaction $v(\lambda)$ and the corresponding configurational partition function $Q^{(c)}(\lambda)$ will also be functions of this λ. Then (15) can be transformed into the following form.

$$F_e = -kT\int_0^1 \frac{\partial \ln Q^{(c)}(\lambda)}{\partial\lambda}\,\mathrm{d}\lambda = \\ = \int_0^1 \mathrm{d}\lambda \left[\sum_{(\beta)} \int \cdots \int \frac{\partial v_e(\lambda)}{\partial\lambda}\,\frac{e^{-v(\lambda)/kT}}{Q^{(c)}(\lambda)}\right]\{\mathrm{d}\mathbf{R}_i\}\,\{\mathrm{d}\mathbf{R}_j\}\,. \tag{31}$$

Here it has been assumed that $v(\lambda) = v_0 + v_e(\lambda)$ where $v_e(\lambda)$ is the potential energy of interaction arising from the charges. Obviously v_0 which represents the potential energy of interaction in the absence of charges is no function of λ. In general it is not possible to separate F_e into independent contributions arising from polyion and small ions respectively.

It is possible to define an internal energy contribution $U_e = U^{(c)} - U_0^{(c)}$ consistently with (31).

$$U_e = \sum_{(\beta)} \int \cdots \int \left[\frac{v\,e^{-v/kT}}{Q^{(c)}} - \frac{v_0\,e^{-v_0/kT}}{Q_0^{(c)}}\right]\{\mathrm{d}\mathbf{R}_i\}\,\{\mathrm{d}\mathbf{R}_j\} = \int_0^1 \mathrm{d}\lambda\,\frac{\partial U^{(c)}(\lambda)}{\partial\lambda}\,, \tag{32}$$

where the following definition has been introduced

$$U^{(c)}(\lambda) = \sum_{(\beta)} \int \cdots \int \frac{v(\lambda)\, e^{-v(\lambda)/kT}}{Q^{(c)}(\lambda)} \{d\mathbf{R}_i\} \{d\mathbf{R}_j\}. \tag{33}$$

Note that $U^{(c)}(\lambda)$ corresponds to the statistical average of the potential energy of the system at a given state of charge characterized by the value of λ. Although it contains a contribution of v_0, in a first approximation it may be assumed that the contribution arising from coulombic interactions are much more important so that $\mathbf{U}^{(c)}(\lambda)$ may also be defined less rigorously by

$$U^{(c)}(\lambda) \sim \sum_{(\beta)} \int \cdots \int \frac{v_e(\lambda)\, e^{-v(\lambda)/kT}}{Q^{(c)}(\lambda)} \{d\mathbf{R}_i\} \{d\mathbf{R}_j\}. \tag{34}$$

From F_e and U_e an entropy contribution S_e can always be calculated.

Expression (31) for the free energy arising from charge interactions is an exact expression so that, together with F_0 as defined previously, it completely defines the total free energy of the system considered here. In the present state of theory an exact evaluation of F_e according to (31) is out of question however for the same reasons that $Q^{(c)}$ cannot be calculated immediately. Therefore a more approximate approach is used analogously to the Debye-Hückel treatment of ordinary ionic solutions. This approach, which we shall call the extended Poisson-Boltzmann approach, will be discussed in the next section.

4. The Extended Poisson-Boltzmann Approach

The fundamental assumption underlying this treatment is the neglect of the explicit presence of the solvent molecules in as far as v_e is concerned. For that purpose the macroion and the small ions are assumed to be immersed in a continuum of electric permittivity D, representing the solvent (The actual value of D is still a difficulty in itself; in the zeroth order approximation it is put equal to the electric permittivity of the pure solvent but it may also be considered to be a variable or adjustable parameter. In this paper we shall confine ourselves to the former approximation). For sake of simplicity we shall also assume that all charges may be represented as point charges located in the center of mass of either small ions or macromolecular units. Then the potential energy of charge interaction in a specified configuration of the system is given by

$$v_e = \frac{1}{D} \left\{ \frac{1}{2} \sum_{j, j'} \frac{e_j e_{j'}}{|\mathbf{R}_j - \mathbf{R}_{j'}|} + \frac{1}{2} \sum_{k, k'} \frac{\zeta_k \zeta_{k'} e_k e_{k'}}{|\mathbf{R}_k - \mathbf{R}_{k'}|} + \sum_{j, k} \frac{\zeta_k e_j e_k}{|\mathbf{R}_j - \mathbf{R}_k|} \right\}. \tag{35}$$

If in any stage of a charging process the charge on a particle or unit is a fraction λ of its final value, it follows from (35) that

$$v_e(\lambda) = \lambda^2 v_e. \tag{36}$$

If (36) is introduced into (31) the following general expression for F_e results,

$$F_e = 2\int_0^1 \frac{\mathrm{d}\lambda}{\lambda}\left[\sum_{(\beta)}\int\cdots\int v_e(\lambda)\frac{e^{-v(\lambda)/kT}}{Q^{(c)}(\lambda)}\{\mathrm{d}\mathbf{R}_i\}\{\mathrm{d}\mathbf{R}_j\}\right]. \tag{37}$$

Let us now define the electrostatic potential arising from the charge distribution in the system as the solution of Poisson's equation. Under the assumption that we deal only with point charges the latter can be written in any specified configuration of the system as

$$\nabla^2\psi(\mathbf{R}) = -\frac{4\pi}{D}\left[\sum_{(k)} e_k\zeta_{k,\beta}\delta(\mathbf{R}_{k,\beta}-\mathbf{R}) + \sum_{(j)} e_j\delta(\mathbf{R}_j-\mathbf{R})\right]. \tag{38}$$

Here e_k represents the charge that unit k of the macromolecule can bear ($e_k = e_{k'} = e_\mathrm{p} = -e$, where e is the elementary positive charge), $e_j = z_j e$ is the charge of ion j, $\zeta_{k,\beta}$ stands for the value of the charge parameter of unit k in the given configuration β of the macromolecule and $\mathbf{R}_{k,\beta}$ for the position of the center of mass of unit k in this configuration β. The Dirac-δ functions $\delta(\mathbf{R}_\mathrm{m}-\mathbf{R})$ express the fact that each point charge m represents a singularity in space. The space charge density due to the macromolecular charges $\varrho_\mathrm{p}(R)$ and to the small ions $\varrho_\mathrm{s}(R)$ have thus been defined by

$$\varrho_{\mathrm{p},\beta}(\mathbf{R}) = \sum_{(k)} e_k\zeta_{k,\beta}\delta(\mathbf{R}_{k,\beta}-\mathbf{R}), \tag{39}$$

$$\varrho_\mathrm{s}(\mathbf{R}) = \sum_{(j)} e_j\delta(\mathbf{R}_j-\mathbf{R}). \tag{40}$$

If both sides of (38) are averaged over all possible configurations of the system at fixed center of mass of the macromolecule the Poisson equation for the average electrostatic potential $\bar{\psi}(R)$ in the system is obtained.

$$\nabla^2\bar{\psi}(\mathbf{R}) = -\frac{4\pi}{D}\left[\bar{\varrho}_\mathrm{p}(\mathbf{R}) + \bar{\varrho}_\mathrm{s}(\mathbf{R})\right]. \tag{41}$$

The average charge densities $\bar{\varrho}_\mathrm{p}(R)$ and $\bar{\varrho}_\mathrm{s}(R)$, due to chain-bound and 'free' charges respectively, are defined by the following equations.

$$\bar{\varrho}_\mathrm{p}(\mathbf{R}) = \sum_{(\beta)}\left[\sum_{(k)} e_k\zeta_{k,\beta}\delta(\mathbf{R}_{k,\beta}-\mathbf{R})\right]P_\beta = e_\mathrm{p}\sum_{(\beta)}\left[\sum_{(k)}\zeta_{k,\beta}\delta(\mathbf{R}_{k,\beta}-R)\right]P_\beta, \tag{42}$$

$$\begin{aligned}\bar{\varrho}_s(\mathbf{R}) &= \sum_{(\beta)}\int\cdots\int\left[\sum_{(j)} e_j\delta(\mathbf{R}_r-\mathbf{R})\right]h\{\mathrm{d}\mathbf{R}_i\}\{\mathrm{d}\mathbf{R}_j\} = \sum_{(j)} e_j p_j(\mathbf{R}) = \\ &= \sum_r e_r N_r p_r(\mathbf{R}) = \sum_r \bar{\varrho}_r(\mathbf{R}).\end{aligned} \tag{43}$$

Here $e_r = z_r e$ is the charge of an ion of species r and the reduced distribution function $p_j(\mathbf{R})$, defined by

$$p_j(\mathbf{R}) = \sum_{(\beta)}\int\cdots\int h\delta(\mathbf{R}_j-\mathbf{R})\{\mathrm{d}\mathbf{R}_i\}\{\mathrm{d}\mathbf{R}_j\} \tag{44}$$

represents the probability density of finding a given ion j at position $\mathbf{R}$ with respect to the center of mass of the macromolecule after averaging over all configurations of the other ions and the polyelectrolyte. Use has been made of the fact that all ions can be split up into groups of N_γ identical ions of species γ. Analogously to $\bar{\varrho}_p(R)$ we can also introduce the average number density of macromolecular units $\bar{C}_p(\mathbf{R})$ at position $\mathbf{R}$.

$$\bar{C}_p(\mathbf{R}) = \sum_{(\gamma)} \left[\sum_k \delta(\mathbf{R}_{k,\gamma} - \mathbf{R})\right] P_\gamma = \sum_{(\gamma)} \varrho_{\gamma,\gamma}(\mathbf{R}) \, P_\gamma \,, \tag{45}$$

where P_γ stands for the probability of finding the macromolecule in a given conformation γ regardless the distribution of ζ-values (cf. Equation (26)). Note that $\int^V C_{p,\gamma} \, d\mathbf{R} = \int^V \bar{C}_p \, d\mathbf{R} = n$. In general there will be no simple relation between $\bar{\varrho}_p(\mathbf{R})$ and $\bar{C}_p(\mathbf{R})$; only if for a given conformation all distributions of ζ-values are equally probable (i.e. characterized by the same potential of average force) an average charge distribution over all macromolecular units may be introduced such that each unit bears the same charge $Ze_p/n = \alpha e_p$. Then $\bar{\varrho}_p(R) = \alpha e_p \bar{C}_p(R)$. In general $\bar{\varrho}_s(R)$, $\bar{\varrho}_p(R)$ and $\bar{C}_p(R)$ will be functions of temperature, composition of the solution and the total charge of the macroion through the different distribution functions.

If certain boundary conditions are satisfied ($\bar{\psi}$ and $\nabla\bar{\psi}$ vanish at the surface of the system) the general solution of the Poisson equation (41) is given by

$$\bar{\psi}(\mathbf{R}) = -\frac{1}{D} \int^V \frac{[\bar{\varrho}_p(\mathbf{R}') + \bar{\varrho}_s(\mathbf{R}')]}{|\mathbf{R} - \mathbf{R}'|} \, d\mathbf{R}' \,. \tag{46}$$

Although in this expression the average charge density due to the macromolecular charges and the average charge density due to free ionic charges contribute independently to the mean electrostatic potential $\bar{\psi}$, both charge densities are interconnected by the averaging procedure and thus cannot be treated as completely independent quantities. In the specific case of a *rigid* macromolecular conformation (independent of charge and ionic strength) and *random* distribution of the chain-bound charges only the situation becomes much more simple as will be discussed below.

The next step in the extended Poisson-Boltzmann treatment consists in evaluating the electric free energy F_e as the reversible work to charge, at constant external variables, the entire system, using the average electrostatic potential and the average charge distribution at any stage of the hypothetical charging process. When all charged groups on the macromolecule and the free ions have the same fraction $0 < \lambda < 1$ of their final charge, $\bar{\varrho}_s(\mathbf{R}, \lambda)$ and $\bar{\varrho}_p(\mathbf{R}, \lambda)$ will be a function of λ and therefore $\bar{\psi}(\mathbf{R}, \lambda)$ as well. As a consequence of (42) and (43) and from the fact that all distribution functions are also dependent on λ, the following expressions hold

$$\bar{\varrho}_p(\mathbf{R}, \lambda) = \lambda e_p f_p(\mathbf{R}, \lambda) \,, \tag{47}$$

$$\bar{\varrho}_s(\mathbf{R}, \lambda) = \sum_{(r)} \bar{\varrho}_r(\mathbf{R}, \lambda) = \lambda \sum_{(r)} e_r N_r p_r(\mathbf{R}, \lambda) \,, \tag{48}$$

where $\bar{f}_{\mathrm{p}}(\mathbf{R}, \lambda)$ is defined by

$$\bar{f}_{\mathrm{p}}(\mathbf{R}, \lambda) = \sum_{(\beta)} \left[\sum_{(k)} \zeta_{k,\beta} \delta(\mathbf{R}_{k,\beta} - \mathbf{R})\right] P_{\beta}(\lambda), \tag{49}$$

and $P_{\beta}(\lambda)$ is defined in a similar manner to (12) with v and $Q^{(c)}$ replaced by $v(\lambda)$ and $Q^{(c)}(\lambda)$. Note that in the case of random distribution of charges on the macromolecule $\bar{f}_{\mathrm{p}}(\mathbf{R}, \lambda) = \alpha \bar{C}_{\mathrm{p}}(\mathbf{R}, \lambda)$.

In the evaluation of F_e we follow now the argument as presented by Marcus [7]. Consider a given volume element $\mathrm{d}\mathbf{R}$ at $\mathbf{R}$ where in a given stage of the charging process the mean electrostatic potential is $\bar{\psi}(\mathbf{R}, \lambda)$. If the total charge in this volume element is now increased by an increase of λ to $\lambda + \mathrm{d}\lambda$ (at constant configuration), the corresponding reversible work δw_e is given by

$$\begin{aligned} \delta w_e &= \bar{\psi}(\mathbf{R}, \lambda) \left[e_{\mathrm{p}} \bar{f}_{\mathrm{p}}(\mathbf{R}, \lambda) + \sum_{(r)} e_r N_r p_r(\mathbf{R}, \lambda)\right] \mathrm{d}\lambda \, \mathrm{d}\mathbf{R} = \\ &= \bar{\psi}(\mathbf{R}, \lambda) \frac{1}{\lambda} \left[\bar{\varrho}_{\mathrm{p}}(\mathbf{R}, \lambda) + \bar{\varrho}_{\mathrm{s}}(\mathbf{R}, \lambda)\right] \mathrm{d}\lambda \, \mathrm{d}\mathbf{R}. \end{aligned} \tag{50}$$

Here use has been made of (46) and (47). The total reversible work of charging the entire system is then

$$F_e = \int_0^1 \frac{\mathrm{d}\lambda}{\lambda} \int^{V} \bar{\psi}(\mathbf{R}, \lambda) \left[\bar{\varrho}_{\mathrm{p}}(\mathbf{R}, \lambda) + \bar{\varrho}_{\mathrm{s}}(\mathbf{R}, \lambda)\right] \mathrm{d}\mathbf{R}. \tag{51}$$

If we may neglect the contribution of v_0 to U_e then the following expression for the internal energy due to the charge interactions is obtained consistently with (50).

$$U_e = \frac{1}{2} \int^{V} \bar{\psi}(\mathbf{R}) \left[\bar{\varrho}_{\mathrm{p}}(\mathbf{R}) + \bar{\varrho}_{\mathrm{s}}(\mathbf{R})\right] \mathrm{d}\mathbf{R}. \tag{52}$$

Making use of (46) this expression can also be put into the following form

$$\begin{aligned} U_e &= \frac{1}{2} \int_0^1 \mathrm{d}\lambda \int^{V} \frac{\partial}{\partial \lambda} \{\bar{\psi}(\mathbf{R}, \lambda) \left[\bar{\varrho}_{\mathrm{p}}(\mathbf{R}, \lambda) + \bar{\varrho}_{\mathrm{s}}(\mathbf{R}, \lambda)\right]\} \mathrm{d}\mathbf{R} = \\ &= -\frac{1}{2D} \int_0^1 \mathrm{d}\lambda \iint^{v} \frac{\partial}{\partial \lambda} \left\{ \frac{\left[\bar{\varrho}_{\mathrm{p}}(\mathbf{R}', \lambda) + \bar{\varrho}_{\mathrm{s}}(\mathbf{R}', \lambda)\right] \left[\bar{\varrho}_{\mathrm{p}}(\mathbf{R}, \lambda) + \bar{\varrho}_{\mathrm{s}}(\mathbf{R}, \lambda)\right]}{|\mathbf{R} - \mathbf{R}'|} \times \right. \\ &\quad \times \mathrm{d}\mathbf{R}' \, \mathrm{d}\mathbf{R} = \int_0^1 \mathrm{d}\lambda \int^{v} \psi(\mathbf{R}, \lambda) \frac{\partial}{\partial \lambda} \left. \left[\bar{\varrho}_{\mathrm{p}}(\mathbf{R}, \lambda) + \varrho_{\mathrm{s}}(\mathbf{R}, \lambda)\right] \right\} \mathrm{d}\mathbf{R}. \end{aligned} \tag{53}$$

With the help of (51) and (53) the entropy contributions S_e due to charge interactions can be derived.

$$TS_e = U_e - F_e = \int_0^1 \mathrm{d}\lambda \int^v \bar{\psi}(\mathbf{R}, \lambda) \left[\lambda e_p \frac{\partial \bar{f}_p(\mathbf{R}, \lambda)}{\partial \lambda} + \sum_{(r)} \lambda e_r \frac{\partial p_r(\mathbf{R}, \lambda)}{\partial \lambda} \right] \mathrm{d}\mathbf{R} . \tag{54}$$

It should be observed that in the expression for F_e as given by (51) in any stage of the charging process the average potential energy $v_e(\lambda)$, which, according to (37) should determine F_e, has been replaced by the product of the average potential and the average charge density. This indicates that (51) can only be correct if certain fluctuations in the average charge density may be neglected, a direct consequence of the definition for δw_e as used in (50). The same observation also holds for the expression for U_e as given by (52) or (53) and S_e as defined by (54). The latter clearly shows a contribution arising both from the distribution of the small ions and from the macromolecular conformation through p_r and $\bar{f}_p$ respectively.

In the extended Poisson-Boltzmann treatment a further assumption is generally introduced in order to relate the average charge density in the system to the mean electrostatic potential. Usually the potential of average force acting on a given ion j at position $\mathbf{R}$, which determines p_j as defined by (44), is replaced by the electrostatic energy $\varepsilon_j \bar{\psi}(\mathbf{R})$. This yields the following expression for $\bar{\varrho}_s(R)$.

$$\bar{\varrho}_s(\mathbf{R}) \sim \sum_{(r)} e_r N_r \frac{e^{-e_r \bar{\psi}(\mathbf{R})/kT}}{\int e^{-e_r \bar{\psi}(\mathbf{R}')/kT} \, \mathrm{d}\mathbf{R}'} . \tag{55}$$

It is clear that, the charge on the macromolecule reduced to zero, ($\bar{\varrho}_p = 0$) all positions in the system become equally probable for any ion so that locally electroneutrality is re-established. This means that for $Z = \alpha = 0$ both $\bar{\varrho}_s$ and $\bar{\psi}$ vanish and implies that in this treatment interactions between small ions independently from their interactions with the polyion are *not* taken into account as already pointed out by Oosawa *et al.* [3 c] and Marcus [7]. (This would no longer be the case if instead of $\bar{\psi}$ a mean electrostatic potential would have been defined at fixed position of the center of mass of the polyion and a specified configuration of the given ion j). The assumption used in defining $\bar{\varrho}_s$ through (55) has the disadvantage that it cannot be used to express $\bar{\varrho}_p$ analogously as neither P_β nor P_γ (for the case where random distribution of charges on the macromolecule is assumed) can be a simple function of $\bar{\psi}$ in the same way as p_j. This is due to the fact that the former quantities are defined in terms of a specified configuration for a large number of charges whereas the latter refers only to the configuration of one single charge.

The so-called Boltzmann equation (analogous to (55)) can however be introduced in an earlier stage of the averaging procedure both for defining the charge density due to the small ions and the charge density arising from the charges on the polyion. For the sake of simplicity we shall assume in what follows that for a given conformation of the polyion all distributions of the Z charges are equally probable.

If both sides of (38) are averaged *at fixed conformation of the macromolecule*, the following expression is obtained

$$\nabla^2 \bar{\psi}^\gamma = -\frac{4\pi}{D}\left[\sum_{(k)} \alpha e_k \delta(\mathbf{R}_{k,\gamma} - \mathbf{R}) + \bar{\varrho}_s^\gamma(\mathbf{R})\right]. \tag{56}$$

Here $\bar{\psi}^\gamma(\mathbf{R})$ and $\bar{\varrho}_s^\gamma(\mathbf{R})$ are defined by the following equations

$$\bar{\psi}^\gamma(\mathbf{R}) = \frac{\sum_{\{\zeta_k\}} \int \cdots \int \psi(\mathbf{R})\, h\{\mathrm{d}\mathbf{R}_i\}\, \{\mathrm{d}\mathbf{R}_j\}}{\sum_{\{\zeta_k\}} \int \cdots \int h\{\mathrm{d}\mathbf{R}_i\}\, \{\mathrm{d}\mathbf{R}_j\}} \tag{57}$$

$$\bar{\varrho}_s^\gamma(\mathbf{R}) = \frac{\sum_{\{\zeta_k\}} \int \cdots \int \left[\sum_{(j)} e_j \delta(R_j - \mathbf{R})\right] h\{\mathrm{d}\mathbf{R}_i\}\, \{\mathrm{d}\mathbf{R}_j\}}{\sum_{\{\zeta_k\}} \int \cdots \int h\{\mathrm{d}\mathbf{R}_i\}\, \{\mathrm{d}\mathbf{R}_j\}} = \sum_{(j)} e_j p_j, \gamma(\mathbf{R}). \tag{58}$$

Of course further averaging of both sides over all possible conformations of the macromolecule with the help of P_γ, defined by

$$P_\gamma = \frac{1}{Q^{(c)}} \sum_{\{\zeta_k\}} h\{\mathrm{d}\mathbf{R}_i\}\, \{\mathrm{d}\mathbf{R}_j\} = \frac{e^{-W_\gamma/kT}}{\sum_{(\gamma)} e^{-W_\gamma/kT}}, \tag{59}$$

where W_γ is the potential of mean force acting on the macromolecule in a specified conformation, exactly yields (41). The mean electrostatic potential $\bar{\psi}^\gamma(\mathbf{R})$ may be used to express $\bar{\varrho}_s^\gamma(\mathbf{R})$ and P_γ. The former may be written by analogy with (55) as

$$\bar{\varrho}_s^\gamma(\mathbf{R}) = \sum_{(\gamma)} e_\gamma N_\gamma \frac{e^{-e_\gamma \bar{\psi}^\gamma(\mathbf{R})/kT}}{\int e^{-e_\gamma \bar{\psi}^\gamma(\mathbf{R}')/kT}\, \mathrm{d}\mathbf{R}'}. \tag{60}$$

The latter may be connected to $\bar{\psi}^\gamma$ by writing for W_γ the following expression

$$W_\gamma \sim W_{0,\gamma} + e_p \alpha \sum_{(k)} \bar{\psi}^\gamma(\mathbf{R}_{k,\gamma}). \tag{61}$$

If the charge on the macromolecule vanishes, so does $\bar{\psi}^\gamma$ and (61) becomes identical with (26). It should be noted that the introduction of $\bar{\psi}^\gamma$ is not necessarily equivalent with the approximation where the average electrostatic potential is introduced in a later stage. This can easily be seen from the fact that averaging both sides of (60) over all conformations of the macromolecule does in general not yield an expression identical with (55). Two particular cases will now be discussed.

(a) If the macromolecule practically only exists in one given (rigid) conformation, (56) and (60) become identical with (41) and (55) respectively. In that particular case P_γ will be zero for all possible conformations except the given rigid one for which it will be unity. No entropy contribution will arise from the macromolecular conformation in F_e and F_0. In expression (51) for F_e, (55) may be used for $\bar{\varrho}_s(\mathbf{R}, \lambda)$ with e_r and $\bar{\psi}(\mathbf{R})$ replaced λe_r and $\bar{\psi}(\mathbf{R}, \lambda)$ respectively. Also $\bar{\varrho}_p(\mathbf{R}, \lambda) = \lambda e_p \alpha C_p(\mathbf{R})$, where C_p completely defined by the fixed conformation will be no function of λ.

(b) Another particular case of more general interest is the one for which a large number of possible conformations are characterized by the same value of W_γ and predominantly determine the average properties of the polyelectrolyte system. Let us assume in this case that out of all conformations γ there are g_m different conformations γ' with the same potential of mean force $W_{\gamma'}$ such that the following condition is satisfied.

$$\sum_{(\gamma)} e^{-W_\gamma/kT} \sim g_m e^{-W'_\gamma/kT}. \tag{62}$$

Then obviously $P_{\gamma \neq \gamma'} \to 0$ and $P_{\gamma'} = g_m^{-1}$. The following expressions will hold

$$\bar{\psi} \sim \sum_{(\gamma')} \bar{\psi}^{\gamma'} P_{\gamma'} = g_m^{-1} \sum_{(\gamma')} \bar{\psi}^{\gamma'}, \tag{63}$$

$$\bar{\varrho}_p(\mathbf{R}) \sim \alpha e_p \sum_{(\gamma')} \left[\sum_{(k)} \delta(\mathbf{R}_{k,\gamma'} - \mathbf{R}) \right] P_{\gamma'} = \alpha e_p g_m^{-1} \sum_{(\gamma')} \sum_{(k)} \delta(\mathbf{R}_{k,\gamma'} - \mathbf{R}). \tag{64}$$

According to (61) a constant value of W_γ, does however not necessarily imply a constant value for $\bar{\psi}^{\gamma'}$. This will rigorously only be the case for $W_{0,\gamma} = 0$, a hypothetical situation already discussed in Section 3. The same would also be true if $W_{0,\gamma}$, in a first approximation, could be neglected with respect to the contribution arising from electrostatic interactions. In both cases $\bar{\psi}^{\gamma'} \sim \bar{\psi}$ and will depend on the charge of the macromolecule. It then also follows that (55) may be used for $\bar{\varrho}_s(\mathbf{R})$. If it is further assumed that condition (62) holds in every stage of the charging process (i.e. if it assumed that for every value of λ the same conformations determine predominantly the average properties depending on charge interactions) the same expression for F_e as in case (a) holds. Here $\bar{\varrho}_p$ will be defined by (64) which again implies that $\bar{C}_p = \bar{\varrho}_p/\alpha e_p$ is independent of λ. Note however that in general $\bar{C}_p$ must be a function of α for such polyelectrolytes of which the average dimensions are known to depend on the degrees of dissociation. The mean squared end-to-end distance $\overline{l^2}$ defined by the following expression

$$\overline{l^2} \sim \sum_{(\gamma')} (\mathbf{R}_{k=n,\gamma'} - \mathbf{R}_{k=1,\gamma'}) \cdot (\mathbf{R}_{k=n,\gamma'} - \mathbf{R}_{k=1,\gamma'}) P_{\gamma'} \tag{65}$$

must depend on α in an analogous way as C_p.

Finally the case may be considered where W_γ, is essentially determined by $W_{0,\gamma'}$. Under these circumstances $\bar{C}_p$ will be independent of α and remain constant throughout the charging process. Here however (63) does not yield $\bar{\psi}^{\gamma'} \sim \bar{\psi}$ so that (55) cannot be used for $\bar{\varrho}_s(R)$ without additional assumptions of rather drastic nature. The same will be true in the most general case for which no simplification is allowed in the expression for W_γ, as defined by (61).

Summarizing it may be stated that the use of the extended Poisson-Boltzmann approach will yield simple expressions for F_e only in the case where the macromolecule is assumed to be in a rigid conformation or to obey conformational statistics determined essentially by conformations characterized by the same value of the electrostatic potentials $\bar{\psi}^{\gamma'}$.

5. Concluding Remarks

It has been shown that the free energy of an (infinitely diluted) polyelectrolyte solution can be represented by the sum of two contributions, F_0 and F_e, defined by (17) or (30) and (31) respectively. All three quantities F, F_0 and F_e will depend on the external variables used to describe the system: V, T and composition. Using certain simplifying assumptions F_e can be related to the average electrostatic potential and average charge distribution in the system as in (51).

$$\begin{aligned} F = N_1 f_1^0 + \sum_{(\gamma)} N_\gamma f_\gamma^0 + n\left[\alpha \mu_{A^-}^0 + (1-\alpha)\,\mu_{AH}^0\right] + n\alpha kT \ln \alpha + \\ + (1-\alpha)\,kT \ln(1-\alpha) + kT \ln V - TS_{s,0} + \Delta F_{p,0} + \\ + \int_0^1 \frac{d\lambda}{\lambda} \int \bar{\psi}(\mathbf{R},\lambda)\left[\bar{\varrho}_p(\mathbf{R},\lambda) + \bar{\varrho}_s(\mathbf{R},\lambda)\right] d\mathbf{R}\,. \end{aligned} \tag{66}$$

Within the assumptions used all contributions arising from the charge interactions are included in F_e which may also be written as a contribution of an internal energy, defined by (52) and of an entropy given by (54). Further simplification by the use of the so-called Boltzmann equation for the total charge density will only be possible in some particular cases.

Expression (66) may be compared to the free energy expression as given by Marcus [7]. It may be seen that the latter is not correct as besides F_e, defined identically to (51), other contributions appear which depend on charge interactions (a configurational entropy of the macromolecule and an entropy of mixing of charged and uncharged groups along the chain). Furthermore in Marcus' treatment it is assumed that $\bar{C}_p$ only depends on a so-called configurational variable, analogous to $\overline{l^2}$. The latter quantity was assumed in any equilibrium state to be determined by a minimizing condition of the free energy, $(\delta F/\delta \overline{l^2})=0$, and therefore it was concluded that in all reversible transformations variations of $\bar{C}_p$ will not contribute to the change in free energy. The argument used by Marcus is however fallacious and leads to an incorrect conclusion as can be seen from the treatment given above. In fact if, as in our present model, $\overline{l_2}$ is to be considered a true average quantity, such as e.g. given by (65), then it will depend on the external variables used to describe the system. Any change in these external variables, such as the temperature and the composition, will modify $\overline{l^2}$ as well as $\bar{C}_p$ and their contribution to the free energy. In other words it will not be possible to change $\overline{l^2}$ independently from these external variables and a change of F at fixed $\overline{l^2}$ cannot be performed (except in the case where all possible conformations would correspond to the same value of $\overline{l^2}$). On the other hand one could treat formally the end-to-end distance of the polyion as an external variable (with all the difficulties pointed out already in Section 2). In this case the minimizing condition of F with respect to l will determine at equilibrium the value of l. In such a case however some external variable used in the treatment above can no longer be a true independent variable but will be defined as an average quantity, function of the real external variables. This would e.g. be the case for V, which could no longer be considered a constant but which should vary with $\overline{l^2}$, T and the composition. Besides it should be pointed out that even in this case a constant value for $\overline{l^2}$ does not necessarily implie a constant value for $\bar{C}_p$ or $\bar{\varrho}_p$ except for the case where the fluctuations in the value of all the $\mathbf{R}_k$ corresponding to fluctuations in the macromolecular conformation at constant $\overline{l^2}$ are negligible. This will in general only be the case for those equilibrium situations where the macromolecule is rather stretched, i.e. characterized by a semi-rigid conformation.

It thus appears that the treatment presented by Marcus [7] will be useful in the particular case only where macromolecules are considered which are either of rigid geometry or for which the conformations contributing predominantly to the average properties may be characterized as negligible small fluctuations around the mean conformation.

References

1. Hermans, J. J. and Overbeek, J. Th. G.: *Rec. Trav. Chim. Pays-Bas* **67**, 761 (1948); *Bull. Soc. Chim. Belges* **37**, 154 (1948).

2a. Fuoss, R. M., Katchalsky, A., and Lifson, S.: *Proc. Natl. Acad. Sci. U.S.* **37**, 579 (1951).
 b. Lifson, S. and Katchalsky, A.: *J. Polymer Sci.* **13**, 43 (1953).
 c. Alexandrowicz, Z. and Katchalsky, A.: *J. Polymer Sci.* **A1** 3231 (1963).

3a. Alfrey, T., Berg, P. W., and Morawetz, H.: *J. Polymer Sci.* **7**, 543 (1951).
 b. Kimbal, G. E., Cutler, M., and Samelson, H.: *J. Phys. Chem.* **56**, 57 (1952).
 c. Oosawa, F., Imai, N., and Kagawa, I.: *J. Polymer Sci.* **13**, 93 (1954).
 d. Oosawa, F.: *J. Polymer Sci.* **23**, 421 (1957).
 e. Wall, F. T. and Berkowitz, J.: *J. Chem. Phys.* **26**, 114 (1957).
 f. Lifson, S.: *J. Chem. Phys.* **27**, 700 (1957).
 g. Kotin, L. and Nagasawa, M.: *J. Chem. Phys.* **36**, 873 (1962).
 h. Gross, L. M. and Strauss, U. P.: in B. E. Conway and R. G. Barradas (eds.), *Chemical Physics of Ionic Solutions* J. Wiley New York, 1966, p. 361.
 i. Manning, G.: *J. Chem. Phys.* **51**, 124 (1969).

4a. Kuhn, W., Kunzle, O., and Katchalsky, A.: *Helv. Chem. Acta* **31**, 1994 (1948).
 b. Katchalsky, A., Kunzle, O., and Kuhn, W.: *J. Polymer Sci.* **5** 283 (1959).
 c. Katchalsky, A. and Lifson, S.: *J. Polymer Sci.* **11** 409 (1953).

5. Rice, S. A. and Harris, F. E.: *J. Phys. Chem.* **58**, 733 (1954).
6. Rice, S. A. and Harris, F. E.: *J. Phys. Chem.* **53**, 725 (1954).
7. Marcus, R. A.: *J. Chem. Phys.* **23**, 1057 (1955).
8. Hill, T. L.: *Introduction to Statistical Thermodynamics*, Addison-Wesley, Reading, 1960.
9. Mandel, M. and Leyte, J. C.: *J. Electronal. Chem.* **37**, 297 (1972).

ION-BINDING PHENOMENA OF POLYELECTROLYTES

MITSURU NAGASAWA
Dept. of Synthetic Chemistry, Nagoya University, Chikusa-Ku, Nagoya, Japan

1. Introduction

The alkali ion salts of carboxylic or sulphonic acid are in general completely dissociated in aqueous solutions if those salts are ordinary simple electrolytes. Even though those groups are fixed on a linear polymer skeleton such as polyacrylate or polystyrene sulphonate, it is reasonable to believe that those salts are completely dissociated in aqueous solutions. Nevertheless, it has often been pointed out that the physical properties of polyelectrolyte solutions can be well explained if we assume that the effective charge density is much lower than the analytical charge density. That is, the counter-ions appear to be bound on fixed charges to decrease the charge density. Such an idea of ion-binding (fixation, condensation etc.) has long been employed by various authors to explain the physical properties of linear polyelectrolyte solutions.

The force acting between fixed charges on polyion and counter-ions is not unique but varies with combination of counter-ion and fixed group. Therefore, the cause for counter-ion binding may depend on the ionic species.

(1) The complex formation between fixed ions and counter-ions: A well known example is the binding of Cu^{++} by $-COO^-$. This type of ion-binding can be detected by the spectroscopic method without ambiguity.

(2) The occlusion of counter-ions inside the polyion coil due to electrostatic force: Then, the polyion behaves like a sphere containing those counter-ions. However, this model is not proper to explain the thermodynamic properties.

(3) The accumulation of counter-ions around the polyion skeleton due to strong electrostatic interaction. This may correspond to ion-pair formation in a simple electrolyte solution. In this case, the polyion is usually assumed to be a rod or a line of beads such as a pearl-necklace.

Obviously, this classification may be too much simplified. There may be various intermediate cases which cannot be classified without ambiguity. For example, the difference between ion-binding of the first class and that of the third class is not clear. The complex formation of Cu^{++}, Ni^{++} etc. with the carboxylic group or the undissociated state of $-COOH$ etc. may be detected without ambiguity but there are many conbinations of fixed group and counter-ion between which the force acting cannot be experimentally determined. In this paper, the ion-binding due to non-electrostatic force is not discussed. Moreover, even if the force acting between ions comprising a strong simple electrolyte is believed to be electrostatic, the force acting between those ions may not always be electrostatic when the electrolyte is a polyelectrolyte. Because of very strong electrostatic attraction of polyion, additional non-electrostatic interactions may occur. For example, if the water structure around

Eric Sélégny (ed.), Polyelectrolytes, 57–77.

the polyion is changed by the approach of counter-ions, the free energy change due to the change in water structure, which is essentially a non-electrostatic force, should be taken into account in addition to pure electrostatic interaction between fixed charges and counter-ions. The force due to charge transfer may be included if two ions approach very closely.

The distinction between the second and third classes is not very clear, either. The model for a polyion in simple salt solutions may be as shown in Figure 1. That is,

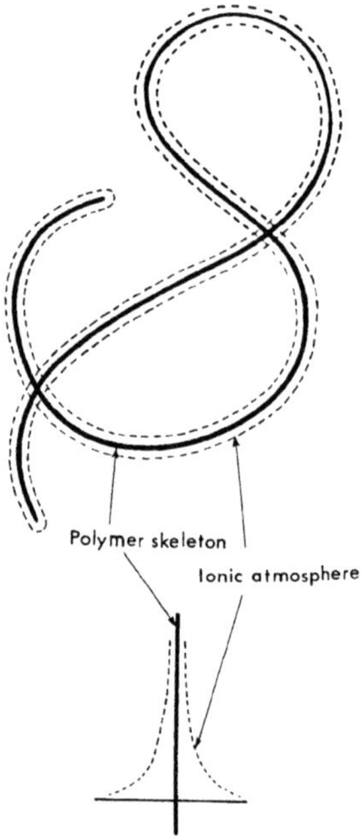

Fig. 1. Model for a polyion in simple salt solutions.

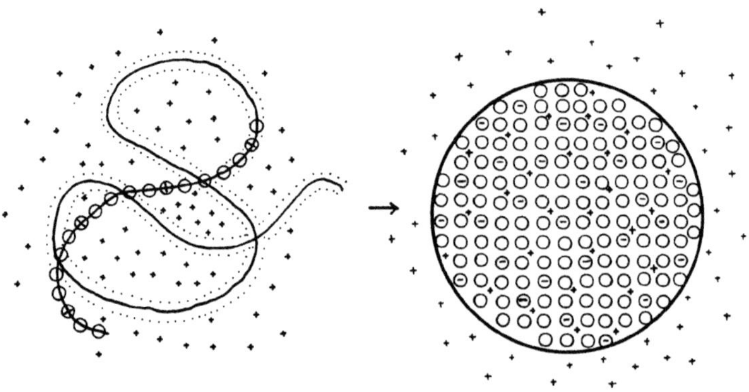

Fig. 2. An equivalent sphere model.

the fixed charges on a polyion are surrounded by their ionic atmosphere and the whole molecule is more or less randomly coiled. The ionic atmosphere around fixed charges may be approximated as cylindrical distribution except at very high and low ionic strengths. When we observe the movement of the whole molecule as in hydrodynamic phenomena, the coiled conformation of polyion is primarily important. In that case, the polyion may be approximated as a sphere in which all fixed charges are uniformly distributed, as shown in Figure 2. The electrostatic distribution and the distribution of counter-ions around the polyion sphere can then be calculated from the Poisson-Boltzmann equation. If the added simple electrolyte is of a uni-uni valence type, the electrostatic potential and ionic distributions calculated are given by [2, 3, 4]

$$-\psi = E_D\left\{1 - e^{-\kappa R}(1+\kappa R)\frac{\sinh(\kappa r)}{(\kappa r)}\right\} \qquad (r < R) \tag{1}$$

$$= \tfrac{1}{2}E_D\{(1+\kappa R)\,e^{-\kappa R} - (1-\kappa R)\,e^{\kappa R}\}\frac{e^{-\kappa r}}{\kappa r}, \qquad (r > R) \tag{2}$$

where

$$E_D = -\frac{kT}{e}\ln\frac{1+\sqrt{1+\left(\frac{8\pi}{3}\frac{1}{z}R^3 n_c\right)^2}}{\left(\frac{8\pi}{3}\frac{1}{z}R^3 n_c\right)} \tag{3}$$

$$n_c = (N_A/10^3)\,C_s \tag{4}$$

and

$$\kappa^2 = \frac{8\pi e^2 N_A C_s}{10^3 DkT} \tag{5}$$

R is the radius of the polyion sphere, Z is the charge number of the polyion, C_s is the concentration of added salt (mol/l), and κ is the reciprocal length of ionic atmosphere. If we calculate the free charge distribution around the polyion sphere, we have Figure 3, and the effective charge of the polyion sphere $[q]$ is [4]

$$[q] = D\frac{E_D}{2\kappa}(1+\kappa R)\,e^{-\kappa R}\{(1+\kappa R)\,e^{-\kappa R} - (1-\kappa R)\,e^{\kappa R}\}. \tag{6}$$

However, this kind of the effective charge is generally inapplicable to the calculation of thermodynamic properties of linear polyelectrolyte solutions. For example, the calculated results depend on molecular weight of the polyelectrolyte, whereas most observed values are independent of molecular weight.

Thermodynamic properties of polyelectrolyte solutions are mostly determined by ionic distribution around the polyion skeleton. To explain thermodynamic properties, therefore, the rod-like model may be effective. It is assumed that the polyion is a rod of infinite length and has smeared charges distributed uniformly over the surfaces of the rod. That is, a real polyion chain consists of a series of discrete charges and each charge is surrounded by its own ionic atmosphere. If the radius of ionic atmosphere is

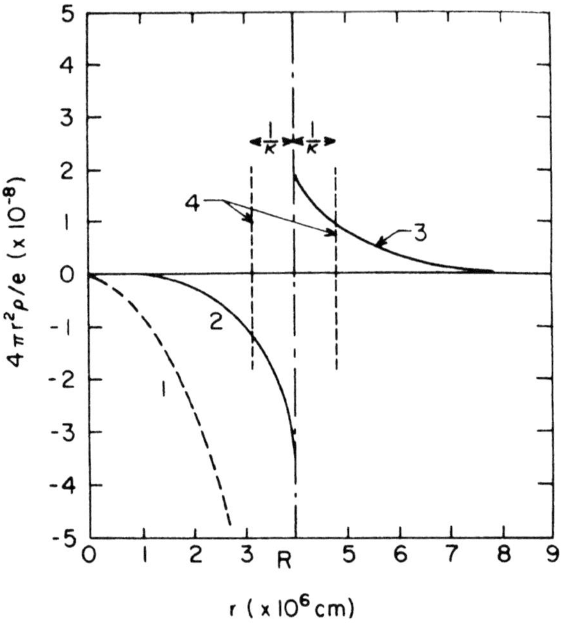

Fig. 3. Charge distribution around a spherical polyion in simple salt solutions. (Reproduced from Reference [4]). Curves 2 and 3 denote the net charge distribution inside and outside the sphere. Lines 4 denote the centres of those distributed positive and negative charges. $Z = 1070$, $R = 3.94 \times \times 10^{-8}$ cm and NaCl conc. $= 1.31 \times 10^{-3}$ N were assumed.

larger than the distance between two neighboring charges so that the ionic atmosphere around the neighboring charges overlaps extensively, the electric field around the series of discrete charges may be well represented by the electric field around a rod having smeared charges on the surface. Therefore the radius may not necessarily be equal to the real radius of the polyion skeleton.

If this rod is located in an infinite volume of a simple electrolyte solution, the distribution of counter-ions may be calculated from the Poisson-Boltzmann equation with proper boundary conditions. To solve the Poisson-Boltzmann equation, the so-called Debye-Hückel approximation $e\psi/kT \ll 1$ can be safely assumed for the porous sphere model but cannot generally be assumed for rod-like model. When the assumption is employed, we have [5]

$$\psi = \frac{2Z_e K_0(\kappa r)}{DL(\kappa a)\, K_1(\kappa a)}, \tag{7}$$

where a is the distance of closest approach of small ions to the polyion and K represent modified Hankel function. Kotin and Nagasawa [6] employed a numerical integration method to obtain the electrostatic potential and ionic distribution around the rod. A few examples of comparison between the values obtained by the numerical integration and the values calculated from Equation (7) are shown in Figure 4. [6] If κ is

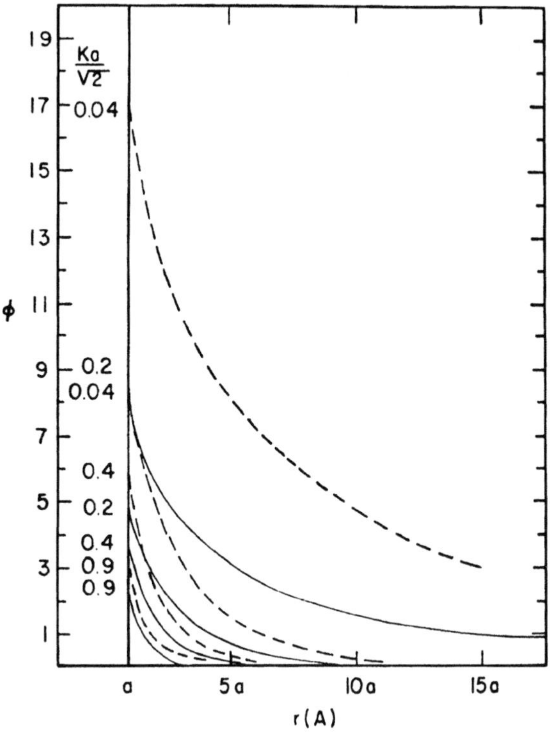

Fig. 4. Comparison between the electrostatic potentials around a rod-like polyion calculated with and without the Debye-Hückel approximation. The broken lines denote the calculated values of Equation (7), while the solid lines denote the values calculated from the Poisson-Boltzmann equation without the Debye-Hückel approximation. (Reproduced from Reference [6].)

large, the difference between both values is not large. As κ decreases, however, the difference becomes larger. A feature of the rod-like model may be that the calculated values are independent of molecular weight. The ion-binding assumption of the third class becomes effective when employing this rod-like model or when employing a more realistic model like a pearl-necklace.

Thus, the model to be employed for polyelectrolytes may be different with the purpose of study, though the most realistic model may be as shown in Figure 1. That is, we should adopt a model considering what we want to clarify or to stress by using the model. However, whichever model we employ, we should remember that a feature of polyelectrolyte is exaggerated by the model and other features are over simplified. In the spherical model, we neglect the distribution of counter-ions around the polyion skeleton. In the rod-like model, we neglect the more or less coiled conformation of the polyion skeleton. In both cases, the simplification of model must cause disagreement in comparison between theory and experiments. When employing the sphere model, the use of the effective charge density may supplement this overly simplified assumption. When employing the rod-like model having smeared charges on the surface, we should remember that the free energy change accompanying the change in the polymer conformation is neglected. It was often pointed out that the neglect of this free energy change prevents agreement between theory and experimental potentiometric titration data. [7, 8, 9]

Moreover, if the ionic strength is so high that the radius of ionic atmosphere around a fixed charge is smaller than the distance between neighboring fixed charges, the rod-like model having smeared charges may fail to be valid and, instead, a discrete charge assembly consisting of beads surrounded by its own ionic atmosphere must be employed.

2. Ion-Binding Due to Electrostatic Interaction

In many discussions on thermodynamic and hydrodynamic properties of linear polyelectrolytes, it was pointed out that a part of counter-ions behaves as if those ions were reacted with fixed charges to form undissociated groups, even though they are believed to be completely dissociated. The best known experiment may be the determination of transport number of counter-ions by Wall and his coworkers [10–12]. That is, a sodium polyacrylate solution is placed in both compartments of the cell as shown in Figure 5 and an amount of electricity is passed through the cell. If there is

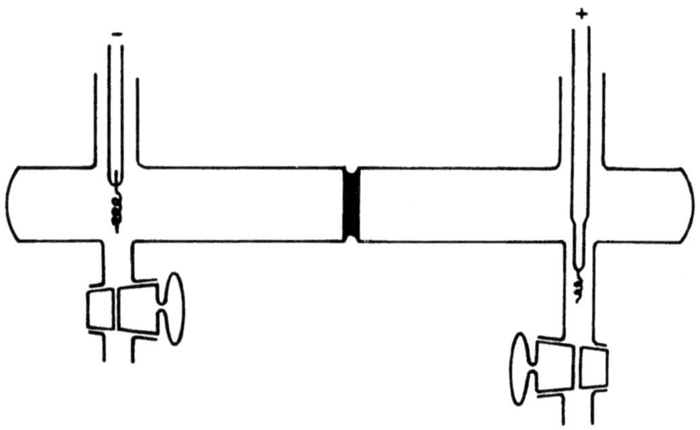

Fig. 5. A cell for transference number determination. (Reproduced from Reference [10].)

no electrostatic interaction between polyion and counter-ions, the concentration of counter-ion in the left compartment and the concentration of polyion in the right compartment should increase with electricity passed. The increases in those concentrations can be calculated if we know the mobilities of those ions. The experimental results, however, show that the increase in the concentration of Na^+ is much lower than the increase expected from the assumption of complete dissociation. At 100% neutralization, the concentration of Na^+ in the left compartment rather decreases with electricity passed. The experimental results can be reasonably explained if we assume that part of the counter-ions are bound with fixed charges to move together with the polyion and the rest of the counter-ions are completely free to have a mobility equal to that in simple electrolyte solutions. In the other transport phenomena such as sedimentation and diffusion, too, the effective charge density of polyion gives a reasonable explanation to experimental results. [13]

In the thermodynamic properties of polyelectrolyte solutions, it is always found that polyelectrolyte solutions are highly non-ideal. That is, the activity coefficient and osmotic pressure coefficient are much lower than unity. [1] Reasonable explanations on these phenomena are usually obtained by assuming that a part of counter-ions are bound on fixed charges.

However, there is ample evidence that this kind of ion-binding is not due to complex formation or other non-ionic force acting between fixed charges and counter-ions.

(1) The pH of polystyrene sulphonic acid solution is much higher than the value expected from the analytical concentration of H^+ if the polystyrene sulphonic acid is completely ionized. This might be interpreted by assuming that sulphonic acid groups are not completely ionized. The apparent degree of ionization of the polyacid would then be about 0.38 independent of polymer concentration. However, it can be proved by a non-thermodynamic method that polystyrene sulphonic acid is completely dissociated in solution. That is, according to the theory of Gutowsky and Saika, [14] the proton magnetic resonance spectrum of H_2O is shifted with increasing concentration of H^+ in the solution. That is, a parameter denoting the chemical shift per unit amount of H^+, p/s, is 11.4 for HCl and 11.8 for nitric acid, [15] which both are believed to be completely ionized in aqueous solutions. The p/s for polystyrene sulphonic acid solution is also found to be 11.5. [16] Therefore, it is certain that the too high pH value of polystyrene sulphonic acid solutions is caused by decrease in activity coefficient of H^+ due to strong electrostatic interaction between polyion and counter-ions. A similar conclusion may be obtained from the Raman spectrum measurements of polystyrene sulphonic acid solutions.

(2) The fact that the interaction between polyacrylate or polymethacrylate ion and sodium ion is purely electrostatic can be confirmed from the IR spectrum of $-COO^-Na^+$ in the solid state and in D_2O solution [16 a]. It was also reported that the interaction between alkali earth ion and carboxylic acid group is not specific [16 a].

(3) In transport phenomena of alkali metal salts of polyacids, the average mobility of counter-ions is much lower than that of the free ion. This is generally interpreted by assuming that a part of counter-ions are bound on fixed groups due to strong electrostatic attraction of the polyion as explained above. However, it is clear that the bound ions cannot be treated as undissociated ions. They have a finite mobility though the mobility may be much lower than the limiting mobility of the ion. This can be concluded by comparing the conductometric titration data of polystyrene sulfonic acid (which is a strong acid) and those of polyacrylic acid (which is a weak acid) with sodium hydroxide. Typical conductometric titration curves of the two polyacids are schematically compared in Figure 6. They are qualitatively similar to those of simple strong and weak acids such as HCl and CH_3COOH. This graph may

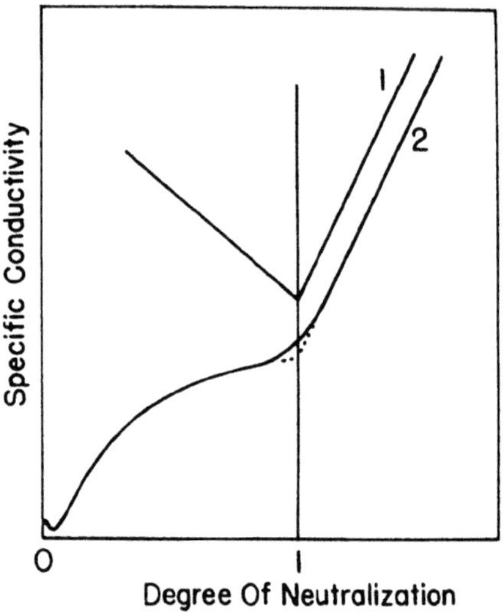

Fig. 6. Schematic comparison between the conductometric titration of polyacrylic acid (2) and polystyrene sulphonic acid (1). (Reproduced from Reference [1].)

be interpreted as follows: the hydrogen ion in the undissociated —COOH must have zero mobility. If the hydrogen ion is replaced with Na^+ (i.e. neutralization), the Na^+ must be completely dissociated and has a finite mobility. As the high charge density of polyion attracts counter-ions around the polyion skeleton, those ions may have mobilities lower than free ions but, at least, the mobilities are not zero. Therefore, the conductivity of the solution increases with degree of neutralization. Since the degree of ion-binding increases with increasing charge density, the increase of conductivity becomes less with increasing degree of neutralization, resulting in the upward concave on the titration curve, as shown in Figure 6. In contrast to that of polyacrylic acid, the titration curve of polystyrene sulphonic acid decreases linearly with increasing degree of neutralization. This is because the charge density of polystyrene sulphonate ion does not change with neutralization. Considerable amounts of both H^+ and Na^+ are bound on polyion. However, bound ions have finite mobilities and the mobility of bound H^+ must be higher than that of bound Na^+, just as the mobility of free H^+ is higher than that of free Na^+. Therefore, replacement of bound H^+ with Na^+ diminishes the conductivity of the solution. It is clear that the bound counter-ions must have a finite value even though the mobilities are lowered by the effect of polyion. The problem may be whether the counter-ions can be classified into two distinct groups discontinuously or not.

(4) Another remarkable example that the bound ions cannot be treated as undissociated groups may be seen in the electrophoretic velocity of polyion in the presence of an added neutral salt. The mobility of polyion can be visually observed in electrophoresis. If a large amount of counter-ions were bound on polyion firmly to decrease the charge density of the polyion, the mobility of the polyion should be much

lower than the mobility of monomer and the mobility should depend on the counter-ion species since the degree of ion-binding depends on the ionic species. In contrast to this prediction, however, there is a theory of Hermans [17], Fujita [18], Overbeek and Stigter [19]. In their theory, it is predicted that the polyion would become like a free-draining sphere at the limit of high ionic strength as if there were no effect from its ionic atmosphere. This conclusion was obtained by using the sphere model for polyion, as shown in Figure 2, and solving the Navier-Stokes equation under the effect of an applied electric field. The basic equation and the method of calculation employed agree with those employed by Debye and Bueche [20] to calculate the frictional coefficients of non-ionic linear polymers if there were no ions around the polyion. At high ionic strengths they obtained

$$U/E = \frac{e_s}{f}\left(1 + \frac{\sigma^2}{3\beta^2}\frac{2 + \sigma/\beta}{1 + \sigma/\beta}\right), \tag{8}$$

$$\sigma^2 = v_m f R^2/\eta_0 , \tag{9}$$

$$\beta = KR , \tag{10}$$

$$v_m = Z/\tfrac{4}{3}\pi R^3 , \tag{11}$$

where f is the frictional coefficient of a segment, and e_s is the charge number of a segment. At the limit of infinite ionic strength where all counter-ions enter into the polyion domain, Equation (8) becomes

$$U/E = e_s/f \tag{12}$$

which is equal to the mobility of a monomer. Therefore, the electrophoretic mobility of polyion at high ionic strengths must be; (1) as high as the mobility of a monomer, (2) independent of molecular weight, (3) independent of ionic species of counter-ion. That is, the polyion should move like a free draining coil at the limit of high ionic strength, as if there were no ion-binding.

This conclusion may be valid independent of the model employed by Hermans, Fujita *et al.*, because the force acting on a segment of polyion has the opposite direction and the magnitude equal to the force acting on a counter-ion and, therefore, the hydrodynamic perturbations arising from the forces acting on both segment and counter-ions cancel each other if the polyion domain is neutral at the limit of high ionic strength.

The speculations were supported by experimental results using sodium salts of polystyrene sulphonic acid or polyacrylic acid. Figure 7 shows that the electrophoretic mobility is independent of molecular weight and as high as the mobility of a monomer [21]. The radius of a monomer calculated from the mobility using the Stokes law is about 3.0 Å which is reasonable compared with the molecular radius of the monomer. Table I shows that the electrophoretic mobility of polyion is independent of ionic species of counter-ion [22].

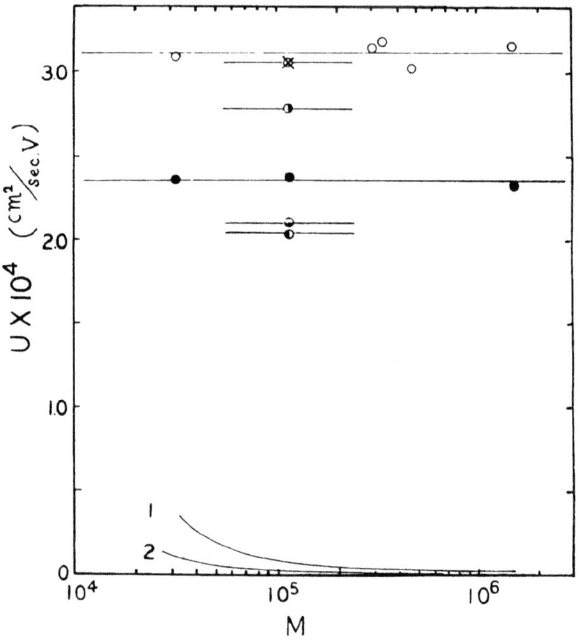

Fig. 7. Molecular weight dependence of electrophoretic mobility of polyacrylate ion. Lines denoted by 1 and 2 are the calculated mobilities from Henry's theory assuming that the polyion sphere is a non-draining sphere having a surface charge of Equation (6). Degree of neutralization of polyacrylate ion is 1.0, 0.6, 0.4, 0.2, 0.1, 0.05 from top to bottom. NaCl concn. is 0.1 N. (Reproduced from Reference [21].)

TABLE I

Electrophoretic mobilities of various salts of poly (styrene sulphonic acid) Salt conc. 0.100 N

Salt	Λ°	$U_p \times 10^{+4}$ ($cm^2/s^{-1}V^{-1}$)
LiCl	115.03	3.4
NaCl	126.45	3.3
KCl	149.85	3.6
CsCl	153.61	3.3
$(CH_3)_4$Ncl	121.27	3.4

(Reproduced from Reference [22].)

(5) It is clear from dielectric constant measurements that even alkaline earth metal ions are movable in the vicinity of polyion. [23]

Thus, it is clear that the counter-ions bound by electrostatic force are different from undissociated molecules. However, the above facts do not mean that the ion-binding of the third class can always be analysed only due to electrostatic interaction. The structure of water, degree of hydration, the conformation of polyion etc., may be, in some cases, changed with a change in the electrostatic force between polyion and counter-ions. Those changes accompanying the change in electrostatic force should be taken into account in discussing the ion-binding of the third class. Here, however, it is pointed that, even when the interaction between polyion and counter-ions is purely electrostatic, the idea of counter-ion binding may be useful.

3. Features of Ion-Binding

If the force acting between fixed and bound ions is a covalent force, the degree of ion-binding may be determined spectroscopically. The counter-ions bound on fixed charges cannot be detected by such methods if the force is electrostatic, though the quadruple magnetic resonance may be used to detect the electrostatic interaction, as was done for ion-pair formation in simple electrolyte solutions [27].

Therefore, the degree of ion-binding due to electrostatic is, in most cases, determined by measuring the amount of free ions. The determination is, moreover, based on the assumption that the counter-ions can be classified into two groups, one being completely free and the other being completely bound. Therefore, the degree of binding obtained varies with the assumption on the behavior of 'free' ions. That is, even the polyion bound with counter-ions has a negative charge and, hence, must have an electrostatic interaction with free ions. Nevertheless, it is often assumed that the activity coefficients of free counter-ions are either unity or the same as that of the ions in simple electrolyte solutions at the same concentration. This assumption is not self-consistent, but may be equivalent to the roughness of the ion-binding assumption. Although it may be possible to take into account the electrostatic interaction between the free ions and the polyion whose charge is diminished by counter-ion binding, in this paper we will adopt the assumption that free ions are the same as the ions in simple electrolyte solution at the same concentration, for convenience.

The ion-binding of the first class, which is due to complex formation or covalent bond formation, may be analysed based on the stepwise dissociation equilibrium, though the effect of charge may be taken into account. In ion-binding of the third class which is caused by electrostatic interaction, however, it is clear that the degree of dissociation cannot be treated by step-wise dissociation equilibrium. A remarkable feature of the ion-binding of the third class may be observed in the fact that most thermodynamic properties such as activity coefficient, osmotic pressure, are approximately additive with respect to the contributions from polyelectrolyte and added salt. For example:

(1) The additivity was first implied in the experiments of Mock and Marshall on the pH of polystyrene sulphonic acid solution containing HCl [25]. That is, if we express pH of pure polystyrene sulphonic acid solution and pure HCl solution by $(\mathrm{pH})_p$ and $(\mathrm{pH})_s$, the pH of the mixture is expressed by

$$\mathrm{pH} = -\log(\gamma^p_{H^+} C^p_{H^+} + \gamma^s_{H^+} C^s_{H^+}), \tag{13}$$

where $(\mathrm{pH})_p = -\log(\gamma^p_{H^+} C^p_{H^+})$, $(\mathrm{pH})_s = -\log(\gamma^s_{H^+} Czx)$ and γ_{H^+} gives the activity coefficients. This means that the activity coefficient of hydrogen ion in the mixture may be expressed by the number average of the contributions from both components.

(2) A similar additivity rule was confirmed for Na^+ activity: The average activity coefficient of Na^+ in the mixture of Na-polyvinylsulphonate and NaCl is given by a number average of the Na^+ activity coefficient of the components [26].

$$\gamma_{Na} = \frac{\gamma^p_{Na^+} C^p_{Na^+} + \gamma^s_{Na^+} C^s_{Na^+}}{C^p_{Na^+} + C^s_{Na^+}}, \tag{14}$$

where $\gamma^p_{Na^+}$ and $\gamma^s_{Na^+}$ are the activity coefficients of Na^+ in the pure aqueous solutions of the polyelectrolyte and NaCl in the absence of the other component, $C^p_{Na^+}$ and $C^s_{Na^+}$ are the concentrations of Na^+ from both salts.

In contrast to the behavior of activity coefficient of counter-ion, the activity coefficient of by-ion, Cl^-, is little affected by the addition of polyanion, such as [26, 27]

$$\gamma_{Cl^-} = \gamma^s_{Cl^-}. \tag{15}$$

Strictly speaking, however, the equalities in Equation (14) and (15) are simply approximate. $\gamma^s_{Cl^-}$ is found to be affected by addition of polyion though slightly, and hence, $\gamma^s_{Na^+}$, too, is believed to be affected by the addition of polyion. The equality in Equation (14) holds if the effect of polyion on $\gamma^s_{Na^+}$ in the mixture is taken into account from the analogy to the effect of polyion on $\gamma^s_{Cl^-}$ [26]. Without such modification for $\gamma^s_{Na^+}$, the values of $\gamma^s_{Na^+}$ calculated from Equation (14) agree with the observed values with an error of several per cent in the system of polyvinyl sulphate and NaCl. This problem will be discussed later.

As is well known, single ion activity coefficients cannot be defined thermodynamically. In the above discussion, $\gamma^2_{NaCl} = \gamma_{Na^+} \cdot \gamma_{Cl^-}$ has a physical meaning in thermodynamics but the separation between γ_{Na^+} and γ_{Cl^-} is carried out based on the assumptions that $\gamma_{K^+} = \gamma_{Cl^-}$ in KCl solutions and the activity coefficient of an ion is the same if the ionic strength is the same. Moreover, the liquid junction potential is assumed to be negligible. However, these ambiguities can be eliminated if we calculate the mean activity coefficient of NaCl in the presence of polyelectrolyte, γ_{NaCl}. The usefulness of the additivity can be observed in the Donnan membrane equilibrium where γ_{NaCl} is discussed [28].

(3) The osmotic pressure of polyelectrolyte in added salt system is given by the sum of the osmotic pressure of each component, such as

$$\pi = \pi^p + \pi^s, \tag{16}$$

where π^p and π^s are the osmotic pressures of polyelectrolyte and the added salt at the same concentrations as in the mixture and in the absence of the other component [28].

(4) A similar additivity is observed for electric conductivity, too. The specific conductivity of Na^+ polystyrene sulphonate in NaCl solutions of various concentrations is shown as a function of polymer concentration in Figure 8 [29]. The observed values in NaCl solutions of various concentrations are almost parallel, showing that

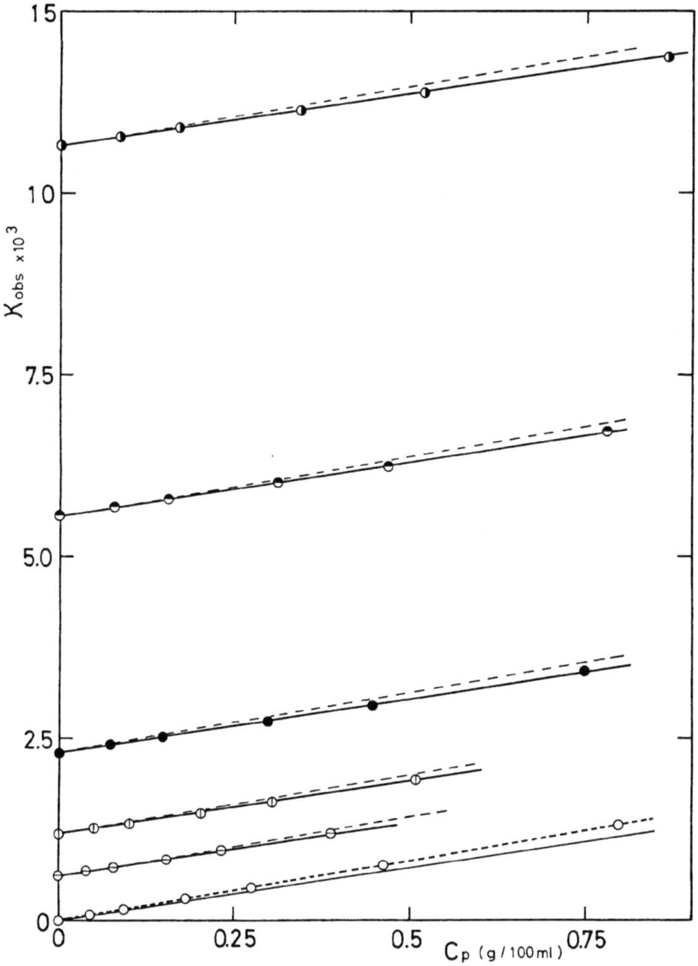

Fig. 8. Specific conductance of sodium polystyrene sulphonate solutions in the presence of NaCl [29]. Number average molecular weight 2.0×10^5, 25°C, 1 kHz. The extrapolation of the specific conductance to infinite high frequency was carried out for the solutions in the absence of added salt, assuming the linearity between specific conductance and reciprocal square root of frequency, but the error was within 0.4 %. NaCl concentrations are 0, 0.005, 0.01, 0.02, 0.05 and 0.1 N from bottom to top. Both solid and broken lines are drawn in parallel, respectively.

the contributions of polyelectrolyte and an added salt to specific conductance are almost additive. All solid lines in Figure 8 are drawn in parallel. Moreover, the polymer concentration dependence of the specific conductance of the pure aqueous solutions shows a feature that the equivalent conductance of polyelectrolyte decreases with decreasing polyion concentration, as the osmotic pressure coefficient and activity coefficient of counter-ion do so [30]. Even this feature appears to be seen in all other experiments with added salt.

TABLE II

Mobilities of polyion, counter-ion and by-ion

No.	NaCl Conc. (N)	Degree of neutralization	Transference numbers			Specific conductance	U_p ($\times 10^4$ cm^2 s$^{-1}V^{-1}$)	Λ_{Na^+}	Λ_{Cl^-}	Mobility of polyion by electrophoresis U_p($\times 10^4$ cm^2 s$^{-1}V^{-1}$)	f
			t_p	t_{Na}	t_{Cl}						
1		1.00	0.14_8	0.51_3	0.33_9	0.01156	4.1_1	$27._7$	$60._4$	3.1_1	0.34
2	0.09823	0.600	0.06_3	0.57_3	0.36_4	0.01129	2.8_6	$33._1$	$65._9$	3.0_8	0.39
3		0.200	0.01_5	0.61_2	0.37_3	0.01078	1.8_8	$37._6$	$67._2$	2.4_0	(0.12)
4		1.00	0.19_6	0.52_6	0.27_7	0.006553	3.0_9	$19._7$	$70._2$		0.29
5	0.04911	0.600	0.13_3	0.51_0	0.35_6	0.006306	3.3_7	$29._9$	$65._5$	3.6_0	0.47
6		0.200	0.04_2	0.53_8	0.42_0	0.005793	2.9_2	$42._1$	$63._5$		0.79
7		1.00	0.73_9	0.29_1	−0.03	0.002385	4.2_4	$-1._2$	$70._6$		0.32
8	0.009824	0.600	0.45_8	0.42_2	0.12_0	0.002134	3.9_2	$7._2$	$91._7$	4.6_5	0.39
9		0.200	0.20_6	0.32_5	0.46_9	0.001598	3.9_6	$40._6$	$52._9$		0.82

Here, $U_p = \Lambda_p/96500$.
Concentration of poly (acrylic acid) in the original solutions is 0.04309. (monomer mol/1000 g).
(Reproduced from Reference [22].)

(5) The specific conductance includes the contributions of counter-ion, byion and polyion. The above additivity holds not only for the total conductivity but also for the equivalent conductance of the counter-ion [22]. The mobility of byion is not changed much with addition of polyelectrolyte, and the change in the mobility of polyion with concentration of added salt is not very large. Table II shows the mobilities of Na^+ and Cl^- in the mixture of NaCl and Na salt of polystyrene sulphonic acid [22]. It is observed that the mobility of Cl^- is little affected by addition of polystyrene sulphonate. These values were determined by measuring the transport numbers of each component using the Hittorff method. Figure 9 shows the degree of

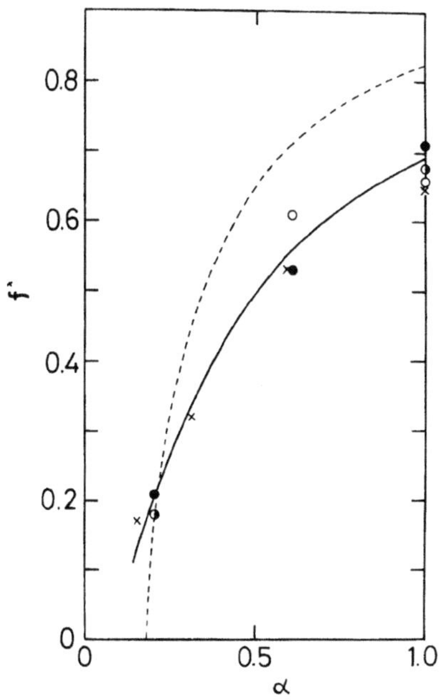

Fig. 9. Degree of ion-binding calculated from transference number in the absence and presence of added salt (NaCl). Crosses denote the data of Wall *et al.* for pure aqueous solutions. Open, half-filled and filled circles denote the data in the presence of NaCl at the concentrations 0.098, 0.049, 0.0098 N respectively. The broken line denotes the calculated values of Equation (24). (Reproduced from Reference [22].)

ion-binding calculated from the observed mobilities of Na^+ assuming that Na^+ in the mixture can be classified into two groups, one having the same mobility as in simple salt solutions and the other being those bound on polyion. The corresponding values obtained for pure aqueous solutions by Wall *et al.* are shown in the same figure [11]. The fact that all observed values fit a single line shows the validity of the additivity in the equivalent conductance of counter-ion.

The reason for this additivity rule has been discussed by various authors, but it is out of the scope of this paper to discuss those theories. Here, it is noted only that the rule is practically important or meaningful because the use of a constant effective charge density in analysing the solution properties of polyelectrolyte may be rationalized from this rule. If the degree of ion-binding were different with ionic strength and polymer concentration much, the assumption of the effective charge density would be useless. In practice, in many papers on thermodynamic properties, the effective charge density was assumed to be independent of ionic strength and, sometimes, of polymer concentration, too.

Strictly speaking, however, it should be stressed that the above rule is only approximate. Let us explain it using the examples of (2) and (4): Even the activity coefficient of byion, Cl^-, is affected by addition of polyion, though slightly (see Figure 10).

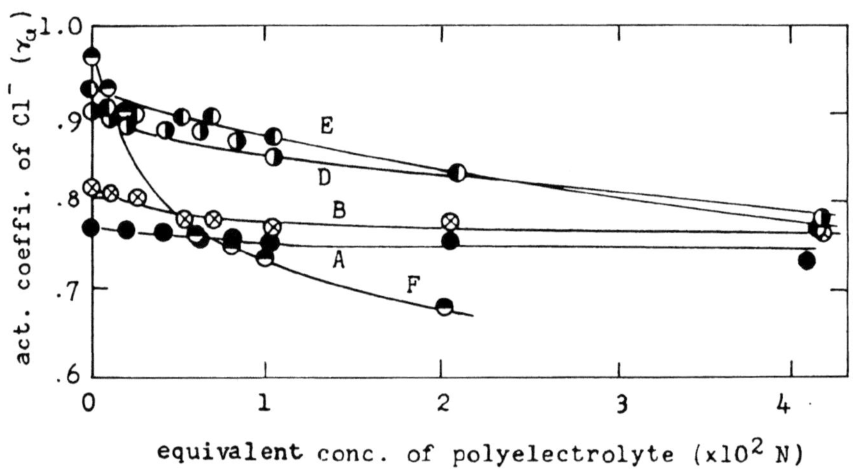

Fig. 10. Activity coefficient of Cl^- in the mixture of sodium polyvinyl alcohol sulphonate and NaCl. NaCl concentration (A) 1.00×10^{-1} (B) 5.00×10^{-2} (D) 1.01×10^{-2} (E) 5.00×10^{-3} (F) 0.991×10^{-3} N. (Reproduced from Reference [26].)

This implies that the activity coefficient of counter-ions constituting the added salt would be affected by the addition of polyelectrolyte. In the comparison between the calculated values of Equation (14) and the observed values, good agreement was found if such a modification was given to $\gamma^s_{Na^+}$. This means that the additivity in its strict meaning may hold only at the limit of zero polyion concentration. If the change in $\gamma^s_{Na^+}$ is neglected, the disagreement between the calculated and observed values would become as large as several percent [31]. Whether this error can be neglected or not depends on the purpose of the research. If we assume that γ_{NaCl} of NaCl added is not affected by addition of polyelectrolytes, the degree of ion-binding in the absence of added salt should be, in some cases, slightly modified by addition of neutral salt. That is, the effective charge of polyion adopted to explain the data over the whole range of salt concentration may be assumed to be slightly different from that in pure aqueous solution. One example is seen in Figure 8. With use of the slightly modified values for specific conductance of pure aqueous solutions of Na-PSS, almost satisfactory agreement between calculated values and observed values is found over the entire range of NaCl concentration used.

Thus, the additivity discussed here is equivalent to the assumption of the effective charge density which is assumed to be constant independent of ionic strength. The assumption may be satisfactory at the limit of infinitely low polymer concentration. If an error of, probably, several percent may be permissible for a purpose, the assumption may be used at finite, but not too high, polymer concentrations.

4. Calculation of the Degree of Ion-Binding

As is clear from the above discussion, the definition of ion-binding is not unique. It depends on our assumption on the state of free ions. Sometimes, it is assumed that the 'free' ions are the same as the ions in simple salt solutions at the same concentration. Sometimes, the electrostatic interaction between polyion and free counter-ions is taken into account. In the case, the so-called Debye-Hückel approximation may be assumed as the charge density is lowered with ion-binding.

If we employ the former assumption, the degree of ion-binding f^* may be calculated from activity coefficient of counter-ion γ_+, osmotic coefficient g or others.

$$f^* = \frac{A_+ \,(\text{total}) - A_+ \,(\text{free})}{A_+ \,(\text{total})} \tag{17}$$

and

$$A_+ \,(\text{free}) / A_+ \,(\text{total}) = \gamma_+ \text{ or } g\,, \tag{18}$$

where A_+ denotes the amount of counter-ion.

The data of ion-binding calculated from osmotic pressure and activity coefficient measurements were accumulated by Katchalsky, Kedem and Alexandrowicz as shown in Figure 11 [37]. Most of those data were obtained in the absence of added salt, but they may be assumed to be applicable in the presence of added salt, too, if the additivity holds. Moreover, the polymer concentration dependence of the values may be assumed negligible. Those data and other data appear to show that the degree of ion-binding is primarily determined by the charge density. In addition, the data published appear to show that: (1) The degree of ion-binding does not depend much on the counter-ion species, if the counter-ions are alkali metal ions and also if the polymer concentration is low. (2) The degree of ion-binding may depend slightly on the radius of polyion rod. (3) The degree of ion-binding calculated from osmotic pressure data is generally half as low as that calculated from counter-ion activity coefficient, as seen in Figure 11.

The degree of ion-binding can also be determined from observed mobility of counter-ion. A typical example is shown in Figure 9. The degree of ion-binding determined in transport phenomena is usually lower than the values determined thermodynamically. That is, the effective charge density of polyion $[q]$ is different

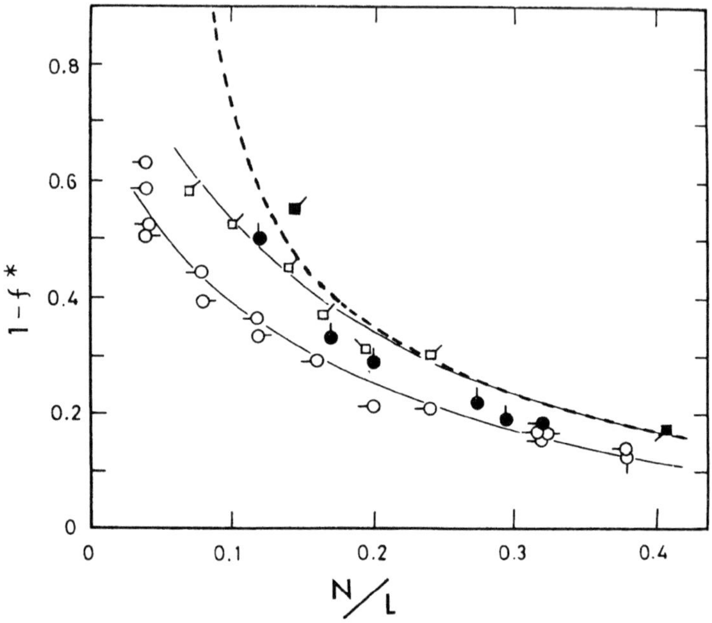

Fig. 11. Degree of ion-binding calculated from thermodynamic data. (Reproduced from Reference [32].) The broken line denotes the calculated values of Equation (24).

with methods for determination such as

$$[\eta]_1 > [\eta]_\gamma > [\eta]_g,$$

where the suffixes show the method for determination. On the other hand, the activity coefficient γ, osmotic coefficient g and conductivity coefficient Λ/Λ° of simple salts are usually in the order of

$$g > \Lambda/\Lambda^\circ > \gamma.$$

If we calculate the effective charge density of a polyion $\{N(\text{effective})/L\}$ from a theory, the degree of ion-binding is related to the effective charge density by

$$f^* = \{N/L - N\,(\text{effective})/L\}/(N/L), \tag{19}$$

where N/L is the analytical charge density.

4.1. Counter-ion condensation theory

Manning [33] concluded that if the charge density is so high that the distance between two neighboring fixed charges b is shorter than a value b_0 which is defined by

$$e^2/DkTb_0 = 1 \tag{20}$$

the counter-ions are condensed on the fixed charges to keep the charge density always at $b = b_0$. The value of b_0 in water at 25 °C is 7.1 Å. This definition of the effective charge is exactly the same as defined by Imai [34] and Oosawa [35]. The difference between the theory of Manning and that of Imai and Oosawa may be in the calculation of the electrostatic interaction between uncondensed ions and the polyion having the effective charge density.

In the theory of Manning, effective charge density is given by

$$N(\text{effective})/L = 1/b_0 \quad \text{for} \quad L/N > 7.1 \text{ Å} \tag{21}$$

and the degree of ion-binding is

$$f^* = 1 - (Ne^2/DkTL)^{-1} \quad \text{for} \quad L/N > 7.1 \text{ Å}. \tag{22}$$

Below the critical charge density, there should be no ion condensation but the activity coefficient or osmotic coefficient of the polyelectrolyte solution is explained on the assumption of the Debye-Hückel approximation. According to this theory, the additivity of osmotic pressure can be proved, though the additivity of counter-ion activity coefficient cannot be proved.

4.2. From Analogy to Ion-Pair Formation of Bjerrum

Kotin and Nagasawa [6] defined the counter-ion binding in analogy to the definition of Bjerrum on ion-pair formation [36]. That is, it is assumed that a polyion is placed in an infinite volume of a neutral salt solution of uni-uni valent type and the polyion is a rod of infinite length having a charge density N/L. Moreover, it is assumed that the ionic distribution around the rod is determined from the Poisson-Boltzmann equation. Then, if one plots the distribution of counter-ions $P_c(r)$ against the distance from the axis of the polyion r,

$$P_c(r)\,\mathrm{d}r = C_s \exp(-e\psi/kT)\,2\pi r\,\mathrm{d}r, \tag{23}$$

where C_s is the bulk concentration of added electrolyte and ψ is the electrostatic potential at the position r, one observes a minimum in the curve which shifts to higher values of r as the salt concentration decreases. The area under each curve is proportional to the number of counter-ions within a specified value of r. If we define the bound ions as the counter-ions which are distributed inside the minimum point r_m, the degree of ion-binding f^* is given by

$$f^* = 1 - (2Ne^2/DkTL)^{-1}. \tag{24}$$

The f^* in Equation (24) is independent of ionic strength. Moreover, we have $f^* = 0$ at $N/L (\equiv b_0^{-1}) = DkT/2e^2$. However, this value for b_0 has no physical importance since the rod-like model having smeared charges is entirely inapplicable at low charge densities.

At finite polymer concentration, Katchalsky *et al.* [32] used the same polyion model and defined the bound ions by the counter-ions inside the distance r_0 which satisfies

$$e\psi/kT = -1 \quad \text{at} \quad r = r_0 \quad \text{for negative macroion}. \tag{25}$$

At moderate polymer concentration, Equation (25) corresponds to

$$\mathrm{d}\phi/\mathrm{d}\ln r = 1.4\,, \tag{26}$$

where $\phi = e\psi/kT$. At infinite dilution, Equation (25) becomes

$$\mathrm{d}\phi/\mathrm{d}\ln r = 1 \tag{27}$$

which also gives Equation (24). Details of the theory of Katchalsky and his coworkers may be referred to reviews by Katchalsky *et al.* [32, 37].

The most important feature of Equation (24) is that the degree of ion-binding calculated from this equation is independent of ionic strength, i.e., the concentration of added salt. Moreover the counter-ions outside r_m or r_0 are assumed to behave in the same manner as those in simple electrolyte solution. Hence, the activity is automatically satisfied.

The values calculated from Equation (24) are shown in Figure 11. The agreement is satisfactory at high charge densities but the observed values deviate from the calculated line as the charge density decreases. It is often observed that agreement between calculated and observed values is poor at low charge densities if a rod-like model having smeared charges on the surface is employed. It is due to the failure of the model. A comparison between the calculated and experimental values estimated from transport phenomena is given in Figure 9. The agreement is fair. It is not reasonable to compare the theory of Manning with experiments in this manner since electrostatic interaction between free ions and polyion is taken into account in his theory. His original papers [33] should be referred to.

The fair agreement between the calculated values and the observed ones in Figure 11 and 9, at high charge densities, does not mean that the additivity was proved by theory, but simply means that the definition of bound ions in Equations (24) and (25) well represents the experimental results so long as the free ions are assumed to be the same as in simple electrolyte solutions. Naturally, it would be more advisable if all solution properties of polyelectrolytes can be explained quantitatively without use of these assumptions.

References

1. Rice, S. A. and Nagasawa, M.: *Polyelectrolyte Solutions*, Academic Press, New York, 1961.
2. Hermans, J. J. and Overbeek, J. T. G.: *Rec. Trav. Chim.* **67**, 761 (1948).
3. Lifson, S.: *J. Chem. Phys.* **27**, 700 (1957).
4. Nagasawa, M. and Kagawa, I.: *Bull. Chem. Soc. Japan* **30**, 961 (1957); Nagasawa, M.: *J. Am. Chem. Soc.* **83**, 300 (1961).
5. Hill, T. L.: *Arch. Biochem. Biophys.* **57**, 229 (1955).
6. Kotin, L. and Nagasawa, M.: *J. Chem. Phys.* **36**, 873 (1962).
7. Nagasawa, M., Murase, T., and Kondo, K.: *J. Phys. Chem.* **69**, 4005 (1965).
8. Nagasawa, M. *Pure Appl. Chem.* **26**, 519 (1971).
9. Muroga, Y., Suzuki, K., Kawaguchi, Y., and Nagasawa, M.: *Biopolymers* **11**, 137 (1972).
10. Huizenga, J. R., Grieger, P. F., and Wall, F. T.: *J. Am. Chem. Soc.* **72**, 2636, 4228 (1950); Wall, F. T. and Doremus, R. H.: *J. Amer. Chem. Soc.* **76**, 868, 1557 (1954).
11. Wall, F. T. and Eitel, M. J.: *J. Am. Chem. Soc.* **79**, 1556 (1957).

12. Wall, F. T. and Hill, W. B.: *J. Am. Chem. Soc.* **82**, 5599 (1960).
13. Alexandrowicz, Z. and Daniel, E.: *Biopolymers* **1**, 447, 473 (1963).
14. Gutowsky, H. S. and Saika, A.: *J. Chem. Phys.* **21**, 1688 (1953).
15. Hood, G. C., Redlich, O. and Reilly, C. A.: *J. Chem. Phys.* **22**, 2067 (1954);
Hood, G. C. and Reilly, C. A.: *J. Chem. Phys.* **27**, 1126 (1957);
Redlich, O. and Hood, G. C.: *Disc. Faraday Soc.* **24**, 87 (1957);
Hood, G. C., Jones, A. C., and Reilly, C. A.: *J. Phys. Chem.* **63**, 101 (1959);
Young, T. F.: *Record Chem. Progr.* (Keresge Hooker Sci., Lib.), **12**, 81 (1951).
16. Kotin, L. and Nagasawa, M.: *J. Am. Chem. Soc.* **83**, 1026 (1961).
16a. Leyte, J. C., Zuiderweg, L. H., and Vledder, H. J.: *Spectrochim. Acta* **23A**, 1397 (1967).
17. Hermans, J. J.: *J. Polymer Sci.* **18**, 529 (1955).
18. Hermans, J. J. and Fujita, H.: *Koninkl. Ned. Akad. Wetensch.*, Proc. **B58**, 182 (1955).
19. Overbeek, J. Th. G. and Stigter, D.: *Rec. Trav. Chim.* **75**, 543 (1956).
20. Debye, P. and Bueche, A. M.: *J. Chem. Phys.* **16**, 573 (1948).
21. Noda, I., Nagasawa, M., and Ōta, M.: *J. Am. Chem. Soc.* **86**, 5075 (1964).
22. Nagasawa, M., Noda, I., Takahashi, T., and Shimamoto, N.: *J. Phys. Chem.* **76**, 2286 (1972).
23. Mandel, M. and Jenard, A.: *Trans. Faraday Soc.* **59**, 2170 (1953);
Mandel, M.: *J. Polymer Sci (C)* **16** 2955 (1967);
Minakata, A.: *Biopolymers* **11**, 1567 (1972).
24. Jardetsky, O. and Wertz, J. E.: *Trans. Faraday Soc.* **49**, 363 (1953).
25. Mock, R. A. and Marshall, C. A.: *J. Polymer Sci.* **13**, 263 (1954).
26. Nagasawa, M., Izumi, M., and Kagawa, I.: *J. Polymer Sci.* **37**, 375 (1959).
27. Kagawa, I. and Kastuura, K.: *J. Polymer Sci.* **9**, 405 (1952).
28. Nagasawa, M., Takahashi, A., Izumi, M., and Kagawa, I.: *J. Polymer Sci.* **38**, 213 (1959).
29. Kato, N. *et al.*: unpublished.
30. Kern, W.: *Makromol. Chem.* **2**, 279 (1948);
Nagasawa, M. and Kagawa, I.: *J. Polymer Sci.* **25**, 61 (1957);
Takahashi, A., Kato, N., and Nagasawa, M.: *J. Phys. Chem.* **74**, 944 (1970).
31. Lyons, J. W. and Kotin, L.: *J. Am. Chem. Soc.* **87**, 1670 (1965).
32. Katchalsky, A.; Alexandrowicz, Z., and Kedem, O.: 'Polyelectrolyte Solutions' in B. E. Conway and R. G. Barradas (eds.), *Chemical Physics of Ionic Solutions*, p. 295, Wiley, New York (1966).
33. Manning, G.: *J. Chem. Phys.* **51**, 924, 934 (1969).
34. Imai, N. and Ohnishi, T.: *J. Chem. Phys.* **30** '1115 (1959).
35. Oosawa, F.: *Polyelectrolytes*, Marcel Dekker Inc., New York, 1971.
36. Bjerrum, N.: *Selected Papers*, Einar Munksgaard, Copenhagen, 1949, p. 108.
37. Katchalsky, A.: *Pure Appl. Chem.* **26**, 327 (1971).

SHORT-RANGE INTERACTIONS BETWEEN POLYIONS AND SMALL IONS

ULRICH P. STRAUSS

School of Chemistry, Rutgers University, New Brunswick, N.J., U.S.A.

List of Symbols

ΔV_D = negative of the differential volume change per mole for the ionization of carboxylic acid groups

α = degree of ionization of polyacid

pK_0 = chemical contribution to the pK of polyacids

Specific short-range interactions between polyions and small ions are gradually becoming recognized for their significant and universal role in the behavior of aqueous polyelectrolyte solutions. These interactions, which are observable even with strong polyelectrolytes, – such as the alkali metal salts of polyacids – characteristically involve changes in the solvation states of the participating species and lead to deviations of the net interionic forces from Coulomb's law. Besides being of intrinsic theoretical interest specific interactions are frequently crucial to the functioning of polyelectrolytes in technological and natural processes [1].

Our current knowledge of solvation phenomena in aqueous solutions of simple electrolytes provides a useful, though limited, background. Several detailed treatises covering this subject are available [2–5]. In brief, the dissolution of ions brings about changes in solvent structure. The region of modified solvent surrounding an ion has been denoted as the *cosphere* [2] of the ion. The degree and manner in which cospheres overlap in the close-range encounter of two ions depends specifically on the nature of both ions and the primary forces between them. The resulting effects on the ions and the solvent structure have been observed by a variety of experimental methods, both equilibrium and kinetic [6]. In many instances several stages of desolvation could be detected, and terms such as 'contact' pairs and 'solvent-separated' pairs have come into use to distinguish the results of complete and partial elimination of solvent molecules from between two interacting ions.

The basic features responsible for the specificities in short-range ionic interactions are, in the case of monatomic ions, their charge, their size, their polarizability, and the availability of electrons and/or orbitals for covalent contributions. Additional features of polyatomic ions are the charge density distribution, and, in some instances, the presence of hydrophobic groups [7]. The ease with which the hydration effects accompanying association-dissociation processes can be observed, depends on the number of ion pairs existent at any instant. Thus the effects are observable more readily with weak electrolytes, with complexes involving covalent bonding or with pairs of multivalent ions, than with strong 1-1 electrolytes. However, even in the last case, solvation changes in the interaction of two ions are recognized as one of the most

Eric Sélégny (ed.), Polyelectrolytes, 79–85.

significant factors in accounting for observed deviations from the Debye-Hückel theory and are now formally included in statistical mechanical treatments of simple electrolyte solutions [8] even though, despite much effort, a generally satisfactory quantitative molecular theory of solvation effects is still lacking.

Since specific interactions are the rule with simple electrolytes, one would expect the same for polyelectrolytes. For example, the order and magnitude of the activity coefficients of alkali metal acetates point to short-range interactions involving solvation changes [9, 10], and consequently, it should not be surprising to see corresponding effects in the interactions of alkali metal ions with polycarboxylates [11]. However, for two reasons such effects might be expected to be much more pronounced with polyelectrolytes. First, the effects will depend not only on the individual properties of the participating ionic group and its counter-ion, i.e. their charges, sizes, charge distributions and polarizabilities, but also on the overall charge of the polyion, as well as on possible cooperating binding sites which might be located close to the ionic group under consideration or far removed from it along the flexible polymer contour. Second, even at high dilution a substantial number of counter-ions is forced into close proximity to the polyion by the long-range electrostatic forces [12, 13], so that, in contrast to dilute strong 1-1 electrolytes, there always exist a large number of ion pairs for which solvation effects should be observable.

It seems paradoxical then, that specific short-range interactions in polyelectrolyte solutions have often escaped detection. Actually, the explanation is quite straightforward. The types of experimental results frequently sought after were intrinsically incapable of revealing the existence of short-range interactions. For example, activity and osmotic coefficients obtained over a wide range of conditions very commonly employed can be shown on theoretical grounds to be insensitive to specific binding effects. These coefficients essentially depend on effective local ion concentrations far away from the polyion, where the specificities of the short-range forces are obscured by compensating effects of the long-range forces [14].

Nevertheless, other polyelectrolyte properties have been shown to be sensitive to specific effects of univalent counter-ions. Such properties include electrophoretic mobility [15–18], molecular dimensions [15, 19], second virial coefficients [20, 21], solubility [15], distribution coefficients of two counter-ions in dialysis equilibrium between solvent and polyelectrolyte solution [22] and potentiometric titration behavior [11].

Depending on circumstances, the specificities can be demonstrated more directly by various methods involving spectroscopic, kinetic or thermodynamic techniques. The application of some of these techniques, such as NMR, ultrasonic absorption, and calorimetry is treated elsewhere. We shall review here results obtained in our laboratory by a dilatometric method [23, 24] and discuss some of the implications of these results to current theoretical treatments of polyelectrolyte solutions. The method involves the measurement of the volume change which occurs when a polyelectrolyte solution is mixed with a solution containing a specifically interacting counter-ion and is especially useful for studying the interaction of polyanions with metal and hydrogen ions. It is based on the observation that the water in the solvation shells of many ions is structured more compactly than it is in the pure liquid state, so that the complete or partial release of such water of solvation upon the association of two ions will produce an overall volume increase. Table I contains the results of dilatometry experiments

TABLE I

Volume increase[a] on mixing solutions of tetramethylammonium polyelectrolytes with solutions of metal chlorides [23]

Polyanion	Li^+	Na^+	K^+	Mg^{++}	Ca^{++}
Poly (styrenesulfonate)	0.9	1.2	1.5	2.1	2.9
Poly (vinylsulfonate)	3.4	4.7	5.6	7.2	10.6
Polyacrylate	4.5	4.1	3.9	11.7	17.3
Polymethacrylate	2.9	3.1	3.0	9.2	13.7
Poly (vinylphosphonate) with					
Singly charged PO_3 groups	6.8	5.9	5.3	15.8	22.2
Doubly charged PO_3 groups	9.6	9.4	7.7	19.6	24.7
Polyphosphate	11.3	11.0	9.6	24.4	25.5
Poly (styrenesulfonate),					
8 % divinyl benzene	5.4				13.0

[a] ml per equivalent of total metal ion present, at 30°C.

involving the mixing of solutions of several polyanions in the tetramethylammonium form with solutions of various alkali and alkaline earth metal ions at constant ionic strength. Tetramethylammonium ion was used as the reference ion because it does not interact strongly with these polyanions and because its volume changes when it does interact are negligibly small [23].

These results clearly show specificities depending on both the polyanion and on the metal ion which would not be expected on the basis of long-range electrostatic forces. The effects of the latter are governed predominantly by the linear charge density of the polyanion, and with the exception of the doubly charged polyphosphonate anion, the linear charge densities are the same for the polyanions in Table I. It follows that the interactions giving rise to the observed volume changes involve specific sites on the polyanions. The short range-interactions have therefore been denoted as 'site-binding', in contrast to the non-specific long-range interactions which have variously been referred to as 'ionic atmosphere binding' or 'domain-binding'.

The data in Table I depict a number of features which are expected to be significant in site-binding. The effect of charge of the polyelectrolyte group can be seen from a comparison of the singly and doubly ionized polyvinylphosphonate; that of the counter-ions, from a comparison of the alkali with the alkaline earth metal ions. The substantial difference between the polyvinyl sulfonate and the singly charged polyvinylphosphonate, both in the magnitude of the effect and in the different order in which the alkali metal ions are affected, has been ascribed to a higher effective field strength of the phosphonate relative to that of the sulfonate group. The effects of cooperation of two or more polyelectrolyte groups have been deduced from a number of comparisons. The polyphosphate with its singly charged groups gives larger volume changes than the doubly ionized polyvinylphosphate, reflecting the ease with which metal ions can be chelated into six-membered rings by adjoining groups in the polyphosphate. The differences in the volume changes observed with the polyvinylsulfonate and the polystyrenesulfonate have been ascribed to the differences in the spacing of the sulfonate groups, making cooperation of adjacent groups in the binding of a cation more difficult for the styrene polyanion. The rather small effects observed for the polystyrenesulfonate indicate that this polyelectrolyte is especially suitable for studying the effects of long-rate electrostatic forces with a minimum of interference from specific short-range interactions. Thus, it is not surprising that heats of dilution obtained with polystyrenesulfonic acid solutions follow the predictions of the general electrostatic theory quite satisfactorily [25]. Such calorimetric measurements would be expected to show deviations from this theory if substantial solvation changes were occurring. In contrast, the large volume changes observed with polyphosphate indicate the outstanding suitability of this polyelectrolyte for studying short-range effects. In fact, earlier, indirect indications of site-binding of alkali metal ions, had been obtained with polyphosphates [15].

Further information concerning the cooperative effects of specific groups of the polyion as well as the influence of its overall charge on the hydration changes was obtained in a subsequent study in which three polyacids of closely related structure were compared [24]. Two of these, polyacrylic acid (PAA) and the alternating copolymer of ethylene and maleic acid (HEMA), were structural isomers which differed only in the spacing of the carboxylate groups along the polymer chain. A third, a copolymer of ethyl vinyl ether and maleic acid (HVMEMA) differed from HEMA only in its methoxy side group. By determining the differential volume changes per mole for the ionization of carboxylic acid groups, $-\Delta V_D$, as a function of the degree of ionization, α, it was possible to follow the course of the hydration with increasing ionization. For the PAA the values of ΔV_D increased continuously from 12.3. ml $mole^{-1}$ $-COOH$ at $\alpha=0$ to 24.6 at $\alpha=1$, indicating a corresponding continuous enhancement of the hydration per ionized carboxylate group with increasing polymer charge. This effect was attributed, following Ikegami [26], to a combination of a cooperative enhancement of hydration of neighboring carboxylate groups and to an overall charge effect on the thickness of the hydration shell of the polymer. Such an interpretation was consistent with the finding that the value of ΔV_D at $\alpha=0$ was very close to corresponding values found for simple mono- and dicarboxylic acids. Quite a different behavior was observed for the maleic acid copolymers. These poly (dicarboxylic acids) show a distinct break at half-neutralization in their potentiometric acid-base titration curves indicating that in all dicarboxylic acid pairs one acid group must have ionized before a significant number of the second acid groups can ionize. As expected, a corresponding break is also seen if the ionization is followed by volume change measurements. Moreover, unlike with polyacrylic acid, the ΔV_D-values showed no continuous increase with α but remained at distinct constant levels within each half-neutralization range. Thus for the first and second ionization ranges, respectively, the ΔV_D-values, were 9.1 and 34.5 for HEMA and 8.8 and 37.4 for HVMEMA, which when compared with the corresponding values of 6.0 and 24.4 for maleic acid shows a very substantial polyelectrolyte charge effect on the hydration in the second ionization region. However, there is no evidence for cooperative hydration enhancement of neighboring dicarboxylate groups. This difference from polyacrylic acid behavior is very likely due to the fact that the dicarboxylate units in the maleic acid copolymers are separated along the polymer chain by two methylene groups whereas the carboxylate units in PAA are separated by only one.

The results obtained with PAA have an important consequence for the interpretation of potentiometric titration results. It is customary to describe such results in terms of pK_0 and a term depending on the change in electrostatic free energy with α. The chemical contribution, pK_0, is conventionally assumed to be constant with varying α. However, if the hydration changes involved in the dissociation reaction are considered to belong to the chemical part of the free energy, then pK_0 will vary with changing α whenever the hydration changes vary, – as they do in the titration of PAA with TMAOH [24, 26], but not with NaOH [26]. By keeping pK_0 constant, the changes in the hydration contribution are, in effect, implicitly included in the 'electrostatic' term.

The interactions of the anions of the three polyacids with several metal ions also showed interesting specificities. At low equivalent ratios of metal ion to polyanion the observed volume changes were about twice as large for HVMEMA as for PAA or HEMA with all alkali and alkaline earth metal ions studied (Li^+, K^+, Mg^{++} and Ba^{++}), indicating a strong effect of the ether group. The results were interpreted as indicating contact chelation of the metal ion by the ether oxygen and an adjacent carboxylate group, resulting in a six-membered ring. Two nearest carboxylate groups in PAA or the maleic acid copolymers, however, could give only solvent-separated chelation because the corresponding contact chelation would lead to unfavorable eight- or seven-membered rings. In contrast, silver ion produced quite large volume changes which were of about equal magnitude for all three polyacids. This finding was ascribed to covalent binding of two carboxylate groups by the silver ion with a valence angle of 180°, thus excluding nearest neighbors and making the binding insensitive to the local polymer structure.

The results reviewed above clearly indicate the features by which site-binding is distinguished from general electrostatic interactions, namely its specificity for one or more chemical groups on the polyion and the critical role played by the solvent. It remains to be considered how the demonstration of the existence of short-range interactions affects the validity of current theoretical treatment of polyelectrolyte solutions. We shall limit ourselves to two of these treatments, both of which apply to the cylindrical rod model, one involving the Poisson-Boltzmann equation, the other the Manning condensation theory.

The classical form of the Poisson-Boltzmann equation takes into account only coulombic interactions. It would be interesting to examine the consequences of including terms for the desolvation and other short-range forces in the Boltzmann factor although, in view of the difficulty of obtaining theoretical *a priori* expressions, such terms would have to depend on empirical estimates. An alternative method of treating the problem is to divide the space around the cylindrical rod into two regions separated by a cylindrical boundary defined so that all solvation changes and other specific effects take place in the interior region. The Poisson-Boltzmann equation may then be applied in the exterior region under the usual boundary conditions, with the modification that the inner boundary condition depends on the *net* linear charge density of the interior region, including all the charges within this region, instead of on the gross linear charge density of the macro-ion alone. A procedure along these lines applied to the interaction of DNA with a number of cations gave results consistent with dialysis equilibrium data which could not be reconciled with solutions of the unmodified Poisson-Boltzmann equation [14].

The condensation theory of Manning which applies at sufficiently low ionic strength has the basic feature that it sets a precisely defined upper limit to the effective linear charge density of the polyion. A sufficient number of counter-ions will 'condense' on the polyion until this limit is reached. The condensation is predicted from a consideration of long-range Coulomb forces alone. Therefore the predictions of the theory contain no information on the characteristic effects of the specific interactions treated in this article. In fact, as long as the extent of specific site-binding is insufficient to reduce the net linear charge density below the critical value, the difference will be made up by condensation, and consequently the number of 'free' counter-ions will be insensitive to the site-binding. Under these conditions, any experimental quantity which depends primarily on the net linear charge density of the polyion and on the number of 'free' counter-ions should not be affected by specific interactions. Such experimental parameters include activity, osmotic and diffusion coefficients in very dilute solutions containing a single counter-ion. Hence, the success of the theory in predicting these parameters can tell us nothing about the presence or absence of specific short-range interactions.

At the present time, a theoretical treatment of polyelectrolytes which adequately takes account of short-range interactions is lacking. Although beginnings have been made in this direction [27, 28], significant progress appears to require a good deal of further information from appropriate experimental studies.

References

1. Armstrong, R. W. and Strauss, U. P.: *Encycl. Polymer Sci. Technol.* **10**, 781 (1969).
2. Gurney, R. W.: *Ionic Processes in Solution*, McGraw-Hill Book Co., New York, 1953.
3. Davies, C. W.: *Ion Association*, Butterworths, London, 1962.
4. Nancollas, G. H.: *Interactions in Electrolyte Solutions*, Elsevier Publishing Co., Amsterdam, 1966.
5. Szwarc, M. (ed.), *Ions and Ion Pairs in Organic Reactions*, Wiley-Interscience, New York, N.Y. 1972. While this book deals with non-aqueous solvents, it is nevertheless relevant.
6. For references to the original literature, see Reference [4]
7. Kauzmann, W.: *Adv. Protein Chem.* **14**, 1 (1959).

8. Friedman, H. L. and Ramanathan, P. S.: *J. Phys. Chem.* **74**, 3756 (1970).
9. See p. 40 in reference [3].
10. Robinson, R. A. and Stokes, R. H.: *Electrolyte Solutions*, Academic Press, New York, 1955, p. 218.
11. Gregor, H. P. and Frederick, M.: *J. Polymer Sci.* **23**, 451 (1957).
12. Fuoss, R. M., Katchalsky, A., and Lifson, S.: *Proc. Natl. Acad. Sci. U.S.* **37**, 579 (1951); Alfrey, T. Jr., Berg, P. W., and Morawetz, H.: *J. Polymer Sci.* **7**, 543 (1951).
13. Manning, G. S.: *J. Chem. Phys.* **51**, 924 (1969).
14. Gross, L. M. and Strauss, U. P.: in B. E. Conway and R. G. Barradas (eds.), *Chemical Physics of Ionic Solutions*, John Wiley & Sons, Inc., New York, 1966, p. 361.
15. Strauss, U. P., Woodside, D., and Wineman, P.: *J. Phys. Chem.* **61**, 1353 (1957).
16. Strauss, U. P. and Ross, P. D. *J. Am. Chem. Soc.* **81**, 5295 (1959).
17. Ross, P. D. and Scruggs, R. L.: *Biopolymers* **2**, 79, 231 (1964).
18. Strauss, U. P., Gershfeld, N. L., and Spiera, H.: *J. Am. Chem. Soc.* **76**, 5909 (1954).
19. Ross, P. D. and Strauss, U. P.: *J. Am. Chem. Soc.* **82**, 1311 (1960).
20. Strauss, U. P. and Wineman, P. L.: *J. Am. Chem. Soc.* **80**, 2366 (1958).
21. Strauss, U. P. and Ander, P.: *J. Phys. Chem.* **60**, 2235 (1962).
22. Strauss, U. P. and Ross, P. D.: *J. Am. Chem. Soc.* **81**, 5299 (1959).
23. Strauss, U. P. and Leung, Y. P.: *J. Am. Chem. Soc.* **87**, 1476 (1965).
24. Begala, A. J. and Strauss, U. P.: *J. Phys. Chem.* **76**, 254 (1972).
25. Skerjanc, J., Dolar, D., and Leskovsek, D.: *Z. Physik. Chem., Frankfurt* **56**, 207 (1967).
26. Ikegami, A.: *J. Polymer Sci., Part A*, **2**, 907 (1964).
27. Eiseman, G.: in A. Kleinzeller and A. Kotyk (eds.), *Membrane Transport and Metabolism*, Academic Press, New York, 1961, p. 163.
28. Ling, G. N.: *A Physical Theory of the Living State*, Blaisdell, New York, 1962.

THE MODES OF SPECIFIC BINDING OF IONS TO POLYELECTROLYTES

HARRY P. GREGOR

Dept. of Chemical Engineering and Applied Chemistry, School of Engineering and Applied Science, Columbia University, New York City, N.Y. 10027, U.S.A.

The contributions to this Advanced Study Institute on Charged and Reactive Polymers represented three different points of view of the same scientific subject. Some authors were concerned with solutions of polymers as whole entities, as for example the density of these solutions as reflecting the electrostriction of solvent by the polymer. A second group of authors was concerned with polymers *per se*, as with their general thermodynamic and dynamic properties as influenced by general considerations of ionic strength, etc., as with the configurational entropy entity of polymers. The third group of authors was not concerned with polymer systems as a whole but rather with a highly localized phenomenon, their binding to ions in solution, with a primary concern on the nature and extent of these bindings reactions.

The binding of ions to polymers is not a necessary phenomenon associated with polyelectrolytes in solution. When it does occur, it has profound influences upon the thermodynamic and dynamic properties of the polymer itself and indirectly upon the nature of the polymer-solvent system. The early investigators of polyelectrolytes in solution, namely Kern, Katchalsky and Fuoss, concerned themselves primarily with the non-specific effects of electrolytes and the general considerations of ionic strength. However, it became apparent that several different modes of binding occurred with certain of these so strong as to manifest themselves in dilute solution. The interest in binding phenomena was enhanced by the availability of ion-exchange resins of synthetic, high polymeric origin. These resins systems showed strong binding effects, ones amenable to relatively simple and direct observation (such as the volume of particles themselves); a body of knowledge on binding phenomena grew with the utilization of these insolubilized polyelectrolytes in technology. With the development of ion-exchange resins in membrane form, one could make measurements of the dynamic properties of these systems also. Data on rates of diffusion and electrical conductivity became readily available and contributed substantially to our understanding of ion-binding phenomena.

We can distinguish five rather different modes of ion binding: coulombic binding; Stern layer binding; hydrophobic binding or adsorption; hydrolytic binding (probably through hydrogen bond formation); ligand binding.

1. Coulombic Binding

Counter-ions may be bound in the annular region near the polymer chain by straightforward coulombic interactions. The usual model is that of Fuoss-Alfrey wherein the

Eric Sélégny (ed.), Polyelectrolytes, 87–95.

polyelectrolyte is considered to be a rigid rod with its fixed charges uniformly dispersed as a continuum along its surface. In the Gregor-Kagawa refinement the counter ions are rigid spheres of a specific size, and the dielectric constant is assigned its bulk value throughout annular region. Under these circumstances the larger counter-ions are excluded from the region close to the rod. The selective uptake of one ion over another can be calculated in the case of polyelectrolytes by selecting an arbitary point of demarkation (by defining the Donnan region) in the case of soluble poly-electrolytes. With ion-exchange resin particles the average annular size can be calculated from the volume of a resin particle. Figure 1 shows the selective binding of a

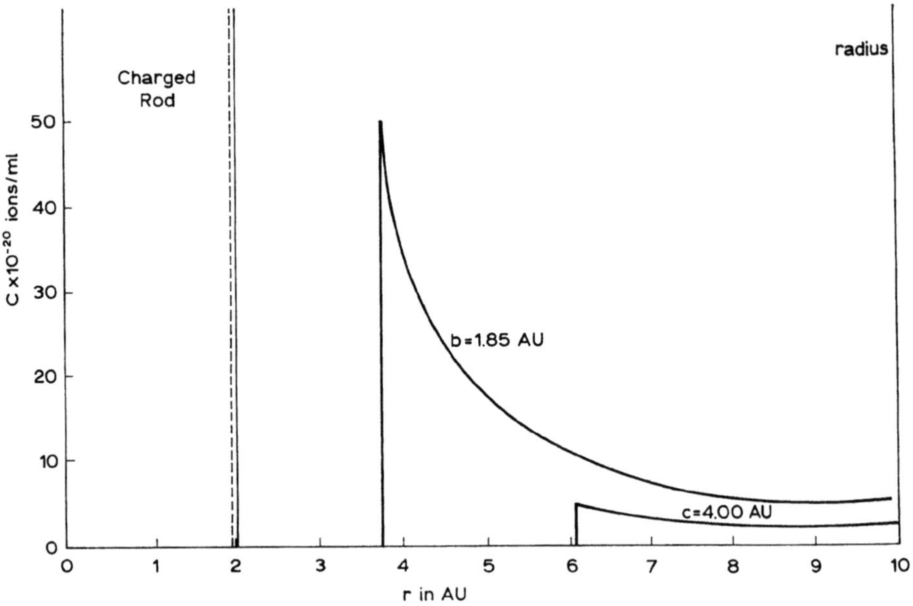

Fig. 1. Concentration of small and large counter-ions about a charged rod of radius 2.1 AU where $\lambda = 2.27$ and solution phase ratio of small to large ions is 2:1; computations of Reference [2] based on data of Reference [1].

small counter-ion over a larger one based on the charged-rod model. The Kagawa-Gregor theory applied to the resin data of Gregor *et al.* [1] showed reasonable agreement, but the computation was a crude one and the resin used was one where adsorptive effects were pronounced, particularly with the quaternary ammonium exchange cations. An exact treatment and computation by Gregor and Gregor [2] and exchange studies on purely aliphatic resins by Greff and Gregor [3] have shown that the charged rod model leads to a considerable discrepancy between theory and experiment. The degree of selective binding observed can be correlated only with a chain potential very much larger than that which can be assigned by a uniformly charged rod model. It is evident that charge localization combined with the use of a local dielectric constant is required. This lack of agreement between theory and experiment with the charged rod model does not vitiate against our view that coulombic binding is an important phenomenon, and that the distance of closest approach of an ion to a polymer chain can be an important parameter in rather profound influences upon the properties of the polymer and of the polymer-solvent system.

Coulombic binding is observed in systems wherein specific interactions between fixed and counter-ions do not occur and where counter-ions may be treated as rigid spheres. This condition is obtained for potassium and quaternery ammonium cations and with carboxyllic and sulfonic acid resins of high charge density and low adsorbability. Pressure-volume and other effects which can influence ion selectivity must be also considered in predicting selectivities.

2. Stern Layer Binding

When chain potentials are so high or counter-ions so small that ions are bound in the Stern layer, as opposed to in the Gouy-Chapman layer in coulombic binding, a different kind of binding occurs. Stern layer binding is not observed in the case of the small alkali metal cations and non-specific fixed charge groups such as the sulphonates. On the other hand, the halogen anions (except the fluoride), thiocyanate and perchlorate anions all show an anomolously strong binding to quaternary ammonium fixed-charge groups. This strong binding of the halogen anions in the sequence of increasing atomic weight can be correlated directly with a small distance of approach which is the result of the high polarizability of these anions. Almost all of the anions show a sufficiently high polarizability to be so bound; the sole exception appears to be the iodate which does not appear to be bound.

The Fuoss-Alfrey charged rod model does not lend itself to treatments of ion binding. Figure 2 shows two calculated concentration-distance functions for counter-ions of finite size about a fixed spherical charge (model of J. Bjerrum) and about a Fuoss-Alfrey charged rod. In the case of the spherical charge, the minimum of the function can be used to distinguish between bound and unbound ions, and it has been shown that the exact position of the minimum is not important in this calculation. In the case of the charged rod model, one can arbitrarily assign a finite thickness to the region adjacent to the rod and call this the Stern layer. This procedure is not only entirely arbitrary but it is also not satisfactory as is evident from Figure 3, computed by Gregor and Gregor. It is evident that a less primitive model will be required to deal with these systems. The use of semispherical charges along a chain of low dielectric constant is probably an essential requirement of this model. The theory of Manning might be modified in this connection.

The binding of the halogen anions to quaternary ammonium polyelectrolytes illustrates the complexity of these systems. It would be convenient if one could consider a counter-ion as being either bound or unbound, giving rise to the simple statement of the selectivity coefficient as,

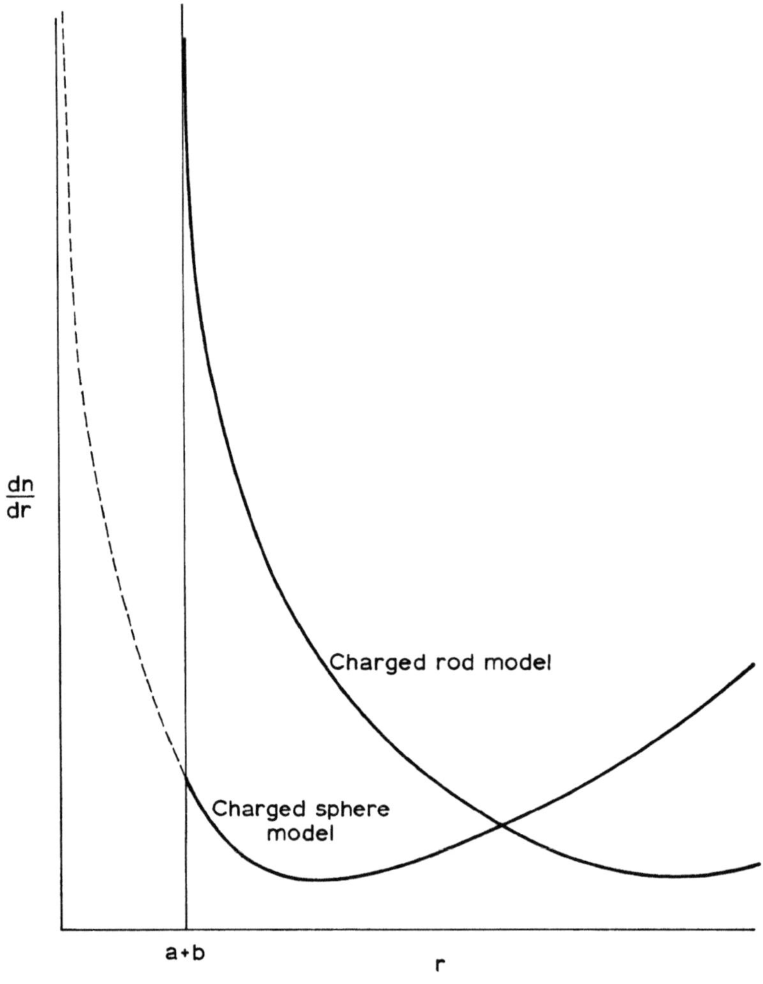

Fig. 2. Comparison of counter-ion concentrations about charged rod and spherical ion models.

$$K_{AB} = \left[\frac{RA + A^-}{RB + B^-}\right]_i \left[\frac{B^-}{A^-}\right]_0,$$

where *RA* and *RB* refer to bound anions and thus obtain a formulation which correlates with measurements of electrical conductivity or diffusion coefficient where the bound ions are immobilized. Such is not the case, because in comparing the chloride and iodide anions, one finds a selectivity coefficient of about 30 in favor of the iodide. This should allow one to predict that the relative self-diffusion coefficients would be in the same ratio, but as Andelman and Gregor [4] have shown, these diffusion coefficients are in the ratio of 5:1. This can be taken to mean that ions in the

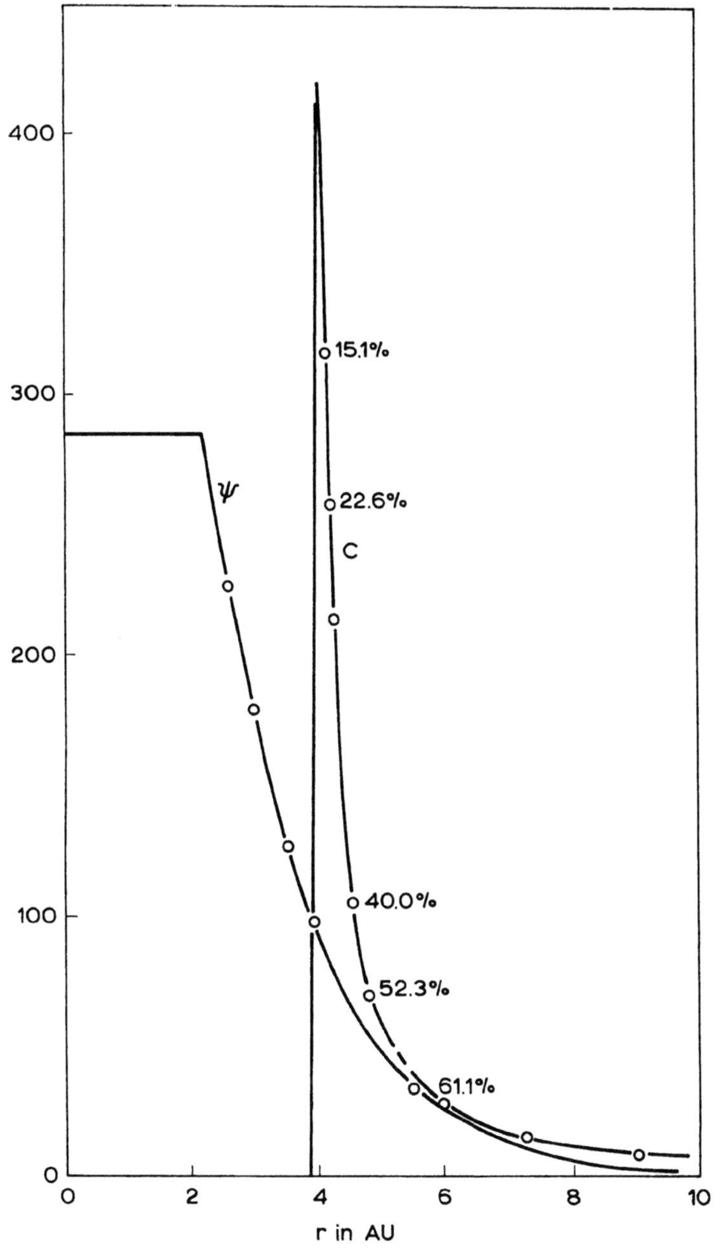

Fig. 3. Computation of chain potential ψ and concentration C of charged rod model. Percentage points refer to fraction of bound ions, assigned to the Stern Layer region at that distance, taken from Reference [2].

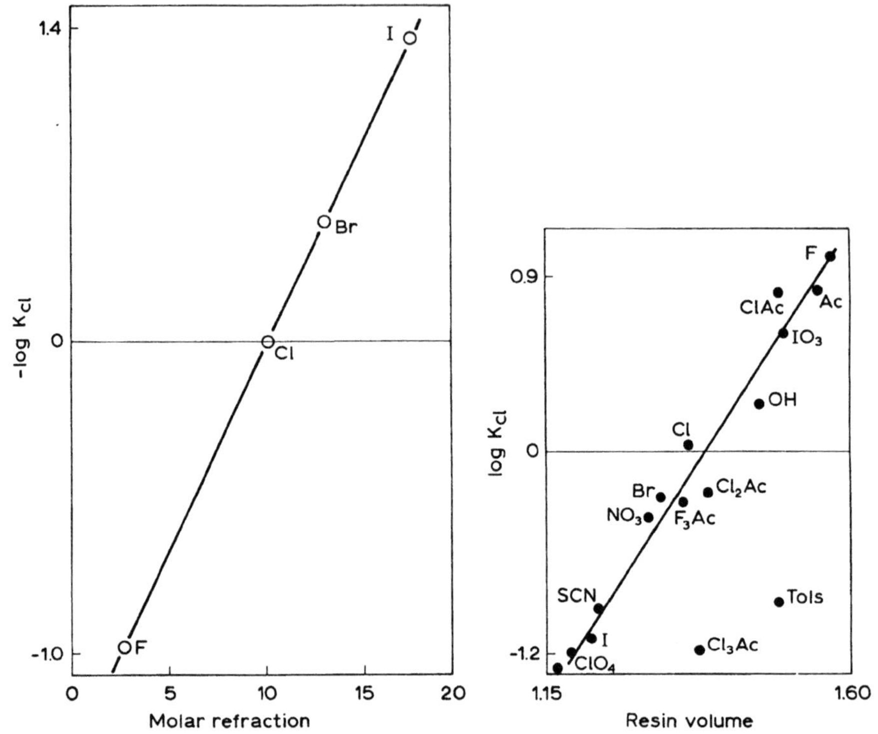

Fig. 4. Selectivity coefficients, resin volumes and ionic molar refraction with benzyltrimethylammonium anion-exchange resins (Gregor, Belle, and Marcus *J. Am. Chem. Soc.* **77**, 2713 (1955).

Stern layer are not completely immobilized but show a relatively high mobility. This concept of diffusion in the adsorbed state is well established in the older colloid literature and obviously has a firm experimental basis. The correlation between binding and the parameters of anions and anion-exchange resins is shown in Figure 4.

3. Hydrophobic Binding or Adsorption

The hydrocarbon nature of polyelectrolytes makes itself felt in the interactions of organic counterions with polyelectrolytes. These phenomena manifest themselves most clearly when one examines the selective uptake of organic ions of different molecular weight and configuration by polyelectrolytes having different charge density.

Figure 5 shows the uptake of a series of tetralkyl-ammonium and the benzyltrimethylammonium cations as compared with potassium by sulfonic acid resins of differing hydrophobic-hydrophilic balance.

As seen in Figure 5, the binding of organic ions to polyelectrolytes is very much a function of the polarity of the polyelectrolyte or of its cohesive energy density and of the organic nature or hydrophobic bonding index of the exchange ion. The binding of

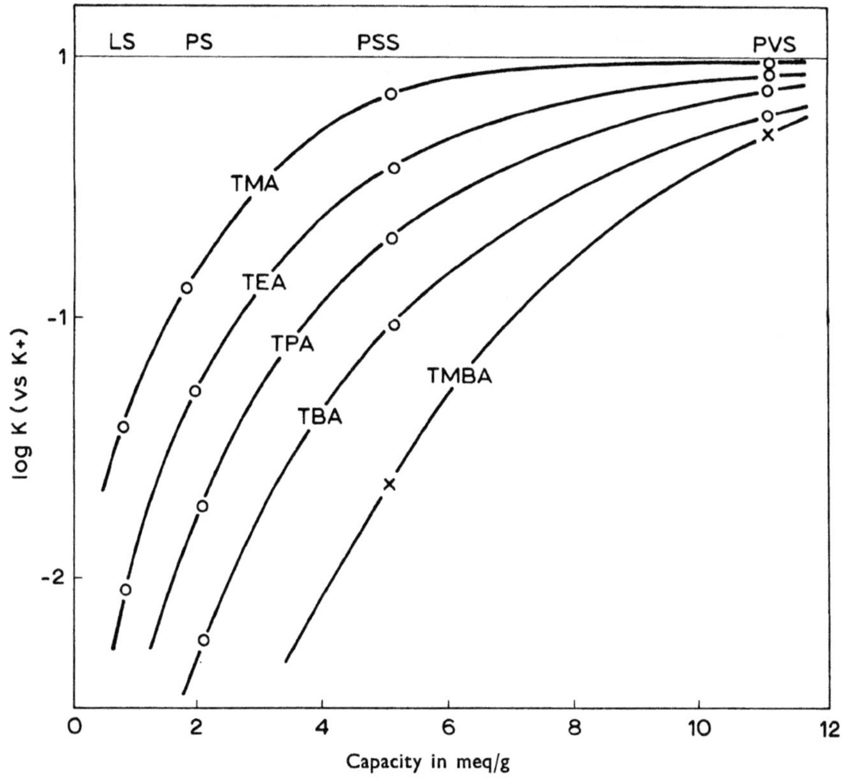

Fig. 5. Selective uptake of tetramethyl, ethyl, propyl, butyl and trimethylbenzylammonium cations by lauroylsulfonic, phenolsulfonic, polystyrenesulfonic and vinylsulfonic acid resins.

aromatic counter-ions by polymers containing the aromatic ring is particularly strong.

Interesting comparisons can be made with the oleophilic ion-exchange polymers [5] whose hydrophilic-hydrophobic balance is altered by increasing or decreasing the length of the aliphatic substituent chain. These polymers swell strongly or weakly depending upon the degree of such substitution and the cohesive energy density of the solvent [6]. By an appropriate combination of polymer and solvent, the specific binding of certain organic ions in the presence of substances which normally interfere can be achieved.

4. Hydrolytic Binding

It was recognized several years ago by Harned that the alkali metal cations of lower atomic weight engage in a specific binding reaction with anionic groups containing a strongly negative oxygen atom. The observations of Scatchard that the binding sequence of the alkali metal cations favored potassium with the monobasic phosphates but lithium as the negative charge increased to the dibasic and tribasic phosphate anions is an excellent case in point.

Harned termed this binding hydrolytic because of indirect evidence that water molecules were involved in the binding mechanism. Several recent studies have given credence to his suggestion that hydrogen-bonded water molecules acted as ligands between the lithium ion and the acetate group to explain the observed low mean activity coefficients of this salt. It can explain also the strong affinity of carboxylic and phosphonic acid resins for lithium and, to a lesser extent, sodium.

5. Ligand Binding

The binding of metallic ions to polymer chains by the formation of a non-dissociated metal ion-ligand complex is well documented from the studies of Wall, Gregor and Morawetz. Since the binding here is quite specific and strong, there is a good correlation between binding constants calculated from potentiometric data and electrical conductivity measurements. Theoretical treatments of these systems involve a consideration of local effects in terms of an intrinsic binding constant to which one adds the effect of the chain potential. One includes in this class of binding phenomena the binding of transition and heavy metals to carboxylic and amine polymers, reactions of complexing polymers which contain groups of the iminodiacetic acid type, of thiolpolymers and the like. Indeed, ligand binding can be evoked to explain a wide range of specific binding phenomena. For example, consider the binding of Ag ions to sulfonic acid resins. Here we observe a very high selectivity coefficient of about 20 when small amounts of Ag are bound (vs K), with a sharp decrease to about 5 when the mole fraction of sites occupied by Ag increases above 0.05. The initial strong binding can be ascribed to the presence of thiol impurities; Stern layer binding can account for the principal phenomena.

6. Discussion

The problems involved in formulating a consistent theory of binding are illustrated when one measures the thermodynamic and dynamic properties of a limited number of polyelectrolytes with different counter-ions and then attempts to formulate a simple theory. For example, Gregor measured the selective uptake, self diffusion coefficients, electrical conductivity and electro-osmotic coefficients of a number of different ions in ion-exchange membrane and resin systems. The measurement of selective uptake is unequivocal and its correlation with binding is straightforward. Data on self-diffusion coefficients are complicated to interpret because the narrow pores of these insolubilized polyelectrolytes place steric and hydrodynamic restrictions upon the diffusive process. These can be overcome, at least in a semi-quantitative manner, by the use of appropriate correction terms [7]. The measurement of the electro-osmotic coefficient is simple and its interpretation is similarly straightforward. Data on electrical conductivity require interpretation because of the steric and hydrodynamic restraints of the pore nature of the system; there is an electro-osmotic correction to the electrical conductivity. Table I tabulates normalized values for different counter-ions with

TABLE I

	Cation-Permeable $\bar{c} = 2.0$, $\bar{m} = 5.1$			Anion-Permeable $\bar{c} = 1.1$, $\bar{m} = 5.0$		
	K	Li	TMA	Cl	I	IO_3
f	0.41	0.45	0.58	0.23	0.21	0.35
$\bar{K}/K$	0.16	0.12	0.078	0.055	0.013	0.046
EO	105	259	267	90	90	169
% Diss.	100	100	100	50	10	100

f – pore volume fraction
$\bar{K}/K$ – eg. conduct in memb./eq. conduct. in solu.
EO – electro-osmotic coef. in ml/Faraday
$\bar{c}$ – molarity; $\bar{m}$ – molality of membrane

sulfonic and quaternary ammonium polyelectrolytes insolubilized in resin or membrane form.

While considerable difference in binding is observed, in the case of the sulfonic acid polymers we can ascribe coulombic binding as being responsible with the alkali metal, quaternary ammonium, alkaline earth and transition metal cations. Coulombic binding can explain the relative uptake of the alkaline earth cations by sulfonic acid resins where the sequence Ba > Sr > Ca > Mg is observed. These phenomena have a higher order of complexity than with the alkali metal cations because binding must occur at two adjacent sites.

In the case of the chloride and iodide Table I shows that Stern layer binding is probably responsible. It is interesting to observe that the electro-osmotic coefficients correlate well with the (hydrated) hydrodynamic size of all ions. It is evident further that in spite of the extremely high molality of these systems (their internal molality is of the order of 5–7 M) there is apparently not a substantial dehydration of the lithium ion. The equal electro-osmotic coefficients of the chloride and iodide ions show that by whatever mechanisms they are transported through the membrane, the volumes transported are very much the same as for the 'free ions themselves.'

In conclusion, we can differentiate between five different binding modes with polyelectrolytes. The complexity of these systems is indeed perplexing to those who attempt to make general, theoretical treatments. However, the existence of these different modes can be used to advantage by those who wish to employ polyelectrolytes for the achievement of practical ends.

References

1. Miller, I., Bernstein, F., and Gregor, H.: *J. Chem. Phys.* **43**, 1783 (1965).
2. Gregor, J. and Gregor, H.: (to be published).
3. Greff, R. and Gregor, H.: (to be published).
4. Andelman, J. and Gregor, H.: *Electrochim. Acta* **11**, 869 (1966).
5. Gregor, H. *et al.*: *J. Am. Chem. Soc.* **87**, 5525, 5534, 5538 (1965).
6. Gregor, H.: *Soc. Chem. Ind.* **436** (1970).
7. Kawabe, H., Jacobson, H., Miller, I., and Gregor, H.: *J. Coll. Int. Sci.* **21**, 79 (1966).

THERMODYNAMIC PROPERTIES OF POLYELECTROLYTE SOLUTIONS

D. DOLAR
Dept. of Chemistry, University of Ljubljana, Ljubljana, Yugoslavia

1. Introduction

During the last two decades the thermodynamic properties of numerous polyelectrolytes in aqueous solution have been investigated. Some research workers were more interested in natural, others again in synthetic polyelectrolytes with a variety of counter-ions. The thermodynamic behaviour of pure polyelectrolyte solutions and of their mixtures with simple electrolytes, was studied. It is impossible to give in one lecture a full account of the efforts in this direction. I shall, therefore, confine myself mainly to the work which has been done at this Department.

In what follows the results will be presented obtained with pure polyelectrolyte solutions to which no salt of a simple electrolyte has been added. As a polyion the negatively charged polystyrenesulphonate ion with the degree of substitution equal to one has been chosen. Strauss and Leung [1] have shown by dilatometry that the extent of the side binding of alkali and alkaline-earth metal ions to this polyion is very small and could be neglected. N.M.R. [2] and Raman spectroscopy [3] of solutions of polystyrenesulphonic acid (HPSS) at different concentrations have revealed that this acid is strong and that the hydrogen ion is not associated with the polyion. Consequently, it seems that this polyelectrolyte is very suitable for testing an electrostatic theory which does not take into account specific interactions between a polyion and its counter-ions.

List of Symbols

a = radius of cylindrical polyion
A = integration constant
b = length of monomeric unit
c_m = monomolarity
e_0 = elementary charge
E_e = electrostatic internal energy
f_+ = activity coefficient of counter-ion
f_- = activity coefficient of polyion
$f_\pm$ = mean activity coefficient of polyelectrolyte
F_e = electrostatic free energy
h = length of stretched polyion
H_e = electrostatic enthalpy
ΔH_D = enthalpy of dilution
k = Boltzmann constant
K = cryoscopic constant
m = molality
m_m = monomolality
m_c = molality of counterion

Eric Sélégny (ed.), Polyelectrolytes, 97–113.

m_p = molality of polyelectrolyte
$n_1{}^0$ = number density of monovalent counter-ions at $\psi = 0$
$n_2{}^0$ = number density of divalent counter-ions at $\psi = 0$
$\bar{n}_1$ = average number density of monovalent counter-ions
$\bar{n}_2$ = average number density of divalent counter-ions
$\bar{N}_1$ = average equivalent fraction of monovalent counter-ions
N_A = Avogadro constant
N_p = number of polyions in volume V
P = pressure
r = cylindrical coordinate
R = gas constant
R = radius of cylindrical cell
T = absolute temperature
ΔT = freezing point depression
u = function related to F_e
v = function related to F_e
V = volume
z_1 = charge number of ionic group on polyion
z_2 = charge number of counter-ions
β = integration constant
ε = dielectric constant of solvent
γ = concentration parameter
ϕ = osmotic coefficient
λ = charge-density parameter
ν = number of ionic groups on polyion
π = osmotic pressure
ψ = electrostatic potential

2. The Cell Model

Since the experimental results will be compared with the calculated ones, it will be convenient to introduce first the basic theory. The model on which the interpretation is based is that of Fuoss *et al.* [4] and of Alfrey *et al.* [5]. Let us imagine the whole volume of a polyelectrolyte solution to be divided into cylindrical cells with one polyion stretched along the axis of each cell. The polyion, too, is represented by a cylinder of radius a and length h, carrying ν negative charges with the charge number z_1 and the length of the monomer unit $b = h/\nu$. The counter-ions with the charge number z_2 are distributed symmetrically around the polyion in the cylindrical cell with the radius R. Supposing that the charges are uniformly smeared over the cylindrical surface of the polyion the Poisson-Boltzmann equation can be applied [4, 5] to the cell. Denoting by ψ the electrostatic potential, and by r the distance from the axis of the cell, and taking into account the boundary conditions

$$\psi(R) = 0, \quad (\mathrm{d}\psi/\mathrm{d}r)_{r=R} = 0 \tag{1}$$

we obtain the following expression for the potential

$$\frac{z_2 e_0 \psi}{kT} = \ln \frac{(1 - \beta^2)\, r^2\, \mathrm{sh}^2\,(\beta \ln Ar)}{\beta^2 R^2}, \tag{2}$$

where e_0 is the elementary charge, k the Boltzmann constant, and T the absolute temperature. The constants β and A may be evaluated from

$$\lambda = \frac{1-\beta^2}{1+\beta \operatorname{cth} \beta\gamma}, \quad \lambda = \frac{z_1 z_2 e_0^2}{\varepsilon k T b}, \tag{3}$$

$$\ln A = -\ln R - (1/\beta) \operatorname{Arth} \beta, \tag{4}$$

$$\gamma = \ln \frac{R}{a} = \tfrac{1}{2} \ln \frac{10^3}{\pi a^2 b N_A} - \tfrac{1}{2} \ln c_m, \tag{5}$$

where λ is the charge-density parameter, ε the dielectric constant of the solvent, γ the concentration parameter, N_A the Avogadro number, and c_m the monomolar concentration. For a more extensive description of the model and of the corresponding calculations the reader is referred to two articles [6, 7] by Katchalsky and his co-workers.

It remains to mention that Lifson and Katchalsky [8] have calculated the electrostatic free energy of an isolated cell by using a charging process [9]. Supposing that the cells form an array of parallelly oriented and symmetrically distributed macromolecular rods, the net force on each rod is zero and the fluctuations are negligible [10]. Thus, the electrostatic free energy of the whole solution may be obtained simply by adding up the contributions of individual cells. This view seems to be supported by the calculations of Onsager [11], who has found that the parallel arrangement of rod-like colloids minimizes the free energy of the system when the repulsive forces between colloidal particles are sufficiently strong.

To prove the applicability of this model to the interpretation of the results obtained with solutions of polystyrenesulphonates we have measured different thermodynamic properties related to different derivatives of the electrostatic free energy.

3. The Activity Coefficient of Counter-ions

There are two thermodynamic properties, that is, the activity coefficient of counter-ions and the osmotic coefficient, which are usually considered as a direct measure for the fraction of the so-called free counter-ions. In Figuie 1, the single activity coefficients of some monovalent [11] and divalent metal ions [12] in salt free polymethylstyrenesulphonate solutions, are presented. The difference in the association of mono- and divalent counter-ions is evident, a fact which is in agreement with the observations on other polyelectrolytes [14]. The explanation of the behaviour of monovalent ions is not easy since the order of sequence of the radii of hydrated ions and that of the activity coefficients do not seem to be correlated. One can ascribe this behaviour either to some 'specific' interactions or to a not well defined meaning of the single activity coefficient. The electrodes, the salt bridge, the calibration and extrapolation procedures can affect appreciably the experimental values of activity coefficients. Thus, it seems safer not to attribute too great a significance to these observations.

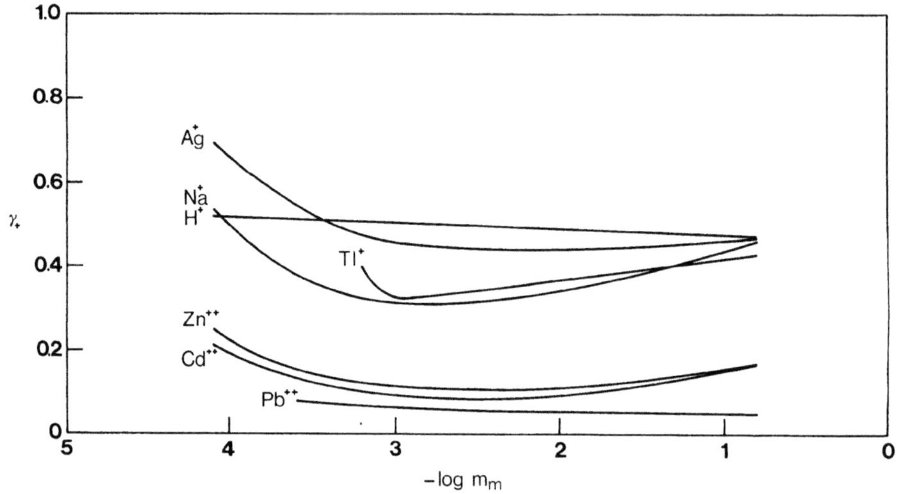

Fig. 1. Single-ion activity coefficients of counterions in saltfree polystyrenesulphonate solutions as functions of monomolality. (Data from References [12] and [13].)

4. The Osmotic Coefficient

The osmotic coefficient ϕ is thermodynamically a well defined property. It can be evaluated from the freezing point depression according to the relation

$$\phi = \Delta T/\Delta T_{\mathrm{id}}, \quad \Delta T_{\mathrm{id}} = Km_{\mathrm{c}}, \tag{6}$$

where ΔT and ΔT_{id} are the freezing point depression obtained by experiment and that of an ideal solution, respectively, K is the cryoscopic constant and m_{c} the molality of counter-ions. We have used a precise apparatus [15] which makes possible the evaluation of the osmotic coefficients of alkali salts of polystyrenesulphonic acid in a wide concentration range down to 0.006 monomolal with an error not exceeding 3% at the lowest concentration.

Inspection of Figure 2 reveals that the osmotic coefficient increases in the order $Cs^+ < K^+ < Na^+ < Li^+ \leqslant H^+$ which corresponds to the order of sequence of the radii of hydrated counter-ions. At concentrations lower than 0.056 monomolal the osmotic coefficient of NaPSS is higher than the coefficients of the other salts, a fact which is not due to experimental error. The same peculiarity in the behaviour of sodium ion has been observed with the heat of dilution data [26]. In the region of higher concentrations extensive measurements on solutions of polystyrenesulphonates with monovalent counter-ions have been made by Chu and Marinsky [16]. They have also found that the osmotic coefficient increases with the increasing size of the hydrated counter-ion. In Figure 3 the data are given for MgPSS and CaPSS [17]. The osmotic coefficient is lower than that for the alkali salts owing to the charge of counter-ions whereas the dependence on the ionic size is the same as for monovalent counter ions.

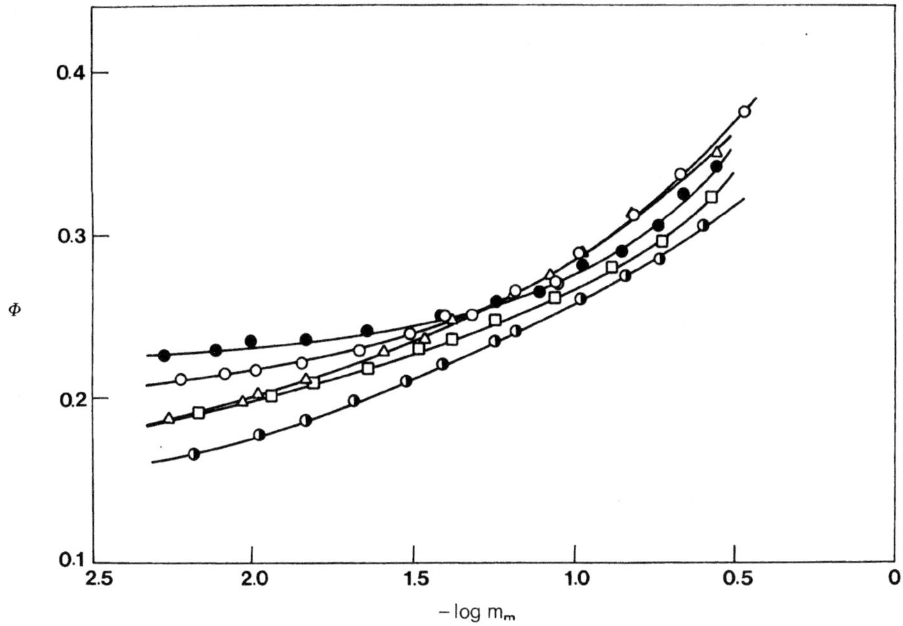

Fig. 2. Osmotic coefficient of solutions of polystyrenesulphonates at 0°C as function of monomolality for HPSS (○), LiPSS (△), NaPSS (●), KPSS (□), and CsPSS (◑). (Reference [15], Figure 1.)

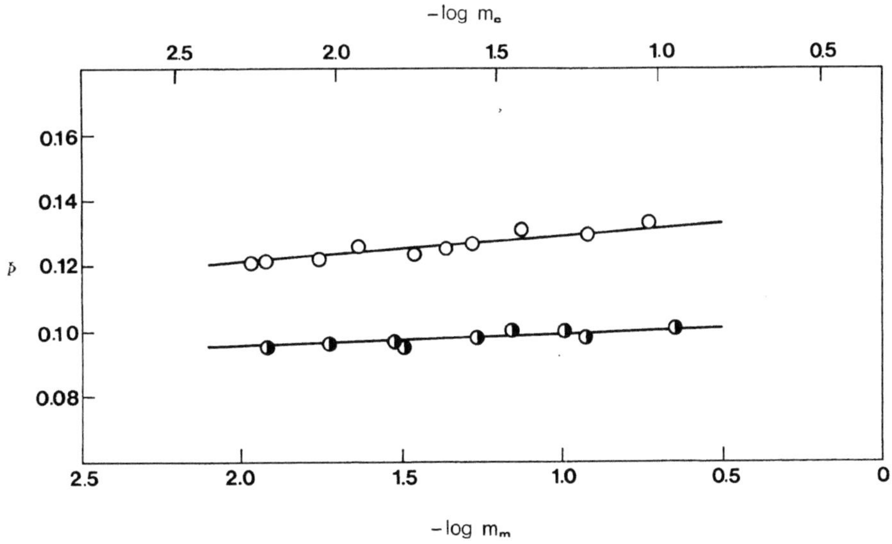

Fig. 3. Osmotic coefficient of solutions of polystyrenesulphonates at 0°C as function of the molality of counter-ions, m_c, and monomolality, m_m, for MgPSS (○) and CaPSS (◑). (Reference [17], Figure 1.)

Considering the osmotic coefficient as equal to the fraction of free counter-ions or $(1-\phi)$ as equal to the fraction of bound counter-ions, we may summarize the experimental results obtained with solutions of polystyrenesulphonates as follows: the degree of binding is higher for divalent than for monovalent counter-ions and increases with the decreasing size of the hydrated ion.

The osmotic coefficient is related [8] to the electrostatic free energy by

$$\phi = 1 - (1/\pi_{\text{id}}) \frac{\partial (N_p F_e)}{\partial V}, \tag{6}$$

where $N_p F_e$ is the electrostatic free energy of the solution containing N_p polyelectrolyte molecules in a volume V. π_{id} is the ideal osmotic pressure,

$$\pi_{\text{id}} = (z_1 v/z_2)\, kT/\pi (R^2 - a^2)\, h, \tag{7}$$

wherein the contribution of the polyion is neglected. The electrostatic free energy F_e, of a cell is, according to the cell model [8],

$$F_e = (z_1 v/z_2)\, kTv/\lambda, \tag{8}$$

$$v = \lambda \ln \frac{(e^{2\gamma} - 1)\,[(1-\lambda)^2 - \beta^2]}{2\lambda} - u, \tag{9}$$

$$u = (1 + \beta^2)\,\gamma + \ln \frac{(1-\lambda)^2 - \beta^2}{1 - \beta^2} + \lambda. \tag{10}$$

From Equations (6) to (10), taking into account Equations (3) and (5), it follows

$$\phi = \frac{1 - \beta^2}{2\lambda} (1 - e^{-2\gamma}). \tag{11}$$

By taking $\varepsilon = 87.9$, $T = 273.15$K, $b = 2.52$ Å, and according to Equation (3), the charge-density parameter $\lambda = 2.76$ which corresponds to the fully extended chain of HPSS in water at 0°C. This is the so-called structural value of the charge-density parameter λ_{str}. In Figure 4, the theoretical curves calculated from Equations (11), (3) and (5), are presented for four different values of the parameter λ and are compared with the experimental values obtained with solutions of HPSS. There is a very good agreement between theory and experiment for $\lambda = 3.8$ which is 1.38 times higher than its structural value. For solutions with divalent counter-ions the osmotic coefficient is almost independent of concentration and the agreement between theory and experiment is good only for a limited concentration range. From Equation (3) it follows that for $z_2 = 2$ the structural value $\lambda = 5.52$. In order to fit the theoretical curve to the experimental points at concentrations of about 0.01 molal, it is necessary for magnesium and calcium ions to multiply λ by a factor of 1.3 or 1.6, respectively.

Considering λ as an adjustable parameter one can find, for each curve in Figure 2, a value of λ which satisfactorily reproduces the experimental results. It is, however,

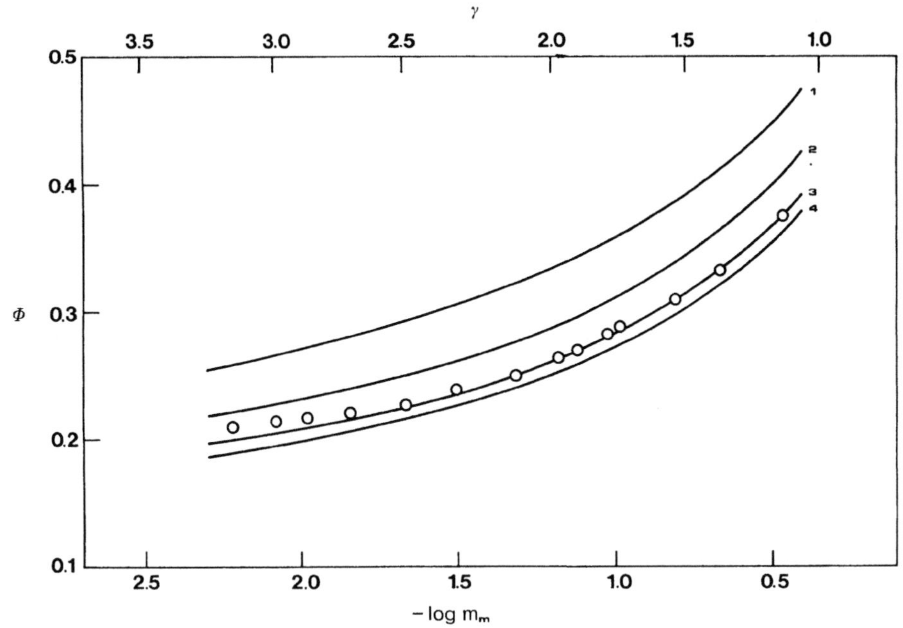

Fig. 4. Osmotic coefficient calculated for the following values of the parameter λ: (1) 2.76, (2) 3.40, (3) 3.80, (4) 4.00. The circles represent the experimental values for solutions of HPSS. (Reference [15], Figure 2.)

sounder to take into account the size of the hydrated counter-ions by changing the polyionic radius a and to keep the parameter λ constant for all counter-ions of the same valency as proposed by Kagawa and Gregor [18]. It is necessary to emphasize that this approach to the osmotic coefficient yields values which are in a qualitative but not quantitative agreement with the results represented in Figure 2.

Recently, Manning [19] has developed a theory in which the real polyelectrolyte chain is replaced by an infinite line charge. Statistical mechanical considerations lead to the conclusion that sufficiently many counter-ions have to condense on the polyion to lower the charge-density parameter to unity. The non-condensed counterions may by treated by the Debye-Hückel approximation. Thus, a limiting law has been developed which is supposed to be valid at high dilution. For the osmotic coefficient this limiting law is

$$\phi = 1/2\lambda, \tag{12}$$

where the charge-density parameter λ, Equation (3), must be calculated for the fully extended polyion.

Exactly the same expression is obtained by extrapolating the right-hand side of Equation (11) to infinite dilution. From the first of Equations (3) it follows for

$$\lambda > 1, \quad \lim_{\gamma \to \infty} \beta = 0 \tag{13}$$

and from Equations (11) and (13)

$$\lim_{\substack{\gamma \to \infty \\ \beta \to 0}} \phi = 1/2\lambda. \tag{14}$$

By taking the structural value for the parameter λ, the limiting value of the osmotic coefficient for solutions with monovalent counter-ions is 0.18. The values of the osmotic coefficient at the lowest concentration are between 0.17 and 0.23, as shown in Figure 2. For solutions having divalent counter-ions $\lambda_{str}=5.52$ and $\phi=0.0906$ which is very close to our experimental observations at low concentrations, see Figure 3.

The extrapolation procedures, based on Equation (13), are applicable to all thermodynamic properties which could be derived from the cell model. It is necessary to emphasize that the limiting expressions obtained in this way are equal to those following from the line charge model.

5. The Mean Activity Coefficient

In the realm of solutions of simple electrolytes the mean activity coefficient has been that basic property which for years has attracted the attention of a large number of investigators. In the realm of polyelectrolytes, however, there are only few papers devoted to this property. The first measurements of the mean activity coefficient of polyelectrolytes were carried out by Ise and Okubo [20]. A collection of data may be found in a review article by Ise [21].

We have measured [22] the electromotive force of the concentration cell with transference having the following scheme:

$$(\text{Pt})\ \text{H}_2\ (1\ \text{atm})\ |\ \text{HPSS}\ (m_0) \vdots \text{HPSS}\ (m)\ |\ \text{H}_2\ (1\ \text{atm})\ (\text{Pt}),$$

where m_0 is the reference and m the variable concentration. In another study [23] the hydrogen electrodes were replaced by Cd-amalgam electrodes and HPSS by CdPSS solutions. From emf and the transference numbers [24] the ratio of the mean activity coefficients $f_\pm/f_\pm^0$, $f_\pm^0$ corresponding to the reference concentration, has been derived. It should be noted that for polyelectrolyte solutions, as contrasted to simple electrolytes, a standard state in which $f_\pm^0=1$ does not exist and therefore only the ratio of activity coefficients has a definite physical meaning. In Figure 5 this ratio is presented for HPSS and CdPSS as a plot of $\log(f_\pm/f_\pm^0)$ versus $\log m_m$ where m_m stands for monomolality. The lowest and reference concentration is 0.002 and the highest concentration is 0.1 monomolal which is a rather wide concentration range.

It is noteworthy that the plot in Figure 5 shows a linear relationship between $\log f_\pm$ and $\log m_m$ and has a considerable slope. If the concentration of HPSS increases from its lowest to its highest value, the ratio $f_\pm/f_\pm^0$ decreases from 1 to 0.066, i.e. the ratio becomes 15-times smaller. On the analogy of simple electrolytes, the mean activity coefficient of the polyelectrolyte is defined by

$$f_+^\nu f_- = f_\pm^{\nu+1}, \tag{15}$$

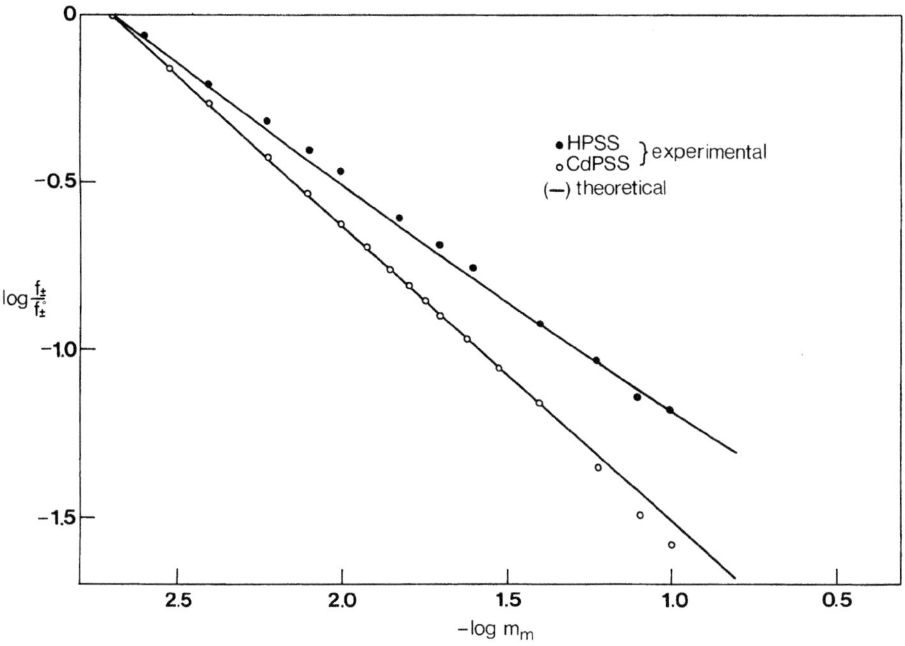

Fig. 5. Ratio of the mean activity coefficients as function of the monomolality. Comparison between theory and experiment. (Data from Reference [22] and [23].)

where f_- and f_+ are the individual activity coefficients of the polyion and its counter-ions, respectively, and ν is the number of ionic groups on the polyion. By taking f_+ as independent of concentration we may write for the ratio of coefficients, corresponding to the lowest and highest concentrations, see Figure 5,

$$(f_-^0/f_-) = (f_\pm^0/f_\pm)^{\nu+1} = 15^{\nu+1}. \tag{16}$$

Since ν is at least 100, the ratio f_-^0/f_- is as high as 10^{120}. This unusual finding has been explained [25] in terms of unequal contributions of the polyion and its counterions to the electrostatic free energy of the system. In other words, the calculations based on the cell model have shown that the contribution of the polyion to the electrostatic free energy of the solution is much more dependent on concentration than the corresponding contribution of the counter-ions.

The mean activity coefficient is related to the electrostatic free energy of the solution as follows

$$\left(\frac{z_1\nu}{z_2} + 1\right) kT \ln f_\pm = \frac{\partial (N_p F_e)}{\partial N_p}. \tag{17}$$

From Equations (5), (8–10) and (17) it follows that

$$\ln f_{\pm} = \frac{z_1 \nu}{z_1 \nu + z_2} \left\{ \frac{v}{\lambda} - \frac{1}{2\lambda} \left(\frac{\partial v}{\partial \gamma} \right)_{\lambda} \right\}, \tag{18}$$

where

$$\frac{1}{2\lambda} \left(\frac{\partial v}{\partial \gamma} \right)_{\lambda} = \frac{e^{2\gamma}}{e^{2\gamma} - 1} - \frac{1 - \beta^2}{2\lambda}. \tag{19}$$

According to Equation (13) we find that the limit for high dilution is

$$\ln (f_{\pm}/f_{\pm}^0) = -(1 - 1/2\lambda) \ln (m_m/m_m^0). \tag{20}$$

The same expression is readily obtained from Manning's limiting law, $\phi = 1/2\lambda$, and the thermodynamic relation

$$\mathrm{d} \ln f_{\pm} = (\phi - 1)\, \mathrm{d} \ln m_p + \mathrm{d}\phi \tag{21}$$

where m_p stands for the molality of the polyelectrolyte.

Figure 5 shows that there is very good agreement between experiment and theory. In the case of HPSS the experimental points are found to reasonably fit the theoretical curve calculated for the structural value of the parameter λ. In the case of CdPSS the best agreement between theory and experiment is obtained at $\lambda = 8$, a value 1.4-times higher than λ_{str}.

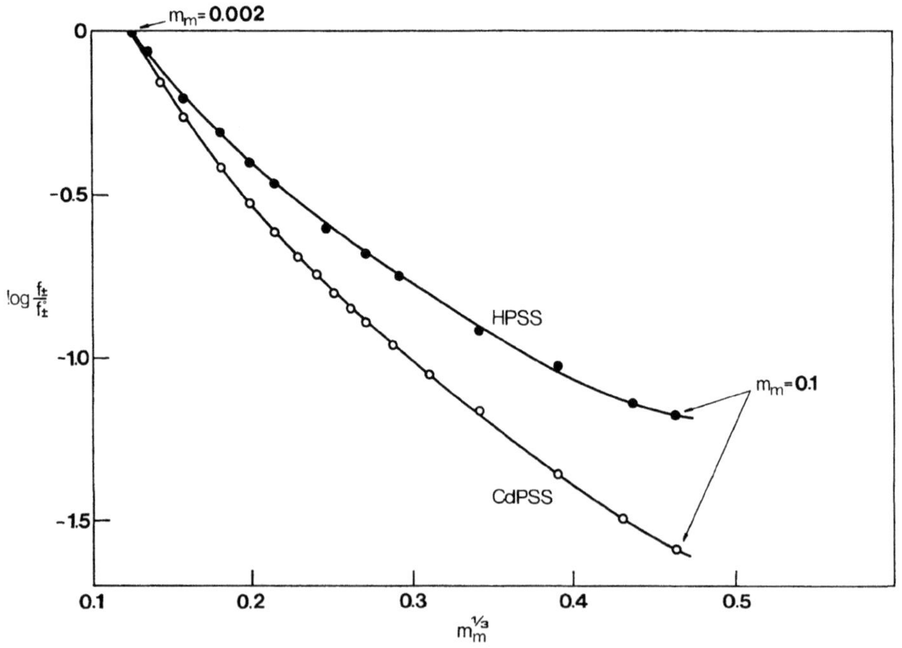

Fig. 6. Comparison of experimental values for the mean activity coefficient with the so-called cube-root relation. (Reference [23].)

It is interesting to compare the slope of the best fit straight line of the experimental points with slope predicted by the limiting law (see Equation (20)). For the former and latter slopes the values −0.68 and −0.823 for HPSS and −0.896 and −0.912 for CdPSS, respectively, have been found. The agreement is good, but in both cases the experimental slope is smaller than the limiting one. It seems that at lower concentrations the agreement would be better.

It remains to see whether the obtained data are in agreement with the so-called cube-root relation. According to Ise [21] the log $f_\pm$ should be linear with $m_m^{1/3}$. The corresponding plots, given in Figure 6, show that the curves for HPSS and CdPSS are not straight lines. The cube-root dependence is based on the idea of an ionic lattice structure of the polyelectrolyte solution in which the polyion-polyion interaction has the same importance as the polyion-counterions interaction. Contrariwise, both the cell model and the line charge model neglect the polyion-polyion interaction and consider that the nonideal behaviour of polyelectrolyte solutions is mainly due to the strong interaction between a polyion and its counterions. The results discussed seem to indicate that the latter view is closer to the real physical situation.

6. Enthalpy of Dilution

The next property worth discussing is the enthalpy of dilution, which, in the field of simple electrolytes, is considered to be a very rigorous test for the theory. Since the heat effect is very small if a polyelectrolyte solution is diluted, it was necessary to build a sensitive calorimeter [26, 27] which makes possible the detecting of temperature differences as small as 2×10^{-6} °C.

The enthalpy of the dilution of a polyelectrolyte solution $\Delta H_{D(m_1\to m_2)}$ is defined as the enthalpy change of the dilution from a concentration m_1 to m_2 and is calculated per monomole of the solute. It may be formally divided in two parts

$$\Delta H_{D\,(m_1\to m_2)} = \Delta H^0_{m_1\to m_2} + \Delta H_{e\,(m_1\to m_2)}, \tag{22}$$

where ΔH^0 is the nonelectrostatic and ΔH_e the electrostatic contribution. According to the cell model, ΔH^0 represents the enthalpy of the dilution of a polyelectrolyte solution in the hypothetical reference state in which all ions are discharged. We suppose that this contribution is equal to zero within the whole concentration range. ΔH_e was calculated [26] from Lifson and Katchalsky's expression for the electrostatic free energy [8]. It is more convenient to reduce their expression per monomole of polyelectrolyte and to write it in the integral form

$$F_e = \frac{z_1 RT}{z_2} \int_0^\lambda \frac{u(\lambda, \beta)}{\lambda^2}\, d\lambda, \tag{23}$$

the function $u(\lambda, \beta)$ being given in Equation (10). By applying the Gibbs-Helmholtz equation to Equation (23) we obtain the electrostatic internal energy E_e

$$E_e = -\frac{z_1 RT}{z_2}\left\{\frac{\partial}{\partial\lambda}\int_0^\lambda \frac{u(\lambda,\beta)}{\lambda^2}\,d\lambda\right\}_\gamma \cdot \frac{d\lambda}{dT} \tag{24}$$

and hence

$$E_e = \frac{z_1 RT}{z_2 \lambda}\, u\left(1 + \frac{d\ln\varepsilon}{d\ln T}\right). \tag{25}$$

The differentiation in Equation (24) has been made at constant volume. In order

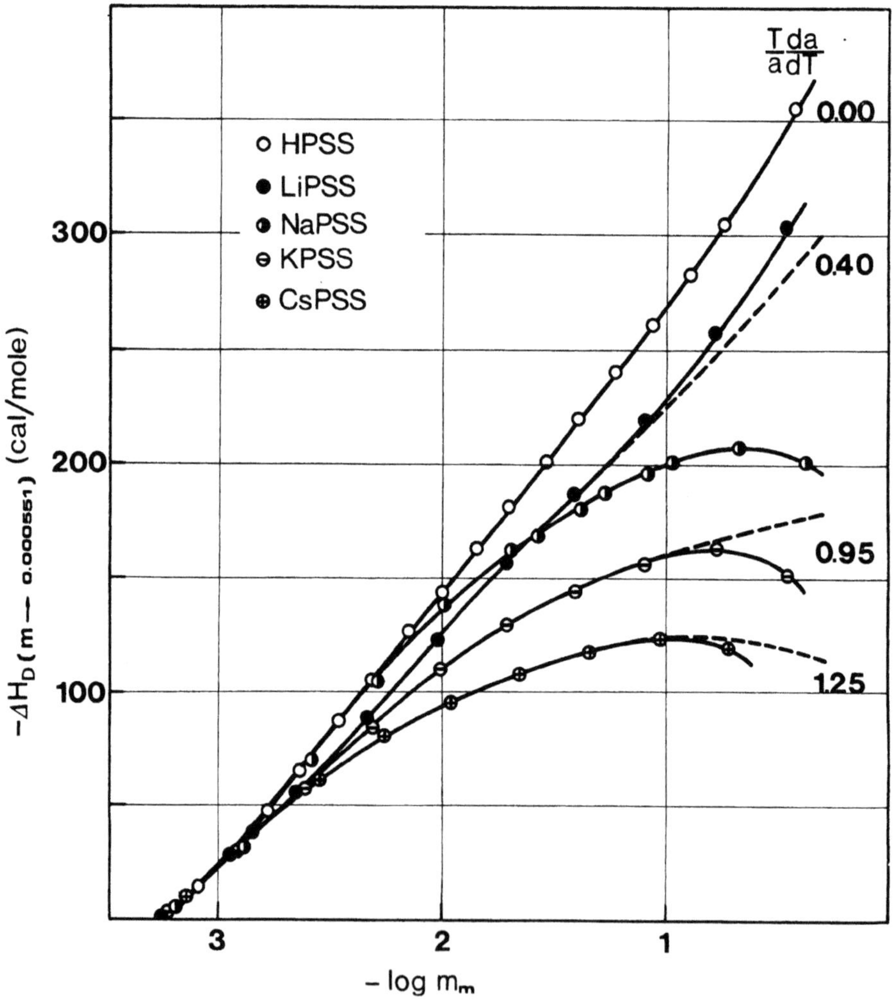

Fig. 7. Enthalpies of dilution of alkali polystyrenesulphonates in water at 25°C. Experimental values (—). Calculated values (---) with d ln a/d ln T as indicated. (Reference [28], Figure 2.)

to obtain the electrostatic enthalpy, H_e, the differentiation in Equation (24) should be made at constant pressure neglecting the small difference between the electrostatic free energy and free enthalpy. By taking into account also the temperature dependence of the concentration parameter γ and of the polyionic radius a, we finally obtain

$$H_e = E_e + \frac{z_1 RT}{2z_2\lambda}\left\{1 - \beta^2 - \frac{2\lambda e^{2\gamma}}{e^{2\gamma} - 1}\right\}\left\{\frac{\mathrm{d}\ln V}{\mathrm{d}\ln T} - 2\frac{\mathrm{d}\ln a}{\mathrm{d}\ln T}\right\}. \tag{26}$$

In Figure 7 the enthalpy of dilution of HPSS and its alkali salts is represented as a function of $\log m_m$. The upper curve plotted according to Equation (25) for the structural value of the parameter λ, is in perfect agreement with the results obtained with solutions of HPSS. At higher concentrations the curves for the alkali salts show negative deviations from the theory. In order to fit the theoretical curves to the experimental points one might, for instance, introduce the short range forces between the polyion and its counterions. A simpler way, however, is to remain in the domaine of electrostatics and to take into account the temperature dependence of the parameter a (see Equation (26)). The values for this dependence, chosen arbitrarily and indicated in Figure 7, have the same order of sequence and magnitude as the corresponding values for alkali chlorides [27].

It is significant that within the range of the lowest concentrations reached in this experiments, the slopes of the curves in Figure 7 seem to approach one and the same value. Thus, the slope of the curve for HPSS, which at the highest concentration is about -150 cal mole^{-1}, changes monotonously with the decreasing concentration and reaches a value of about -100 cal mole^{-1} at the lowest concentration. From Equations (26) and (25) we obtain in the limit for very low concentrations

$$\Delta H_{\mathrm{D}\,(m_1 \to m_2)} = \frac{z_1 RT}{2z_2\lambda}\left(1 + \frac{\mathrm{d}\ln\varepsilon}{\mathrm{d}\ln T}\right)\ln\frac{m_1}{m_2} \tag{27}$$

which could be also derived from Manning's expression for the excess free energy [19]. By taking $\lambda = 2.83$, $T = 298$ K, and $\mathrm{d}\ln\varepsilon/\mathrm{d}\ln T = -1.372$ one obtains for the limiting slope of ΔH_D versus $\log m_m$ a value of -90 cal mole^{-1}, which is very close to the experimental observation.

The apparent molal volume belongs, from a formal point of view, to the same group of thermodynamic properties as the enthalpy of dilution. The expression for the concentration dependence of the apparent molal volume of a polyelectrolyte can be derived from the electrostatic free enthalpy in a way similar to that shown above. In the final expression, which is analogous to Equation (24), the derivative $\mathrm{d}\lambda/\mathrm{d}T$ is replaced by $\mathrm{d}\lambda/\mathrm{d}P$. Recently, Škerjanc [28] has studied the apparent molal volume of solutions of polystyrenesulphonates and shown that the cell model is applicable within a wide concentration range.

7. The Polyelectrolyte Solutions with a Mixture of Mono- and Divalent Counterions

Let us apply the cell model to a polyelectrolyte solution containing a mixture of mono- and divalent counterions. Such solutions frequently occur in living matter and for that reason the present study might be of some interest to the field of biopolymers.

The Poisson-Boltzmann equation applied to this system reads [29]

$$\frac{1}{r}\frac{\mathrm{d}}{\mathrm{d}r}\left(r\frac{\mathrm{d}\psi}{\mathrm{d}r}\right) = -\frac{4\pi e_0}{\varepsilon}\left\{n_1^0 \exp\left(-e_0\psi/kT\right) + \right.$$
$$\left. + 2n_2^0 \exp\left(-2e_0\psi/kT\right)\right\}, \quad a \leqslant r \leqslant R \tag{28}$$

where n_1^0 and n_2^0 are the numbers of mono- and divalent counterions per unit volume at $\psi = 0$. The boundary conditions are

$$\psi(R) = 0, \quad (\mathrm{d}\psi/\mathrm{d}r)_{r=R} = 0 \tag{29}$$

and the normalization conditions

$$\left.\begin{aligned} \bar{n}_1 V &= n_1^0 \int \exp\left(-e_0\psi/kT\right)\mathrm{d}V \\ \bar{n}_2 V &= n_2^0 \int \exp\left(-2e_0\psi/kT\right)\mathrm{d}V \end{aligned}\right\} \tag{30}$$

with

$$v = (\bar{n}_1 + 2\bar{n}_2)\, V, \quad V = \pi(R^2 - a^2)\, h, \tag{31}$$

where $\bar{n}_1$ and $\bar{n}_2$ are the average values, corresponding to n_1^0 and n_2^0. The values of n_1^0 and n_2^0 in Equation (28) are not known in advance but are determined by Equations (30). Thus it is necessary to solve numerically the simultaneous Equations (28) and (30). The details of the calculation are given in Reference [29]. A simpler procedure is to solve the differential Equation (28) by any of the numerical methods. By introducing the dimensionless quantities in Equation (28) a new parameter equal to $(\bar{n}_1 + 2\bar{n}_2)/(n_1^0 + 2n_2^0)$ appears. For it a guess-value has to be found which would satisfy the boundary condition at the surface of the polyion derived from the Gauss's theorem.

From the solution of the potential problem defined above we have calculated the degree of ion binding as a function of the equivalent fraction of monovalent counterions. It has been found that the degree of ion binding of the less bound monovalent ions increases with its equivalent fraction. The more bound divalent ionic species obeys the opposite rule: the degree of binding increases with the decrease of its equivalent fraction. A similar rule has been found experimentally by studying the selectivity of ion exchange resins.

According to the cell model the osmotic coefficient is given by

$$\phi = \frac{n_1^0 + n_2^0}{\bar{n}_1 + \bar{n}_2}. \tag{32}$$

In Figure 8, the theoretical values of ϕ are plotted as functions of the equivalent

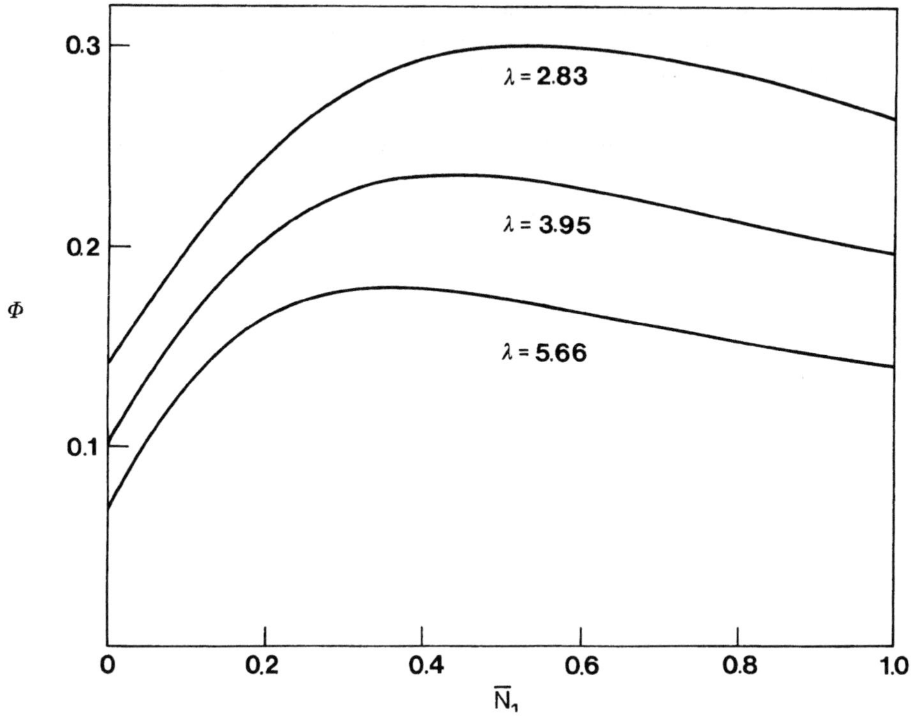

Fig. 8. Osmotic coefficient as function of the equivalent fraction of monovalent counter-ions. Theoretical curves calculated for the indicated values of the charge-density parameter λ and for $\ln(R/a) = 3.00$. (Reference [30], Figure 3.)

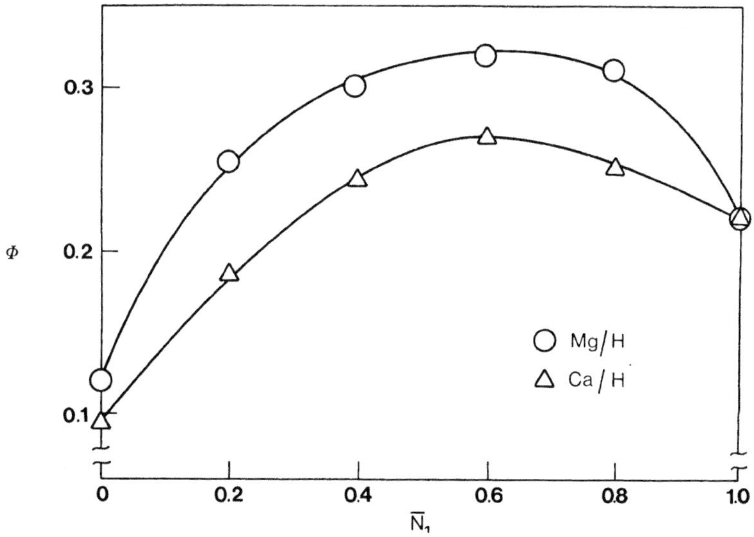

Fig. 9. Osmotic coefficient of solutions of polystyrenesulphonates with a mixture of counter-ions as function of the equivalent fraction of monovalent counter-ions at 0°C for indicated mixtures of counterions. (Reference [30], Figure 1.)

fraction of the monovalent ions defined as $\bar{N}_1 = \bar{n}_1/(\bar{n}_1 + 2\bar{n}_2)$. These curves as well as those found by experiment [30] and presented in Figure 9 show distinct maxima. The same general behaviour was observed with other pairs of mono- and divalent counter-ions.

Finally, to satisfy our curiosity, some additional experiments have been made with solutions of polystyrenesulphonates having a mixture of counter-ions. We have found that the dependence of the enthalpy of dilution and of the apparent molal volume on the equivalent fraction of monovalent ions can also be satisfactorily interpreted in the light of the cell model [31].

8. Concluding Remarks

Before concluding this lecture I would like to make some additional remarks concerning the cell and the line charge models. From the electrostatic free energy, which was calculated for the cell model, it is possible to derive expressions giving the concentration dependence of the thermodynamic properties of polyelectrolyte solutions. By extrapolation of these expressions to infinite dilution we obtain formulas predicting the limiting values or limiting slopes which the thermodynamic properties should approach at high dilution. We come to exactly the same results by applying the line charge model. Thus, we can say that the line charge model is the limiting case of the cell model at high dilution.

The applicability of both models for the interpretation of the experimental results has been illustrated on a variety of thermodynamic properties related to different derivatives of free energy. It is clear that such an agreement cannot be just accidental. In this review, it is true, we have confined ourselves only to the data obtained with polystyrenesulphonates. Unfortunately, there is no complete set of thermodynamic data of other polyelectrolyte systems. It seems, however, that in the polyelectrolyte solutions discussed above the polymeric chain consists of rather long fully extended segments. Between segments belonging to the same or to a neighbouring polyion, there should be a considerable volume where the electric field intensity is close to zero. This justifies the application of a model with cylindrical symmetry. Moreover, the absence of specific interactions permits the application of a pure electrostatic theory.

It is evident that the idea of the cell model with cylindrical symmetry, introduced years ago by Prof. Katchalsky to the field of polyelectrolytes, has been very successful. It has promoted the experimental research and initiated the development of other models having a cylindrical symmetry. Today, we would like to have a deeper understanding of the reasons for the wide applicability of this model. It seems that at this stage no satisfactory explanation can be expected. We believe, however, that the line charge model will be the first and the cell model the second approximation of a future more elaborate theory of polyelectrolyte solutions.

References

1. Strauss, U. P. and Leung, Y. P.: *J. Am. Chem. Soc.* **87**, 1476 (1965).
2. Kotin, L. and Nagasawa, M.: *J. Am. Chem. Soc.* **83**, 1026 (1961).
3. Lapanje, L. and Rice, S. A.: *J. Am. Chem. Soc.* **83**, 496 (1961).
4. Fuoss, R. M., Katchalsky, A., and Lifson, S.: *Proc. Natl. Acad. Sci. U.S.* **37**, 579 (1951).
5. Alfrey, T., Berg, P. W., and Morawetz, H.: *J. Polymer Sci.* **7**, 543 (1951).
6. Katchalsky, A., Alexandrowicz, Z., and Kedem, O.: in B. E. Conway and R. G. Barradas (eds.), *Chemical Physics of Ionic Solutions*, John Wiley & Sons, Inc., New York, 1966. p. 295.
7. Katchalsky, A.: *Pure Appl. Chem.* **26**, 327 (1971).
8. Lifson, S. and Katchalsky, A.: *J. Polymer Sci.* **13**, 43 (1954).
9. Verwey, E. J. W. and Overbeek, J. Th. G.: *Theory of the Stability of Lyophobic Colloids*, Elsevier, Amsterdam, 1948, p. 56.
10. Philip, J. R.: *J. Chem. Phys.* **52**, 1387 (1970).
11. Onsager, L.: *Ann. N.Y. Acad. Sci.* **51**, 627 (1949).
12. Oman, S. and Dolar, D.: *Z. Physik. Chem., Frankfurt* **56**, 1 (1967).
13. Oman, S. and Dolar, D.: *Z. Physik. Chem., Frankfurt* **56**, 13 (1967).
14. Armstrong, R. W. and Strauss, U. P.: in H. F. Mark and N. G. Gaylord (eds.), *Encyclopedia of Polymer Science and Technology*, vol. 10, Interscience, New York, 1969, p. 781.
15. Kozak, D., Kristan, J., and Dolar, D.: *Z. Physik. Chem., Frankfurt* **76**, 85 (1971).
16. Chu, P. and Marinsky, J. A.: *J. Phys. Chem.* **71**, 4352 (1967).
17. Kozak, D. and Dolar, D.: *Z. Physik. Chem., Frankfurt* **76**, 93 (1971).
18. Kagawa, I. and Gregor, H. P.: *J. Polymer Sci.* **23**, 477 (1957).
19. Manning, G. S.: *J. Chem. Phys.* **51**, 924 (1969).
20. Ise, N. and Okubo, T.: *J. Phys. Chem.* **69**, 4102 (1965).
21. Ise, N.: *Adv. Polymer Sci.* **7**, 536 (1971).
22. Dolar, D. and Leskovšek, H.: *Makromol. Chem.* **118**, 60 (1968).
23. Vesnaver, G. and Dolar, D.: in press.
24. Dolar, D., Špan, J., and Pretnar, A.: *J. Polymer Sci., Part C* **16**, 3557 (1968).
25. Dolar, D.: *Z. Physik. Chem., Frankfurt* **58**, 170 (1968).
26. Škerjanc, J., Dolar, D., and Leskovšek, D.: *Z. Physik. Chem., Frankfurt* **56**, 207 (1967).
27. Škerjanc, J., Dolar, D. and Leskovšek, D.: *Z. Physik. Chem., Frankfurt* **70**, 31 (1970).
28. Škerjanc, J.: *J. Phys. Chem.* **77**, 2225 (1973).
29. Dolar, D. and Peterlin, A.: *J. Chem. Phys.* **50**, 3011 (1969).
30. Dolar, D. and Kozak, D.: *IUPAC Symposium on Macromolecules*, Leiden, 1970, *Book of Abstracts*, Vol. 1, p. 363.
31. Dolar, D. and Škerjanc, J.: in preparation for press.

THERMOCHEMISTRY OF SYNTHETIC AND NATURAL POLYELECTROLYTES IN SOLUTION

VITTORIO CRESCENZI

Laboratory of Macromolecular Chemistry, Institute of Chemistry, University of Trieste, Trieste, Italy

Abstract. The importance of experiments which may enable to describe accurately how free energy changes accompanying a variety of typical equilibrium processes in polyelectrolyte solutions are built up by enthalphy and entropy contributions is emphasized.

In particular, the usefulness of accurate microcalorimetric data is pointed out.

In this context, a number of results of studies recently carried out in the Laboratory of Macromolecular Chemistry of the Instituto di Chimica of the University of Trieste are illustrated.

These studies have dealt with the physico-chemical characterization of aqueous solutions of both synthetic and natural polyelectrolytes.

From the experimental standpoint, working with synthetic polyelectrolytes attention has been focussed on:

(a) The measurement of the enthalpy of dissociation of a class of polycarboxylic acids in aqueous solution [1–4].

(b) The direct determination of the enthalpy of pH induced conformational transitions of vinyl polyelectrolytes and of poly-α-aminoacids in aqueous solution [1–3, 5].

(c) The measurement of the enthalpy of binding of transition metal ions by polycarboxylates in dilute aqueous solution.

In (a) and (c) polyions exhibiting only relatively small differences in their structure have been chosen so that more would be learned about short range interactions in polyelectrolytes systems.

In (b) conformational transitions of the type compactglobular coils $\underset{\text{pH}}{\rightarrow}$ expanded-solvated coils (for polymethacrylic acid and the maleic butyl vinyl ether copolymer, respectively) as well as ordered (β) form → coil transitions (for poly-*L*-histidine) [5] have been considered.

The useful joint application of accurate microcalorimetric and potentiometric measurements is illustrated by all cases mentioned above.

The research on biopolymers has dealt with a few interesting problems concerning a number of proteins and nucleic acids [6, 7]. In particular, the results obtained investigating:

(i) the interaction of spermine and spermidine with DNA and t-RNA in dilute aqueous solution [6]; and

(ii) the association reaction of *S*-protein (from RNase-A) with *S*-peptide and *S*-peptide synthetic analogs [7] are discussed in some detail.

List of Symbols

ΔH_{diss}	differential enthalpy of dissociation
$\alpha =$	degree of neutralization
$\text{pK}_a \equiv -\lg$	K_a, where K_a is the apparent dissociation constant of a polyelectrolyte
$\Delta G_{\text{diss}} =$	2.303 RT pK_a, free energy of dissociation
ΔS_{diss}	entropy of dissociation
ΔH_c	enthalpy of conformational transition
ΔS_c	entropy of conformational transition
ΔG_c°	standard free energy of conformation transition (uncharged states)
ΔH_c°	standard enthalpy of conformational transition
ΔH_p	differential enthalpy of protonation
ORD	optical rotatory dispersion
ΔH_P	enthalpy of the interaction (in kcal per mole of phosphate groups) between nucleic acids and spermine or spermidine.

Eric Sélégny (ed.), Polyelectrolytes, 115–133.

1. Motivation and enumeration of the Researches

The direct, accurate determination of changes in thermodynamic state functions accompanying certain typical reversible processes taking place in polyelectrolyte solutions is, needless to emphasize, a valuable achievement.

Such processes include, for example: the dissociation of interdependent ionizable groups along the chains, the binding or chelation of counterions by macroions, the change in shape or conformation of the chains brought about by changing one of the parameters controlling the equilibrium in the polymer solvent system, etc.

All these pheonomena have been studied extensively by many workers for a number of synthetic and natural polyelectrolytes using a variety of different techniques leading, in some cases, to a more or less straightforward evaluation of the associated free energy changes. However, the important goal of describing accurately how such free energy changes are built up by enthalpic and entropic contributions, e.g. with the aid of direct microcalorimetric measurements, has been achieved in relatively few instances only.

Because of our interest in equilibrium properties of polyelectrolyte solutions, it appeared worthwhile, in the light of what said above, to start a series of studies with an emphasis from the experimental standpoint on microcalorimetric measurements, dealing with both synthetic and natural polyelectrolytes. In particular, working with synthetic polyelectrolytes, we have focussed attention on:

(a) The measurement of the enthalpy of protonation of a class of polycarboxylic acids in water [1–4].

(b) The measurement of the enthalpy of binding of divalent transition metal ions by a few polycarboxylates in dilute aqueous solution.

(c) The direct determination of the enthalpy of: (i) pH-induced conformational transitions of vinyl polyelectrolytes [1–3] and of poly-α-aminoacids in aqueous solution [5] and (ii) poly-α-aminoacids order $\rightarrow$ disorder transitions in mixed organic solvents [8].

In (a) and (b) polyions exhibiting only relatively small differences in their structure have been chosen. The polycarboxylate considered by us so far include: poly (acrylic acid) [1, 4], poly (methacrylic acid) [1, 3] and four different maleic acid copolymers [4].

Polyaminoacids studied are: poly-*L*-histidine (pH-induced, $\beta \rightarrow$ coil transition in aqueous solution) [5] and poly-*L*-tyrosine (α-helix $\rightarrow$ coil transition in mixed organic solvents) [8].

Our research on biopolymers has dealt with the thermodynamic characterization of a few interesting features of a number of proteins and nucleic acids in aqueous solution, as for instance:

(α) The association reaction of *S*-protein (from RNAse-A) with *S*-peptide and with *S*-peptide synthetic analogs [7].

(β) The dilution-denaturation of DNA (and of t-RNA) and the binding of spermine and spermidine by the nucleic acids [6].

Obviously, the various problems mentioned above have been tackled with the aid of a number of experimental techniques besides microcalorimetric ones.

In the presentation of only a few of the results of the different studies mentioned above, I shall limit myself to a brief survey of what may be new, worthwhile, informations. Full details on the techniques employed and on the experimental procedures may be found in the articles quoted above.

2. Enthalpy of Dissociation of Synthetic Polyelectrolytes

2.1. Polycarboxylic acids

Passing immediately to the experimental results let us first examine the case of five polycarboxylic acids, namely; poly (acrylic acid) (PAA), maleic acid-ethylene

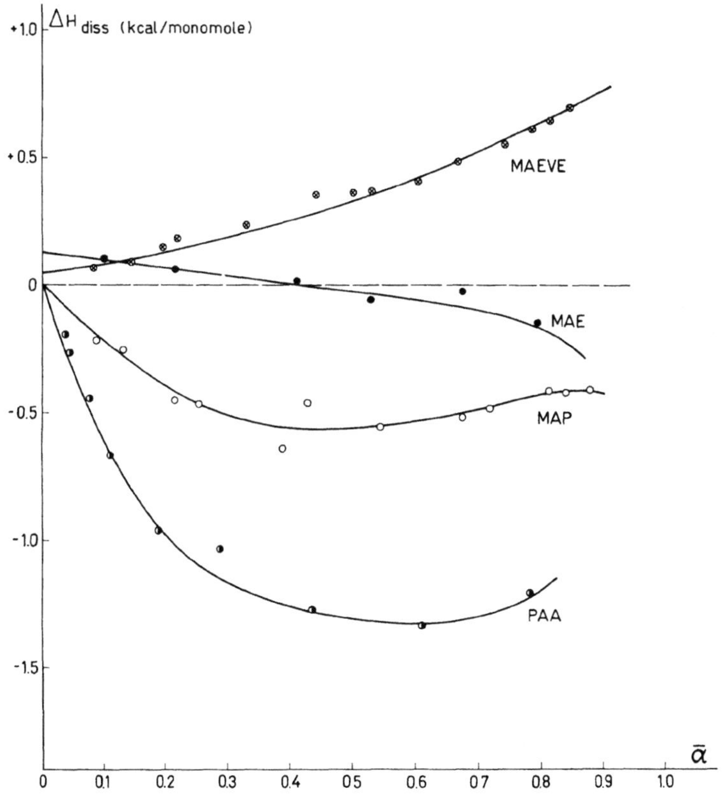

Fig. 1. Dependence of the enthalpy of dissociation, ΔH_{diss} upon the degree of neutralization α for aqueous solutions of:

◑ poly(acrylic acid), PAA; $m_p = 6.5 \times 10^{-2}$

○ maleic acid-propylene copolymer, MAP; $m_p = 2.0 \times 10^{-2}$

● maleic acid-ethylene copolymer, MAE; $m_p = 2.0 \times 10^{-2}$

⊗ maleic acid-ethyl vinyl ether copolymer, MAEVE; $m_p = 2 \times 10^{-2}$.

Neutralization with NaOH. Experiments carried out with LKB batch type and flow-type calorimeters. (Reproduced from Reference [4].)

copolymer (MAE), maleic acid-propylene copolymer (MAP), maleic acid-isobutilene copolymer (MAiB) and maleic acid-ethyl vinyl ether copolymer (MAEVE), whose chains are supposed to gradually expand and solvate during the charging process (neutralization with NaOH in water, and 25°). The enthalpy of dissociation data ΔH_{diss} in kcalories per monomole, are reported in Figure 1 against the degree of neutralization $\alpha(\alpha=1$ corresponds to half neutralization in the case of the maleic acid copolymers).

The enthalpy of dissociation of the polyelectrolytes is seen to become more negative upon increasing α, with the notable exception of the MAEVE copolymer for which $\Delta H_{diss}>0$ in the whole range of α values studied so far.

Data of Figure 1 clearly indicate, furthermore, that the trend of ΔH_{diss} with degree of neutralization markedly depends on the nature of the side chains and/or on the structure of the comonomers regularly alternating with the maleic acid residues in the

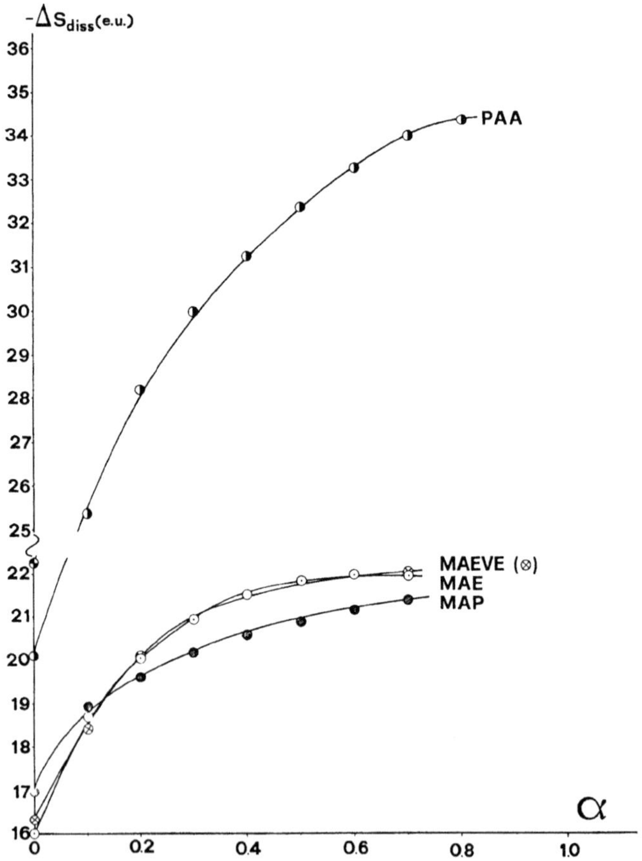

Fig. 2. The entropy of dissociation of poly (acrylic acid), PAA, and of the maleic acid copolymers of Figure 1 in water at 25°. (Reproduced from Reference [4].)

polyelectrolyte chains. With the aid of a number of potentiometric titration pK_a data,

$$pK_a \equiv pH + \lg \frac{1-\alpha}{\alpha} \tag{1}$$

the total free energy for the removal of a mole of protons from the different polycarboxylate chains at each given value, $\Delta G_{diss} = 2.303\ RT \cdot pK_a$, was readily calculated and, finally, the associated entropy change, ΔS_{diss}, estimated on the basis of both calorimetric and potentiometric data:

$$\Delta S_{diss} = (\Delta H_{diss} - \Delta G_{diss})/T.$$

The results are shown in Figure 2.

It is seen that the entropy of ionization is distinctly more negative for PAA than for the maleic acid copolymers at least for $\alpha \leqslant 1$. In our opinion, this should be at least in part connected with a stronger immobilization of water molecules (solvation) during the charging process of the former polymer than of the latters.

This hypothesis is in agreement with the results of dilatometric experiments recently reported by Begala and Strauss [9].

> In summary our calorimetric and potentiometric data provide original evidence that both the enthalpy and the entropy of dissociation of polymeric weak acids are sensitive functions mainly of the charge density and of the nature of non-ionic substituent along the chains.

For a given polyelectrolyte in water the overall change in heat content of the system for the removal of a proton from a chain at a given α value, may be though as the algebraic sum of different energy terms stemming, for example, from the built up of the charge density (corresponding to the α value considered), the associated changes in average chain dimensions and in the hydration of ionizable groups and of chain backbone, (all of which contribute to determine the value of the local effective dielectric constant) as well as from changes in the extent of interaction and eventually in the extent of site binding of counterions (Na^+ ions in our case).

Likewise the overall entropy of dissociation will reflect all the above mentioned phenomena, among which the more or less steadely increasing solvation of the polyelectrolyte with increasing α would play a dominant role.

For an attempted, more detailed, interpretation of the set of data reported in this section, the reader is referred to the original articles in the literature [1–4], as it would take too long here to summarize the approaches utilized even though these have been, necessarily, qualitative ones.

Let us now proceed to consider the case of two different synthetic polyelectrolytes, namely: poly(methacrylic acid), PMA and the maleic acid-butyl vinyl ether copolymer, MABVE, which are known to assume in water (when uncharged) tightly globular conformations which expand only within a critical range of α values to yield open solvated conformations. Such unwinding process has for both PMA and MABVE the

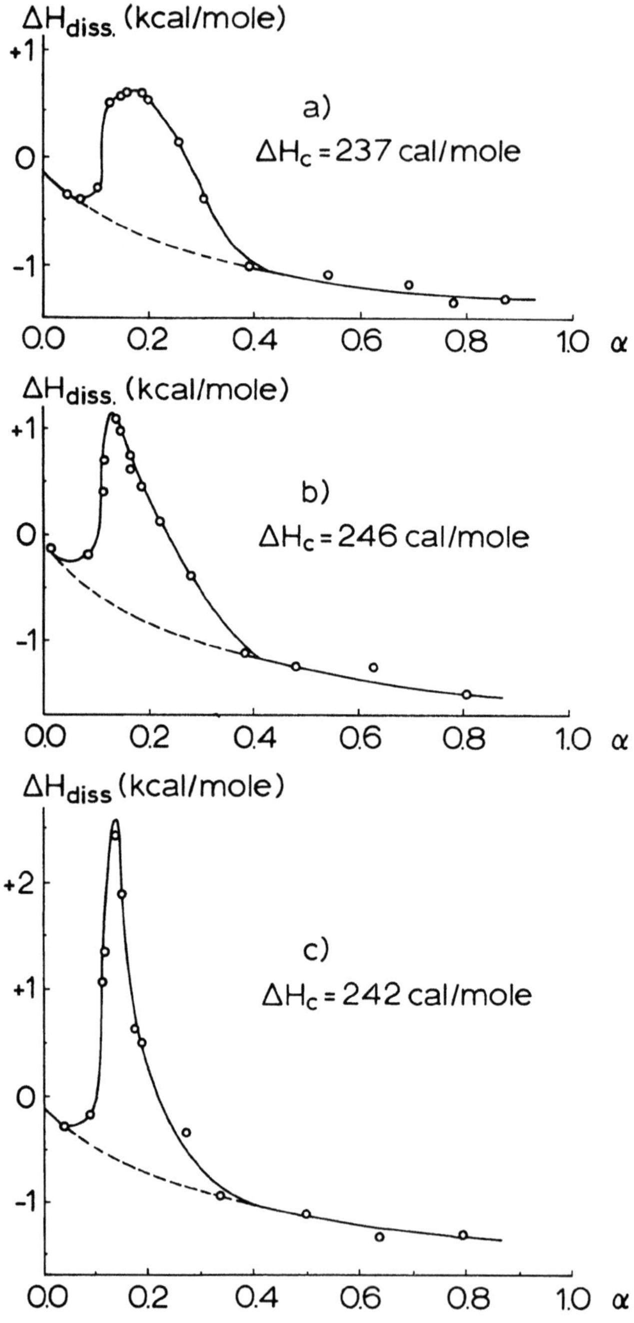

Enthalpy of dissociation of Poly(methacrylic acid) in water at 25°C

Fig. 3.

character of a cooperative conformational transition, usually schematized as:

$$\text{'globule'} \xrightarrow{\text{pH}} \text{expanded coils.}$$

Hydrophobic interactions are considered one of the factors responsible for tight globular state of the uncharged chains.

A systematic study of such relatively simple systems should prove useful for a transfer of information to the more complicate case of biopolymers transitions, for which hydrophobic forces do play a dominant role.

Passing to the experimental results already accumulated, the calorimetric ΔH_{diss} data for PMA in water at 25° reported in Figure 3 clearly show that dissociation behaviour of this polymer is quite anomalous. The anomalies in the ΔH_{diss} against α plots for PMA must be evidently connected with the occurrence of the globule → coil transition of the polymer.

Would this pH-induced transition not occur, the ΔH_{diss} values had in fact to vary smoothly and continuously with α as we have seen before in the case of five different polycarboxylic acids (Figure 1). The range of α values within which the transition is

TABLE I

Poly (methacrylic acid). pH-induced conformational transition: thermodynamic data. Summary of microcalorimetric and potentiometric data for the conformational transition of poly (methacrylic acid) in aqueous solution

		ΔH_c (cal mole^{-1})	Conc. (equiv lt^{-1})	ΔG_c° (cal mole^{-1})	Conc. (equiv lt^{-1})	ΔS_c (e.u.)
(1) 25°						
in water	C	246	6.4×10^{-2}	184	$1.8–6.5 \times 10^{-2}$	0.20
	C	242	2.1×10^{-2}			
	S	237	6.4×10^{-2}	185	$3.3–6.5 \times 10^{-2}$	0.17
in 0.5M NaCl	C	127	2.1×10^{-2}	165	3.4×10^{-2}	−0.13
	S	121	6.3×10^{-2}	165	$2.7–6.5 \times 10^{-2}$	−0.15
(2) 45°						
in water	C	364	5.9×10^{-2}	164	$3.2–5.6 \times 10^{-2}$	0.60
	S	344	5.9×10^{-2}	157	$3.2–3.4 \times 10^{-2}$	0.59

C = Conventional PMA, $M_w = 3.4 \times 10^5$
S = Syndiotactic PMA, $M_w = 9 \times 10^3$

←

Fig. 3. Dependence of the enthalpy of dissociation of PMA on the degree of neutralization α in water at 25°. (Reproduced from Reference [2].)
(a) Syndiotactic PMA ($M_w = 9 \times 10^3$) polymer concentration, 6.4×10^{-2} equiv lt^{-1};
(b) conventional PMA ($M_w = 3.4 \times 10^5$) polymer concentration, 6.46×10^{-2} equiv lt^{-1};
(c) conventional PMA ($M_w = 3.4 \times 10^5$) polymer concentration, 2.09×10^{-2} equiv lt^{-1}.
The ΔH_c values have been calculated from the areas under the curves as delimited by the dotted base line (see text).

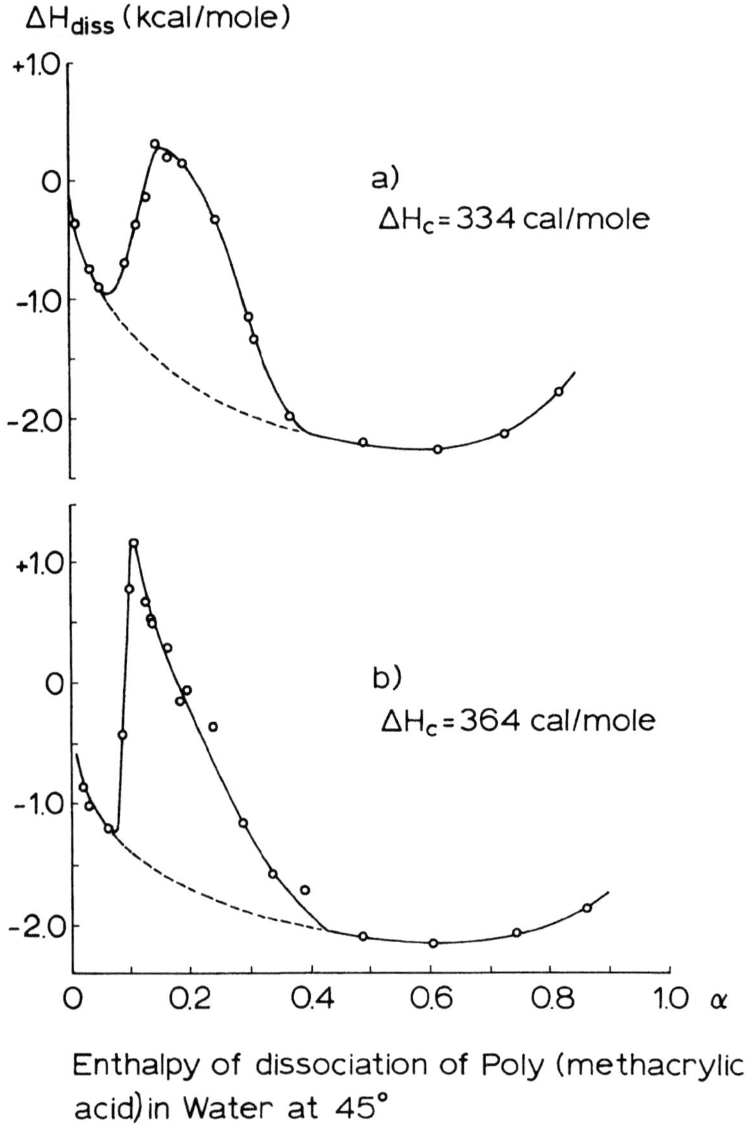

Fig. 4. Dependence of the enthalpy of dissociation of PMA on α in water at 45°. (Reproduced from Reference [3].)

(a) Conventional polymer concentration, 5.94×10^{-2} equiv lt^{-1};

(b) syndiotactic polymer concentration, 5.94×10^{-2} equiv lt^{-1}.

sought to occur is also clearly, and concordantly with the calorimetric data, defined by the potentiometric titration plots of PMA [1]. Therefore, we take the normalized areas under the peaks in the curves of Figure 3 (as delimited by the dotted base lines) to represent the enthalpy of conformational transition of PMA, ΔH_c. It is interesting to notice that the shapes of the ΔH_{diss} against α curves for PMA in water appear to depend on polymer stereoregularity and/or concentration. However, as indicated in Figure 3, the ΔH_c values are nearly independent of the above mentioned variables within experimental errors.

Data for PMA in water at 45° are illustrated in Figure 4: comparison of these data with those reported in Figure 3 clearly shows that increasing temperature increases the ΔH_c of PMA.

A summary of the microcalorimetric data for PMA is reported in Table I.

In this table the values of the free energy change ΔG_c^0 associated with the globule → → coil transition of PMA for experimental conditions identical to those employed in the calorimetric experiments, and evaluated by means of potentiometric titration data are also shown [1].

On the assumption that our ΔH_c values are not significantly different from the standard state ones and with the necessary confidence in our ΔG_c^0 data, the ΔS_c values reported in Table I have been finally estimated.

I wish to draw your attention particularly on the fact that ΔH_c for PMA increases with temperature. Naïve expectation might have been that, were hydrophobic forces relevant in the conformational transition of PMA, as we do believe they are, both ΔH_c and ΔS_c had to result negative in all cases. This expectation disregards of course that other factors (of very difficult evaluation in the case of PMA whose initial 'compact-globular' state is ill defined) are also important. In particular, the change of *configurational free energy of PMA chains* passing from a compact state to an 'open' state should be considered, in addition, of course, to the change in free energy of the water molecules engaged in solvating the 'open' PMA chains.

We believe that the observed increase of ΔH_c with temperature for PMA is an indication that we are dealing with a hydrophobic forces driven phenomenon, as we had already evidences from earlier studies.

TABLE II

Maleic acid – bulit vinyl ether copolymer pH-induced conformational transition: thermodynamic data.

Summary of microcalorimetric and potentiometric data for the conformational transition of the maleic acid – butyl vinyl ether copolymer in aqueous solution

Solvent	ΔG_c° (cal mole⁻;)	ΔH_c (cal mole ⁻;)	ΔS_c (e.u.)
water	330 (310)	$-560(-720 \pm 300)$	-3.0
25° 0.04M NaCl	~ (310)	-345	-2.2
0.20M NaCl	~ (310)	<0	<0
45° water	(330)	~93 (>0)	-0.7

Figures given in parenthesis are those obtained by P. Dubin and U. P. Strauss by the interpolation of potentiometric data (U.P. Strauss: private communication).
The uncertainty in our calorimetric ΔH_c data should not exceed $+5\%$ at 25° and about $+20\%$ at 45°.

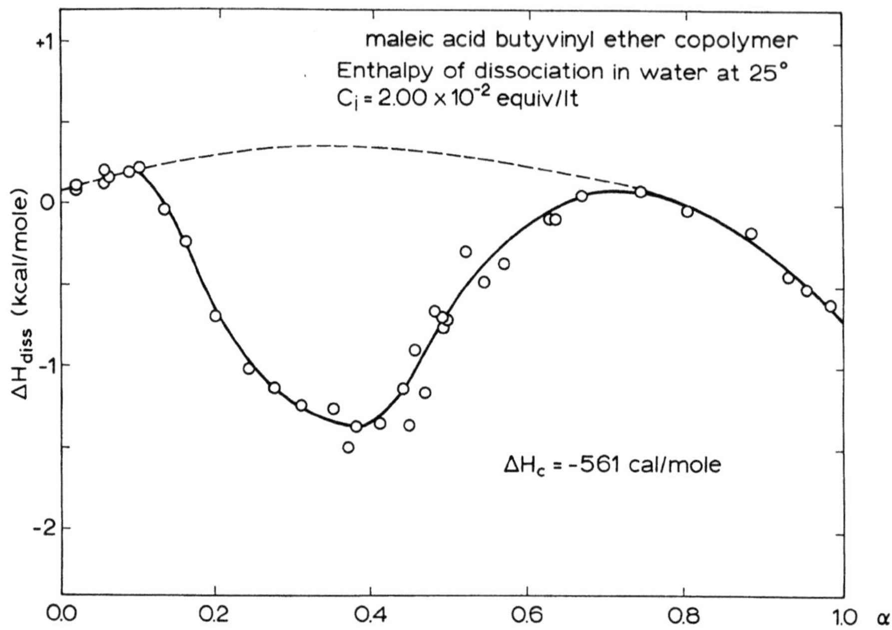

Fig. 5. Dependence of the enthalpy of dissociation of the maleic acid-butyl vinyl ether copolymer on the degree of neutralization α, in water at 25°. (Reproduced from Reference [2]).
Polymer concentration, 2.0×10^{-2} equiv lt^{-1}.

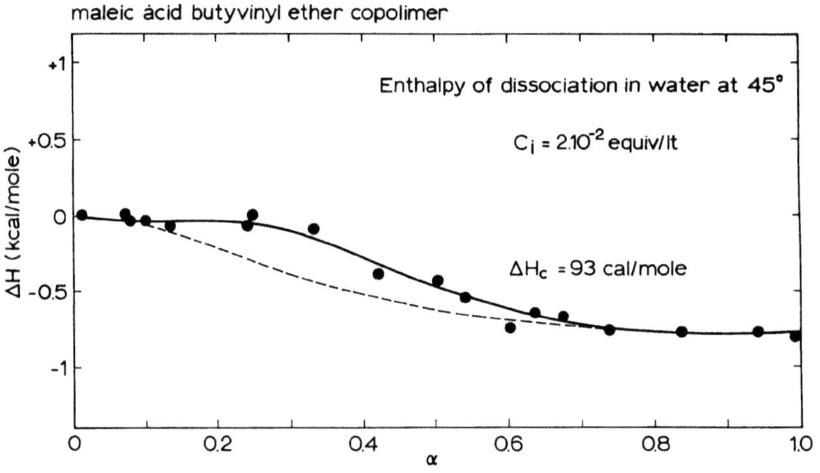

Fig. 6. Dependence of the enthalpy of dissociation of the maleic acid-butyl vinyl ether copolymer on the degree of neutralization α, in water at 45°. (Reproduced from Reference [2]).
Polymer concentration, 2.0×10^{-2} equiv lt^{-1}.

In this connection, and in view of their own interest, the results for the MABVE copolymer appear particularly worth mention [2]. A few microcalorimetric data for this copolymer are reported in Figures 5 and 6.

The line of reasoning for the evaluation of the ΔH_c values for the copolymer is identical to that outlined before in the case of PMA. A summary of the more important data is given in Table II.

In particular I wish to emphasize that for the conformational transition of the MABVE copolymer, which is doubtless a hydrophobic forces driven process, ΔH_c is negative at 25° but is *positive* at 45°.

2.2. Poly-*L*-histidine

As it is well known, in the case of polymers of naturally occurring α-*L*-aminoacids, steric regularity and chemical constitution of the backbone permit the development of highly regular ordered conformations of the chains (under appropriate conditions) also in dilute aqueous solution. In particular, the conformational properties of poly-*L*-histidine (Poly His) in water have been the subject of a number of investigations. It is presently well known that this polymer may undergo a conformational transition induced by changing the degree of protonation of the side-chains. The circular dichroism (CD) spectra of pure poly(His) and of random copolymers of *L*-lysine and *L*-histidine have shown unambiguously that pure poly (His) assumes the random coil conformation at high degrees of protonation of the imidazole groups, while at low degrees of protonation an ordered structure is formed.

On the basis of CD data one can safely exclude the righthanded α-helix from the possible ordered forms. Myer and Barnard [10] interpreted the CD spectra of the poly(His) ordered conformations in terms of a β-structure, this assignment however being complicated by the influence of side-chain chromophores [11]. In our study we have collected more evidences in favour of Myer and Bernard's proposal and investigated the thermodynamics of the $\beta \rightarrow$coil conformational transition [5]. Among the different types of data and informations derived from this study, I shall here collect and illustrate to you a few results more directly connected with the energetics of the transition of poly(His) and which, in my opinion, afford a further example of the useful joint applications microcalorimetry and potentiometry may find in the field of our concern.

Typical examples of potentiometric titration data for poly(His) in 0.02M KCl obtained working at 15° and 35° are reported in Figure 7. Potentiometric data were also collected at other two different temperatures. The curves were drawn according to the usual polyelectrolyte potentiometric equation. In this case α is the degree of *deprotonation* of poly(His), given by:

$$\alpha = 1 - \frac{(\mathrm{HCl}) - (\mathrm{KOH}) - (\mathrm{H}^+)_{\mathrm{free}}}{\mathrm{poly}\,(\mathrm{His})_{\mathrm{total}}},$$

where (HCl) is the molarity of HCl added before the titration, (KOH) is the molarity of added titrant and $(\mathrm{H}^+)_{\mathrm{free}}$ refers to the molarity of free H^+ ions. The plots in the Figure 7 exhibit marked inflection points which reflect the onset of poly(His) con-

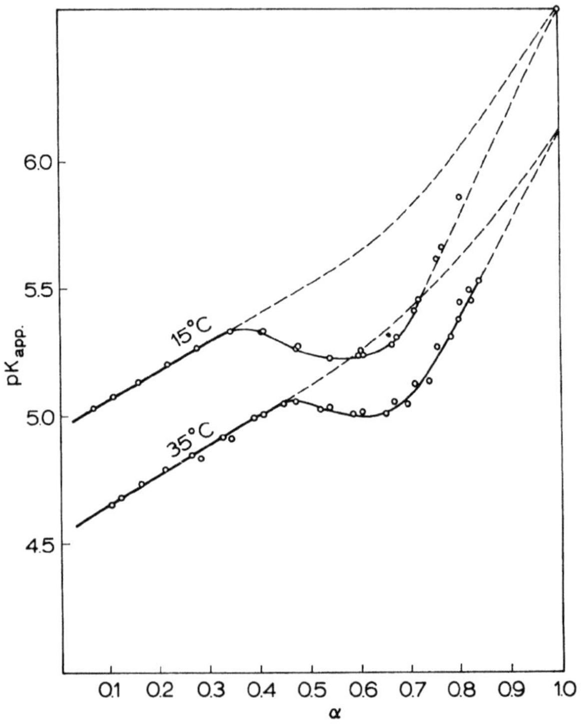

Fig. 7. Potentiometric titration curves of poly-*L*-histidine in 0.02M KCl at 15° and 35°. (Reproduced from Reference [5].)

formational transition, as concordantly monitored within the same narrow range of α values also by CD measurements.

The area limited by the extrapolated and experimental curves of the random coil and of the ordered form in the pK_a against α plots is proportional to the standard free energy, ΔG_c^0, per residue for the transition from *uncharged* ordered form to uncharged coil. There are however some difficulties in the extrapolation of the coil titration curves, since the extrapolation is very long and somewhat arbitrary. In principle, the pK_0 of the coil and of the ordered form could even not be the same. In order to determine more accurately the pK_0 values to which the extrapolations had to be made, we have carried out potentiometric titrations on the model compound

$$Z-\text{Leu}-\text{His}-\text{OCH}_3$$

in 0.02M KCl and at the same temperatures at which the titration curves of poly(His) have been obtained. From the results of such measurements, it was possible to determine the pK_0 of an isolated imidazole group and its temperature dependence. Assuming that the pK_0 of the imidazole group in the model compound is the same as in the coiled polymer at $\alpha = 1$, the final points of extrapolation of the coil titration curves such as those of Figure 7 have been fixed. From the same figure it appears that the portions of the titration curves corresponding to the ordered from of poly(His) are almost linear and extrapolate with good approximation to the same pK_0 of the coil titration curve. In the evaluation of the ΔG_c^0 values for the conformational transition of poly(His) from the potentiometric plots, the main point of uncertainty is in the *shape* of the extrapolation curves for the coil regions. As a reasonable compromise we have drawn curves on all potentiometric plots having approximately the same shape, but in any case the limit of uncertainty is relevant. In consequence, our ΔG_c^0 values must be considered approximate ones. Such ΔG_c^0 values are reported in Figure 8

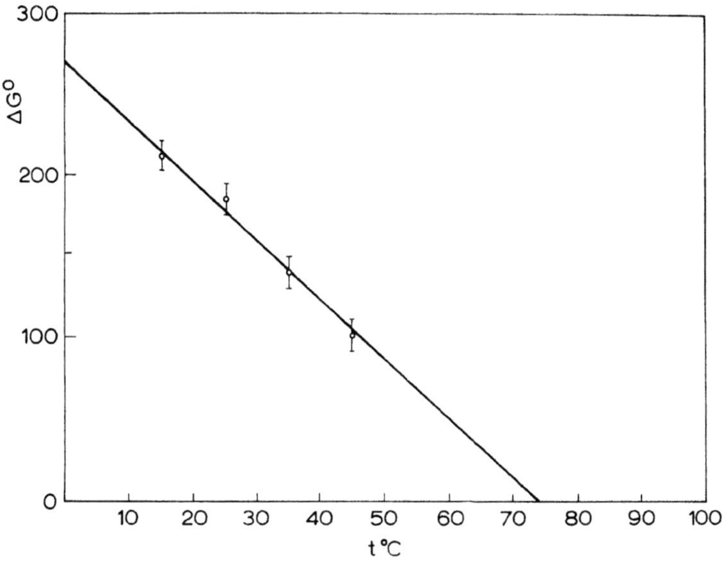

Fig. 8, Free energy (calories per residue) relative to the uncharged ordered-from→uncharged-coil transition of poly-*L*-histidine in 0.02M KCl as a function of temperature. (Reproduced from Reference [5].)

as a function of temperature: from these data the associated ΔH_c^0 and ΔS_c^0 data may finally be evaluated. The results for 25° are:

$$\Delta H_c^0 = 1.16 \text{ kcal mol}^{-1}; \qquad \Delta S_c^0 = 3.65 \text{ e.u.}$$

It was considered worthwhile to confirm these data by means of direct microcalorimetric measurements. The results of these measurements are given in Figure 9 as a plot of the differential heat of protonation against the degree of deprotonation of poly(His). We assume the heat of the conformational transition to be measured by the area between the curve and a suitable base-line. The impossibility of measuring the heat of protonation of the imidazole ring when the polymer is in its ordered state prevents to draw a good base line. In this case, however, it is possible to derive the most probable base line according to various methods each of which will be discussed in what follows. All methods are based on the extrapolation of the microcalorimetric

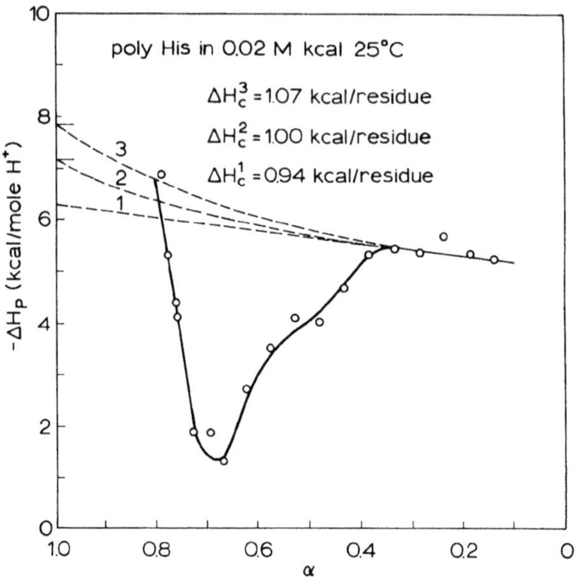

Fig. 9. Differential heat of protonation of poly-*L*-histidine in 0.02M KCl at 25°. (Reproduced from Reference [5].)

data obtained in the range of α values where the polymers is entirely random to $\alpha=1$, where the polymer is completely ordered. The assumption which underlines these methods is that the heat of protonation of the histidyl group at $\alpha=1$ is independent on the conformational state of the polymer. On the other hand this assumption is analogous to that made in the case of the titration curves for the pK_0 value.

The first method is that of extrapolating directly with a straight line the data at low α values. In this case, a heat of protonation of the imidazole ring at $\alpha=1$ (ΔH_p, $\alpha=1$) of 6.30 kcal mole^{-1} and a ΔH_c of 0.94 kcal residue^{-1} is obtained. This extrapolation is represented by curve 1 in Figure 9. Base line 2 is obtained extrapolating to a value of $\Delta H_p(\alpha=1)$ of -7.14 kcal mole^{-1}. This value was obtained by direct calorimetric measurements on *L*-histidine in 0.1M KCl at 25°C by Christensen *et al.* [12]. With this extrapolation a ΔH_c value of 1.00 kcal residue^{-1} is obtained.

Base line 3 is drawn extrapolating to a value of ΔH_p ($\alpha=1$) of -7.85 kcal mole^{-1} obtained by potentiometric measurements at different temperatures on the model compound mentioned above. This extrapolation leads to a ΔH_c of 1.07 kcal residue^{-1}.

The values of ΔH_c obtained by microcalorimetry have to be compared with the value of 1.16 kcal residue^{-1} obtained by the temperature dependence of potentiometric data. Taking into account the errors connected with the potentiometric method, the errors of the calorimetric one (essentially due to the uncertain choice of the baseline) and the possible difference between ΔH_c and ΔH_c^0, the agreement between the two sets of data is rather good indeed. This fact is most gratifying in view of the eventually large, unexplained, difference in thermodynamic data as derived by means of potentiometry and calorimetry for polyaminoacids conformational transitions reported by different authors.

For example, the enthalpy values for the $\beta \rightarrow$ coil transition of poly-*L*-lysine reported by Pederson *et al.* [13] (potentiometry) and by Chou and Scheraga [14] (calorimetry) differ by nearly 2 kcal mole^{-1}.

The only other case of a polyaminoacid $\beta \rightarrow$ coil transition which has been studied recently in details is that of poly-*L*-tyrosine for which Senior *et al.* [15] conclude that $\Delta G_c^0 \sim 0$ at 25° C!

3. Biopolymers

3.1. Ribonuclease S′: the interaction of S-protein with S-peptide and with S-peptide synthetic analogs

Ribonuclease. A(RNase) affords a case of a globular protein whose chemical constitution and structures are well established. RNase-A can be split at position 20 by subtilisin and the two pieces (*S*-protein and *S*-peptide) brought together again to yield RNaseS' with *full* recovery of enzymatic activity.

A detailed investigation on the thermal denaturation of such a protein and of the thermodynamics of association of *S*-protein with *S*-peptide and with synthetic *S*-peptide analogs seemed therefore particularly interesting. We decided to study the thermal denaturation of intact RNase-A by means of differential scanning calorimetry, which as far as I know, we were the first to apply to the study of concentrated biopolymers solutions [16] using a commercially available apparatus, and then to study with the LKB microcalorimeter the enthalpy of recombination of *S*-protein with *S*-peptide as well as with *S*-peptide analogs [7]

I shall here limit myself to the latter type of study. In brief, the main motivations of this research were (1) to find out which aminoacids in the *S*-peptide chain are indeed essential in the recombination process with *S*-protein and (2) to investigate the energetics of the process. To this end different peptides have been considered, in particular *S*-peptide, Orn^{10} *S*-peptide; Gly^8, Orn^{10} *S*-peptide and Cpg^8, Orn^{10} *S*-peptide.

Besides microcalorimetry, different types of experimental approaches have been utilized, namely: (a) difference spectroscopy, (b) differential ORD spectroscopy and (c) melting curves (thermal denaturation measurements).

Measurements of type (a) and the microcalorimetric ones, have permitted direct evaluation of the stability constants for the different *S*-protein – *S*-peptide adducts and their enthalpy of formation. ORD difference spectroscopy allowed us to estimate comparatively the increase in the helix content in the formation of partially synthetic RNase-*S*' analogs with respect to RNase-*S*′.

In all cases, moreover, the biological activity of RNase-*S*′ and its analogs was monitored in a systematic fashion by means of standard catalytic activity essays. The main results of such essays are: the Orn^{10} *S*-peptide is able to regenerate high ribonuclease activity in the presence of *S*-protein, while, on the contrary, the Cpg^8, Orn^{10} – and the Gly^8, Orn^{10} *S*-peptide are unable to activate *S*-protein.

Without going into any detail about the various interesting and consistent informations obtained by the different experimental approaches mentioned above (this would in fact take too much time here) let us focus attention on what microcalorimetric measurements in particular, have added to this piece of research.

The results of these measurements are summarized in Figure 10. It is evident (from such data that interaction of *S*-protein with *S*-peptide, as well as with the Orn^{10} *S*-peptide in dilute aqueuos solution is accompanied by a relatively large decrease in

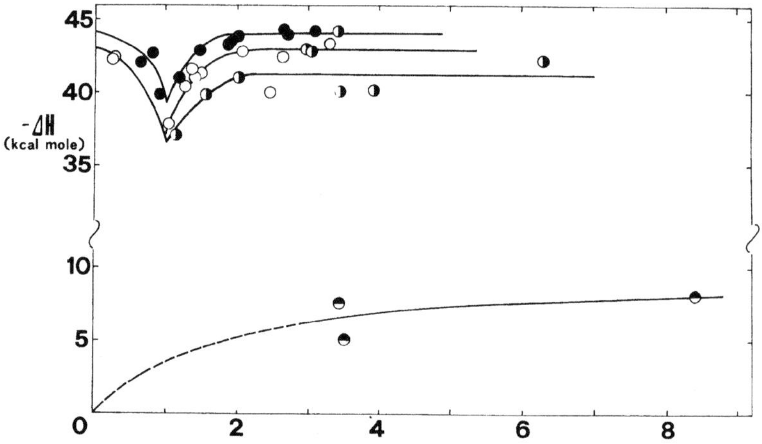

Fig. 10. Enthalpy change for the recombination of *S*-protein with *S*-peptide and two of its analogs as a function of the peptide: *S*-protein molar ratio, *r*. Unless states otherwise the measurements were carried out in 0.01M sodium phosphate buffer (pH = 6.8) containing 0.9% NaCl. (Reproduced from Reference [7].)

ΔH is in kcalories per mole of RNase-*S'* or RNase-*S'* analog. The enzyme concentration is calculated on the basis of the peptide of *S*-protein concentration assuming complete recombination.

— — ● — — S-peptide
— — ◑ — — Orn^{10} S-peptide
— — ◒ Cpg^8, Orn^{10} S-peptide
— — ○ S-peptide (in water)

enthalpy of the system. The *ΔH* attains a limiting value of approximately −44 kcal $mole^{-1}$ (kcalories per mole of the maximum stoichiometric amount of RNase-*S'* present in solution) at a molar ratio $r \sim 2$ and at 25°.

Our data also show, however, that the *ΔH* values in the case of the *S*-peptide are systematically higher than in the case of the Orn^{10} *S*-peptide.

With the Cpg^8, Orn^{10} *S*-peptide the microcalorimetric results indicate a much weaker interaction with *S*-protein, since only ca. 8 kcal $mole^{-1}$ are released when this peptide is mixed with *S*-protein at molar ratios higher than 8.

From our *ΔH* data for the *S*-peptide – *S*-protein system it is possible to derive for the equilibrium constant, K_d, of the process:

$$\text{RNase-}S' \rightarrow S\text{-protein} + S\text{-peptide} \tag{1}$$

value of 6.2×10^{-6} (standard deviation of the mean 1.5×10^{-6}) at 25°. This leads to a free energy change, ΔG_d^0 for the dissociation reaction of ca. 7.1 kcal mole $^{-1}$ at 25°.

Neglecting the small heat of dilution effects, and thus assuming as a first approximation that our ΔH data are standard state values, one readly estimates for the above mentioned process an entropy change of 124 e.u.

The marked increase in entropy connected with RNase-S' dissociation must be largely determined by the increased conformational freedom of S-protein and, especially, of S-peptide when separate in dilute aqueous solution. In the case of the Orn^{10} – S-peptide – S-protein system it is difficult to evaluate a reliable K_d value on the basis of our calorimetric data. However, the order of magnitude of this K_d should not be too different from that obtained for reaction (1).

The selective increase of *order* upon association of S-protein with S-peptide as compared with S-peptide analogs is also clearly demonstrated by ORD data. These data confirm the hypothesis that S-protein can act only through very specific effects and is thus able to recognize, to bind and to change the conformation of only some S-peptide analogs, although all peptides examined prossess the structural requirements which are needed to undergo a solvent-induced coil → helix transition according to their CD spectra in water and in organic solvents, respectively.

To grasp the real, original kind of information derivable from these results, in particular from the microcalorimetric ones, we are lead to conclude that:

(1) the small difference in behaviour between S-peptide and the Orn^{10} S-peptide may be due to a lack in the latter of a proper site, i.e. $arginine^{10}$, which is supposed to interact with glutamic acid-2 in the S-peptide;

(2) More important, phenylalanine in position 8 has to play a very specific role in the binding process of S-peptide with S-protein *and* in controlling the local structure of the reconstituted enzyme and hence its biological activity.

What I have given here is a very short summary of a research which is of course far from being complete in the spirit of its motivations indicated in the outset.

We are continuing this type to study using different synthetic polypeptides with the aim of finally arriving at a sort of 'thermodynamic mapping' of the important binding sites which control native conformation and function of RNase-A.

3.2. Interaction of DNA and t-RNA with spermine and spermidine

Studies being carried out in our laboratory on the nucleic acids include: (1) the measurement of the enthalpy of DNA and t-RNA isothermal denaturation promoted by the addition of different strong acids or by simple dilution in salt-free solution [17]; (2) the study of the selective interaction of DNA with different histones in salt solutions and; (3) the thermodynamic characterization of the binding of low molecular weight compounds and of certain divalent counterions by DNA [6].

I shall here limit myself to a brief report of a few data concerning point (3) above, more precisely dealing with the interaction of spermine and of spermidine with DNA and t-RNA respectively.

As it is well known DNA as well as t-RNA interact strongly and selectively with both spermine and spermidine. A number of studies on such interactions have been carried out in the past and models have been proposed to account for the specific nature of the binding processes in terms of the conformational states of both DNA and the two biological polyamines [18].

One of results of the binding of such compounds onto DNA chains is that the so called 'melting-temperature' of the biopolymer is markedly increased. We have found that the stabilizing influence of spermine and spermidine against thermal denaturation of DNA and of t-RNA is, qualitatively, very similar. The results of equilibrium dialysis measurements have shown that also the free energy of binding of

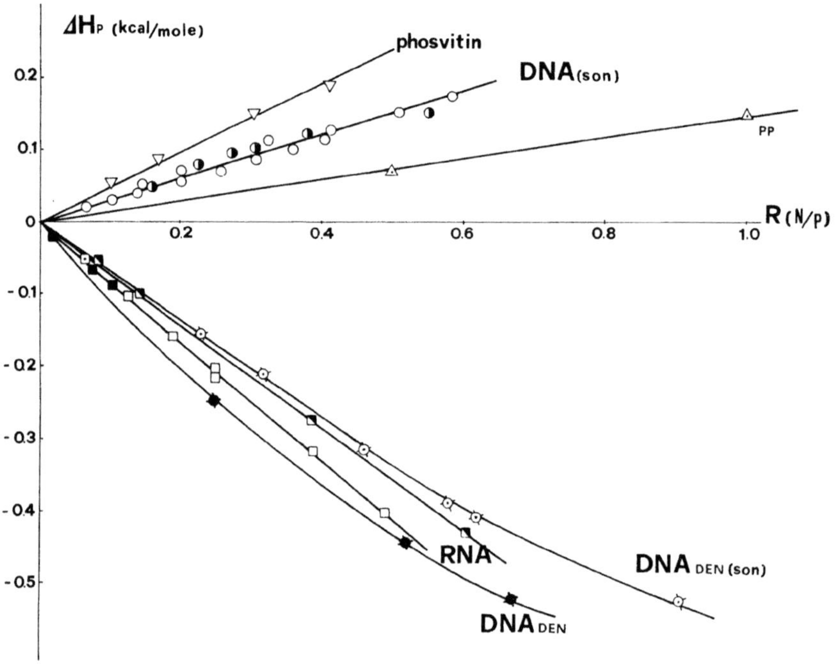

Fig. 11. Enthalpy of interaction (ΔH_P) of spermine (sp.) and of spermidine (spd.) with calf thymus DNA and baker's yeast t-RNA, and with other polyelectrolytes. (Reproduced from Reference [6].)

Measurements at 25° with an LKB batch-type microcalorimeter; solvent: 0.02M NaCl, 0.002 M cacodilate (pH = 5.5).

□ s-RNA ($6.7 \times 10^{-3} M_P$) + sp.
◩ s-RNA ($8.4 \times 10^{-3} M_P$) + spd.
■ t-RNA ($1.8 \times 10^{-2} M_P$) + sp.
∇ Phosvitin ($6.2 \times 10^{-3} M_P$) + sp.
△ Na polyphosphate ($7.0 \times 10^{-3} M_P$) + sp. (ph = 6.5)
○ DNA, sonicated ($6.4–9.0 \times 10^{-3} M_P$) + sp.
◑ DNA, sonicated ($9.0 \times 10^{-3} M_P$) + spd.
-○- DNA, sonicated-denaturated ($4.1 \times 10^{-4} M_P$) + sp.
-●- DNA, alkalidenaturated ($7.4 \times 10^{-4} M_P$) + sp.
s-RNA was a Sigma product; t-RNA was a Miles product.

the two amines by DNA and t-RNA should not be too different. The microcalorimetric data on the heat of mixing the nucleic acids with spermine and spermidine do show however that the enthalpy of binding clearly differentiates the behaviour of the two biopolymers. These data are reported in Figure 11, together with the results of calorimetric measurements carried out using other anionic polyelectrolytes such as sodium polyphosphate and phosvitin [6]. The calorimetric data also show that, interesting enough, when alkali-denaturated DNA is used the change of enthalpy upon binding of spermine is of opposite sign to that found working with native DNA, either sonicated or not, and is quite close to that found with t-RNA.

For 'native' t-RNA in aqueous solution it is commonly assumed that about half of the bases are exposed to solvent or are not engaged in intramolecular hydrogen bonding, while for alkali-denaturated DNA the percentage of bases in a similar state is certainly greater.

Our results might thus be taken as an indication that in these cases interactions between the nucleic acids and spermine should involve the purine and pirimidine moieties. As it is difficult to speculate on the sign of an enthalpy of binding, a less qualitative picture may be only gained by taking advantage of additional informations on the systems considered using also different experimental approaches. It is relevant however to point out that possible thermal effects arising from an even partial renaturation of DNA, consequent to interaction with spermine and/or spermidine, seem to be, even though quite indirectly, ruled out by CD measurements.

References

1. Crescenzi, V., Quadrifoglio, F., and Delben, F.: *J. Polymer Sci.* **A-2**, **10**, 357 (1972.)
2. Crescenzi, V., Quadrifoglio, F., and Delben, F.: *J. Polymer Sci.* **C**, **39**, 241 (1971); paper presented at the 9th IUPAC Microsymposium on Macromolecules – Prague, 1971.
3. Delben, F., Crescenzi, V., and Quadrifoglio, F.: *European Polymer J.* **8**, 933 (1972).
4. Crescenzi, V., Delben, F., Quadrifoglio, F., and Dolar, D.: *J. Phys. Chem.* **77**, 539 (1973).
5. Terbojevich, M., Cosani, A., Peggion, E., Quadrifoglio, F., and Crescenzi, V.: *Macromolecules* **5**, 622 (1972).
6. Crescenzi, V., Quadrifoglio, F., Cesàro, A., and Ciancotti, V.: in H. Peeters (ed.), Proceedings of the XXth Annual Colloquium on *Protides of the Biological Fluids* – Brugge, 1972; Pergamon Press, 1972, p. 483.
7. Rocchi, R., Borin, G., Marchiori, F., Moroder, L., Peggion, E., Scoffone, E., Crescenzi, V., and Quadrifoglio, F.: *Biochemistry* **11**, 50 (1972).
8. Giancotti, V., Quandrifoglio, F., and Crescenzi, V.: *Makromol. Chem.* **158**, 53 (1972).
9. Begala, A. J. and Strauss, U. P.: *J. Phys. Chem.* **76**, 254 (1972).
10. Myer, F. and Barnard, D.: *Arch. Biochem. Biophys.* **143**, 116 (1971).
11. Peggion, E., Cosani, A., Terbojevich, M., and Scoffone, E.: *Macromolecules* **4**, 725 (1971).
12. Christensen, J. J., Izatta, R. M., Wrathall, D. P., and Hansen, L. D.: *J. Chem. Soc.* (**A**), 1212 (1969).
13. Pederson, D., Gabriel, D., and Hermans Jr., J.: *Biopolymers* **10**, 2133 (1971).
14. Chou, P. J. and Scheraga, H. A.: *Biopolymers* **10**, 657 (1971).
15. Senior, M. B., Garrell, S. L. H., and Hamori, E.: *Biopolymers* **10**, 2387 (1971).
16. Delben, F. and Crescenzi, V.: *Int. J. Protein Res.* **3**, 57 (1971).
17. Cesàro, A. and Crescenzi, V.: (in preparation).
18. Liquori, A. M., Constantino, L., Crescenzi, V., Elia, V., Giglio, E., Puliti, R., Savino, M., and Vitagliano, V.: *J. Mol. Biol.* **24**, 113 (1967).

THERMODYNAMIC PROPERTIES OF STRONG ELECTROLYTE – STRONG POLYELECTROLYTE MIXTURES AT 25°C [1]

G. E. BOYD

Oak Ridge National Laboratory, Oak Ridge, Tenn. 37830, U.S.A.

Abstract. Isopiestic vapor pressure comparison measurements were performed with aqueous mixtures of sodium chloride (*B*) with sodium polystyrenesulfonate, NaPSS, (*C*) and with polyvinylbenzyltrimethyl ammonium chloride, PVR_4NCl, (*C*) to determine molal osmotic coefficients, ϕ, for typical strong electrolyte-strong polyelectrolyte systems. The McKay-Perring method was applied to compute the stoichiometric mean molal activity coefficients, γ_B, of NaCl and γ'_C of NaPSS or PVR_4NCl, respectively, in the mixtures as a function of the osmolal concentration, **m**, and of the osmolal fraction of polyelectrolyte, y_C.

The activity coefficient of the NaCl was strongly decreased by the addition of polyelectrolyte at constant **m**, while the activity coefficients of the polyelectrolytes were increased or decreased only slightly by the addition of salt. The mean molal activity coefficients, Γ'_B, of the polyelectrolytes in their binary solutions also were computed relative to an arbitrary concentration of 0.001 equiv./kg H_2O where the activity coefficient was defined as unity. The values of $\log \Gamma'_B$ decreased linearly with increasing $\log m$. The measured ϕ values were employed to test the predictions of the Additivity Rule for mixtures at constant water activity. Small but significant departures from additivity dependent on y_C were found at all concentration. The variations of $-\log \gamma_B$ and $-\log \gamma'_B$ with y_C in mixtures of constant osmolality showed that Harned's Rule was not obeyed. The electrolyte-polyelectrolyte interaction coefficients, α_B, α_C, β_B, and β_B computed from the variations of $\log \gamma_B$ and $\log \gamma'_B$ with y_C were much larger than for simple electrolyte-electrolyte mixtures, especially in dilute solutions.

The thermodynamic properties of aqueous electrolyte-polyelectrolyte mixtures are quite unusual, and therefore, a subject of interest because apparently they can be described empirically as a superposition of the individual properties of the pure components [2–6]. For example, the osmotic pressure, exerted by a mixture is equal to the sum of the osmotic pressure of the salt-free polyelectrolyte solution, π_p, and of the polyelectrolyte-free salt solution, π_s, each at the same concentration as when it was present in the mixture:

$$\pi = \pi_p + \pi_s . \qquad \text{(A)}$$

The activity of the counter-ions, a_+, in a mixture likewise appears to be given as the sum of the independent contributions of the counter-ions from the polyelectrolyte, a_+^p, and from the salt, a_+^s, respectively:

$$a_+ = a_+^p + a_+^s . \qquad \text{(B)}$$

Additivity of thermodynamic properties is not generally shown by aqueous mixtures of simple strong electrolytes, and, while it might be expected for dilute solutions of non-interacting solutes, this behavior is surprising for mixtures of salts with highly charged polyions.

Several attempts have been made to supply a theoretical explanation for the Additivity Rule. Katchalsky and Alexandrowicz [7] have demonstrated that (A) and (B) above can be derived thermodynamically from the Gibbs-Duhem equation for a

Eric Sélégny (ed.), Polyelectrolytes, 135–155.

ternary polyelectrolyte-electrolyte-water mixture on the basis of the single assumption that the osmotic activity of the coions of the electrolyte is not significantly influenced by the presence of the polyion. More specifically, it is assumed that the activity coefficient of the coion, f_-, (i.e., in systems where polyanions are present) is virtually unity over a wide range of concentration and ionizations. Only counter-ions are attracted by the polyion, and their activity coefficient, f_+, is strongly decreased. Coions are slightly repelled by the screened polyion and are attracted by the counterion atmosphere of the latter. Both effects cancel each other and f_- is left largely unchanged. The approximate nature of this assumption, however, is shown by experimental measurements [3] of f_- wherein the coion activity coefficient is found to depart perceptibly from unity, and also to vary with polyelectrolyte concentration.

An independent justification has been given by Oosawa [8] who has shown that additivity is a consequence of the properties of the integrated coulomb potential of the macroion. The approximate nature of the Rule is recognized in Oosawa's treatment, and it would seem, by analogy with the development of the present understanding of low molecular weight electrolyte mixtures, that departures from additivity when carefully assessed would be more truly revealing of the complex interactions between a polyelectrolyte and added salt.

Thus, far few experimental tests have been conducted on the scope and limitations of the Additivity Rule. The approximate character of relation (B) for counter-ion activities has been demonstrated by Lyons and Kotin [9]. Departures from (B), defined $\Delta a_+ = a_+ - a_+^p - a_+^s$, were greatest when the contributions from each component in the electrolyte-polyelectrolyte mixture were equal, and Δa_+ approached zero when either component was present in excess. The deviation also was significant for large concentrations, but decreased and vanished with increasing dilution. Alexandrowicz [6] has performed an extensive series of measurements of the osmotic pressure of electrolyte-polyelectrolyte mixtures in an attempt to establish the general validity of strict additivity as expressed in (A). An analysis of his data shows, however, that when the electrolyte and polyelectrolyte are present in approximately equal amounts and the osmolality of the mixture exceeded 0.5, the difference from additivity was as much as ten percent. Additivity was obeyed, however, whenever salt or polyelectrolyte was present in large excess.

The Additivity Rule has been assumed to describe the Donnan equilibrium between polyelectrolyte gels and aqueous solutions of low molecular weight electrolytes. [10] The mean molal activity coefficient, $\bar{\gamma}_\pm$, for the salt which 'invades' the gel may be computed from experimental measurements of the equilibrium partition of electrolyte with the thermodynamic Donnan equation, and its value compared with that given by the Rule. According to the Rule, $\bar{\gamma}_\pm$ should be constant and independent of salt concentration; in fact, $\bar{\gamma}_\pm$ was observed to increase with dilution. [11]

I now wish to report new measurements on the osmotic properties of aqueous strong electrolyte-strong polyelectrolyte mixtures conducted using the isopiestic vapor pressure comparison technique. [12] This method for determining osmotic coefficients is not as sensitive as the membrane osmometer, but it is much more accurate. It is not subject to complications (i.e., leaky membranes, etc.) nor are slightly questionable assumptions necessary as in the 'concentration method' developed by Alexandrowicz, which uses polyethylene glycol (PEG) to vary the water activity in a membrane osmometer. Activity coefficients for the electrolyte and for the polyelectrolyte in the mixtures may be computed by applying thermodynamic theory to the variation of the measured isopiestic concentration ratios with composition at constant water activity. These thermodynamic properties in turn may be employed in the consideration of mixture rules other than simple superposition, and in the evaluation of free energies of mixing of salt with polyelectrolyte.

1. Experimental Section

Osmotic coefficients of aqueous sodium polystryrenesulfonate (NaPSS) and polyvinylbenzyltrimethylammonium chloride (PVBTMACl) solutions and their mixtures with sodium chloride solutions were measured with the gravimetric isopiestic vapor pressure comparison technique of Sinclair and Robinson [12]. Two units of the design shown in Figure 1 were employed. Aliquots of solutions of polyelectrolyte, of NaCl, and of NaCl-polyelectrolyte mixture all of predetermined composition were weighed into the 4-ml capacity platinum sample cups shown which were held firmly pressed against a one-inch thick, gold-plated copper block inside a heavy-walled glass desiccator vessel. The large central dish contained pure NaCl solution of predetermined concentration with which the solutions in the outer twelve Pt dishes come to isopiestic

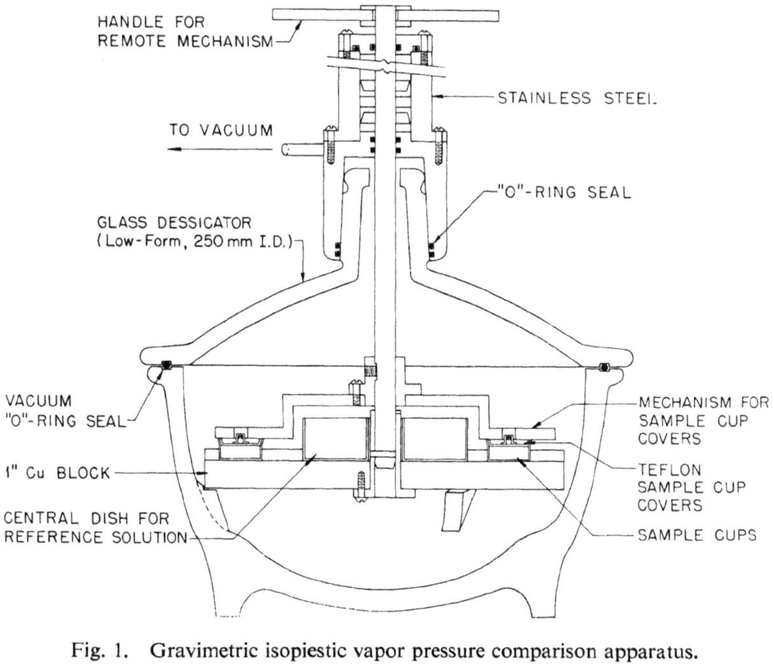

Fig. 1. Gravimetric isopiestic vapor pressure comparison apparatus.

vapor pressure equilibrium. The unit was submerged in a large water bath, maintained for long periods at 25.00±0.01 °C, and evacuated with a water pump to remove all air from the chamber in-so-far as possible. The units were gently rocked in the bath to hasten the attainment of vapor pressure equilibrium. The essentials of the apparatus and the procedure are: good heat conduction from one dish to another, a good thermal buffer to minimize temperature fluctuations, efficient stirring to assure mixing and heat conduction without heating, evacuation sufficient to remove the air, and provision (i.e. Teflon caps) for closing the dishes when the unit is removed from the bath so that accurate weighings of the dishes plus solution can be performed.

At equilibrium the partial pressure of water vapor through-out the chamber and above all the solutions will be constant, and the water activity in each solution will be the same as in any other. The isotonic concentrations of the solutions will be related by:

$$\nu_r m_r \phi_r = \nu_x m_x \phi_x, \quad (1)$$

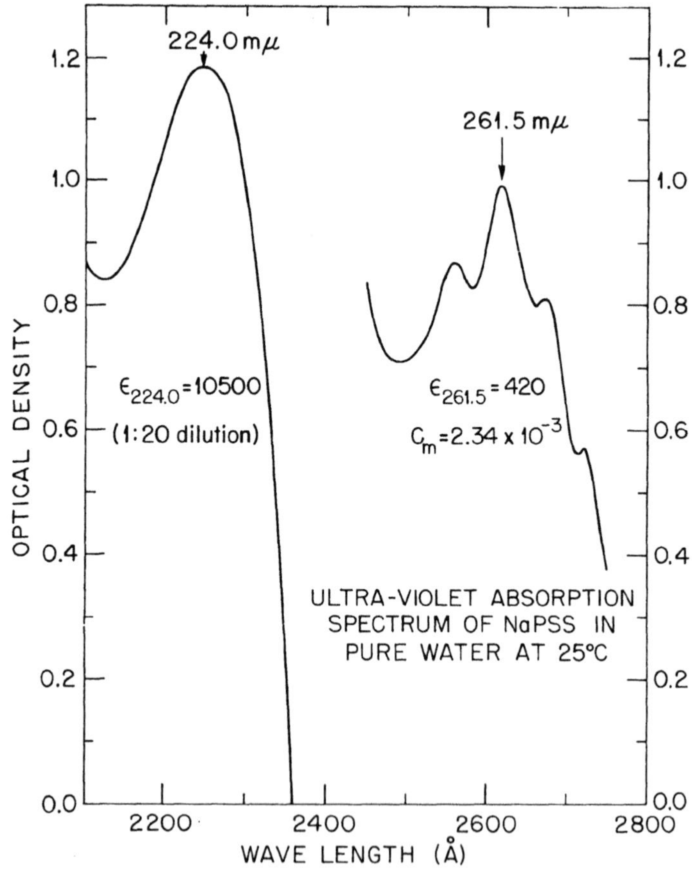

Fig. 2. Ultra-violet absorption spectrum of purified NaPSS in water at 25°C (Cary Model 14P spectrophotometer, 1 cm quartz cells).

where ν_r, m_r, and ϕ_r are the number of ions per mole, the molality, and the practical osmotic coefficient of the reference electrolyte, r, and ν_x, m_x, ϕ_x are the corresponding quantities, respectively, of the polyelectrolyte, or of the polyelectrolyte-electrolyte mixture. Pure NaCl solution was taken as the reference electrolyte because its osmotic coefficient is well established as a function of its molality.

The solid NaPSS and PVBTMACl samples used were obtained from the Dow Chemical Company, Midland, Michigan, and possessed nominal weight average molecular weights of ca. 500000 and ca. 140000 respectively. The compounds were dissolved and dialyzed against pure water with a deacetylated cellulose triacetate hollow-fiber dialysis cell [13] until the electric conductivity of the dialyzate decreased to approximately five micromho per cm. The purified polyelectrolyte solution was concentrated with a rotary evaporator and pure, dry compound was obtained by vacuum freeze drying. The equivalent weight of NaPSS, determined by acidimetric weight titration of the HPSS produced by cation exchange, was 206.5 corresponding to ca. 100% degree of substitution. The equivalent weight of PVBTMACl, determined by weight titration of chloride ion with standard $AgNO_3$, was 228.8 which corresponds to ca 93% degree of substitution. The absence of degradation products in the purified polyelectrolytes was established by measurements of the ultra-violet absorption spectra shown in Figures 2 and 3. The value of the molar absorption coefficient,

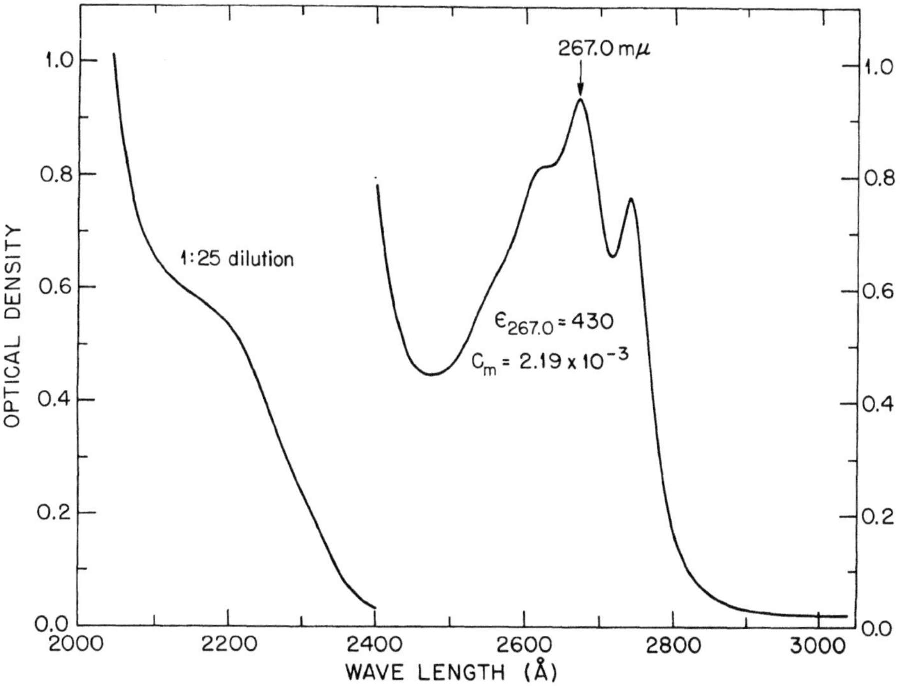

Fig. 3. Ultra-violet absorption spectrum of purified polyvinylbenzyltrimethylammonium chloride (PVBTMACl) in water at 25°C.

$\varepsilon_{261.5} = 420$, for pure NaPSS is in good agreement with recent measurements by others. [14]

Solutions of polyelectrolyte were made up from known weights of the pure compound and triply-distilled water. The aqueous polyelectrolyte-electrolyte mixtures were prepared by weight from the pure polyelectrolyte solutions and standardized NaCl solution. The compositions of the mixtures were expressed in terms of the 'osmolal' fraction of polyelectrolyte as defined below.

2. Experimental Results and Treatment of Data

The concentration dependence of the osmotic coefficients for the binary aqueous NaPSS and PVBTMACl solutions at 25° is shown in Figure 4 where a comparison can be made with previously published measurements. The data for the NaPSS solutions are in good agreement with those of Reddy and Marinsky [15] but not with those of Bonner and Overton [16], nor of Ise and O'Kubo [17], The value of the osmotic coefficient appears to be approaching $\phi = 0.18$ for very small concentrations predicted by the theories of Lifson and Katchalsky [18] and of Manning [19]. It is also of interest to note (Figure 4) that the osmotic coefficients measured with a lightly cross-linked polyelectrolyte gel [20] (nominal 0.5% divinylbenzene, Na-form) agree reasonably well with the osmotic coefficients determined with polyelectrolyte solutions. The general concordances of the measurements with the gel and with linear NaPSS with $M_v = 40000$ and 150000, respectively, supports the view that the colligative properties

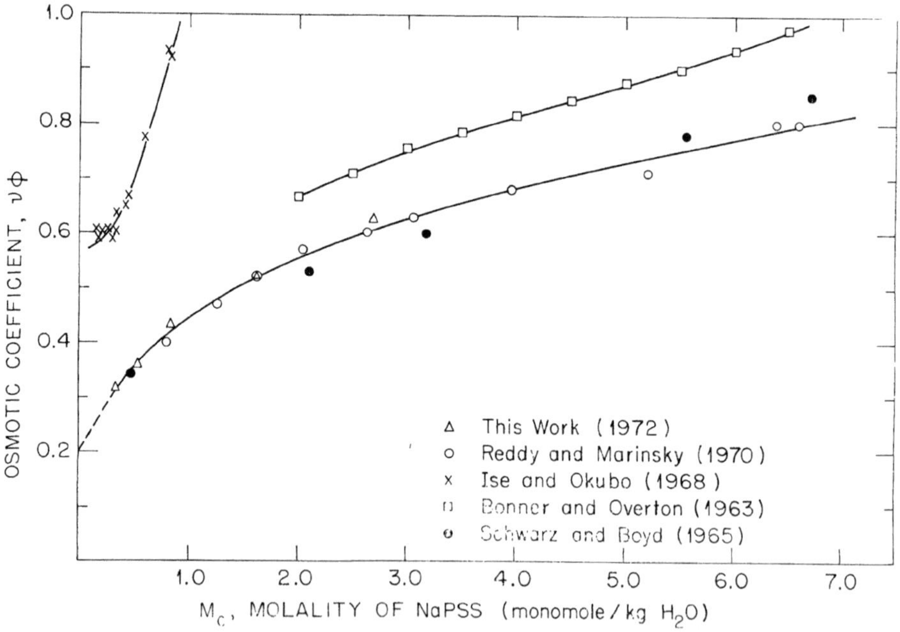

Fig. 4. Concentration dependence of osmotic coefficient for NaPSS at 25°C.

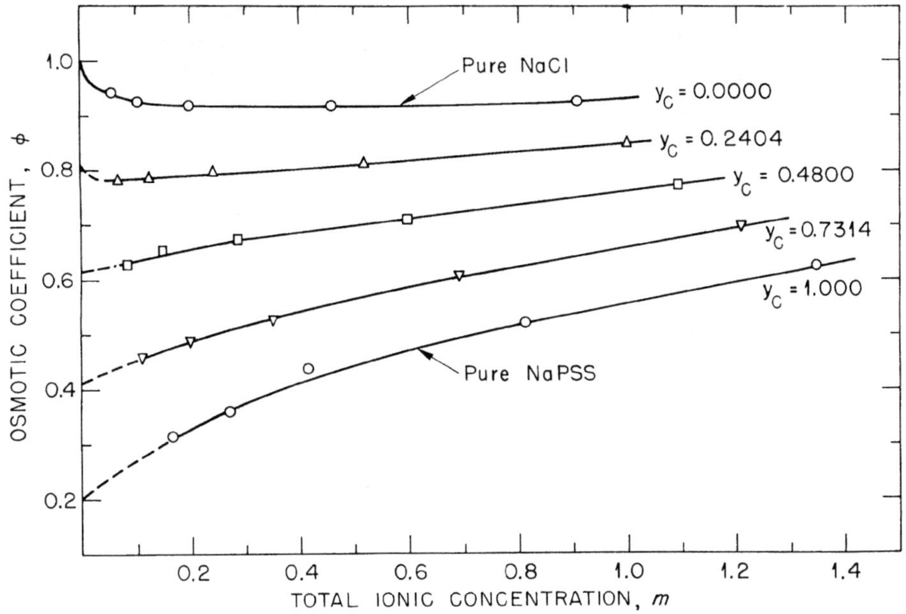

Fig. 5. Concentration dependence of osmotic coefficients of aqueous NaCl + NaPSS Mixtures at 25°C.

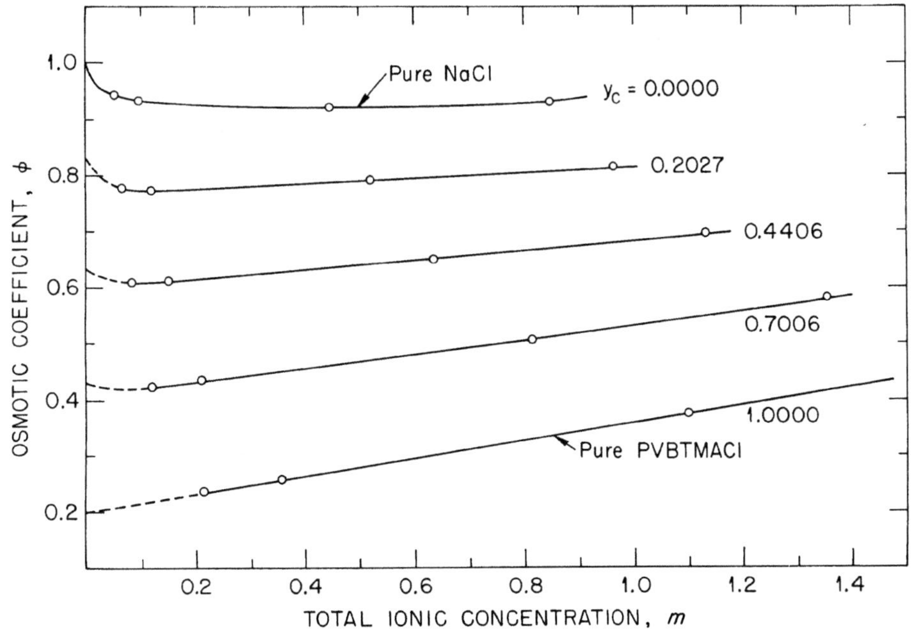

Fig. 6. Concentration dependence of osmotic coefficients of aqueous NaCl + PVBTMACl mixtures at 25°C.

of polyelectrolyte solutions are independent of the molecular weight. The concentration dependence of the osmotic coefficients for the salt-polyelectrolyte mixtures is shown in Figures 5 and 6 where ϕ is plotted as a function of the total ionic concentration, or osmolality, **m**, defined by, $m = v_B m_B + v_C m_C$, with $B \equiv$ NaCl and $C \equiv$ NaPSS or PVBTMACl and v_B and v_C defined as the number of ions per mole and per monomole, respectively.

The measurements shown in Figures 5 and 6 may be employed to test the Additivity Rule given in (A). This relation may be restated in terms of osmotic coefficients using the well-known equation relating the latter to the osmotic pressure. Hence,

$$m\phi_{B/C} = v_B m_B \phi_B^0 + v_C m_C \phi_C^0, \tag{2}$$

where $\phi_{B/C}$ is the osmotic coefficient for the salt-polyelectrolyte mixture and ϕ_B and ϕ_C are the coefficients for the pure salt and polyelectrolyte when present alone at concentrations m_B and m_C, respectively equal to their concentrations in the mixture. Taking $v_B = 2$, $v_C = 1$, Equation (2) may be reduced to:

$$\phi_{B/C} = y_B \phi_B^0 + y_C \phi_C^0, \tag{3}$$

where y_B and y_C are osmolal fractions defined by:

$$y_B = v_B m_B / (v_B m_B + v_C m_C) = m_B / (m_B + 0.5\ m_C), \tag{4a}$$

$$y_C = v_C m_C / (v_B m_B + v_C m_C) = 0.5\ m_C / (m_B + 0.5\ m_C). \tag{4b}$$

The function, $\Delta\phi$, defined by:

$$\Delta\phi = \phi_{B/C} - y_B \phi_B^0 - y_C \phi_C^0 \tag{5}$$

will be used as a measure of the departure from the Additivity Rule. A plot of $\Delta\phi$ derived from the data (Table IA) on NaCl-NaPSS mixtures at constant water activity (i.e., $m\phi_{B/C}$=constant) as a function of y_C is given in Figure 7. A $\Delta\phi$, y_C plot for constant osmolality, m, may be constructed by interpolation of $\phi_{B/C}$ values from Figure 5.

Activity coefficients for the electrolyte and for the polyelectrolyte in the mixtures may be derived by application of the method of McKay and Perring [21] to the experimental results presented in Table I. Thus, the mean molal activity coefficient, γ_B, of sodium chloride is:

$$\ln \gamma_B = \ln \Gamma_B + \ln R_B + y_B^2 \int_0^{M_B \phi_B^0} (b/M_B)\, \mathrm{d}\,(M_B \phi_B^0). \tag{6a}$$

The corresponding equation for the mean molal activity coefficient, γ_C', of the polyelectrolyte is:

$$\ln \gamma_C' = \ln \Gamma_C' + \ln R_C + y_C^2 \int_0^{M_B \phi_B^0} (b/M_B)\, \mathrm{d}\,(M_B \phi_B^0) \tag{6b}$$

TABLE I

Isopiestic ratios for McKay-Perring calculations of activity coefficients

M_B	M_C	y_C	R_B	R_C
A. $NaCl + NaPSS + H_2O$				
0.0556	0.3299	0.0000	1.0000	2.9662
		0.2404	0.8309	2.4644
		0.4800	0.6689	1.9839
		0.7314	0.4995	1.4595
		1.0000	0.3371	1.0000
0.1039	0.5380	0.0000	1.0000	2.5891
		0.2404	0.8436	2.1835
		0.4800	0.7051	1.8250
		0.7314	0.5250	1.3593
		1.0000	0.3862	1.0000
0.4593	1.6242	0.0000	1.0000	1.7681
		0.2404	0.8856	1.5659
		0.4800	0.7725	1.3658
		0.7314	0.6629	1.1722
		1.0000	0.5655	1.0000
0.9065	2.6938	0.0000	1.0000	1.4858
		0.2404	0.9136	1.3575
		0.4800	0.8301	1.2335
		0.7314	0.7495	1.1136
		1.0000	0.6730	1.0000
B. $NaCl + PVR_4NCl + H_2O$				
0.0537	0.4278	0.0000	1.0000	3.9836
		0.2166	0.8217	3.2733
		0.4514	0.6453	2.5709
		0.7163	0.4509	1.7961
		1.0000	0.2510	1.0000
0.0981	0.7115	0.0000	1.0000	3.6253
		0.2166	0.8285	3.0037
		0.4514	0.6560	2.3777
		0.7163	0.4662	1.6899
		1.0000	0.2758	1.0000
0.4467	2.1993	0.0000	1.0000	2.4568
		0.2027	0.8581	2.1078
		0.4406	0.7046	1.7312
		0.7006	0.5494	1.3498
		1.0000	0.4070	1.0000
0.8473	3.1674	0.0000	1.0000	1.8691
		0.2027	0.8789	1.6426
		0.4406	0.7479	1.3979
		0.7006	0.6247	1.1675
		1.0000	0.5350	1.0000

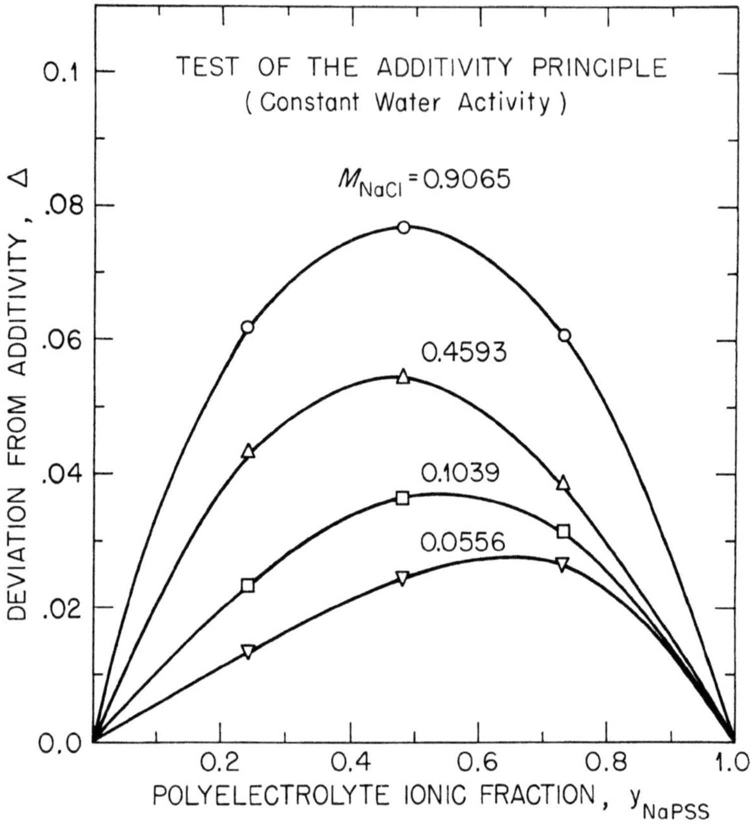

Fig. 7. Test of the additivity principle (Constant water activity, 25°C). Scale for Δ to be divided by 10.

In Equations (6) Γ_B and Γ'_C are the activity coefficients of the salt or of the polyelectrolyte, respectively, in their own pure solutions (i.e., two-component) at molalities M_B and M_C in isopiestic equilibrium with the mixed electrolyte-polyelectrolyte (i.e., three-component) solutions. The required values of Γ_B are tabulated [12] but those for Γ'_C must be estimated from the measurements on the pure polyelectrolyte solution as will be shown below. The quantities R_B and R_C are isopiestic ratios defined by:

$$R_B = \nu_B M_B/(\nu_B m_B + \nu_C m_C) = M_B/(m_B + 0.5\, m_C), \tag{7a}$$

$$R_C = \nu_C M_C/(\nu_B m_B + \nu_C m_C) = 0.5\, M_C/(m_B + 0.5\, m_C), \tag{7b}$$

where m_B and m_C are the molalities of components B and C in the mixtures. The dependence of R_B on y_C is shown in Figures 8 and 9 for the NaCl-NaPSS and the NaCl-PVBTMACl mixtures, respectively.

The coefficient **b** which appears in the integral in Equations (6) was determined by a least-squares fit of R_B to the equation:

$$R_B = 1 - ay_C - by_C^2. \tag{8}$$

The values of the parameters **a** and **b** and their standard errors (S.E.) are given in Table II. No simple polynominal could be found relating (b/M_B) to $M_B\phi_B$, hence, the integral in Equations (6) was evaluated by numerical integration with a digital computer.

The definition, $m = m_B + 0.5\, m_C$, employed in the activity coefficient calculations deserves a special comment as an implicit assumption was made concerning the dissociation of the polyelectrolyte. It was assumed that the polyelectrolyte supplies one osmotically active ion (i.e., the counter-ion) per monomole. Many of the properties of

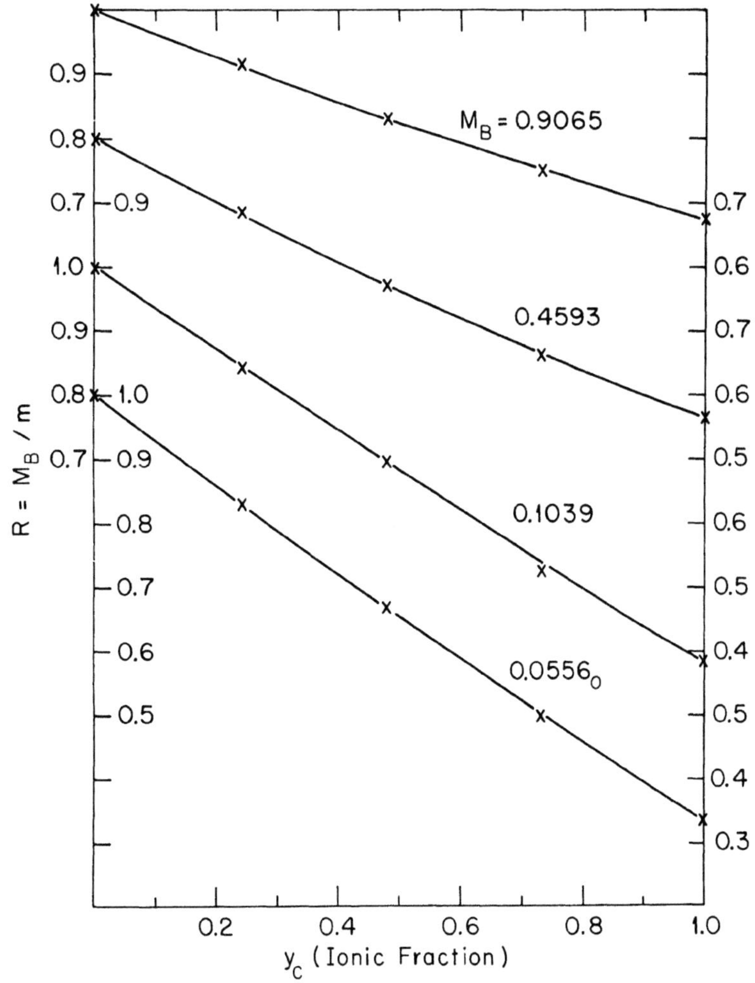

Fig. 8. Composition dependence of isopiestic ratios for aqueous NaCl + NaPSS mixtures at 25°C.

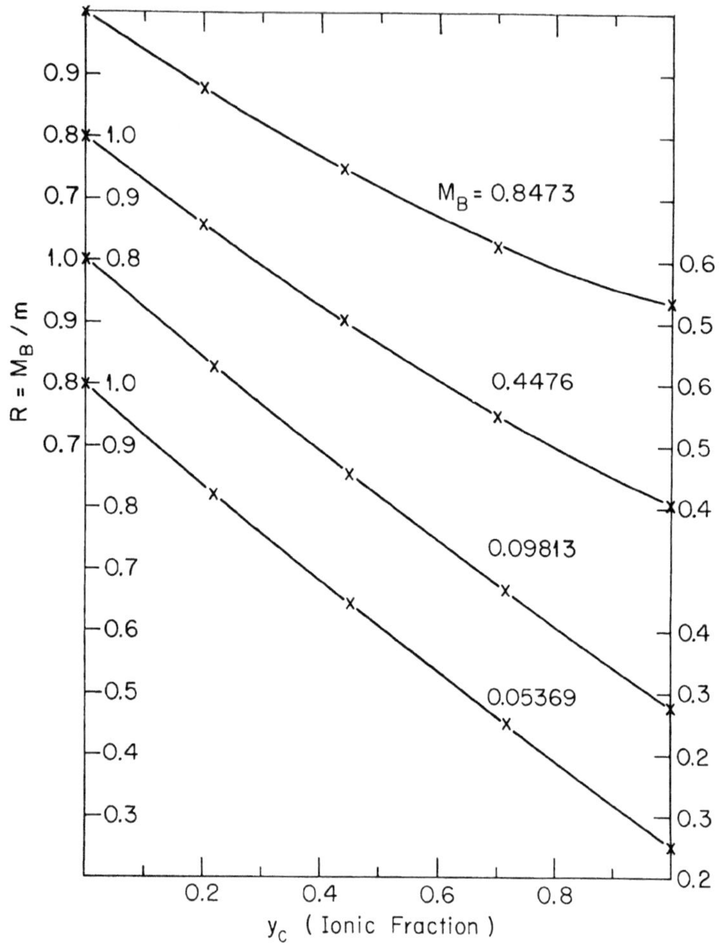

Fig. 9. Composition dependence of isopiestic ratios for aqueous NaCl + PVBTMACl mixtures at 25°C.

strong polyelectrolytes, however, suggest that a fraction of these mobile ions is strongly associated with the polyion. Thus, an alternative definition might be:

$$m = m_B + \beta m_C, \tag{9}$$

where $\beta = \alpha/z$ is the charge fraction, α is the net polyion valency and z is the stoichiometric polyion valency. The number of ions per monomole, ν_C, is then,

$$\nu_C = (\alpha + 1)\, z \approx \beta. \tag{10}$$

This alternate approach to an estimation of the osmolal concentration was not adopted however, because there was experimental evidence that β varied with m_C, the mono-

TABLE II

Values of the parameters of Equation (8) and of the integral in Equation (6)

M_B	a	S.E.	b	S.E.	$M_B\phi°_B$	$\int$[a]
A. $NaCl + NaPSS + H_2O$						
0.0556	0.7221	0.0054	0.0583	0.0060	0.0524	0.02821
0.1039	0.6869	0.2000	0.0703	0.0221	0.0969	0.04475
0.4593	0.5099	0.0096	0.0742	0.0108	0.4229	0.09310
0.9065	0.3777	0.0033	0.0503	0.0038	0.8448	0.11118
B. $NaCl + PVR_4NCl + H_2O$						
0.0537	0.8211	0.0054	0.0727	0.0061	0.0506	0.03694
0.0981	0.7990	0.0032	0.0749	0.0035	0.0916	0.05408
0.4476	0.7394	0.0061	0.1454	0.0069	0.4121	0.12298
0.8473	0.6643	0.0157	0.1975	0.0175	0.7880	0.16631

[a] This column gives values of the integral, $0.4343 \int_0^{M_B\phi°_B} (b/M_B)\, d(M_B\phi°_B)$, obtained by numerical integration.

molality of the polyelectrolyte. The values of γ_B and γ'_C calculated in Equations (6), therefore, are stoichiometric activity coefficients, while those that might be estimated, knowing β from non-thermodynamic measurements, with Equations (9) and (10) and the experimental data would be the 'species' activity coefficients, γ_B^* and γ_C^*. The relationship between γ and γ^* has been discussed by Robinson and Stokes [12].

The activity coefficients, Γ'_C, for the polyelectrolyte present alone in pure (i.e., binary) aqueous mixtures were computed by applying the Gibbs-Duhem to the osmotic coefficient data.

$$\mathrm{d}\ln\gamma_C = \mathrm{d}\phi + (\phi - 1)\,\mathrm{d}\ln M_C. \tag{11}$$

Because of the unusual concentration dependence of ϕ at high dilution and because no simple limiting behavior such as the Debye-Huckel law for electrolytes was available, the integration of Equation (11) was conducted from an arbitrarily chosen concentration, $M_C = 10^{-3}$ equiv./kg H_2O, where the activity coefficient, γ_C, was defined as unity and the polyelectrolyte osmotic coefficient was approximated by its limiting value $\phi_0 = 0.2$. Thus,

$$\int_{1.0}^{\Gamma'_C} \mathrm{d}\ln\gamma_C = \int_{0.2}^{\phi} \mathrm{d}\phi + \int_{10^{-3}}^{M_C} (\phi - 1)\,\mathrm{d}\ln M_C \tag{12a}$$

or,

$$\ln\Gamma'_C = (\phi - 0.2) + \int_{10^{-3}}^{M_C} (\phi - 1)\, d\ln M_C. \tag{12b}$$

An empirical equation of the type

$$\phi_C = \phi_0 + k_1 M_C + k_2 M_C^2/(1 + k_3 M_C) \qquad (13)$$

which was fitted to the experimental data (see Figure 4) by least-squares methods, was employed to evaluate the integral on the right hand side of Equation (12b). The values of the constants in Equation (13) were, $k_1 = 0.430$, $k_2 = -0.340$ and $k_3 = 0.848$, respectively, for the NaPSS solution. For the PVBTMACl solution $\phi_0 = 0.202$, $k_1 = 0.0776$, $k_2 = 0$, 0, and $k_3 = 0.0$ for osmolalities less than 3.5. An interesting empirical fact appears to be that $\log \Gamma'_C$ may be expected to vary linearly with $\log M_C$ over a surprisingly wide range in concentration. This has been explained by Katchalsky [22] who has derived the following equation from thermodynamics:

$$(\mathrm{d} \ln \Gamma'_C / \mathrm{d} \ln M_C) = -(2 - \phi_C) + \mathrm{d}\phi_C / \mathrm{d} \ln M_C .$$

In dilute solutions $\phi_C = \phi_0 + k_1 M_C$ to a fair approximation, hence

$$(\mathrm{d} \ln \Gamma'_C / \mathrm{d} \ln M_C) = -(1 - \phi_0) .$$

The data for the NaPSS solutions could be fit with good accuracy to the relation:

$$-\log \Gamma'_C = 2.4104 - 0.804 \log M_C$$

so that $\phi_0 \sim 0.2$, as required. An analogous behavior was noted for the PVBTMACl

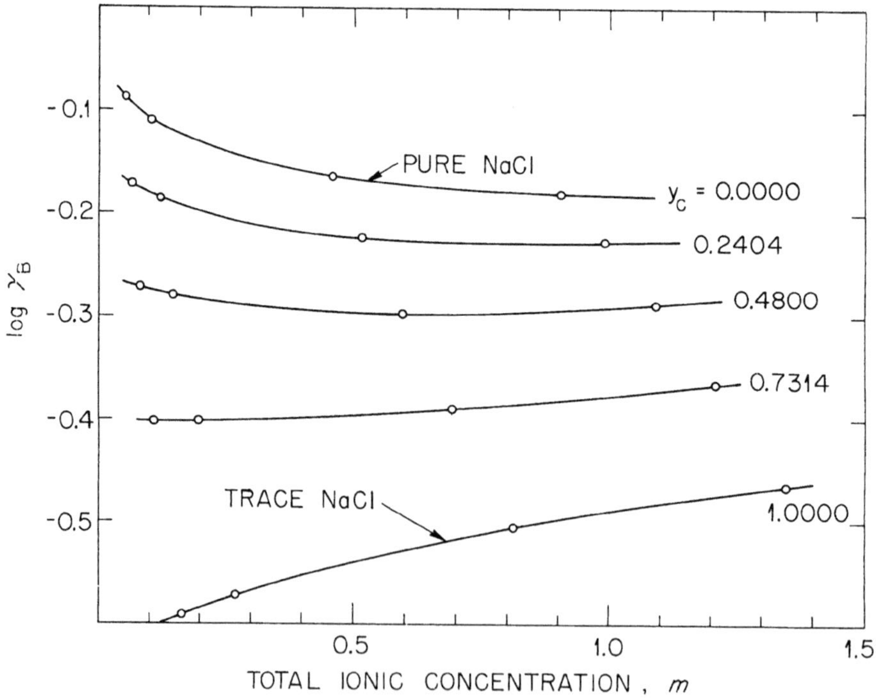

Fig. 10. Concentration dependence of activity coefficients of NaCl in the presence of NaPSS.

solutions. Results from a typical activity coefficient calculation are given in Figure 10 for the NaCl-NaPSS mixtures where the concentration dependence of γ_B for various osmolal fractions, y_C, of polyelectrolyte is shown.

3. Discussion and Conclusions

3.1. The Additivity Rule

The data plotted in Figure 7 for the NaCl+NaPSS mixtures show that additivity is not strictly obeyed. The deviation as measured by $\Delta\phi$ is the largest when the electrolyte and polyelectrolyte are present in equal amounts ($y_C=0.5$) and it varies from ca. four to ten percent from the lowest to the highest concentration. Whenever the salt or the polyelectrolyte is present in large excess, Figure 7 shows that the Additivity Rule is approximately valid. Because of the decrease of $\Delta\phi$ with the osmolality of the mixture it seems reasonable to assume that additivity would hold at high dilution. The general dependence of $\Delta\phi$ on y_C shown in Figure 7 also was observed with the NaCl+ +PVBTMACl mixtures although the deviation was smaller (i.e., 7.8%) at the highest and, within experimental error, vanished at the lowest concentration.

The magnitudes of $\Delta\phi$ for both types of mixtures examined are significantly larger than those observed with mixtures of simple low molecular weight electrolytes. Thus, the maximum departure from additivity in six mixtures of NaCl, NaBr, KCl, and KBr was 3.2% for NaBr with KCl at 3.977 molal. [23] A slightly different use of Equation (5) above was made, however. The values of ϕ_B^0 and ϕ_C^0 were those for solutions of B and C, respectively, at the same osmolality, m, as the *total osmolality* of the mixed solution rather than at m_B and m_C, respectively, in a mixture of constant water activity. When the $\Delta\phi$ values shown in Figure 7 are recalculated following this convention, the departure from additivity is 3.0% for $m=0.1$ and 1.4% for $m=1.0$.

The dependence of $\Delta\phi$ for the electrolyte-polyelectrolyte mixtures on y_C is of the same type as with electrolyte mixtures, namely, $\Delta\phi=y_By_C\beta_0$. The value of β_0 for the NaCl+NaPSS mixture, for example, was 0.0409 at $m=1.0$ compared with 0.0042 for a NaCl-NaBr mixture at $m=2.9$. The excess free energy of mixing, ΔG^E, may be estimated with Equation (14) if it is assumed that β_0 is a linear function of the concentration:

$$\frac{\Delta G^E}{RT} = y_By_Cm\beta_0\,. \tag{14}$$

Thus, for the NaCl-NaPSS mixture, $\Delta G^E=6$ cal/mole of solute while for the NaCl-NaBr mixture $\Delta G^E=0.6$ cal/mole solute. This ten-fold difference in part at least reflects the difference between the strength of the anion-anion interaction in the two mixtures. The excess free energy of a mixture containing a salt A with cation M and anion X, and salt B with cation N and anion Y is, according to Guggenheim [24]:

$$\frac{\Delta G^E}{RT} = y_By_Cm^2(A_{MY}+A_{NX}+\delta_{MN}+\delta_{XY}-A_{MX}-A_{NY})\,. \tag{15}$$

The A terms in Equation (15) allow for specific ionic interaction and the δ terms for interaction between ions of like charge. For a mixture with a common cation (e.g., NaCl+NaPSS):

$$\frac{\Delta G^E}{RT} = y_B y_C m^2 \delta_{XY}, \tag{16}$$

where δ_{XY} is β_0/m in Equation (14). Thus, for the NaCl+NaPSS and the NaCl-NaBr mixtures, ΔG^E depends on the magnitude of the coefficient δ_{XY} which characterizes the interaction between the anions. Not surprisingly, there is a much stronger interaction between Cl^- ion and PSS^- ion than between Cl^- and Br^- ions.

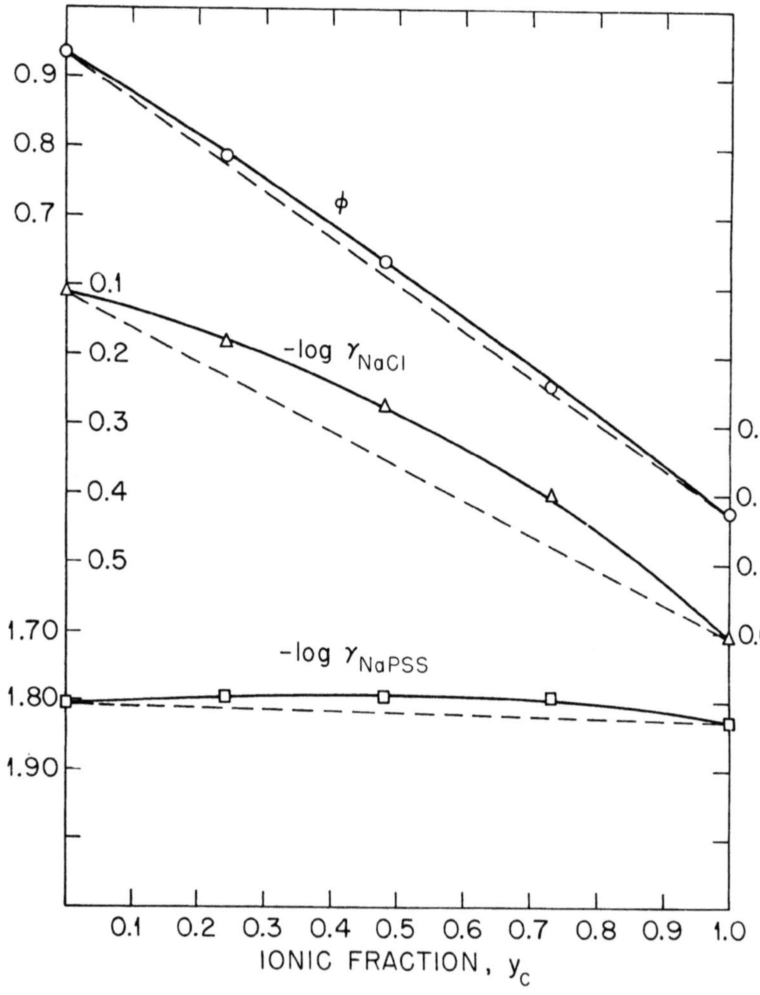

Fig. 11. Composition dependence of osmotic and activity coefficients for aqueous NaCl + NaPSS mixtures at constant $m = 0.10$.

3.2. The activity coefficients of the solutes in electrolyte-polyelectrolyte mixtures

It is clear from the results presented in Figure 10 that the activity coefficient of sodium chloride is strongly lowered by the addition of increasing amounts of sodium polystyrenesulfonate. An analogous plot also holds for the concentration and composition dependence of $-\log \gamma_{NaCl}$ in the presence of the cationic polyelectrolyte, polyvinylbenzyl-trimethylammonium chloride. On the other hand, as is shown in Figures 11–14, the activity coefficients of the polyelectrolytes are virtually unaffected by the addition of NaCl. In addition, the figures show that Harned's Rule, indicated by the broken lines, is not obeyed. Strong ionic interactions are unmistakably present in electrolyte-polyelectrolyte mixtures as reflected by the unusually large values of the coefficients, α and β in the equations:

$$\log \gamma_B = \log \gamma_B^0 - \alpha_B m y_C - \beta_B m y_C^2, \tag{17a}$$

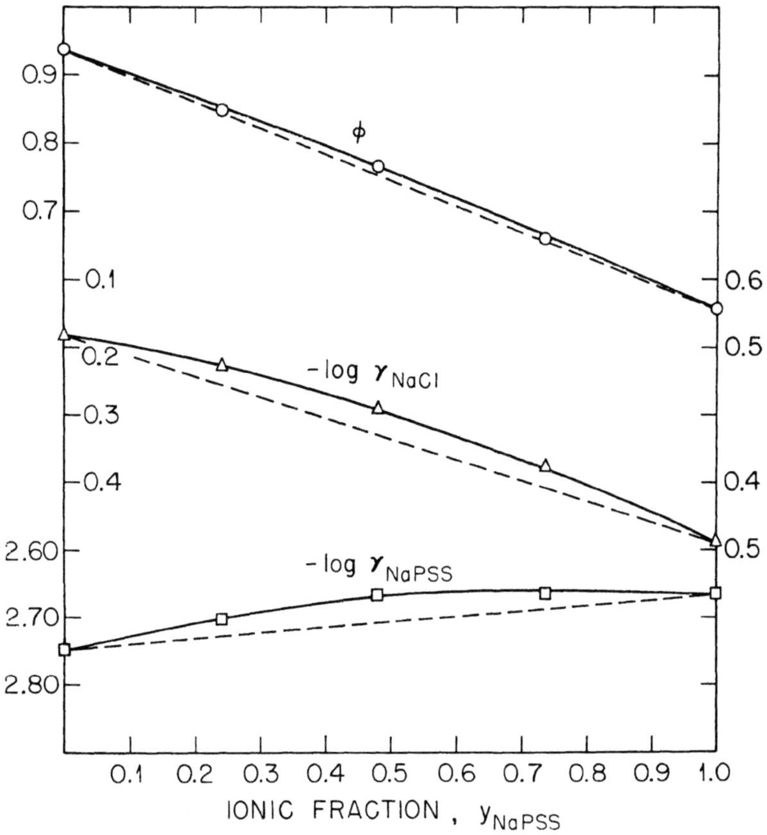

Fig. 12. Composition dependence of osmotic and activity coefficients for aqueous NaCl + NaPSS mixtures at constant $m = 1.0$.

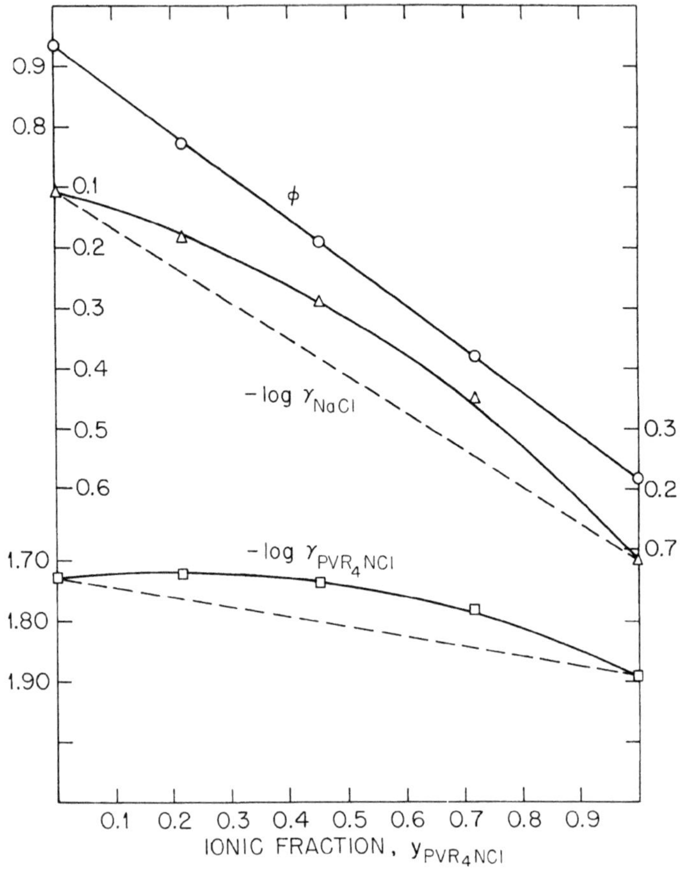

Fig. 13. Composition dependence of osmotic and activity coefficients for aqueous NaCl + PVBTMACl mixtures at constant $m = 0.10$.

$$\log \gamma_C' = \log \gamma_C^{0'} - \alpha_C m(1 - y_C) - \beta_C m(1 - y_C)^2\,, \tag{17b}$$

where γ_B^0 and $\gamma_C^{0'}$ are the activity coefficients, respectively, for NaCl in its own pure solution at a concentration equal to that for the mixture, m, and for NaPSS or PVR_4NCl in their own pure solutions of the same concentration. Values of α_B, α_C, β_B, and β_C estimated from a least-squares fit of Equations (17) to the data in Figures 11–14 are summarized in Table III. The magnitudes of these quantities for solutions of unit concentration are not much larger than those found with mixtures of simple electrolytes at $m = 1.0$. However, at $m = 0.1$, the large values of the coefficients in Equations (17) indicates a strong interaction between the polyelectrolyte and electrolyte. No mixtures of 1-1 type electrolytes are known which show values of α exceeding 0.1, and, almost, invariably, β is vanishingly small.

The origin of the unique behavior of polyelectrolyte-electrolyte mixtures lies in the

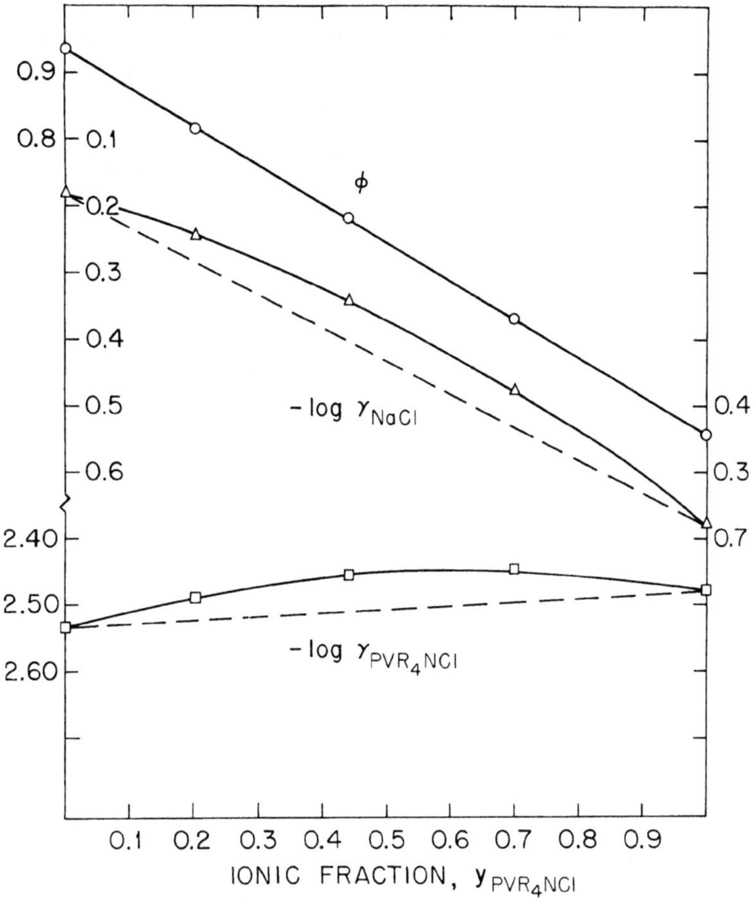

Fig. 14. Composition dependence of osmotic and activity coefficients for aqueous NaCl + PVBTMACl mixtures at constant $m = 1.0$.

TABLE III

Interaction coefficients for aqueous electrolyte-polyelectrolyte mixtures at 25°

M_B	α_B	$-\alpha_C$	β_B	β_C
A. NaCl + NaPSS				
0.1	0.988	1.367	7.40	11.29
1.0	0.0760	0.0456	0.0384	0.1337
B. NaCl + PVR_4NCl				
0.1	1.102	4.370	9.62	27.77
1.0	0.125	0.170	0.0613	0.226

strong association of Na^+ ion with the polystyrenesulfonate anion (or, of Cl^- ion with the polycation). The activity coefficient of sodium ion, γ_{Na^+}, is strongly decreased from its value in pure NaCl solutions of the same concentration as has been demonstrated by measurements with single-ion electrodes. [3, 25] These measurements indicate that the sodium ion activity coefficient is ca. 0.3 [3] or 0.435 [25] for solutions of sodium polyvinylsulfate or sodium polymethylstyrenesulfonate at $m=0.1$, respectively. The smaller of these two γ_{Na^+} values is almost the same as $\gamma_\pm$ for 'trace' concentrations of NaCl in the presence of 0.1 m NaPSS (Figure 11). This near equality would appear to require that the activity coefficient of chloride ion be approximately the same as that for Na^+ ion, or, $\gamma_{Cl^-} \sim 0.3$, in marked disagreement with the basic assumption of the Additivity Rule. The estimated value, $\gamma_{Cl^-} \sim 0.3$, is also much lower than the value of ca. 0.7 derived from measurements with an electrochemical cell with an electrode reversible to Cl^- ion. [3, 26] However, the cell measurements cannot be accepted unreservedly because of the well-known, strong interaction of the polyelectrolyte with Ag^+ ion from a Ag, AgCl electrode. It is essential that the isopiestic measurements of $\gamma_\pm$ for NaCl in the presence of polyelectrolyte be confirmed by e.m.f. measurements. Fortunately, studies of mixtures of NaCl with PVBTMACl appear to be feasible.

3.3. Comparison of the interactions of NaPPS and PVR_4NCl in mixtures with NaCl

Small but significant differences appear to exist between the interactions of the anionic polyelectrolyte, NaPSS and of the cationic polyelectrolyte, PVR_4NCl, with NaCl. Although the activity coefficient of NaCl is strongly reduced in mixtures with either polyelectrolyte, the reduction in $\gamma_\pm$ by PVR_4NCl is slightly greater. This difference appears to be a consequence of the substantially lower osmotic coefficient of PVR_4NCl compared with NaPSS at all concentrations except at high dilution. The nature of the structurally-bound ionic groups in the two compounds may be the cause: in PVR_4NCl the positive charge resides on a quaternary nitrogen atom which is well-shielded by large hydrophobic groups, while in NaPSS the negative charge is distributed over the sulfonate group which is hydrated. The difference in the hydration of the two polyions may be responsible for the difference in the strength of their interactions with NaCl. Another small difference is in the sign and magnitude of the like-ion interactions, δ_{XY} for Cl^- with PSS^- ion and δ_{MN} for Na^+ with PVR_4N^+ ion. The value of $\delta_{XY}=\beta_0/m=+0.0409$ whereas $\delta_{MN}\approx=0.01$, respectively, at $m=1.0$. The magnitude of δ_{MN}, which is virtually zero, suggests that the large, hydrophobic PVR_4N^+ cation interacts extremely weakly with Na^+ ion.

References

1. Research sponsored by the U.S. Atomic Energy Commission.
2. Mock, R. A. and Marshall, C. A.: *J. Polymer Sci.* **13**, 263 (1954).
3. Nagawasa, M., Izumi, M., and Kagawa, I.: *J. Polymer Sci.* **37**, 375 (1959).
4. Nagawasa, M., Takahashi, A., Izumi, M., and Kagawa, I.: *J. Polymer Sci.* **38**, 213 (1959).

5. Katchalsky, A., Cooper, R. E., Upadhyoy, J., and Wasserman, A.: *J. Chem. Soc. London*, 5198 (1961).
6. Alexandrowicz, Z.: *J. Polymer Sci.* **43**, 325, 337 (1960); *J. Polymer Sci.* **56**, 97, 115 (1962).
7. Katchalsky, A. and Alexandrowicz, Z.: *J. Polymer Sci.* [A] **1**, 2093 (1963).
8. Oosawa, F.: *J. Polymer Sci.* [A] **1**, 1501 (1963).
9. Lyons, J. W. and Kotin, L.: *J. Am. Chem. Soc.* **87**, 1781 (1965).
10. Marinsky, J. A.: *J. Phys. Chem.* **71**, 4349 (1967).
11. Boyd, G. E. and Bunzl, K.: *J. Am. Chem. Soc.* **89**, 1776 (1967).
12. Robinson, R. A. and Stokes, R. H.: *Electrolyte Solutions*, 2nd ed., Butterworths, London, 1959, p. 177ff.
13. (Beaker Dialyzer, b/HFD-1), Bio-Rad Laboratories, Richmond, California.
14. Kozak, D., Kristan, J., and Dolar, D.: *Z. Physik. Chem. N.F.* **76**, 85 (1971).
15. Reddy, M. and Marinsky, J. A.: *J. Phys. Chem.* **74**, 3884 (1970).
16. Bonner, O. D. and Overton, J. R.: *J. Phys. Chem.* **67**, 1035 (1963).
17. Ise, N. and Okubo, T.: *J. Phys. Chem.* **72**, 1361 (1968).
18. Lifson, S. and Katchalsky, A.: *J. Polymer Sci.* **13**, 43 (1954).
19. Manning, G. S.: *J. Chem. Phys.* **51**, 924 (1969).
20. Schwarz, A. and Boyd, G. E.: *J. Phys. Chem.* **69**, (1965).
21. McKay, H. A. C. and Perring, J. K.: *Trans. Faraday Soc.* **49**, 163 (1953).
22. Katchalsky, A.: *Pure Appl. Chem.* **26**, 371 (1971).
23. Covington, A. K., Lilley, T. H., and Robinson, R. A.: *J. Phys. Chem.* **72**, 2759 (1968).
24. Guggenheim, E. A.: *Trans. Faraday Soc.* **62**, 3446 (1966).
25. Oman, S. and Dolar, D.: *Z. Physik. Chem.* **56**, 1 (1967).
26. Leszko, M. and Gregor, H. P.: *Kunstharz-Ionenaustauscher*, Akad. Verlag, Berlin, 1970, p. 681.

COMPARISON BETWEEN EXPERIMENTAL RESULTS OBTAINED WITH HYDROXYLATED POLYACIDS AND SOME THEORETICAL MODELS

M. RINAUDO

Centre de Recherches sur les Macromolécules Végétales (C.N.R.S.) – 38 – Grenoble, France

1. Introduction

We intend to present first a set of experimental data concerning the equilibrium properties (pK, activity of counterions) of polyelectrolytes in salt free solutions; in a second part, these results are compared with theoretical values obtained when a rod-like model is adopted in the Lifson-Katchalsky, Manning and Oosawa treatments. The polyelectrolytes tested are respectively polysaccharides (alginic and pectic acids), carboxymethylated derivatives of cellulose (CMC), dextran (CMD), amylose (CMA), and sulphonic derivatives of polyvinyl alcohol. The essential difference between them is local rigidity. It is possible to vary by synthesis the degree of substitution and consequently the linear charge density; our data are given for dilute aqueous solutions (2×10^{-3} N–5×10^{-3} N) at 25°C.

2. Characterization of the Polyelectrolytes

The fundamental parameter is the linear charge density λ proposed by Lifson and Katchalsky [1]; λ is given by the relation:

$$\lambda = \frac{\nu\varepsilon^2}{DhkT} = \frac{\alpha}{b}\cdot\frac{\varepsilon^2}{DkT} \tag{1}$$

with ν number of ionic sites on a polymeric chain whose length is h; D is the dielectric constant ($\simeq 80$), ε the electron charge, kT the Boltzman term; α/b means a number of ionic sites per unit length.

In the particular case of polysaccharide derivatives characterized by their $\overline{DS}$ (the mean number of carboxyl groups per monomeric unit of length b) and degree of neutralization α', relation (1) can be written:

$$\lambda = \frac{\alpha\overline{DS}}{b}\,\frac{\varepsilon^2}{DkT} = \alpha\cdot\overline{DS}\cdot\lambda_0,$$

λ_0 is a constant for a given polysaccharide; for example, with carboxymethylcellulose, $\lambda_0 = 1.38$ at 25°C and for the polyvinyl derivative, $\lambda_0 = 2.83$ [2].

Figure 1 gives the different monomeric units studied in this work. Briefly, they are:

Eric Sélégny (ed.), Polyelectrolytes, 157–193.

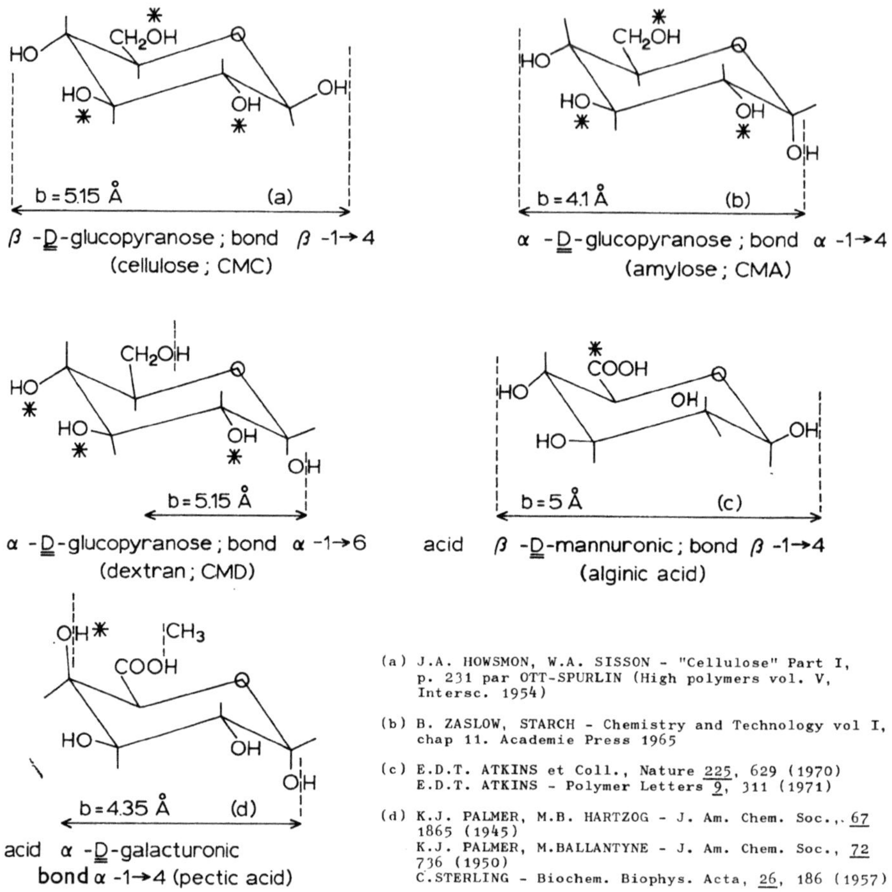

(a) J.A. HOWSMON, W.A. SISSON - "Cellulose" Part I, p. 231 par OTT-SPURLIN (High polymers vol. V, Intersc. 1954)

(b) B. ZASLOW, STARCH - Chemistry and Technology vol I, chap 11. Academie Press 1965

(c) E.D.T. ATKINS et Coll., Nature 225, 629 (1970)
E.D.T. ATKINS - Polymer Letters 9, 311 (1971)

(d) K.J. PALMER, M.B. HARTZOG - J. Am. Chem. Soc., 67 1865 (1945)
K.J. PALMER, M.BALLANTYNE - J. Am. Chem. Soc., 72 736 (1950)
C.STERLING - Biochem. Biophys. Acta, 26, 186 (1957)

Fig. 1. Monomeric units of the different polysaccharides studied
* (substitution position or —COOH site).

(a) *Polycarboxylics*

• pectic acids with different degree of esterification kindly donated by Doctor Kohn (Bratislava):

• alginic acid $\overline{DS}=1$ is a commercial but purified sample (Touzart and Matignon);

• carboxymethylated polysaccharides (amylose, cellulose, dextran) obtained by the following reactions [3]

$$R-OH \xrightarrow{Na/NH_3} R-ONa \xrightarrow{ClCH_2-COONa} R-O-CH_2-COONa$$

Their $\overline{DS}$ varies from 0.5 to 3 for soluble samples

(b) *Polysulphonics*

• Sulphopropylic derivatives of polyvinyl alcohol (PVS) with different $\overline{DS}$ have been obtained through the reactions [4]:

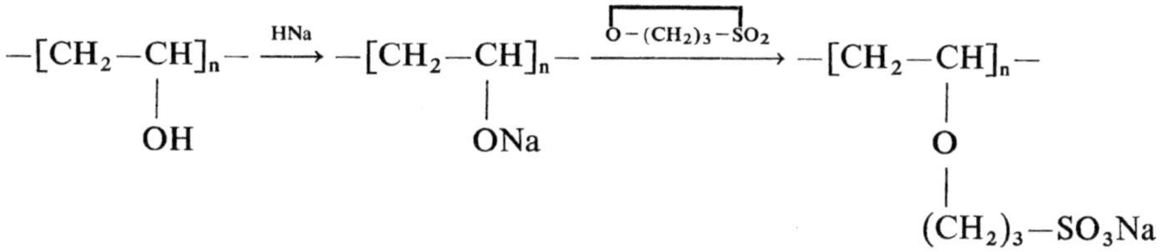

- Polyethylenesulphonic acid (PES) is a commercial sample from Hercules Powder Co.

Each polyelectrolyte is characterized by its λ value calculated for a fully extended chain; the essential differences between these polyelectrolytes allow us to study the influence of:

- the charge density,
- the rigidity of the backbone,
- the chemical type of the ionic site ($-COOH$; $-SO_3H$),
- the chemical structure of the chain.

3. Experimental Techniques and Results

3.1. Potentiometry

3.1.1. *Activity Coefficients of Counterions*

The free fraction of counterions is determined with specific electrodes for Na^+, Ca^{2+}, Ba^{2+}, Sr^{2+}, Mg^{2+}. The measurements are sensitive to H^+ ions, so the data are obtained for complete neutralization ($7<pH<8$). For each counterion, a calibration curve is drawn which relates the electrochemical potential to the concentration; its independance of the anion is assumed. The activity coefficient γ assimilated to the free fraction of counterions is given by:

$$\gamma = C_{exp}/C_{tot} \tag{2}$$

with C_{exp} the experimental value of the concentration and C_{tot} the concentration of the polyelectrolyte (in equivalent.l^{-1}) or of counterions investigated.

We define then the effective ionisation i by

$$i = \alpha \cdot \gamma \equiv \alpha\phi$$

if Katchalsky's conclusion $\gamma \equiv \phi$ is retained [5, 11]. The activities of H^+ and Na^+ are measured with a Tacussel TS 40 (France) potentiometer, a calomel reference electrode and respectively a Tacussel H^+ or a Jena glass Na^+ electrode; the Ca^{2+} activity is measured with a Sargent S 30000 potentiometer and an Orion electrode (model 92.32).

Free fraction of monovalent counterions – The results obtained with Na^+ as counterions and different CMC are plotted on Figure 2a; the experimental points (not represented)

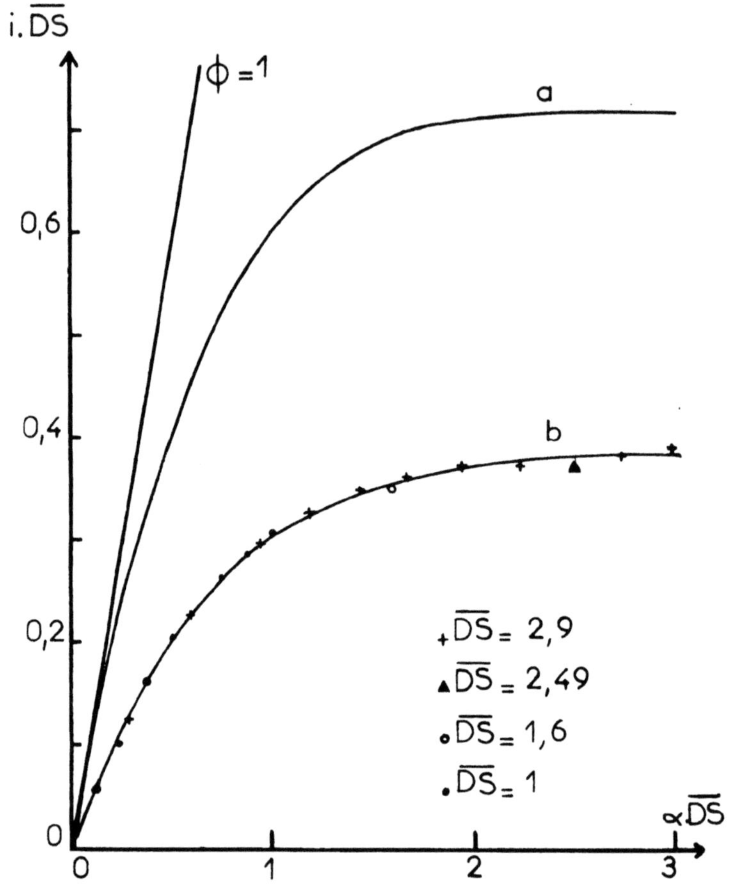

Fig. 2. Effective charge ($i \cdot \overline{DS}$) versus structural charge ($\alpha \cdot \overline{DS}$) of CMC [10]. (a) monovalent counter-ions; (b) divalent counter-ions.

are all on the same curve which limits up at $i \cdot \overline{DS}$ 0.72 when represented as a function of the structural charge density ($\alpha \cdot \overline{DS}$); this limit corresponds to $\lambda_{eff} = 1$; such a behaviour is in favour of an electrostatic mechanism.

Figure 3 gives the results for the two families of polyelectrolytes; a normalization is adopted with ($i \cdot \overline{DS}/b$) as a function of ($\alpha \cdot \overline{DS}/b$) the charge density per unit length; the behaviour of the polysaccharide is quite different from the polyvinylic one and can be attributed to the difference in rigidity; the assumption of a maximum extension of the vinylic chain of 75%, is concordant with previous data of Katchalsky [5, 11].

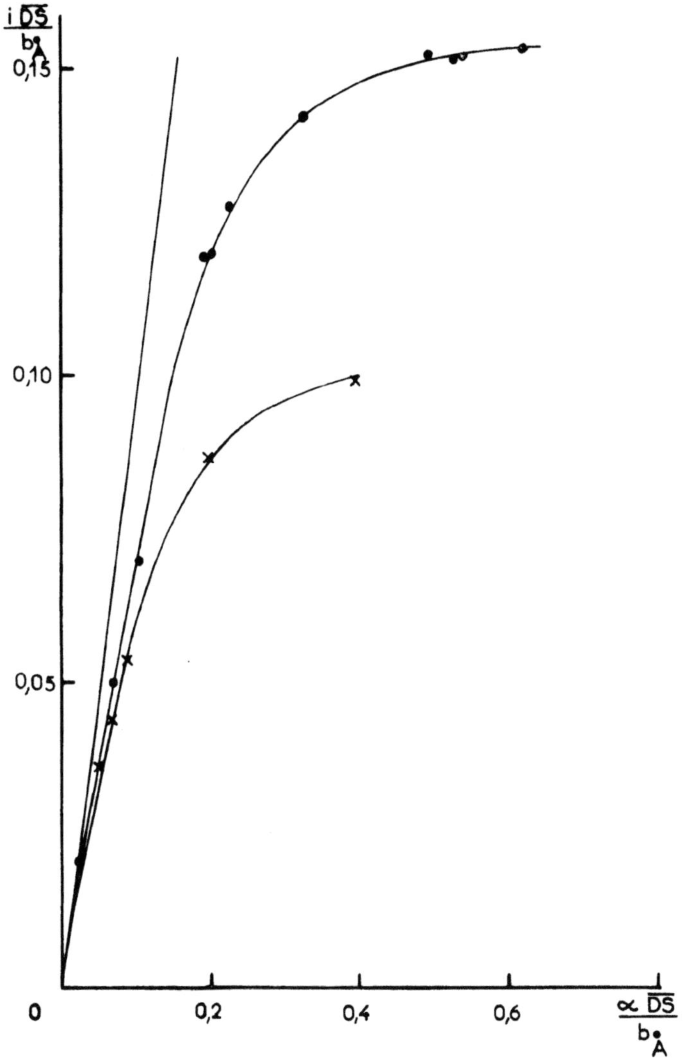

Fig. 3. Effective charge density obtained with monovalent counter-ions: × polysulphonic derivatives; ● polysaccharides.

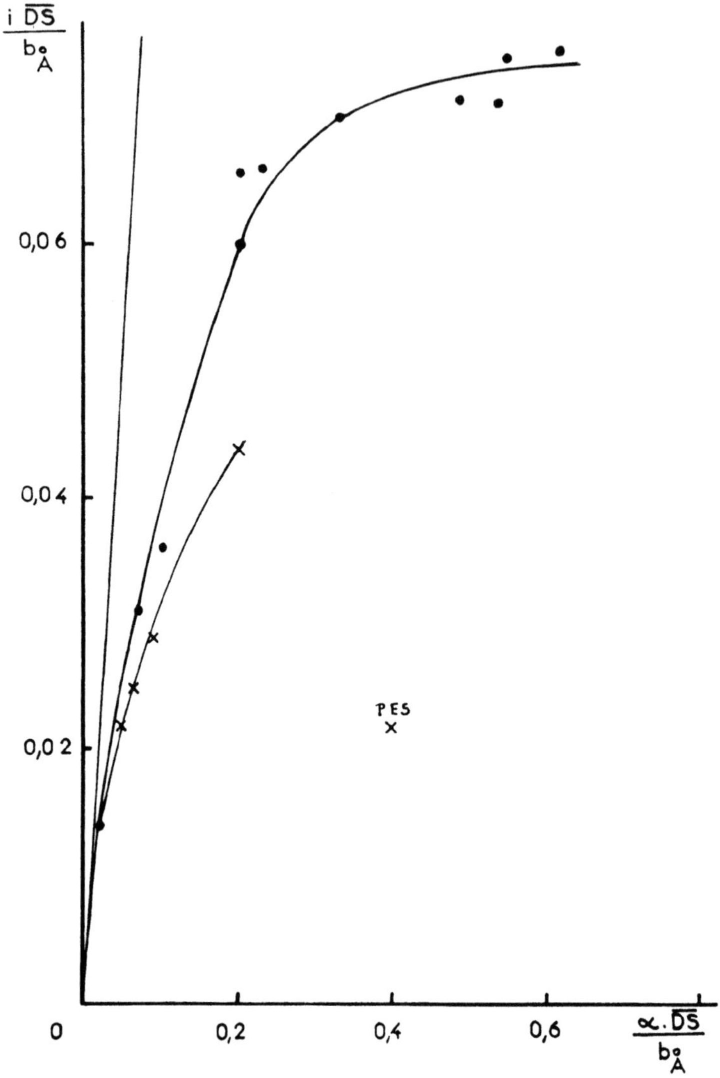

Fig. 4. Effective charge density obtained with divalent counter-ions: × polysulphonic derivatives; ● polysaccharides.

Free fraction of divalent counterions – With CMC, the results obtained give in Figure 2b a single curve; one can also see that for $\lambda > 1$, $\gamma_{Na}/\gamma_{Ca} = 2$. Figure 4 compares the carboxylic and sulphonic families; the particularly low value characterizing PES could well be attributed to the formation of intramolecular bridges with Ca^{++}. The experimental data are collected in Table I.

TABLE I

Potentiometric data [6, 8]

Samples	$\overline{DS}$	λ	γ_{Na}	γ_{Ca}	pK_0
CMC	1	1.38	0.630	0.345	3.00
	1.7	2.35	0.410	0.225	2.60
	2.49	3.44	0.330	0.145	2.90
	2.77	3.83	0.280	0.120	–
CMA	2.2	3.80	0.285	0.140	3.20
	2.5	4.32	0.265	0.126	3.10
CMD	1.15	1.59	0.575	0.295	2.60
PA	0.10	0.16	0.910	0.630	3.30
	0.30	0.48	0.725	0 450	3.25
	0.45	0.72	0.675	0.340	3.25
AA	1	1.38	0.600	0.300	–
PVS	0.11	0.307	0.870	0.500	
	0.16	0.447	0.700	0.390	
	0.21	0.587	0.650	0.350	
	0.43	1.20	0.560	0 330	
	0.50	1.397	0.440	0.220	
PES	1	2.79	0.230	0.055	

Free fraction of monovalent and divalent counterions in mixture – CMC is completely neutralized by a series of mixed solutions of sodium and calcium hydroxides, the equivalent ionic fraction X_{Na} varying from zero to one, $(X_{Na} + X_{Ca} = 1)$; in the resulting polysalt solutions γ_{Na} and γ_{Ca} are determined separately. The total dissociation X_T calculated following Equation (3) is plotted vs. X_{Na} in Figure 5.

$$X_T = \gamma_{Na} \cdot X_{Na} + \gamma_{Ca}(1 - X_{Na}). \tag{3}$$

The results clearly show the marked influence of the charge density. Note that with a CMC $\overline{DS}$ 2.49 and $X_{Na} > 0.8$, X_T is constant and $\gamma_{Ca} = 0$ thus showing that with high charge density and small fractions of divalent counterions each Ca^{++} excludes two Na^+ quantitatively.

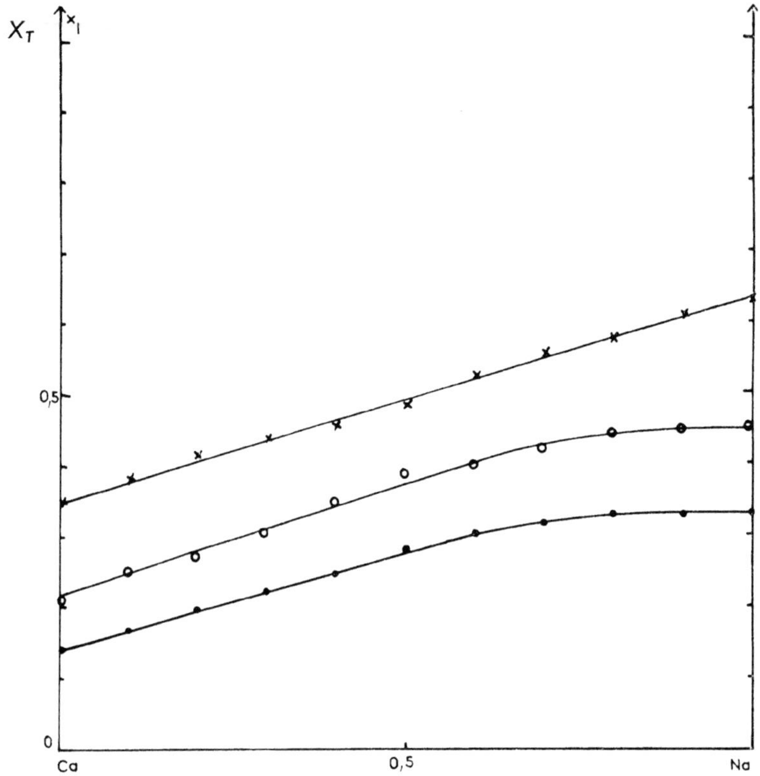

Fig. 5. Equivalent free fraction of counterions (X_T) with variable Na^+/Ca^{2+} ratios in salt free solution of CMC [9] ($\alpha' = 1$; $c = 2 \times 10^{-3}$ N) ● $\overline{DS} = 2.49$; ○ $\overline{DS} = 1.7$; × $\overline{DS} = 1$.

3.1.2. *Neutralization curves – Intrinsic pK*

Polyacid solutions are neutralized progressively by a monovalent or divalent hydroxide and the pH measured for each degree of neutralization α'; the apparent pK_a is obtained from (4):

$$pK_a = pH + \log \frac{1 - \alpha'}{\alpha'}. \tag{4}$$

Introducing pK_0, the characteristic intrinsic dissociation constant of the polyacid at nil charge density, the variation of pK_a can be written:

$$pK_a = pK_0 + \Delta\, pK(\alpha). \tag{5}$$

The theoretical interpretation of $\Delta pK(\alpha)$ representing the contribution of electrostatic interactions will be widely discussed in the second part of this paper. Figures 6, 7, 8 give the experimental results obtained with CMC and polysulphonic derivatives.

The ion-selectivity is clearly shown:

for $-COOH: Ca \gg Li > Na > K > Cs$
for $-SO_3H: Ca \gg Cs > K > Na > Li$.

Due to the contribution of autodissociation of polycarboxilic acids to the effective charge density, lower values of α must be corrected and pK recalculated in function of α_{tot}:

$$\alpha_{tot} = \alpha' + \alpha_{H^+} \quad \text{with} \quad \alpha_{H^+} = \frac{[H^+]}{C}.$$

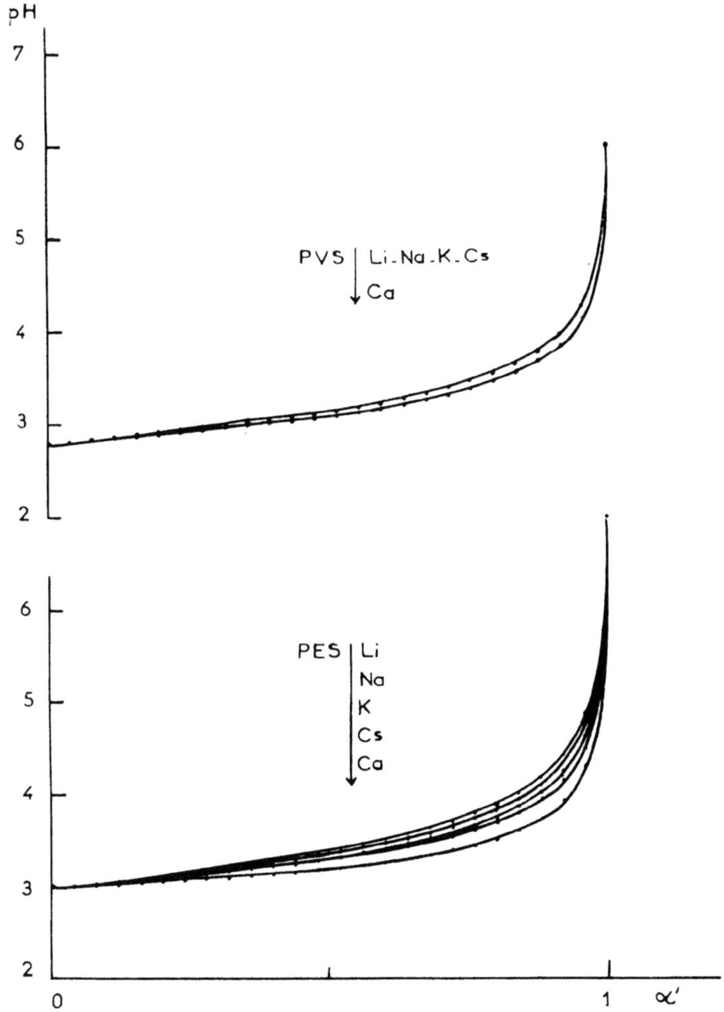

Fig. 6. Potentiometric neutralization: polysulphonic acids (PVS, $\overline{DS}=0.16$; PES) [8].

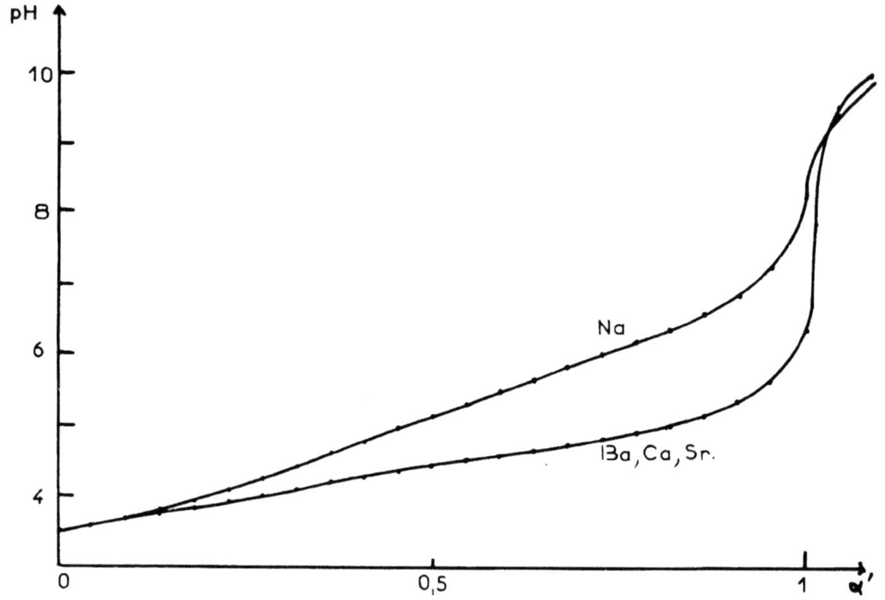

Fig. 7. Potentiometric neutralization: CMC $\overline{DS} = 1.25$ $c = 2, 1 \times 10^{-3}$ N [10a].

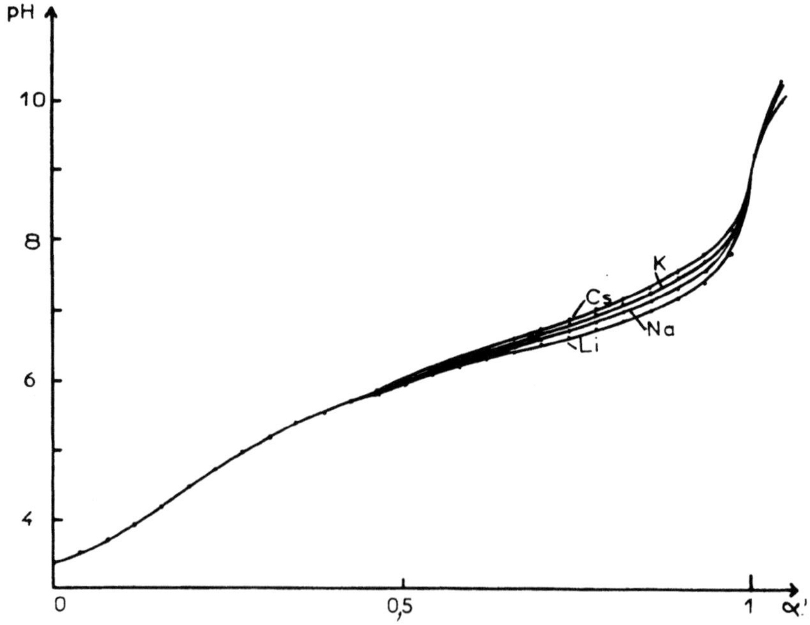

Fig. 8. Potentiometric neutralization by different X OH of a CMC $\overline{DS} = 2.9$; $c = 5 \times 10^{-3}$ N [10a].

Extrapolation to $\alpha_{tot}=0$, gives pK_0. The experimental values used to determine pK_0 were obtained on neutralization by $N(ET)_4^+$ which is free of site binding or in other terms interference of non-electrostatic terms.

pK_0 values are given in Table I; for CMC, pK_0 is independent of $\overline{DS}$ and equal to 3 ± 0.2; the same value is obtained for a monomer [6] as well as for oligomers of glutamic acid [7]; thus it is clearly shown that pK_0 is independent of the degree of polymerization.

3.2. Conductimetry

Figures 9 and 10 give some examples corresponding to the neutralization of CMC and sulphonic acids.

The behaviour is directly related to the nature of the ionic site ($-SO_3H$ or $-COOH$). In the particular case of divalent counterions, one can assume that the free fraction of

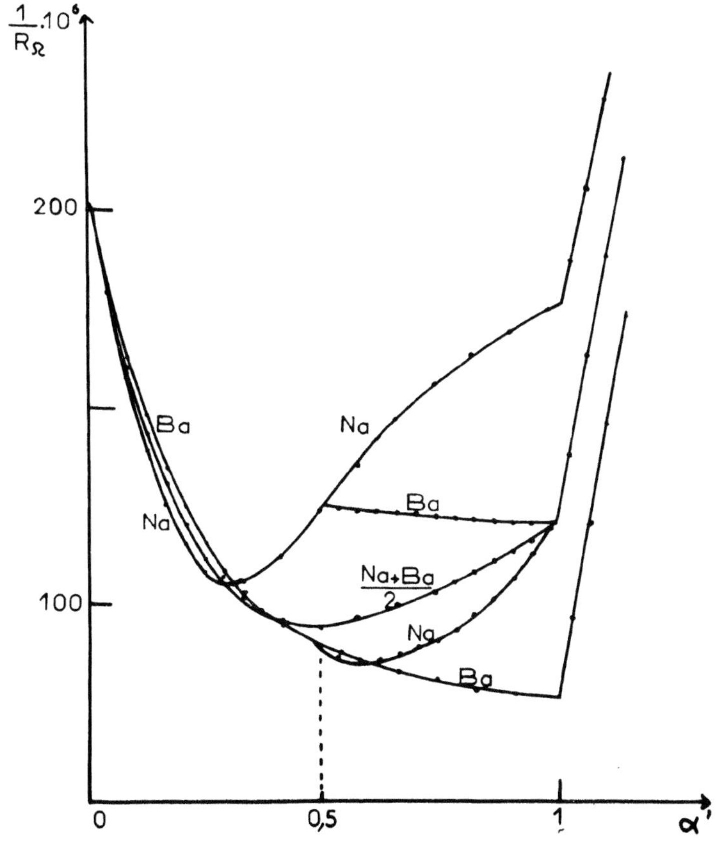

Fig. 9. Conductimetric neutralization by NaOH, $Ba(OH)_2$ and their mixture of a CMC solution $\overline{DS}=1.25$; $c=2.3\times10^{-3}$ N [10a]].

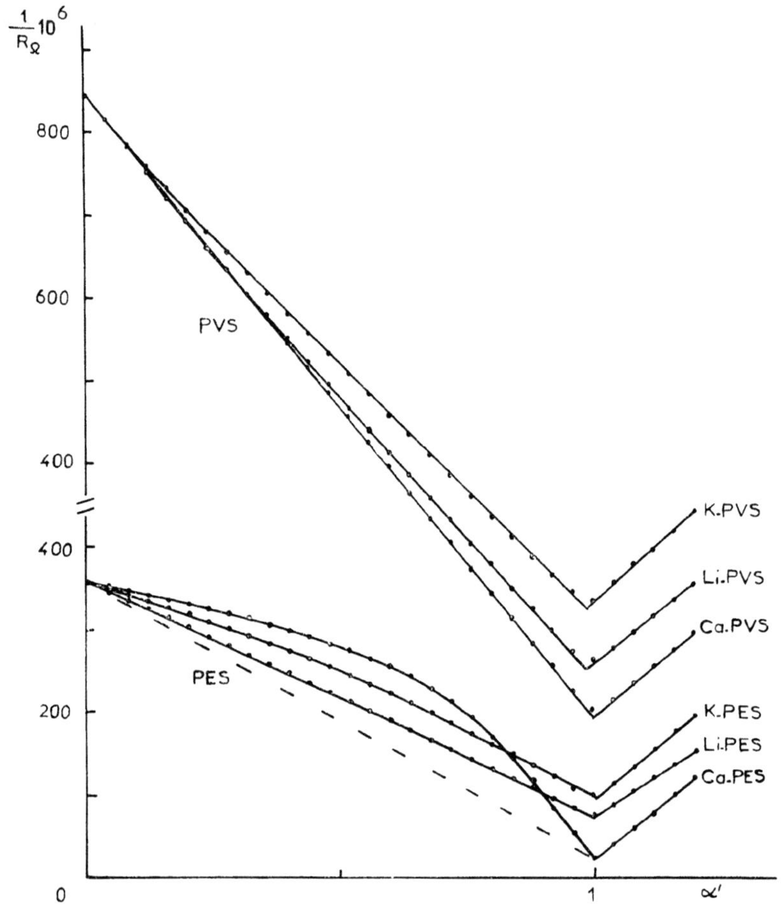

Fig. 10. Conductimetric neutralization by LiOH, KOH, $Ca(OH)_2$ of polysulphonic derivatives (PVS $\overline{DS} = 0.16$; PES) [8]

counterions ϕ is directly obtained from

$$\Lambda = \phi(\lambda_p + \lambda_c) \tag{6}$$

with Λ the equivalent conductance, and λ_p, λ_c respectively the equivalent ionic conductances of the polyion and of the counterion. The values of λ_p are taken from an experimental study with two monovalent counterions [10].

Experimental values of the free fraction of counterions in a CMC solution, obtained by potentiometry and conductimetry are given in Table II; one can conclude that both techniques give similar results for divalent ions.

To conclude, the principal properties of these polyelectrolytes in aqueous solution are:

TABLE II

Free fraction of divalent counterions obtained by potentiometry and conductimetry [10b]

$\overline{DS}$	counter-ions / methods	Mg^{+2}	Ca^{+2}	Sr^{+2}	Ba^{+2}	estimated error
2.9	conductim.	–	0.134	0.134	0.129	±0.005
	potentiom.	0.140	0.140	0.135	0.130	±0.010
2.49	conductim.	–	0.145	0.150	0.145	±0.005
	potentiom.	0.135	0.145	0.145	0.140	±0.010
2 1	conductim.	–	–	–	–	–
	potentiom.	0.165	0.170	0.180	0.180	±0.010
1.6	conductim.	–	0.205	0.210	0.210	±0.005
	potentiom.	0.190	0.190	0.190	0.190	±0.010
1	conductim.	–	0.323	0.334	0.323	±0.010
	potentiom.	0.300	0.305	0.280	0.297	±0.015

– The free fraction of counterions ϕ is essentially dependent on the linear charge density; the role of the rigidity of the backbone is clearly shown. The phenomenon is purely electrostatic with an exception for Ca-PES.

– ϕ is independant of $\overline{DP}$ (with $\overline{DP} > 30$) and of the chemical nature of the chain (experiments on different polysaccharides with different $\overline{DP}$ from 30 to 600).

– ϕ is practically independent of the concentration and of the nature of cations of equal valence (example Table II).

– the effective ionisation goes to a limit when the charge density increases (Figure 2); this limit attains $\lambda_{\mathrm{eff}} = 1$ for CMC.

– with CMC the ion-selectivity appears for a structural value $\lambda_{\mathrm{struct}} = 0.75$ as shown by potentiometry (and ultrasonic absorption) and corresponds to ion pair formation. The sequence and the magnitude of selectivity is directly related to the chemical nature of the ionic site ($-SO_3H$ or $-COOH$).

4. Summary of Theoretical Treatments and Applications

From a general point of view, the distribution of counterions is imposed by that of the electrostatic potential. Our purpose is to treat the thermodynamic behaviour of a cylindrical system without excess added salt; the polyelectrolyte is always considered as a thin rod characterized by its linear charge density. Two approaches are investigated: the first one needs the resolution of a Poisson-Boltzmann equation without the Debye approximation, as proposed by Lifson and Katchalsky [1]; the second one considers a two phase system such as the one treated by Oosawa [14].

4.1. Lifson-Katchalsky Theory [1]

In the cell model, a polyelectrolyte solution is divided in subvolumes with a radius R, a length h, parallel to each other; Figure 11 represents the system. If there are N

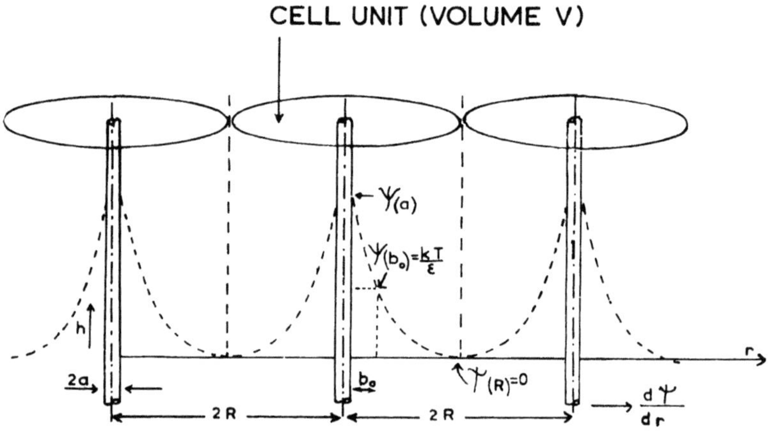

Fig. 11 Cylindrical model from Katchalsky and Alexandrowicz [11].

polyelectrolyte molecules per unit volume,

$$N\pi R^2 h = 1 .$$

The radius 'a' is introduced and means a minimum distance of approach to the axis, so that the volume fraction φ occupied by the polymer is given by

$$\varphi = a^2/R^2 .$$

4.1.1. *Osmotic Coefficient*

The thermodynamic properties are described by ϕ which is equal to the ratio of the effective osmotic pressure to the ideal calculated one:

$$\pi_{\text{ideal}} = \alpha \cdot n_m \cdot kT ,$$
$$\pi_{\text{real}} = \phi \cdot \alpha \cdot n_m \cdot kT ,$$
$$\phi = \pi_{\text{real}}/\pi_{\text{ideal}} . \tag{7}$$

In these expressions, $n_m = N \cdot P$ (P degree of polymerization) represents the concentration in monomeric units.

Assuming that there is only one site per monomeric unit on a chain, the following relation holds:

$$\alpha = \nu/P$$
$$n_c = N\nu = n_m \cdot \alpha .$$

α is the degree of dissociation and n_c the monovalent counterion concentration.

The effective ionisation is then introduced as:

$$i = \alpha \cdot \phi \quad \text{or} \quad \pi_{\text{real}} = i \cdot n_m \cdot kT.$$

The thermodynamic treatment [11] leads to:

$$\gamma = n_R/\bar{n},$$
$$\gamma = \phi, \tag{7'}$$

γ is the activity coefficient, $\bar{n}$ the mean counterion concentration, and n_R the counterion concentration at the distance R.

Writing this relation involves that the activity coefficient is independent of the exact nature of counterions of the same valence.

4.1.2. *Electrostatic Potential*

Due to the symmetry of the system (see Figure 11) calculation of the electrostatic potential ψ can be limited to the subvolume of radius $a \leqslant r \leqslant R$.

The Poisson-Boltzmann equation for a monovalent counterion writes

$$\nabla^2 \psi(r) = -\frac{4\pi \varrho(r)}{D} = -\frac{4\pi \varrho_0}{D} \exp - \varepsilon\psi/kT \tag{8}$$

ϱ_0 is the charge density corresponding to the condition $\psi = 0$; D is the dielectric constant, ε the electron charge and kT the Boltzmann term. The resolution of (8) is realized easily [12] with the initial conditions:

$$\left(\frac{\partial \psi(r)}{\partial r}\right)_{r=R} = 0; \quad \left(\frac{\partial \psi(r)}{\partial r}\right)_{r=a} = -\frac{4\pi\sigma}{D} = \frac{2\alpha\varepsilon}{Dab} \tag{9}$$

(σ = superficial charge density of the cylinder at 'a'). The electric field on the polyion's surface $(\partial\psi(r)/\partial r)_{r=a} = E$ can be written as:

$$E = \frac{kT}{\varepsilon} \cdot \frac{2\lambda}{a}. \tag{9'}$$

In addition, to that, the condition $(\partial\psi(r)/\partial r)_{r=a} = -(4\pi\sigma/D)$ is equivalent to the electroneutrality convention for the system

$$\frac{\alpha\varepsilon}{b} = 2\pi n_0 \varepsilon \int_a^R \exp - \frac{\varepsilon\psi}{kT}\, r\, \mathrm{d}r \tag{10}$$

(n_0 is the counterion concentration for $\psi = 0$). The general solution can be written in different simple forms depending on the respective values of λ and λ_0.

$$\psi(r) = \frac{kT}{\varepsilon} \ln\left\{\frac{K^2 r^2}{2\beta^2} \sinh^2(\beta \ln Ar)\right\} \quad \lambda < \lambda_0$$

$$\psi(r) = \frac{kT}{\varepsilon} \ln \left\{ \frac{K^2 r^2}{2} (\ln Ar)^2 \right\} \qquad \lambda = \lambda_0 \tag{11}$$

$$\psi(r) = \frac{kT}{\varepsilon} \ln \left\{ \frac{K^2 r^2}{2|\beta|^2} \sin^2 (|\beta| \ln Ar) \right\} \quad \lambda > \lambda_0 .$$

In expressions (11), both β and A are integration constants and K^2 is defined by

$$K^2 = \frac{4\pi n_0 \varepsilon^2}{DkT} .$$

The set of relations (11) has two more useful expressions depending on the introduction of a new condition for ψ_R.

(a) Lifson-Katchalsky expressions [1]: the question is to define n_0 ($\varrho_0 = n_0 \varepsilon$) in the general expression (8). With conditions (9), n_0 the number of counterions per unit volume can be obtained through the relation:

$$\alpha = \pi b (R^2 - a^2) \quad \text{so} \quad n_0 = \bar{n}$$

introducing R deduced from the monomole per unit volume polyelectrolyte concentration n_m given by:

$$n_m = (\pi \mathcal{N} R^2 b)^{-1} .$$

($\mathcal{N}$ Avogadro's number; b length of monomeric unit).

Then, $K^2 = 4\lambda/(R^2 - a^2)$ and it follows for $\lambda < \lambda_0$:

$$\psi(r) = \frac{kT}{\varepsilon} \ln \left\{ \frac{2\lambda}{\beta^2} \frac{r^2}{R^2 - a^2} \sinh^2 (\beta \ln Ar) \right\} . \tag{12}$$

Figure 12 shows the variation of β with λ at different polyelectrolyte concentrations; note that the latter is also expressed by $\gamma = \ln R/a$.

The final expression for $\psi(r)$ depends on λ compared to $\lambda_0 = \gamma/1 + \gamma$ corresponding to $\beta = 0$.

β has limiting values at infinite dilution:

$$\begin{aligned} &\lambda < 1 \quad \beta = \lambda , \\ &\lambda \geqslant 1 \quad \beta = 0 . \end{aligned} \tag{13}$$

For the condition $\lambda > \lambda_0$, the expression of ψ is

$$\psi(r) = \frac{kT}{\varepsilon} \ln \left\{ \frac{2\lambda}{|\beta|^2} \frac{r^2}{R^2 - a^2} \sin^2 (|\beta| \ln Ar) \right\} . \tag{14}$$

(b) Condition $\psi_R = 0$: More recently [5], a treatment was proposed with this additional condition involving $n_R = n_0$; the Poisson-Boltzmann equation becomes

$$\nabla^2 \psi = - \frac{4\pi n_R \varepsilon}{D} \exp - \frac{\varepsilon \psi}{kT} \tag{15}$$

with $n_0 < \bar{n}$.

From the initial conditions:

$$\left(\frac{\partial\psi}{\partial r}\right)_{r=R} = 0 \quad \text{and} \quad \psi_R = 0 \tag{16}$$

n_R is deduced and introduced in (15) together with the electroneutrality condition

$$\alpha = 2\pi b \int_a^R n_R \exp - \frac{\varepsilon\psi}{kT} r \, \mathrm{d}r .$$

The analytical expression for (11) and $\lambda < \lambda_0$ becomes:

$$\psi(r) = \frac{kT}{\varepsilon} \ln \left\{ \frac{1-\beta^2}{\beta^2} \frac{r^2}{R^2} \sinh^2(\beta \ln Ar) \right\} . \tag{17}$$

4.1.3. *Expression of* ϕ

From $\psi(r)$, Lifson and Katchalsky [1] have calculated the free electrostatic energy and deduced the osmotic coefficient:

$$\phi \simeq \frac{1-\beta^2}{2\lambda} \quad \text{when} \quad \lambda < \lambda_0 ,$$

$$\phi \simeq \frac{1+|\beta|^2}{2\lambda} \quad \text{when} \quad \lambda > \lambda_0 .$$

Using relation (7′) it is easy to calculate ϕ from the $\psi(r)$ function [5]; this treatment leads to the result:

$$\phi_K = \frac{1-\beta^2}{2\lambda}$$

and as $\lambda = \alpha\lambda_0$, it follows that:

$$i = \alpha\phi = \frac{1-\beta^2}{2\lambda_0} , \qquad \text{when } \lambda < \lambda_0$$

λ_0 is constant for a given type of polyelectrolyte; thus the behaviour of $i(\lambda)$ follows that of β^2; from Figure 12 one can justify the fact that i reaches a limit for high values of λ. At infinite dilution the next values of ϕ follow from (13):

$$\begin{array}{ll} \lambda < 1 & \phi = 1 - \lambda/2 \\ \lambda = 1 & \phi = 1/2 \\ \lambda > 1 & \phi = 1/2\lambda ; \end{array} \tag{18}$$

they are identical to those given by the Manning theory [15]. The calculation of the ΔpK(α) function (5) can be made from the law giving the variations of $\psi(r)$ with α if it is assumed that relation (19) is valid.

$$\Delta \mathrm{pK} = -\, 0.434\, \varepsilon \Delta\psi / kT \tag{19}$$

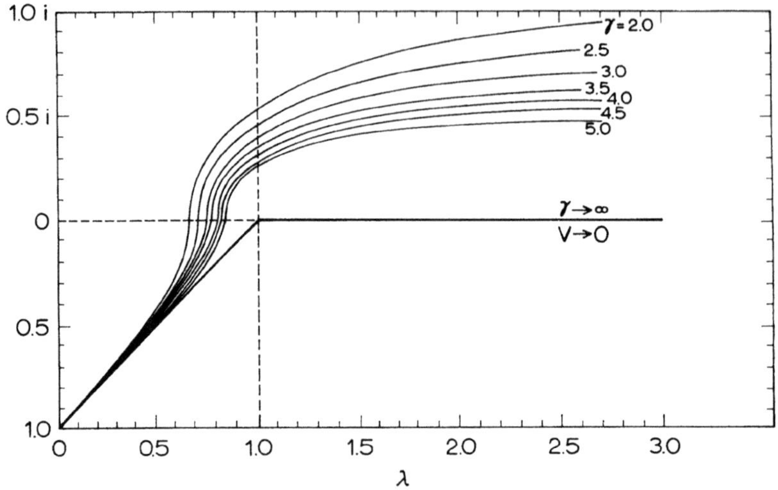

Fig. 12. Integration constant β as a function of charge density parameter λ for different concentrations (γ) [5].

($\Delta\psi$ potential difference between 'a' and R) in order to interpret the potentiometric titration curves.

4.1.4. *Extension of the Treatment and Application*

(1) *Extension* – From the knowledge of $\psi(r)$ with monovalent counterions, the treatment may be applied to

- divalent counterions [25],
- mixtures of monovalent and divalent counterions [23, 19].

Different approaches can be proposed to evaluate the fraction of free counterions; Dolar and Peterlin [23] defined three sorts of bound counterions (f):

- $f = 1 - n_R/\bar{n}$ as in Lifson-Katchalsky treatment
- f corresponds also to the surface area below the curve representing the probability of distribution of counterions ($d\pi(r)$, Figure 16) taken between its minimum and a (this recalls Bjerrum's theory).
- f is the fraction contained in a region where $\varepsilon\psi/kT \geqslant 1$; in our study, we call $\bar{\alpha}_1 = 1 - f$ the free fraction. We calculate also $\bar{\alpha}_2$ giving another free fraction for which $\varepsilon\psi/kT \leqslant 2$ selected because of the energetical limit of ion-pairing given by Bjerrum.

The experimental values are also compared to the diffusion coefficient ratio (D/D_0) defined by Jackson and Coll [17, 18].

For a diffusion perpendicular to the axis of the cylinder, D is related to ψ

$$D = D_0 \{\langle \exp\varphi \rangle_v \langle \exp - \varphi \rangle_v\}^{-1} \qquad (20)$$

with $\varphi = \varepsilon\psi/kT$, D_0 diffusion coefficient in absence of electrostatic field. By symmetry, it follows

$$\langle \exp \pm \varphi \rangle_s \equiv \langle \exp \pm \varphi \rangle_v$$

so the mean value on a cylindrical section has to be calculated.

From relations (12), (14) or (17), D/D_0, independent of the reference potential is obtained; for $\lambda > \lambda_0$ and $R \gg a$:

$$D_0/D = \frac{\lambda}{2} \cdot \frac{13 + |\beta|^2}{(|\beta|^2 + 1)(|\beta|^2 + 4)}. \tag{21}$$

In the work of Marcus [24], (21) equals the activity coefficient of the counterions in the absence of added salt.

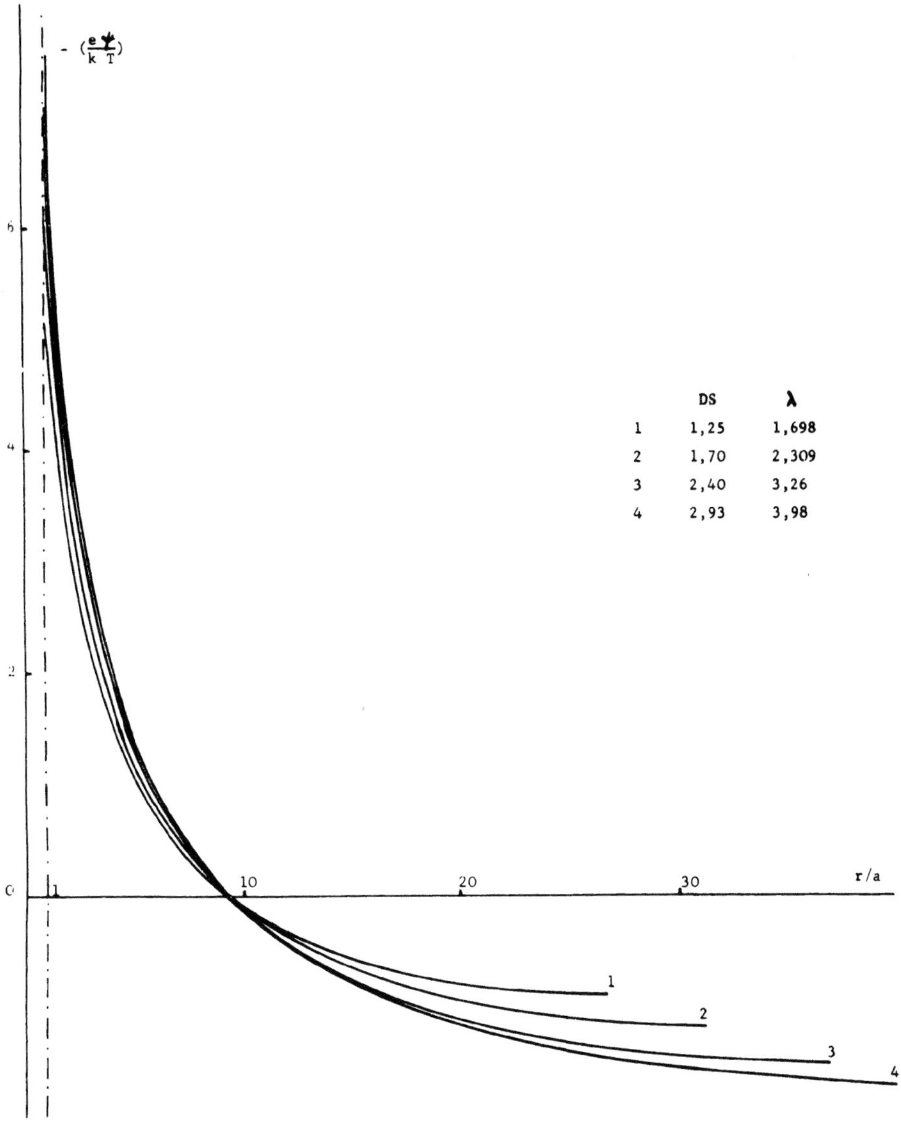

Fig. 13. Variation of the electrostatic potential $\varepsilon\psi/kT$ as a function of the distance r/a for CMC ($c = 5 \times 10^{-3}$ N) [13].

(2) *Application*

Activity coefficient of monovalent counterions – On Figure 13, $\varepsilon\psi/kT$ vs. (r/a) obtained from (12) is plotted; on Figure 14, the representation $\varepsilon\Delta\psi/kT$ vs. (r/a) from (17) is given ($\psi_R=0$). If $\psi(r)$ is known, the distribution of counterions is expressed by:

$$n(r) = n_0 \exp\left(-\frac{\varepsilon\psi(r)}{kT}\right)$$

which is independent of the reference potential; $\bar{\alpha}(r)$ the fraction of counterions contained between r and R is

$$\bar{\alpha}(r) = 1 - f(r)$$

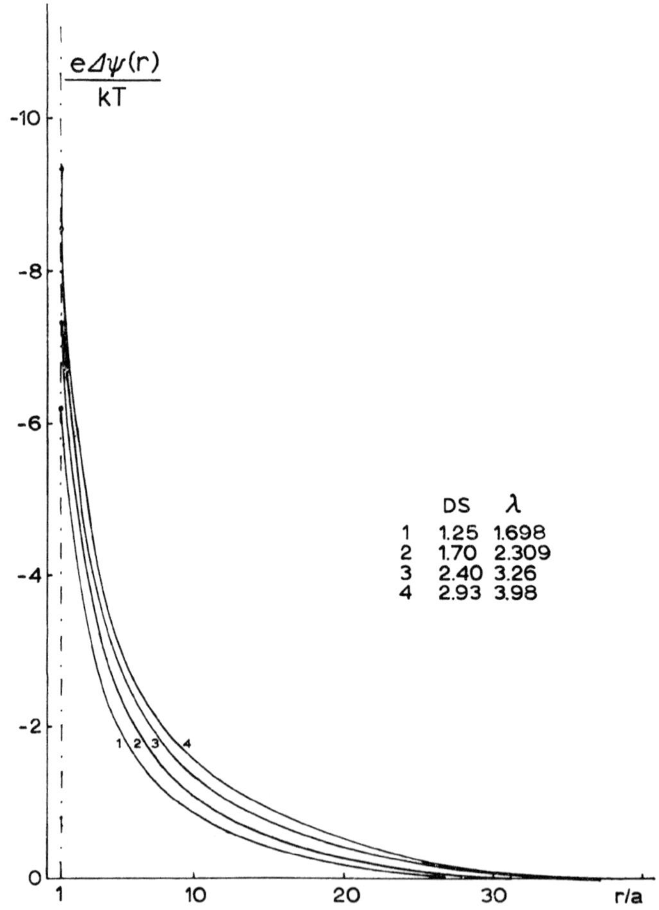

Fig. 14. Variation of the difference of potential $\varepsilon\Delta\psi/kT$ as a function of r/a for CMC ($c = 5 \times 10^{-3}$ N) [13].

with

$$f(r)=\frac{2\pi h n_0 \int_a^r r\exp(-\varepsilon\psi/kT)\,\mathrm{d}r}{2\pi h n_0 \int_a^R r\exp(-\varepsilon\psi/kT)\,\mathrm{d}r} \tag{22}$$

or

$$f(r)=\frac{2}{R^2-a^2}\int_a^r r\exp(-\varepsilon\psi/kT)\,\mathrm{d}r. \tag{23}$$

Results are represented in Figure 15. The probability of the distribution can be computed from (23); the function has a minimum at $r=r_0$; in Figure 16, a typical result is presented. The fraction of counterions contained between r_0 and R is $\bar{\alpha}=1/2\lambda$ for all values of λ/λ_0. This is also found in the Manning theory when $\lambda>1$.

Figure 17 gives the results for a series of CMC; experimental values are compared

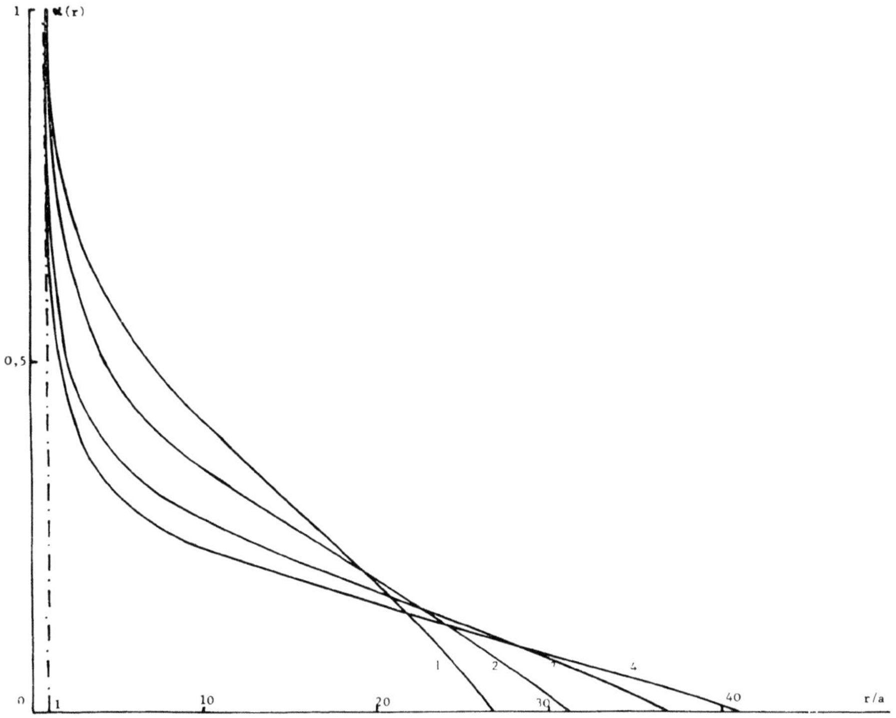

Fig. 15. Calculated fraction of monovalent counterions $\bar{\alpha}(r)$ outside a cylindrical volume of radius r[13]. $\overline{DS}$ values: 1(1.25); 2(1.70); 3(2.40); 4(2.93).

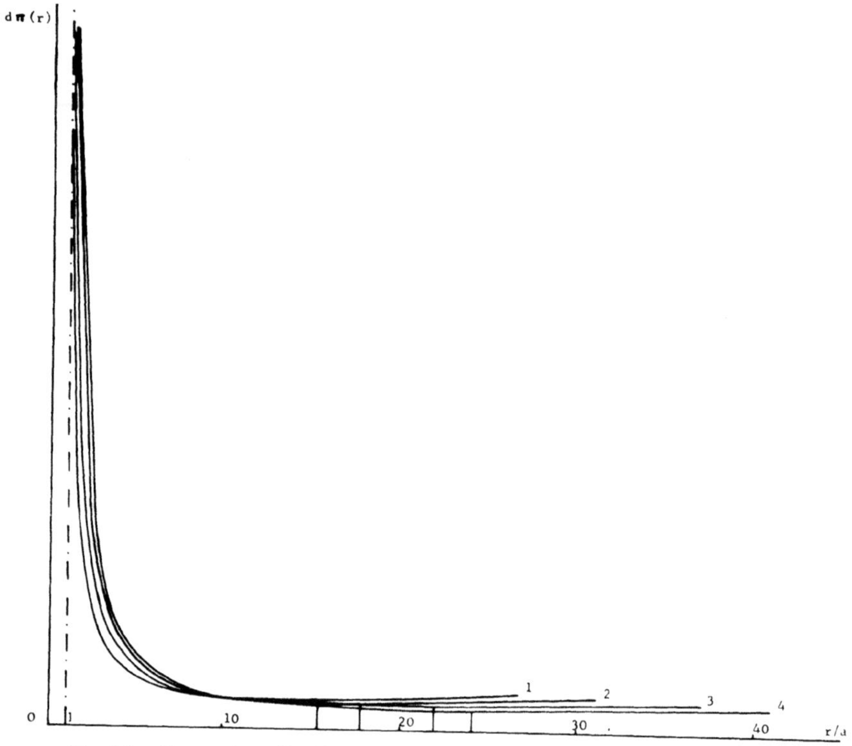

Fig. 16. Counterions distribution $d\pi(r)$ as a function of the distance r/a [13]. $\overline{\text{DS}}$ values: 1 (1.25); 2 (1.70); 3 (2.40); 4 (2.93).

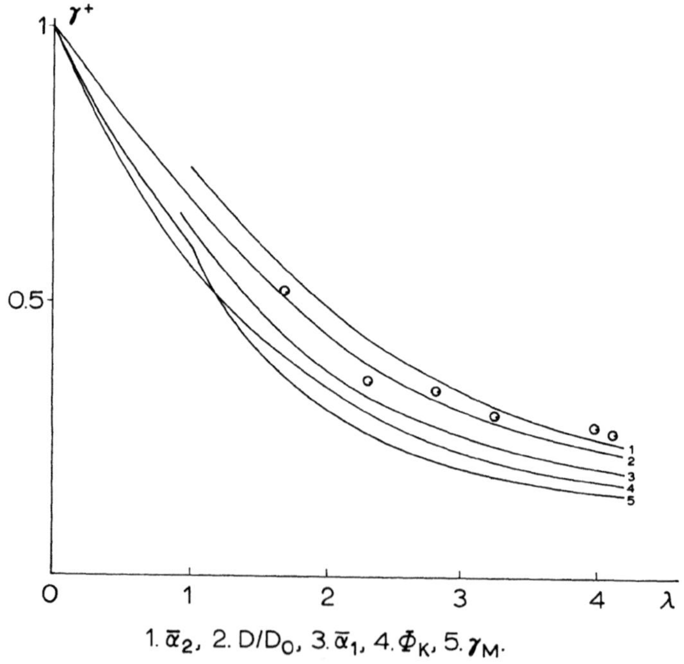

Fig. 17. Free fraction of counterions obtained with different definitions as function of the charge parameter λ [13] ——— calculated values; ○ experimental data.

TABLE III

Experimental and theoretical values compared [6]

data / samples	$\overline{DS}$	Monovalent counter-ions					Divalent counter-ions			
		γ_{Na}exp	$\gamma^+ = D/D_0$	ϕ_K	γ_M	γ_O	γ_{Ca}exp	$\gamma^{2+} = D/D_0$	ϕ_K	γ_O
1-CMC	1.1	0.630	0.574	0.447	0.440	0.725	0.345	0.327	0.249	0.362
2-CMC	1.7	0.410	0.366	0.281	0.258	0.425	0.225	0.193	0.147	0.212
3-CMC	2.49	0.330	0.253	0.194	0.176	0.291	0.145	0.130	0.099	0.145
4-CMA	2.2	0.285	0.228	0.175	0.159	0.263	0.140	0.116	0.089	0.131
5-CMA	2.5	0.265	0.200	0.154	0.140	0.232	0.126	0.102	0.078	0.116
6-CMD	1.15	0.575	0.515	0.399	0.380	0.630	0.295	0.286	0.217	0.315
7-P.A	0.10	0.910	1	0.935	0.925	–	0.630	0.982	0.872	–
8-P.A	0.30	0.725	0.933	0.803	0.785	–	0.450	–	–	–
9-P.A	0.45	0.675	0.858	0.710	0.695	–	0.340	–	–	–
10-A.A	~ 1	0.600	0.574	0.447	0.438	0.725	0.300	0.327	0.149	0.362
11-CMC	2.77	0.280	0.228	0.175	0.159	0.263	0.120	0.116	0.089	0.131

ϕ_K osmotic coefficient from Katchalsky's treatment.
γ_M activity coefficient from Manning's treatment.
γ_O apparent degree of dissociation from Oosawa's treatment.

to calculated ones: $\bar{\alpha}_1$, $\bar{\alpha}_2$, (D/D_0), ϕ_K; it appears that $\bar{\alpha}_2$ and (D/D_0) reflect with a good approximation the experimental behaviour. On Figure 21 and Table III, theoretical (D/D_0) values are compared to experimental ones for different polysaccharides.

Activity coefficient of divalent counterions – The treatment is the same as above; the results are given in Figures 18–21; replacement of ε by 2ε takes care of the increase of the valence [25]. Just as for monovalent counterions for $\lambda > 1$, the value of γ_{Ca} is close to that of (D/D_0) and also $\gamma_{Na}/\gamma_{Ca} = 2$ (Figure 21). The rigidity of the chain surely plays a very important role in this case.

Variation of pK(a) – From $\psi(r)$, $\Delta\psi(a)$ is easily obtained and represents the difference between the surface potential and the reference one at a distance R [26].

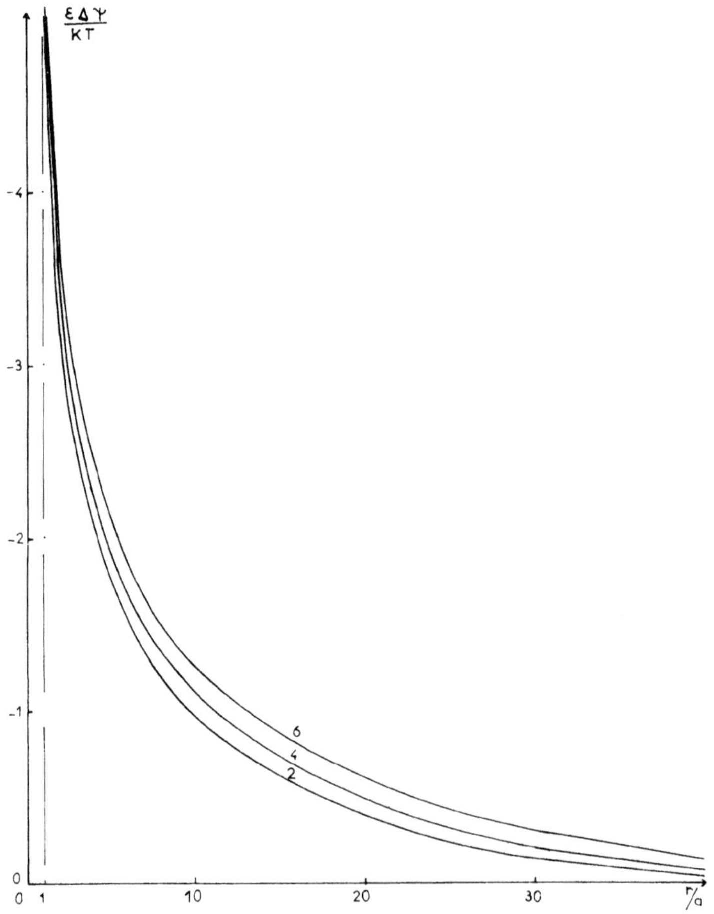

Fig. 18. Electrostatic potential $\varepsilon\Delta\psi/kT(r)$ as a function of the distance for divalent counterions [25] (CMC; $c = 2 \times 10^{-3}$ N; 2: $\overline{DS}$ 1.25; 4: $\overline{DS}$ 2.1; 6: $\overline{DS}$ 2.9).

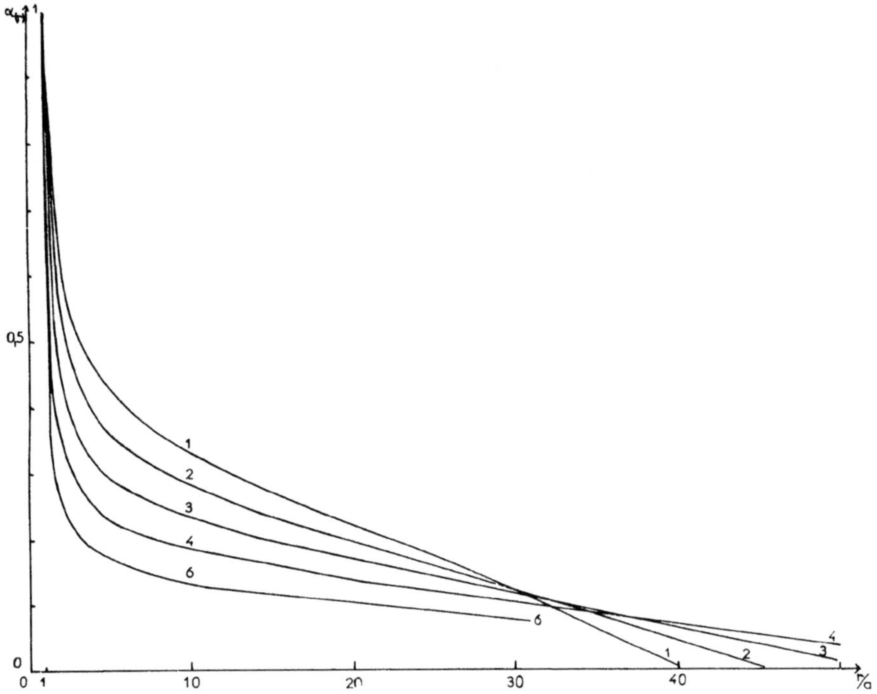

Fig. 19. Fraction of divalent counterions $\bar{\alpha}(r)$ outside a cylindrical volume of radius r [25] (CMC; $c = 2 \times 10^{-3}$ N; 1: $\overline{DS}$ 1; 2: $\overline{DS}$ 1.25; 3: $\overline{DS}$ 1.6; 4: $\overline{DS}$ 2; 5: $\overline{DS}$ 2.49; 6: $\overline{DS}$ 2.9).

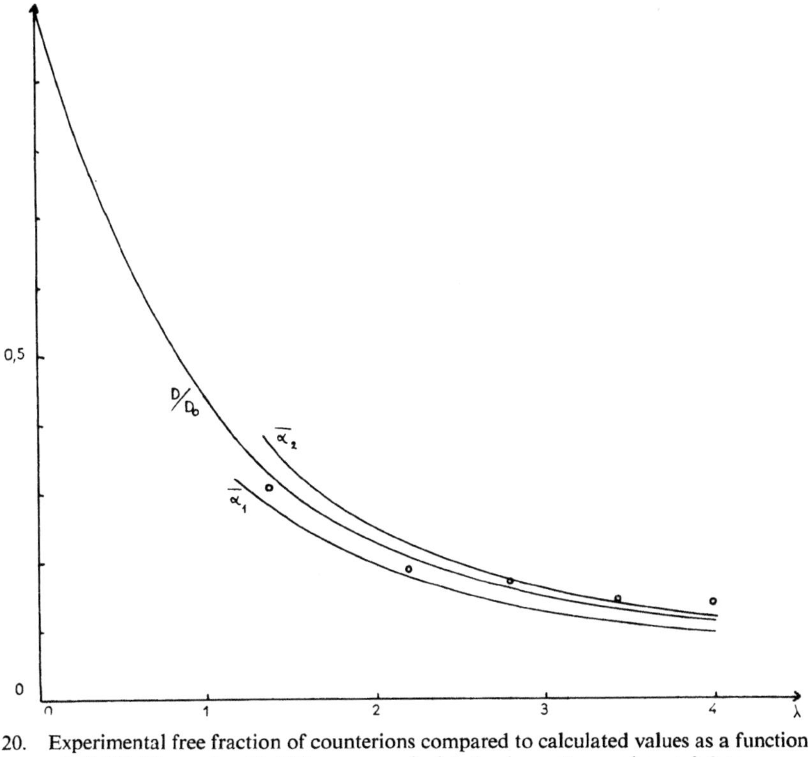

Fig. 20. Experimental free fraction of counterions compared to calculated values as a function of λ [25] (CMC; $c = 2 \times 10^{-3}$ N) ——— calculated values; ○ experimental data.

Examples of calculated variations of ΔpK (related to $\Delta\psi(a)$ by (19)) with α are given in Figure 22 for different concentrations of a CMC $\overline{DS}=1$; in Figure 23 the influence of $\overline{DS}$ at the same equivalent concentration of CMC is illustrated. pK_0 can be obtained by superimposing these curves on the experimental $pK_a(\alpha)$ one (R_4N^+ as counterion). By this means no hazardous extrapolation is needed (Figure 24) and a characteristic pK_0, independent of $\overline{DS}$ and the concentration equal to 3 ± 0.2, is found for COOH in CMC.

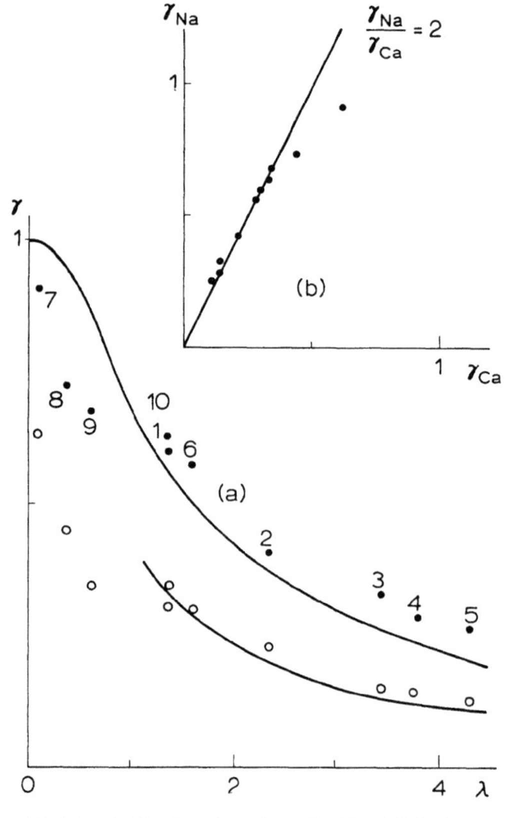

Fig. 21. (a) Experimental data obtained with polysaccharides [6] (points) and compared to calculated (D/D_0) (full lines) (characteristics of samples 1 to 11 are given in Table III). (b) Ratio γ_{Na}/γ_{Ca} obtained on the previous sample ($\gamma_{Na}/\gamma_{Ca}=2$ for $\lambda > 1$).

4.2. Oosawa's treatment

4.2.1. *Theoretical Data* [14]

The model is a two phases system separated by a potential difference $\delta\psi$; a first region (radius a) of the sub-volume V countains the bound counterions, in a second region the ions are free ($a<r<R$); the representation is given in Figure 25.

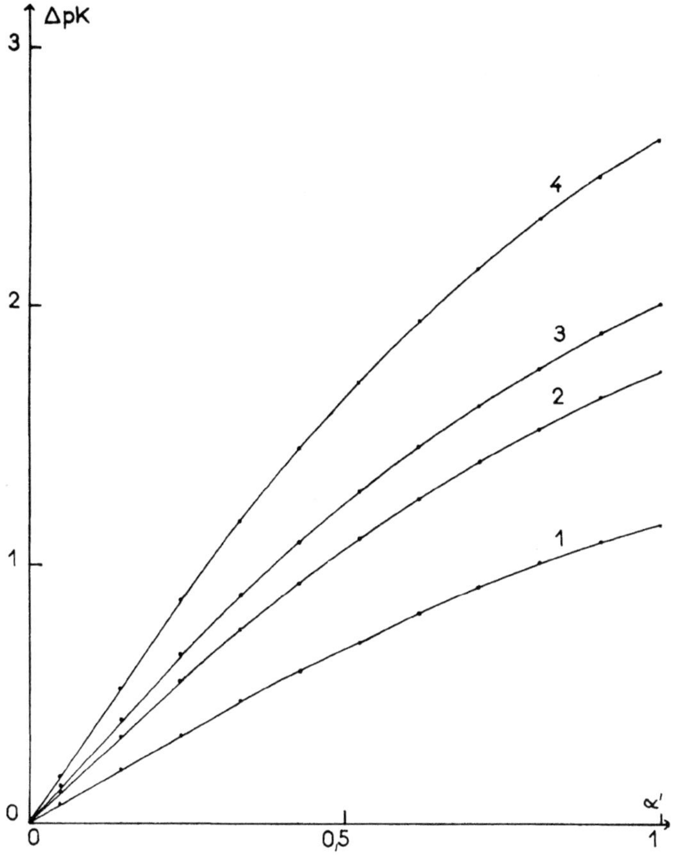

Fig. 22. Calculated values of ΔpK (α) as a function of α for various concentrations (CMC $\overline{DS}=1$; 1:9.2 × 10^{-2} N; 2:1.84 × 10^{-2} N; 3:9.2 × 10^{-3}; 4:1.84 × 10^{-3} N) [26].

(a) *Monovalent Counterions*

The activity coefficient of free fraction of counterions γ is given by

$$\ln\left(\frac{1-\gamma}{\gamma}\right) = \ln\left(\frac{\varphi}{1-\varphi}\right) - \frac{\varepsilon\delta\psi}{kT} \tag{24}$$

φ is the volume fraction occupied by the polyelectrolyte. In a cylindrical system, $\partial\psi$ is related to φ by:

$$\partial\psi = -\frac{\gamma\varepsilon}{Db}\ln(1/\varphi)$$

thus

$$\ln\left(\frac{1-\gamma}{\gamma}\right) = \ln\left(\frac{\varphi}{1-\varphi}\right) + \gamma\lambda\ln(1/\varphi). \tag{25}$$

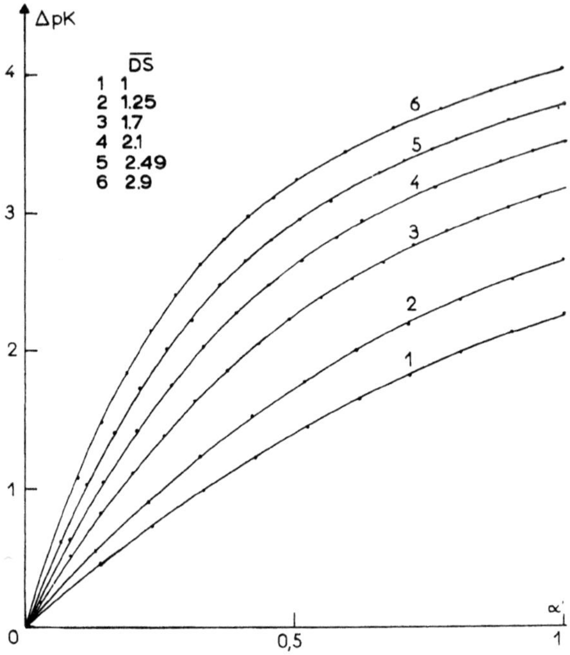

Fig. 23. Calculated values of $\Delta pK(\alpha)$ as a function of α for different $\overline{DS}$ [26] (CMC $c = 5 \times 10^{-3}$ N; $1{:}\overline{DS} = 1$; $2{:}\overline{DS} = 1.25$; $3{:}\overline{DS} = 1.7$; $4{:}\overline{DS} = 2.1$; $5{:}\overline{DS} = 2.49$; $6{:}\overline{DS} = 2.9$).

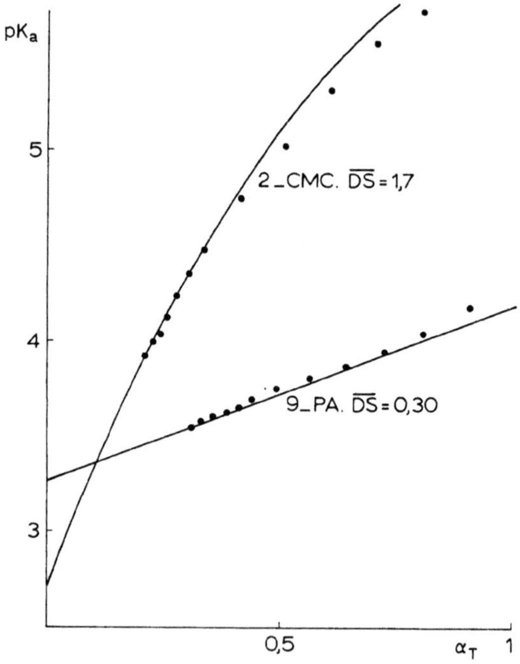

Fig. 24. Determination of pK_0 [6, 26] ——— calculated $\Delta pK(\alpha)$ function
. . . experimental values.

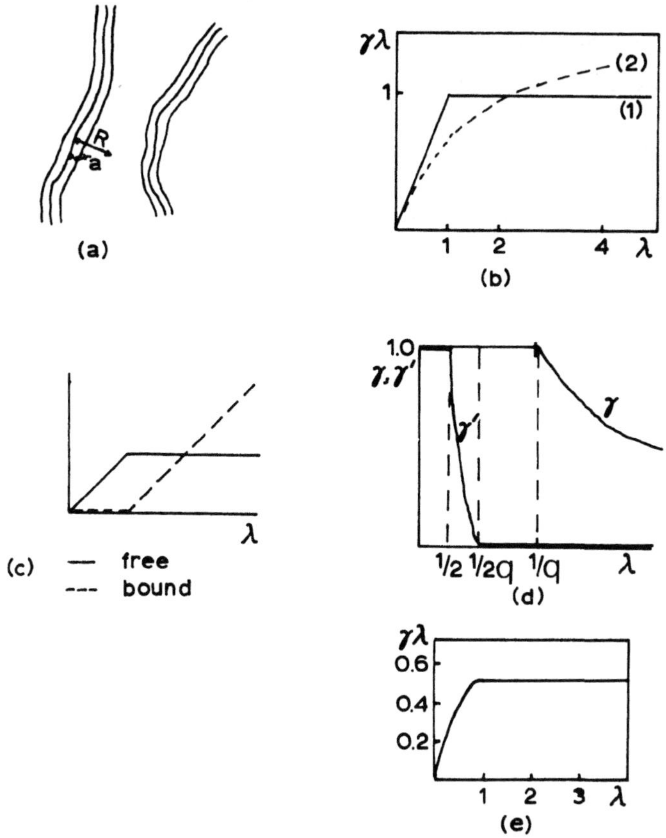

Fig. 25. Oosawa model [14]
(a) two phases model;
(b) effective values of $\gamma\lambda$ as a function of λ: (1) infinite dilution; (2) $\varphi = 5 \times 10^{-2}$;
(c) ——— free counterions ----- bound counterions;
(d) γ, γ' values as a function of λ for mixtures of monovalent and divalent counterions;
(e) effective values of $\gamma\lambda$ as a function of λ. Katchalsky data for infinite dilution.

For infinite dilution, the limit of (25) is:

$$0 \leqslant \lambda \leqslant 1 \qquad \gamma \to 1,$$
$$\lambda > 1 \qquad \gamma \to 1/\lambda.$$

Expression (25) predicts two types of variations of γ with concentration (Figure 25b); when $\lambda > 2$, γ increases with φ but for $\lambda < 2$ the opposite is true. At infinite dilution, one has to consider a limiting $\gamma\lambda = 1$ which can be compared to a 'condensation'. In Figure 25c the fraction of bound counterions is plotted; in Figure 25e the limiting behaviour following the treatment of Lifson and Katchalsky is illustrated in order to compare both conclusions.

(b) *Divalent Counterions*

The activity coefficient γ' is obtained by

$$\ln\left(\frac{1-\gamma'}{\gamma'}\right) = \ln\left(\frac{\varphi}{1-\varphi}\right) + 2\lambda\gamma' \ln(1/\varphi) \tag{26}$$

when $\lambda > 1/2$, γ' decreasing with the concentration tends towards $1/2\lambda$ in opposition with the case of $\lambda < 1/2$ where $\gamma'_{\text{lim}} \to 1$ at infinite dilution.

(c) *Mixture of Monovalent and Divalent Counterions*

With both types of counterions mixed in the ratio q to q' ($q + q' = 1$), the γ and γ' values are given by a set of two expressions:

$$\ln\left(\frac{1-\gamma}{\gamma}\right) = \ln\left(\frac{\varphi}{1-\varphi}\right) + (\gamma q + \gamma' q')\,\lambda \ln(1/\varphi), \tag{27}$$

$$\ln\left(\frac{1-\gamma'}{\gamma'}\right) = \ln\left(\frac{\varphi}{1-\varphi}\right) + (\gamma q + \gamma' q')\,2\lambda \ln(1/\varphi). \quad (28$$

The values at infinite dilution are diagrammatically represented in Figure 25d;

$$0 \leqslant \lambda \leqslant \tfrac{1}{2} \qquad \gamma, \gamma' \to 1$$

$$\tfrac{1}{2} \leqslant \lambda \leqslant \frac{1}{2q} \qquad \gamma \to 1;\ \gamma' \to \left(\frac{1}{2q'\lambda} - \frac{q}{q'}\right)$$

$$\frac{1}{2q} \leqslant \lambda \leqslant 1/q \qquad \gamma \to 1;\ \gamma' \to 0$$

$$1/q \leqslant \lambda \qquad \gamma \to 1/q\lambda;\ \gamma' \to 0$$

4.2.2. *Application of the Oosawa's Treatment* [27]

Applying this treatment to CMC we have already shown that the value of 'a', the only parameter which is not defined, has only limited influence on γ. Thus for example with monovalent ions and CMC $\overline{DS}$ 2.93 at $c = 5 \times 10^{-3}$ N:

$a = 6$ Å	$a = 48$ Å	Experimental value	Infinite dilution
$\gamma = 0.283$	$\gamma = 0.310$	$\gamma = 0.275$	$\gamma = 0.251$

(a) *Activity Coefficient of Monovalent Counterions*

The points of Figure 26 were calculated from Equation (25) using $a = 12$ Å and assimilating the potentiometric results to the free fraction γ. Oosawa's treatment best fits the experimental data for $\lambda > 1$. The ion-selectivity [20] (Figure 25) or the ultrasonic absorption [21] increases with the number of counterions whilst the effective ionisation approaches the $\gamma\lambda \simeq 1$ limit.

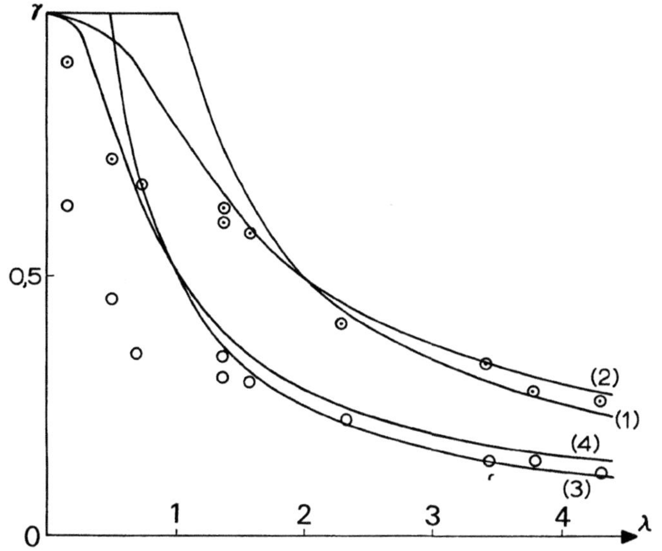

Fig. 26. Comparison between experimental and calculated values of γ in a function of λ using the Oosawa model [27]

1 – monovalent counterions; $\overline{DS}$ 2.93; $c = 5 \times 10^{-3}$ N; $a = 12$ Å (function of α);
2 – monovalent counter-ions infinite dilution;
3 – divalent counter-ions; $\overline{DS}$ 2.49; $c = 2 \times 10^{-3}$ N; $a = 12$ Å (function of α);
4 – divalent counter-ions infinite dilution

○ divalent counter-ions } experimental data on polysaccharides [6].
○ monovalent counter-ions }

(b) *Activity Coefficient of Divalent Counterions*

The data obtained from (26) plotted in Figure 26 show good agreement for polysaccharides, but never for polyvinyl derivatives.

(c) *Mixture of* Na^+ *and* Ca^+ *Counterions*

The polyacid is exactly neutralized as described on p. 163. γ_{Na} and γ_{Ca} are both measured and also calculated for each X_{Na} and φ using the sets of relations (27) and (28).

The experimental and theoretical values are given in Table IV and in Figure 27. The agreement is better than ever before and namely better than that obtained by the resolution of the Poisson-Boltzmann Equation [23, 19].

4.3. The Manning Theory [15]

This treatment is proposed for dilute solutions. The phase integral diverges as soon as $\lambda > 1$ and implies an effective charge density limit $\lambda_{eff} = 1$ with monovalent counterions.

The characteristic $\lambda = 1$ is a particular value in different treatments (Oosawa, Lifson-Katchalsky at infinite dilution) and is explained here by a new approach.

TABLE IV

Activity coefficients for Na^+ and Ca^{2+} in salt free solution with various stoechiometric ratios [27] (Oosawa's treatment)

X_{Na}		0	0.1	0.2	0.3	0.4	0.5	0.6	0.7	0.8	0.9	1
CMC	γ_{Na}th	–	0.931	0.923	0.911	0.894	0.869	0.832	0.787	0.740	0.692	0.651
	γ_{Na}exp	–	1	0.980	0.900	0.870	0.852	0.840	0.800	0.730	0.680	0.630
$\overline{DS}=1$ $\lambda=1.38$	γ_{Ca}th	0.390	0.343	0.290	0.231	0.169	0.110	0.065	0.037	0.011	0.0025	–
	γ_{Ca}exp.	0.320	0.310	0.280	0.237	0.174	0.126	0.075	0.033	0	0	
CMC	γ_{Na}th	–	0.907	0.854	0.741	0.622	0.530	0.461	0.408	0.366	0.332	0.325
	γ_{Na}exp	–	0.850	0.792	0.696	0.613	0.564	0.508	0.453	0.414	0.363	0.330
$\overline{DS}=1.7$ $\lambda=2.35$	γ_{Ca}th	0.178	0.098	0.037	0.009	0.003	0.0014	0.00083	0.00045	0.00034	0.00012	–
	γ_{Ca}exp	0.145	0.091	0.050	0.015	0	0	0	0	0	0	–
CMC	γ_{Na}th	–	0.922	0.901	0.863	0.797	0.716	0.642	0.577	0.523	0.478	0.440
	γ_{Na}exp	–	0.930	–	0.790	0.787	0.740	0.668	0.607	0.556	0.500	0.455
$\overline{DS}=2.49$ $\lambda=3.44$	γ_{Ca}th	0.275	0.187	0.122	0.062	0.024	0.01	0.0016	0.0013	0.0003	0.00005	–
	γ_{Ca}exp	0.200	0.175	0.128	0.097	0.058	0.040	0	0	0	0	–

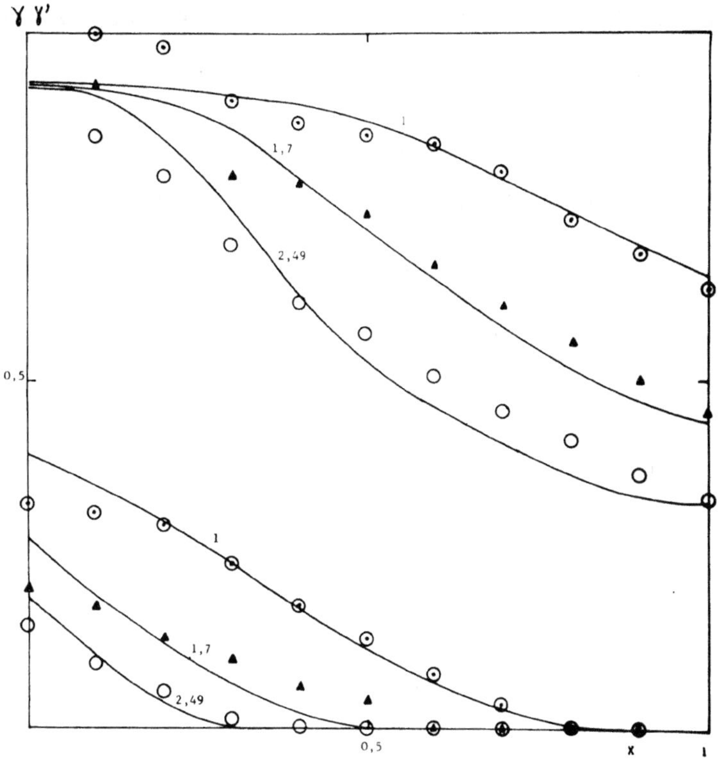

Fig. 27. Comparison between experimental data (γ, γ') and theoretical ones (Oosawa's treatment); CMC $c = 2 \times 10^{-3}$ N ○ $\overline{DS}$ 1 ▲ $\overline{DS}$ 1.7 ○ $\overline{DS}$ 2.49.

The consequence is:

– For $\lambda < 1$, there is no condensation and the total field interacts with counterions.

– For $\lambda > 1$, there is condensation for a fraction $(1 - 1/\lambda)$ of counterions to the limit $\lambda_{eff} = 1$; the fraction $1/\lambda$ is then in the electrostatic residual potential.

In both cases, the Debye-Huckel approximation can be used.

At infinite dilution, the predictions are:

$$\lambda < 1 \qquad \phi = 1 - \lambda/2$$
$$\ln \gamma = -\lambda/2$$

thus

$$\phi = 1 + \ln \gamma .$$
$$\lambda > 1 \qquad \phi = 1/2\lambda$$
$$\gamma \cong 1.21\ \phi$$
$$D/D_0 = 0.87 \cdot \lambda^{-1}$$

D/D_0 is the coefficient of free diffusion.

Conclusions. The essential difference with the Lifson-Katchalsky treatment is that

$$\gamma \neq \phi$$

and

$$\phi < \gamma < D/D_0 \text{ is predicted.}$$

Manning's values for ϕ are equal to those of the Lifson-Katchalsky theory at infinite dilution.

This treatment implies a 'condensation' of counterions for a critical charge density.

Application of the Treatment

(a) *Activity Coefficient of Monovalent Counterions*

On Figure 28, the curves for γ, ϕ are given; the γ_{Na} from Table I are plotted. Two regions appear:

$\lambda < 1$, the concordance is correct and this treatment seems to be the only convenient one in this region.

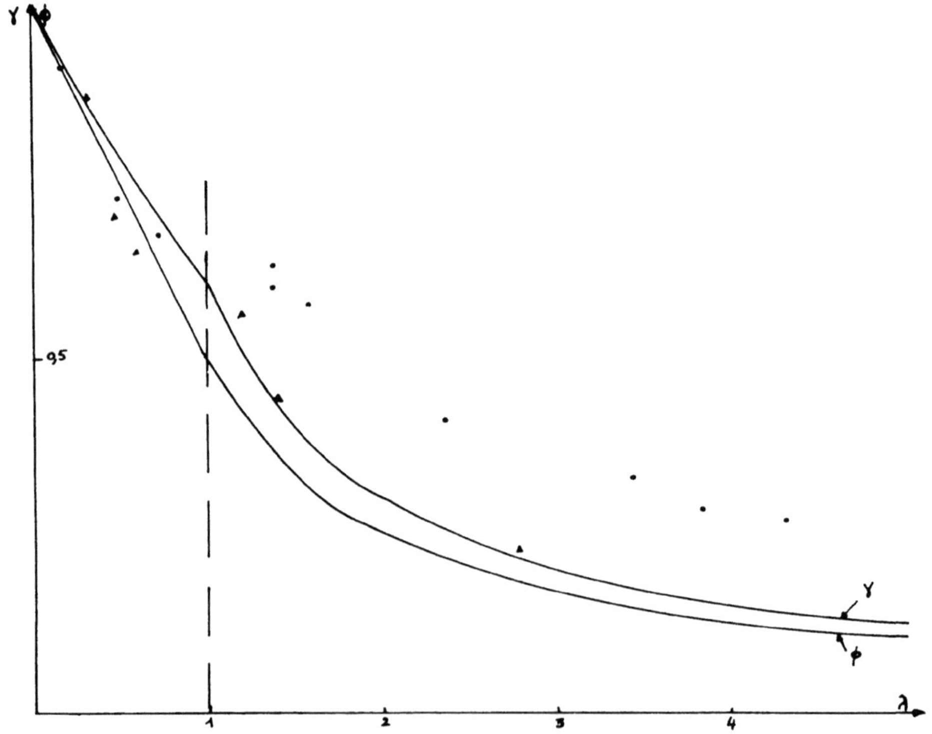

Fig. 28. Manning's treatment [15]
——— calculated values of γ, ϕ as a function of λ;
●●● polysaccharides
▲▲▲ polysulphonic derivatives } experimental data (Table I).

$\lambda > 1$, the experimental values obtained with rigid polysaccharides are higher than the theoretical ones.

The limit level of effective ionisation is found to be $\lambda \cong 1$ for polysaccharides and in favour of the Manning hypothesis. The ionic selectivity obtained by potentiometry or ultrasonic absorption begins for $\alpha \cdot \overline{DS} \cong 0.6 (\lambda_{\text{struct}} \simeq 0.85)$ for CMC and for $\alpha \mp 0.35 (\lambda_{\text{struct}} \simeq 1)$ for vinylic.

(b) *Coefficient of Free Diffusion*

Free diffusion of Cs^+ was measured [22] and calculated; the values are given just under and compared with the free fraction of counterions obtained by conductimetry (as proposed by Manning [16])

$\overline{DS}$ *CMC*	λ	γ_{Na}	$(D/D_0)_{Cs^+}$ [a]	$(D/D_0)_M = 0.87 \cdot \lambda^{-1}$	(D/D_0) [22]	ϕ_X [+b]
1.25	1.698	0.520	0.665	0.512	0.572	0.615
1.70	2.309	0.410	0.650	0.377	0.444	0.530
2.40	3.260	0.290	0.523	0.267	0.321	0.440
2.90	3.980	0.275	0.565	0.218	0.266	0.395

[a] $C = 2 \times 10^{-2}$ N.

[b] $\phi_X = \dfrac{A}{(\lambda_p + \lambda_c)}$.

The agreement is poor; experimentally $(D/D_0) > \gamma$ and the theory predicts the opposite.

5. Conclusions

In this paper, we have tried to apply the theories proposed by Lifson and Katchalsky, by Manning and by Oosawa and compare the calculated data with the results of potentiometric measurements on polyanions:

- rigid polysaccharides which constitute good models to test these theories and also
- polyvinylalcohol sulphonic derivatives.

A set of experimental results was discussed first in terms of charge density; the essential laws of polyelectrolytes are shown. The systematic theoretical investigation permits to discuss the validity of the different theories. From the resolution of the Poisson-Boltzmann equation, we have shown that only (D/D_0) or $\bar{\alpha}_2$ is a good approximation for the free fraction of counterions measured by potentiometry.

By opposition, Oosawa's treatment gives very good agreement between experimental and calculated values when $\lambda > 1$.

In Table III, the values are given together for polysaccharides and to conclude Figure 29 compares the theoretical values for infinite dilution. From a general point of view, it seems that $\lambda = 1$ is a critical value in the different treatments; for $\lambda < 1$, the assumption of a regular charge density is perhaps not at all convenient.

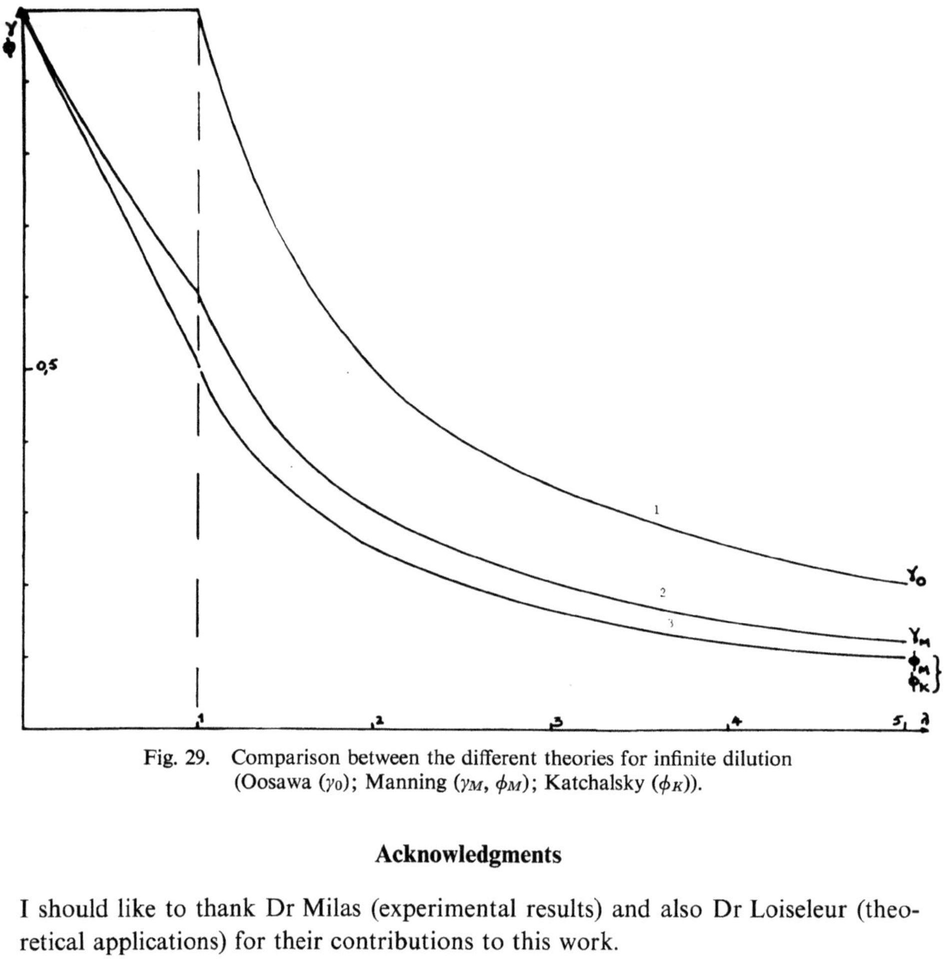

Fig. 29. Comparison between the different theories for infinite dilution (Oosawa (γ_0); Manning (γ_M, ϕ_M); Katchalsky (ϕ_K)).

Acknowledgments

I should like to thank Dr Milas (experimental results) and also Dr Loiseleur (theoretical applications) for their contributions to this work.

References

1. Lifson, S. and Katchalsky, A.: *J. Polymer Sci.* **13**, 43 (1954).
2. Oman, S. and Dolar, D.: *Z. Physik. Chem. Neue Folge* **56**, 1 (1967).
3. Rinaudo, M. and Hudry-Clergeon, G.: *J. Chim. Phys.* **64**, 1753 (1967).
4. Dolle, F., Le Moigne, J., and Gramain, P.: *Eur. Polymer J.* **6**, 1227 (1970).
5. Katchalsky, A.: *International Symposium on Macromolecules*, Leiden, Sept. 1970.
6. Rinaudo, M. and Milas, M.: *International Symposium on Macromolecules*, Helsinki, 1972, vol. **2**, p. 115.
7. Domard, A.: Thèse Spécialité Grenoble, 1971; Domard, A. and Rinaudo, M.: *Biopolymers* 2211 (1973).
8. Rinaudo, M., Milas, M., Le Moigne, J., and Gramain, P.: *Eur. Polymer J.* **8**, 1073 (1972).
9. Rinaudo, M. and Milas, M.: *Eur. Polymer J.* **8**, 737 (1972).

10a. Milas, M.: Thèse Spécialité Grenoble, 1969.
b. Rinaudo, M. and Milas, M.: *Compt. Rend. Acad. Sci.* **271**, 1170 (1970).

11. Katchalsky, A., Alexandrowicz, Z., and Kedem, O.: in B. E. Conway and R.G. Barradas (eds.), *Chemical Physics of Ionic Solutions*, Wiley, New York, 1966.

12. Fuoss, R. M., Katchalsky, A., and Lifson, S.: *Proc. Nat. Acad. Sci. U.S.* **37**, 579 (1951).
13. Rinaudo, M. and Loiseleur, B.: *J. Chim. Phys.* **68**, 882 (1971).
14. Oosawa, F.: *Polyelectrolytes*, Dekker, 1971, Chap. 2, 3, 4.
15. Manning, G. S.: *J. Chim. Phys.* **51**, 924 (1969).
16. Manning, G. S.: *Biopolymers*, **9**, 1543 (1970).
17. Lifson, S. and Jackson, J. L.: *J. Chem. Phys.* **36**, 2410 (1962).
18. Jackson, J. L. and Coriell, S. R.: *J. Chem. Phys.* **38**, 959 (1963).
19. Rinaudo, M. and Loiseleur, B.: *Bull. Soc. Chim.* **124**, 1 (1973).
20. Rinaudo, M. and Milas, M.: *Compt. Rend. Acad. Sci.* **269**, 1190 (1969).
21. Zana, R., Tondre, C., Rinaudo, M., and Milas, M.: *J. Chim. Phys.* **68**, 1258 (1971).
22. Rinaudo, M., Loiseleur, B., Milas, M., and Varoqui, R.: *Compt. Rend. Acad. Sci.* **272**, 1003 (1971).
23. Dolar, D. and Peterlin, A.: *J. Chem. Phys.* **50**, 3011 (1969).
24. Marcus, R. A.: J. *Chem. Phys.* **23**, 1057 (1957).
25. Rinaudo, M., Loiseleur, B., and Milas, M.: *Compt. Rend. Acad. Sci.* **273**, 1235 (1971).
26. Rinaudo, M., Loiseleur, B., and Milas,M.: *Compt. Rend. Acad. Sci.* **273**, 1148 (1971).
27. Rinaudo, M. and Loiseleur, B.: *J. Chim. Phys.* **69**, 1606 (1972).

SOME TYPICAL PROPERTIES OF A WEAKLY BASIC POLYELECTROLYTE WITH NON HYDROPHYLIC CHAINS

G. MULLER

Laboratoire de Chimie macromoléculaire. U.E.R. Sciences exactes et naturelles, E.R.A. 471, Université de Rouen, 76130 Mont Saint Aignan, France

1. Introduction

The physicochemical behaviour of polyelectrolyte solutions is strongly dependent on electrostatic interactions between the ionized groups located along the chains. Considering that a polyelectrolyte molecule consists of a hydrocarbon backbone, its solubility in water is due to the presence of ionizable groups, their hydration, electrostatic repulsion and to the contribution of counterions to the entropy of mixing. In fact many properties are well explained by the variation of the electrostatic potential energy. However, many experimental observations show that the conformation of polyelectrolytic chains is also dependent on others factors e.g. the nature of side groups, the structure of water surrounding the polyions, etc....

Besides those purely electrostatic in nature, other kinds of interactions may play a role in controlling chain conformation. Many investigators have reported the pH-induced transition of the molecular configuration of PMA in relation to the intramolecular hydrophobic or Van der Waals interactions due to the methyl groups at low degrees of neutralization [1]. The term 'hydrophobic interactions' describes the tendency of non polar groups to associate leading them to a decreased contact with neighbouring molecules of solvent [2]. The stabilization of the conformation of proteins and polysoaps [3] has been attributed to the existence of such interactions.

These examples show that in certain cases (and under certain conditions) non electrostatic interactions can become a predominant factor controlling the state in solution of polyelectrolyte molecules. A very weak polyelectrolyte with a non hydrophylic chain can allow the illustration of such effects. A great deal of work has been carried out on polyanions but only a relatively limited number of studies is concerned with the properties of polycations. In this paper, we would like to present some properties concerning the polyelectrolytic behavior of aqueous solutions of (non quaternized) isomolecular atactic or isotactic poly-2-vinylpyridine (PVP), which have already been the subject of some of our previous investigations [4]. The experiments were performed on solutions without added salt with a few monovalent or bivalent counter-ions; the effect of ionization and concentration on viscosimetric and dielectric properties of PVP was investigated.

2. Ionization of poly-2-vinylpyridine

When neutralized by an acid, PVP insoluble when unionized is protonated according

Eric Sélégny (ed.), Polyelectrolytes, 195–205.

to (1)

$$PVP + H_3O^+ \rightleftarrows PVP \cdot H^+ + H_2O \tag{1}$$

and behaves as a cationic polyelectrolyte.

However, due to the low basicity of pyridinic groups, this equilibrium is shifted to the left (hydrolysis). Therefore the degree of protonation (α: ratio of protonated sites, dissociated or not, to polyelectrolyte concentration C_p) is lower than the theoretical degree of neutralization ($\bar{\alpha}$ = added H^+/C_p). A part of the added acid which has not reacted with the polybase always remains in solution; the presence of this free acid (the concentration of which is $C_{free} = (\bar{\alpha} - \alpha)\, C_p$) acts as an added salt and explains most, if not all, of the properties of PVP. Due to the low basicity of pyridinic groups, it was difficult to obtain the degree of ionization from the usual potentiometric titration. The dependence of the degree of protonation on the degree of neutralization has been determined for various acids from spectroscopic and conductivity measurements, as already stated [4].

The absorption coefficients ε were determined at 2620 Å and the degree of protonation was calculated from the relation (2)

$$\alpha = \frac{\varepsilon - \varepsilon_n}{\varepsilon_i - \varepsilon_n} \tag{2}$$

in which ε_i and ε_n are respectively the absorption coefficients for the fully ionized form ($\varepsilon_i = 5600$, in 2N hydrochloric acid) and for the neutral form ($\varepsilon_n = 3000$, in water by extrapolation to C_p and $\bar{\alpha} = 0$). The conductance due to the free acid was calculated by assuming that the polyion and its counter-ions did not affect the properties of the excess acid. Owing to experimental accuracy and the approximations, the best agreement between the two techniques is expected and found for high polymer concentration and high degree of neutralization. More recently, a value of the degree of protonation of PVP has also been determined from dialysis equilibrium between a solution of the polyelectrolyte and a solution of the monovalent electrolyte (the latter permits the determination or the control of the activity of the acid) [5]. The values of α obtained from dialysis equilibrium are always higher than those calculated from conductivity measurements with best agreement when the degree of neutralization is above 60% (this difference is not unexpected owing to the conductimetric approximations). The comparison between the three methods is reported in Figure 1.

In the concentration range of our measurements, the degree of protonation determined from conductivity is practically independent of the nature of the counter-ion with a very slight tendency however of the polybase to be somewhat more protonated in presence of nitrate, at the same degree of neutralization; this tendency is confirmed by the results of Donnan equilibria which show that the difference although small is consistent (Figure 2).

The dissociation process of PVP, treated as a polyacid, is PVP. $H^+ \rightarrow PVP + H^+$. The apparent ionization constant K_a is defined in the usual manner by $(H^+) \times$ $\times (PVP/PVPH^+)^n = K_a$. K_a varies with α since the interaction of the polyion with the

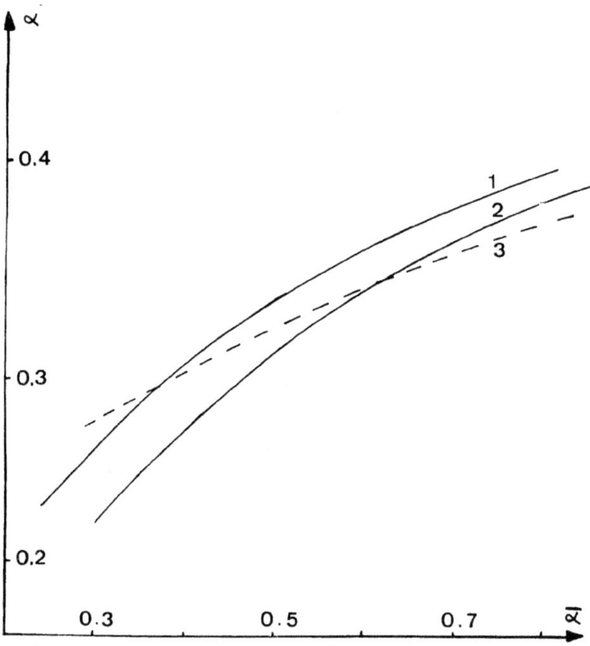

Fig. 1. Dependence of the degree of protonation (α) on the degree of neutralization (ᾱ). (1) from dialysis equilibrium; (2) from conductivity; (3) from UV absorption.

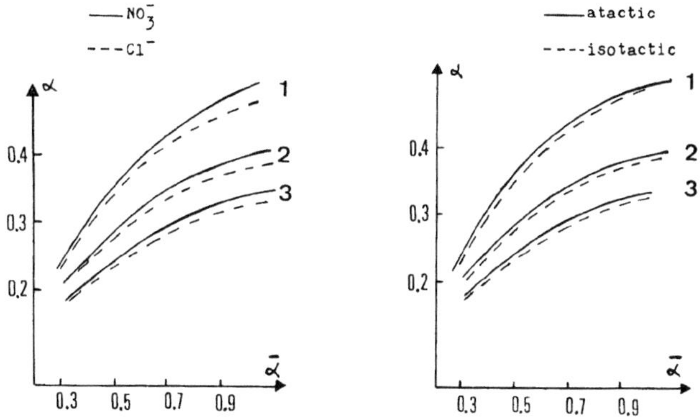

Fig. 2. Dependence of the degree of protonation (α) on the λ the nature of λ counter-ion and on the tacticity. Polymer concentration: $C_p = 1.5 \times 10^{-2}$ (1); 5×10^{-3} (2) and 2.5×10^{-3} (3).

protons is a function of α. Expressing this interaction by means of the mean electric potential we obtain $pK_a = pK_0 + 0.43e\bar{\psi}(\alpha)/kT$. $\bar{\psi}$ is easily determined by measuring pK_a at various values of α and using the condition that for $\alpha = 0$, $\bar{\psi} = 0$. Some empirical formulae have been proposed to express the dependence of $\bar{\psi}$ on α. The calculated parameters of the Henderson-Hasselbach equation are given in Table I.

TABLE I

Adjustable parameters of the Henderson-Hasselbach equation for PVP.

PVP	Counter-ion	pK_a	n
atactic	chloride	2.3_0	2.0_9
	nitrate	2.1_5	2.1_2
	bromide	2.2_1	2.2_0
isotactic	bromide	2.3_6	2.0_0

The results reported here illustrate clearly the weakness of PVP which is only partially ionised in aqueous solution (hydrolysis at high dilution). The protonation is only slightly dependent on the nature of counterions and on the nature of the chain (the isotactic chains are less protonated but the difference is very small). Only an excess of electrolyte (salt or acid) can, by reducing the charge repulsions, increase the apparent basicity of the last groups sufficiently and allow a higher degree of protonation.

3. Hydrodynamic Behaviour of Poly-2-Vinylpyridine

The effect of the degree of neutralization and of the concentration on the viscosity of aqueous solutions of PVP was investigated. Dilution was carried out by addition of solvent and not by a dialysis technique; therefore the chemical potential of all diffusible components is not constant.

The following features were observed:

– from a partial degree of neutralization (30 to 40%) depending on the exact nature of the counter-ion, the reduced viscosity decreases until the ionizable groups are completely neutralized.

– for each degree of neutralization, the reduced viscosity of PVP.Cl solutions is always lower than that of PVP.NO_3 (or PVP.Br) solutions ($\eta_{NO3}/\eta_{Cl^-}=1.3$ ($\bar{\alpha}=0.3$) or 1.4 ($\bar{\alpha}=0.44$);

– the concentration dependence of the reduced viscosity of PVP shows a maximum at low concentration. The position of this maximum depends on the degree of neutralization and on the nature of the counter-ion. The observed maximum is shifted to lower concentration when $\bar{\alpha}$ is increased and with NO_3 as a counter-ion (Figure 3).

The decrease in viscosity with increasing neutralization (from $\bar{\alpha}=0.35$) can be readily explained by the weakness of the polybase. The effect of increasing 'charge screening' (leading to a contraction of polymeric coils) is more important than the increase in the charge density (leading to an extension); some of the unreacted acid produces a salt effect and causes a reduction of the repulsion forces between the segments. In presence of bromide and nitrate the reduced viscosities are different from those in the presence of chloride; this can be due to changes in the binding of counter-ions. Counter-ion binding for the halide ions should decrease in the order $I^- > NO_3^- = Br^- > Cl^- > F^-$ (sequence observed for the selectivity of strongly basic anion ex-

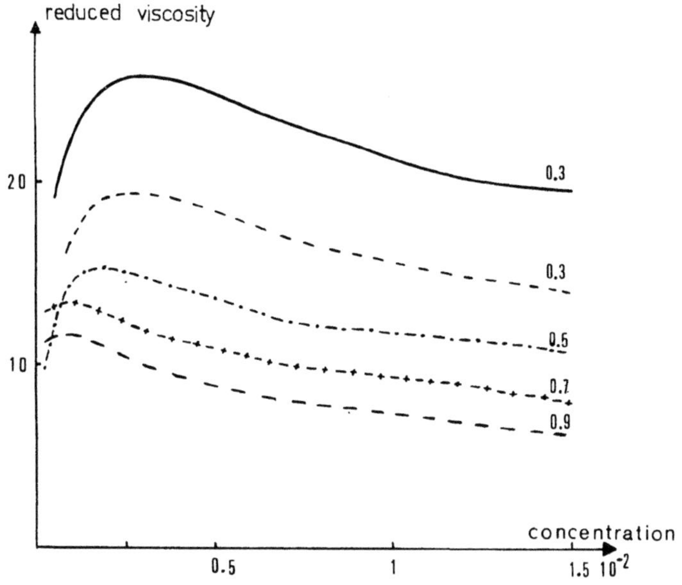

Fig. 3. Reduced viscosity plotted against concentration for aqueous solutions of PVP neutralized with hydrochloric acid (broken line). The upper curve refers to PVP in presence of counterion nitrate (or bromide) for a degree of neutralization 0.3.

change resins for the same anions) [6]. Therefore the viscosity should decrease from chloride to nitrate unless chloride ions interacte with specific sites. The selectivity coefficient of the polyelectrolyte towards chloride and nitrate has been determined from Donnan equilibrium measurements at $\bar{\alpha}=0.50$ [5]. The experimental value found for the selectivity constant $K_d=(\overline{NO_3^-})\,(Cl^-)/(NO_3^-)\,(\overline{Cl^-})$ (where $\overline{NO_3^-}$, $\overline{Cl^-}$, NO_3^- and Cl^- are respectively the concentrations of ions in the polyelectrolyte phase and in the electrolyte phase) shows that the polyion has a stronger affinity for NO_3^-.

The observed difference in viscosity between the two kinds of counter-ions can probably be attributed to the slight difference in ionization. The 'salt effect' (arising from the unreacted acid) is somewhat smaller in the presence of nitrate; as the charge is always low it can be sufficient to explain the reported behaviour.

The existence of a discontinuity in the concentration dependence of the reduced viscosity has already been reported for many polyelectrolytes of different natures which indicates that the Fuoss relation is no longer valid at high dilution. Maxima have been reported for both weak and strong polyelectrolytes and many explanations have been proposed e.g. absorption of carbon dioxyde [7], hydrolysis of ionized groups [8], purity of the solvent [9], effect of rate of shear [10]. If the presence of carbon dioxyde can lead to a decrease of viscosity in the case of a weak polyacid, it can be reasonably excluded for the precise case of PVP (pH$\sim$4). Owing to the relatively low molecular weight of PVP ($M_w\sim 8\times 10^4$), the effect of rate of shear can also be neglected. Moreover it has been reported that although greatly influenced by the purity of water, the maximum does not disappear after purification of the solvent [9]. In the case of PVP this behaviour can be explained by the combined effects of two factors: firstly the weak electrolyte character of the ionizable groups and secondly, to a lesser extent, the nature of the chain. The hydrolysis occurring when the dilution increases leads to a decreasing charge; when the latter is sufficiently low the polymer-polymer interactions become of significant magnitude. The chains of partially ionized PVP are constituted of 'hydrophylic portions' (carrying a charge, with affinity for the solvent) and 'hydrophobic portions' (uncharged, with no affinity for the solvent). The relative proportion of each of them depends on a number of parameters. At high dilution and decreasing ionization, the relative proportion of hydrophobic portions increases: the mutual Van der Waals attractions between them and the insolubility of these chain segments can explain the decrease in viscosity (the hydrophobic portions tend to cluster together avoiding the contact with neighbouring molecules of solvent; in this sense we can consider that these interactions are hydrophobic). On the other hand the remaining hydrophylic groups carrying a residual charge exert an attraction upon the molecules of the solvent and the resultant hydration contributes to the maintenance in solution. At relatively higher concentrations and degrees of ionization, the electrostatic interactions overbalance all other kinds of interactions. It is worth mentioning that the magnitude of the viscosity is largely influenced by the stereoregularity of chains. The much lower viscosity observed for isotactic PVP, over the whole range of ionization, shows a limited extension for such a structure. The small

TABLE II

Polymer	Solvent	$K' \times 10^2$	a
Atactic	0.1N HCl	2.37	0.75
–	methanol	1.13	0.67
Isotactic	0.1N HCl	3.89	0.67
–	methanol	2.52	0.64

difference in ionization between the two forms is not sufficient to explain adequately such a behaviour only in terms of electrostatic interactions. One may anticipate that the contribution of non electrostatic forces to the conformation of chains is of greater importance in the case of the isotactic configuration favouring the intrapolymer interactions acting in an opposite direction to the repulsions of charges. In the presence of an excess of acid, PVP behaves like a neutral polymer (the reduced viscosity is a linear function of the concentration and the values of the Huggins constant in IN hydrochloric acid is of the same order of magnitude as that for the uncharged polymer). The values of parameter a of the Mark-Houwink equation confirm the limited extension of isotactic PVP even at high concentration in acid (Table II).

4. Dielectric Behaviour of Poly-2-Vinylpyridine

The electric permittivity of solutions of macromolecules is an interesting method for the study of molecular properties and has been often used in the field of polyelectrolytes to obtain information about the molecular dimensions and association phenomena.

Polyelectrolytes generally have large dielectric increments which are due to the behaviour of counter-ions associated to a polyion; the associated counter-ions have restricted movements and give rise to a large polarizability. Therefore it seemed interesting to study the effect of ionization and concentration on the dielectric properties of PVP.

The dielectric properties of aqueous solutions of PVP were studied over the frequency range 2.5 KHz–100 MHz using bridge techniques [11, 12]. The experimental values of the real part of the frequency-dependent complex electric permittivity were calculated by a least-squares fitting procedure according to the analytical expression proposed by Cole [13] with two separated dispersion regions. Each dispersion region is characterized by a mean relaxation time τ_1 and τ_2 and by a dielectric increment $\Delta\varepsilon_1 = \varepsilon_s - \varepsilon_2$ and $\Delta\varepsilon_2 = \varepsilon_2 - \varepsilon_\infty$ (ε_s and ε_∞ were obtained by extrapolating the experimental curves to the corresponding frequency regions). All results could be fitted by a curve with a Cole-Cole parameter $\beta = 0.8$.

The dielectric dispersion profile of aqueous solutions of PVP consists of two separated dispersion regions, in the kilocycle/second range and in the megacycle/second range respectively. Figure 4 shows the dispersion pattern of PVP partially neutralized

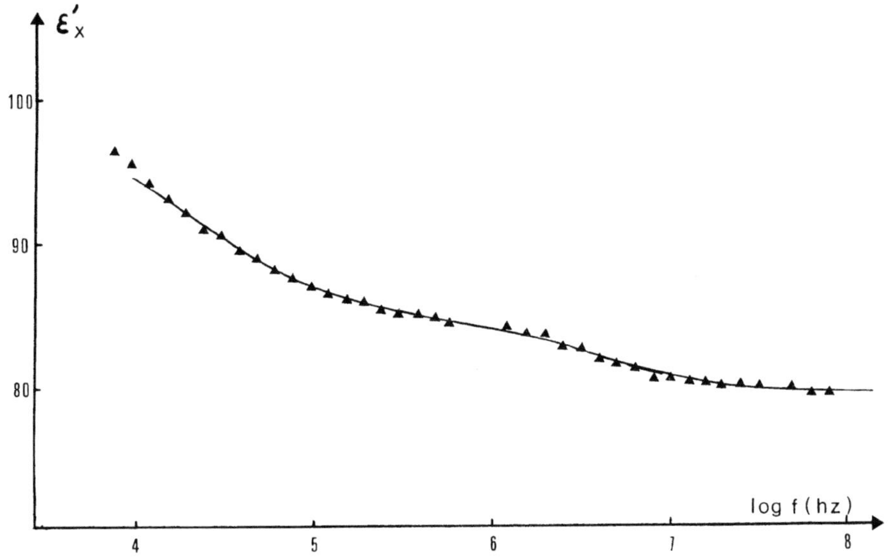

Fig. 4. Electric permittivity as function of the frequency f. PVP-NO_3^- (DP:2850), $C_p = 2.5 \times 10^{-3}$ eq. l^{-1}, degree of ionization 0.30 (relaxation time distribution parameter = 0.80).

by HNO_3 ($C_p = 2.5 \times 10^{-3}$ eq.l^{-1}, atactic form, degree of ionization $\bar{\alpha} = 0.30$). For all solutions investigated the dielectric increment $\Delta\varepsilon_1$ is larger than $\Delta\varepsilon_2$.

Owing to the nature of pyridinic groups the presence of free acid which has not reacted limits the measurements as the method is limited with respect to the conductivity range.

We have investigated the dependence of dielectric parameters on the degree of ionization for monovalent and bivalent ions. The experimental results are reported in Table III.

TABLE III

Effect of ionization on dielectric parameters of atactic PVP

Counter-ion	C_p(eq. l^{-1})	$\bar{\alpha}$	$\Delta\varepsilon_s$	$\Delta\varepsilon_2$	$\tau_1 \times 10^6$ s	$\tau_2 \times 10^7$ s
Cl^-	10^{-3}	0.3	13.3	4.0	8.0	1.1
		0.4	10.8	4.3	4.0	0.7
		0.5	10.0	3.4	1.3	0.4
NO_3^-	10^{-3}	0.3	12.1	3.8	8.9	1.3
		0.4	13.0	4.6	5.6	1.3
SO_4^{--}	10^{-3}	0.3	18.7	7.0	7.9	2.6
		0.4	14.8	4.8	2.8	1.7
		0.5	13.4	3.4	1.6	0.4

The effect of the concentration on the dielectric properties of PVP was also investigated (at a value of the degree of ionization $\bar{\alpha} = 0.3$). The measurements were carried out with two samples of PVP of different stereoregularity. The results are summarized in Tables IV and V.

TABLE IV

Effect of concentration on dielectric parameters of PVP

PVP atactic	DP ~ 2850	degree of ionization:0.3			counter-ion NO_3^-
$C_p (g\,l^{-1})$	0.26	0.16	0.10_5	0.08	0.05_2
$\Delta\varepsilon_s\, C_p^{-1}$	66	100	115	240	200
$\Delta\varepsilon_2\, C_p^{-1}$	21	36	36	60	63
$\tau_1 \times 10_6$(s)	5.6	8.9	8.9	8.6	5.0
$\tau_2 \times 10^7$(s)	0.4_6	1.0	1.2	3.0	2.3

TABLE V

Effect of concentration on dielectric parameters of PVP

PVP isotatic	DP ~ 2900	degree of ionization:0.3		counter-ion NO_3^-
$C_p (g\,l^{-1})$	0.52	0.31_5	0.21	0.10_5
$\Delta\varepsilon_s\, C_p^{-1}$	27	52	65	91
$\Delta\varepsilon_2\, C_p^{-1}$	11	16.5	26	35
$\tau_1 \times 10^6$(s)	2.5	4.5	5.0	7.11
$\tau_2 \times 10^7$(s)	0.3_6	0.2_5	0.4_4	1.1_5

From the experimental results reported in Tables III, IV and V it can be seen that:

– In contradiction with the behaviour generally observed for other stronger polyelectrolytes, the dielectric increment does not increase with increasing degree of neutralization. Moreover the relaxation times τ_1 and τ_2 are found to decrease.

– Concerning the effect of the concentration, it is observed that the dielectric increment is only slightly dependent on the concentration. Both the reduced increments $\Delta\varepsilon_s/C_p$ and $\Delta\varepsilon_2/C_p$ are found to decrease with increasing concentration. The second relaxation time is always increasing with decreasing concentration whereas the low frequency relaxation time τ_1 is found to be less dependent on the concentration. These features are reported in Figure 5.

It must also be mentioned that no significant difference in the values of dielectric

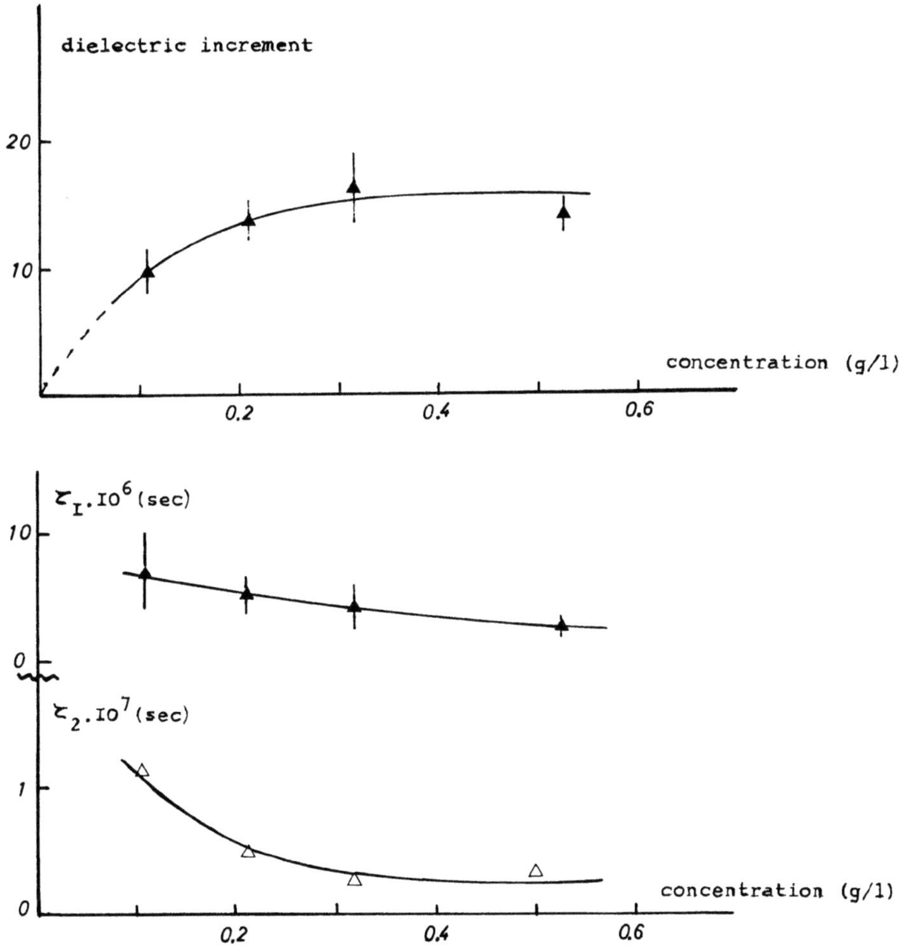

Fig. 5. Dependence of the dielectric increment ($\Delta\varepsilon_s$) and the relaxation time on the concentration of isotactic PVP partially neutralized by HNO_3 (degree of neutralization $\bar{\alpha} = 0.30$).

parameters between the atactic and isotactic species is detected whereas the reduced viscosity of isotactic PVP has been found to be always smaller than that of the atactic.

– In the presence of bivalent ions it was experimentally observed that the dielectric increment is somewhat larger than in the presence of monovalent ions.

Schwarz [14] and Mandel [15] have derived equations for the polarizability of the counter-ion atmosphere in the case of spheres and rods respectively. For an ellipsoid it can be shown that the dielectric increment $\Delta\varepsilon_s$ is roughly proportional to the square of the length of the polyion and can be expressed by the relation (3):

$$\Delta\varepsilon_s = \frac{1}{3\varepsilon_0}\varrho\sigma\frac{x(z_e)^2\langle L^2\rangle}{12kT} \tag{3}$$

in which: ε_0 is the electric permittivity of vacuum (8.854×10^{-12} F m^{-1}). ϱ is the number of monomeric units per unit volume; σ is the ratio of internal field to external field x is the fraction of associated counter-ions with charge z_e; and L is the length of the polyion.

On the other hand the relaxation time is found to be proportional to L^2/ukT (u being the mobility of counter-ions along the chain).

Usually with the neutralization of ionizable groups the number of charged groups of polyions and their counterions increases and $\Delta\varepsilon_s$ is also expected to increase during the ionization. The experimental results reported here show that the observed decrease in $\Delta\varepsilon_s$ with increasing $\bar{\alpha}$ may be attributed to a decrease in the dimensions of polyions (according to 3) which is in accordance with the decreasing reduced viscosity.

This unusual behaviour can be attributed to the very weak electrolyte character of PVP and confirms that the 'salt effect' is the predominant factor governing the properties of PVP.

In the presence of bivalent counter-ions there are two opposite effects. Firstly the bivalent ions tend to increase $\Delta\varepsilon_s$ through a charge effect and secondly the reduction of the viscosity proves that the polyions contract in the presence of SO_4^{--}. The experimental values of dielectric parameters show that the two effects cancel each other out and can explain the observed behaviour.

The variation of reduced increments versus the concentration is probably a consequence of the polyion-polyion interactions which reduce the extension of polyions and their ionic atmosphere.

In conclusion there are three regions to be distinguished. For very low charge ($\bar{\alpha}$ lower than 0.15) hydrophobic or Van der Waals interactions between uncharged parts of the polymer overbalance the effect of electrostatic interactions. Above that charge until $\bar{\alpha}$ is about 0.35, as usual the chains extend as $\bar{\alpha}$ increases. Higher protonation is only possible if the net charge of the polyelectrolyte is decreased by counter-ion-site binding (by addition of an excess acid or of salt); in this last domain the chains contract with increasing α. The first domain is inexistent with hydrophylic soluble chains. The last domain is restricted to the region of nearly complete ionization with less weak ionizable groups; it is extended with isotactic PVP as compared to the atactic one. The reported results can be explained on this basis. The drop of the reduced viscosity at low concentration (and low degree of neutralization) can be due to hydrolysis of pyridinic groups thus diminishing the degree of ionization of the polyelectrolyte. PVP probably does not form fully extended chains under electrostatic interactions because of attractive interactions and this may be the reason for the observed dependence of the reduced viscosity on the concentration. All the experimental facts show that the 'salt effect' arising from the very weak electrolyte character of the polybase is the predominant factor governing the behaviour of aqueous solutions of PVP. The observed decrease in viscosity and in dielectric parameters during the neutralization show that the length of polyions is maximum for an only partial neutralization of ionizable groups.

Acknowledgments

I should like to thank Professor Sélégny for his constant interest and help in the development of these investigations and Professor Mandel for introducing the field of electric permittivity to me.

References

1. Liquori, A. M., Barone, G., Crescenzi, V., Quadrifoglio, F., and Vitagliano, V.: *J. Macromol. Chem.* **1**, 291 (1966).
   Mandel, M. and Leyte, J. C.: *J. Polymer Sci.* **A2**, 1879 (1964).
   Anufrieva, E. M., Birshtein, T. M., Nekrasova, T. N., Ptitsyn, O. B., and Sheveleva, T. V.: *J. Polymer Sci.* **C16**, 3519 (1968).
2. Nemethy, G. and Scheraga, H. A.: *J. Chem. Phys.* **36**, 3382 (1962).
3. Kauzmann, W.: *Adv. Protein Chem.* **14**, 1 (1959).
   Dubin, P. L. and Strauss, U. P.: *J. Phys. Chem.* **74**, 2842 (1970).
4. Muller, G., Ripoll, C., and Sélégny, E.: *European Polymer J.* **7**, 1373 (1971).
5. Chabot, C.: Thesis, Paris, 1973.
6. Helfferich, F.: *Ion Exchange*, McGraw-Hill, 1962.
7. Flory, P. J. and Osterheld, J. E.: *J. Phys. Chem.* **58**, 653 (1954).
8. Alfrey, T. and Morawetz, H.: *J. Am. Chem. Soc.* **74**, 436 (1952).
9. Butler, J. A. V., Robins, A. B., and Schooter, K. V.: *Proc. Roy. Soc.* **A241**, 299 (1957).
10. Eisenberg, H. and Pouyet, J.: *J. Polymer Sci.* **13**, 85 (1954).
11. van der Touw, F. and Mandel, M.: *Trans. Faraday Soc.* **67** 1336 (1971).
12. van Beek, W. M., van der Touw, F., and Mandel, M.: (in preparation).
13. Cole, K. S. and Cole, R. H.: *J. Chem. Phys.* **9**, 341 (1941).
14. Schwarz, G.: *J. Phys. Chem.* **66**, 2636 (1962).
15. Mandel, M.: *Mol. Phys.* **4**, 489 (1961).

POLYSOAPS AT INTERFACES
AS MODELS FOR POLYELECTROLYTES

I: *Ion Exchange and Surface Pressure*

L. TER-MINASSIAN-SARAGA and S. J. ABITBOUL
Laboratoire de Physico-Chimie des Surfaces et des Membranes C.N.R.S. and U.E.R. Biomédicale, Université René Descartes, 45 rue des Saints Pères – 75270 Paris

1. Introduction

The effect of calcium on several properties of acidic monolayers has been studied in the hope that the ionic behaviour of these systems may approximate that of the biological membranes. Calcium ions are important in muscle contraction, erythrocyte deformability, enzyme reactions, nerve impulse transmission and propagation, and membrane permeability.

It has been found that the axon membrane may have negatively charged sites at its surface [1]. The variation of the membrane permeability to cations in the presence of calcium ions has been related to their adsorption on these sites [2] the chemical nature of which is ignored yet.

Recent studies of the binding of calcium ions by erythrocyte ghosts have shown that the largest amount of these ions was fixed by the free carboxyl groups of glutamic and aspartic residues of the membrane proteins [3, 4].

Monolayers of stearic acid have been used as models to study the interaction of negatively charged surfaces with calcium ions. It is known since 1937 that these monolayers are more condensed and are rigid when calcium ions are present in the aqueous substrate [5]. This behaviour may not be reproduced by the monolayers of the unsaturated fatty acid or of the acidic phospholipids [6]. Furthermore, at a given surface density of the lipids, two very similar carboxylic acids: oleic and stearic acids, bind different amounts of calcium ions [7].

In the present study a relation was sought for between the ionisation of an anionic monolayer and calcium binding. A hydrophobic polyacid: the alternate copolymer of maleic acid and hexadecylvinylether has been utilised.

Under definite conditions of hydration and of ionisation, this hydrophobic polyelectrolyte may present several phases: lamellar, cylindric or micellar [8]. They correspond to different specific areas [8] and conductivities [9].

The HPA was spread at the surface of aqueous solutions at various pH, ionic strengths and concentrations of calcium ions. The surface pressure, the amount of calcium binding (Part I) and the surface potential (Part II) were measured. In Part II the degree of ionisation of the monolayers was deduced from the values of their surface potentials. The interpretation of the results has shown that:

– the binding of calcium corresponds to an ion exchange reaction by the carboxylate groups of the polyacid film;

Eric Sélégny (ed.), Polyelectrolytes, 207–218.

– the electrostatic free energy of this reaction contributes to the difference of the electric potential between the film and the substrate and accordingly to the measured surface potential.

2 Materials and Methods

The HPA has the following chemical formula:

$$\mathrm{HPA} = \left[\begin{array}{ccccccc} - \mathrm{CH} & - & \mathrm{CH} & - & \mathrm{CH_2} & - & \mathrm{CH} - \\ | & & | & & & & | \\ \mathrm{COOH} & & \mathrm{COOH} & & & & \mathrm{O} - \mathrm{R} \end{array}\right]_n$$

The molecular weight of the monomer is 384 g mole^{-1}. Stereomodels led to a projected area equal to 45–50 A^2 for a monomer laying flat on a surface (side on position).

The samples used were polydisperse. They were kindly provided by R. Varoqui*. Their preparation was described in references [10, 11]. The samples were used without further purification or fractionation.

The spreading solvents were mixtures of chloroform with different organic polar liquids: acetone, methyl, ethyl or amyl alcohol. They are of analytical grade (Merck pro analysis).

The water was distilled three times as follows: distillation – distillation on $KMnO_4$ – distillation. The last two distillations were carried on in an all glass apparatus.

The mineral salts were reagent grade (Merck pro analysis). Their aqueous solutions were purified by foaming in a stream of N_2 before use. Radioactive $^{45}CaCl_2$ was supplied by Amersham (England). Its original specific activity was of the order of 100–200 mCi g^{-1}. It has been diluted isotopically to a specific activity equal to 3–4 mCi g^{-1}.

The films were formed by delivering various volumes of the polyacid solution from an Agla microsyringe. The concentration of the solution was 0.5 mg ml^{-1}. The delivered volumes were known with an accuracy of ± 0.2 μl.

The substrates were contained in glass dishes or in a teflon trough. The spreading areas were equal to 73 cm^2, for the glass dishes, or to 221 cm^2 for the trough.

The surface pressures were deduced from the surface tension measurements performed by the method of Wilhelmy using a Sanborn force transducer and a sand blasted Pt plate, 2 cm wide. The accuracy was ± 0.5 dyne cm^{-1}.

The amount of Ca^{2+} ions bound by the film was measured following the method described elsewhere [12]. The HPA was spread at the surface of aqueous solutions containing radioactive ^{45}Ca ions. The film area was equal to 10 cm^2. A teflon ring with an inner area equal to 7 cm^2 was placed above the film. Through it, β-rays (0.254 MeV) were emitted by the ^{45}Ca ions either bound to the film or located just below it. The total emission was detected by a counter placed above the ring. For each calcium concentration in the substrate, two measurements of this emission were performed

* R. Varoqui, Centre des Recherches sur les Macromolecules, C.N.R.S., Strasbourg 67.

and the surface density of the bound calcium ions was deduced following the method described in reference [12].

The phosphate buffers contained various mixtures of salts:
PO_4NaH_2 (3.3×10^{-3} M) + NaCl (y)+NaOH (x). The pH of the buffer was adjusted by varying x. The ionic strength was modified by changing y. The concentration of the free calcium ions of the substrate was calculated by the method of Caldwell [13], using an equilibrium constant K of the complex $HCaPO_4$ formation equal to $10^{1.5}$ (see ref. [14]).

The experiments were performed at the room temperature = 22±2 °C. Small temperature variations did not change significantly the amount of bound Calcium ions.

3. Results

3.1. Choice of the spreading solvent and of the amount of spread polyacid

The polyacid (HPA) was dissolved in mixed polar-non polar solvents. The solutions were quite stable after a certain period of ageing.

A given amount of HPA was spread at the surface of an aqueous substrate at a given pH and ionic composition and the amount of bound calcium ions was determined. It is inferred that the best spreading solvent led to the highest amount of bound Ca ions.

The obtained results are shown in Table I.

Out of the four studied mixtures the highest amount of Calcium binding was obtained with the solvent: methanol chloroform (1/9). This mixture was used all throughout the present study.

To find out the effect of the amount of the spread HPA on the efficiency of spreading various volumes of the solution were delivered on the surface of a given substrate contained in different dishes. The binding of calcium by the film was then measured.

TABLE I

Effect of solvent on the amount of bound Ca^{2+} ions by a HPA film

Solvent composition *v/v*		Bound Ca *r*/2	Solvent composition *v/v*		Bound Ca *r*/2
($CH_3OH/CHCl_3$)	(3/7)	31 ± 1	($C_5H_{12}O/CHCl_3$)	(3/1)	20 ± 2
	(1/9)	35 ± 5		(1/9)	24 ± 5
	(1/1)	30 ± 3		(2/8)	23 ± 6
($C_2H_6O/CHCl_3$)	(1/9)	32 ± 5	($(CH_3)_2CO/CHCl_3$)	(3/1)	19
	(2/8)	28 ± 4		(6/4)	19
	(1/1)	14		(1/1)	16
				(4/6)	25 ± 2

Area of spreading: 10 cm². Surface density of the HPA: 9.65×10^{14} (residues cm^{-2}).
pH = 8; $c_{CaCl_2} = 8.15 \times 10^{-5}$ M.
Specific activity of Ca: 3.6 mCi g^{-1}. r: ratio of Ca ions/monomer.

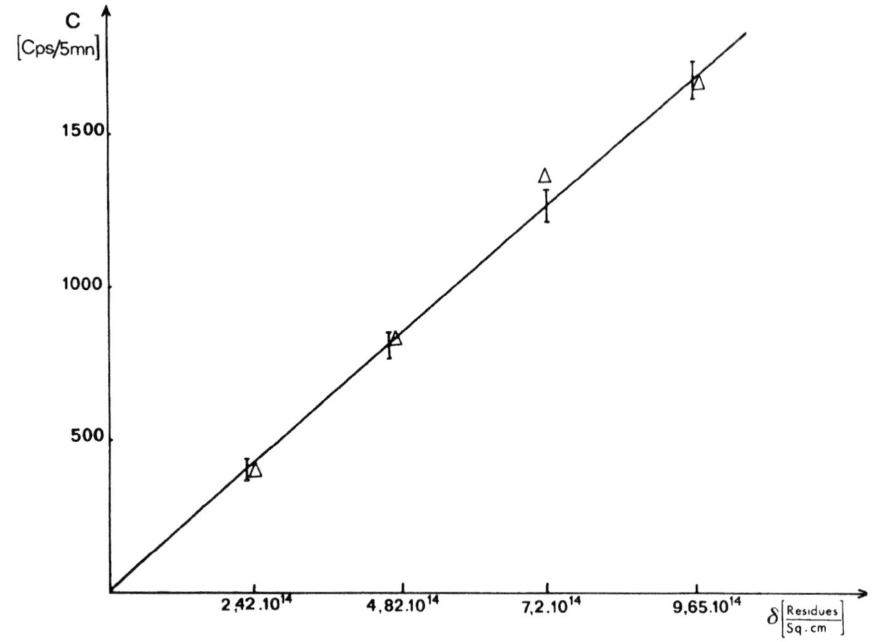

Fig. 1. Increase of the counting rate with the surface concentration of HPA. Phosphate buffer. Concentration of the spread HPA solution: 0.5 mg ml^{-1}. δ = surface density of HPA. C = counting rate. I: error. Specific activity of ^{45}Ca: 3.9 mCi g^{-1}. r = 50%.

Figure 1 shows that the intensity of the β-radiation emitted by the film is proportional to the amount of the spread HPA. The last point on this figure corresponds to a volume equal to 12 μl. Larger spread volumes led to erratic results. In the subsequent experiments, the amounts of the HPA spread and studied films has not exceeded the maximum value of 1.85×10^{15} COOH groups cm^{-2}, or 9.25×10^{14} copolymer residues cm^{-2}.

3.2. Surface pressure measurements

These were performed by two techniques.

– Increasing amounts of a HPA solution were spread on a given area, equal to 73 cm^2, of substrate. Each amount has been delivered on a fresh surface.

– A given quantity of spread HPA film was compressed.

The two techniques led to stable films.

The compression curves are shown in the Figures 2, 3 and 4. They show the effect of the pH of the substrate and of Ca binding by the films, on their surface pressure.

The results of the surface pressure determination by the first technique have been shown in Figure 5. Only the results obtained at pH = 8 have been reproduced. They were similar to those corresponding to pH = 6.

In order to compare the two techniques the corresponding results of the surface

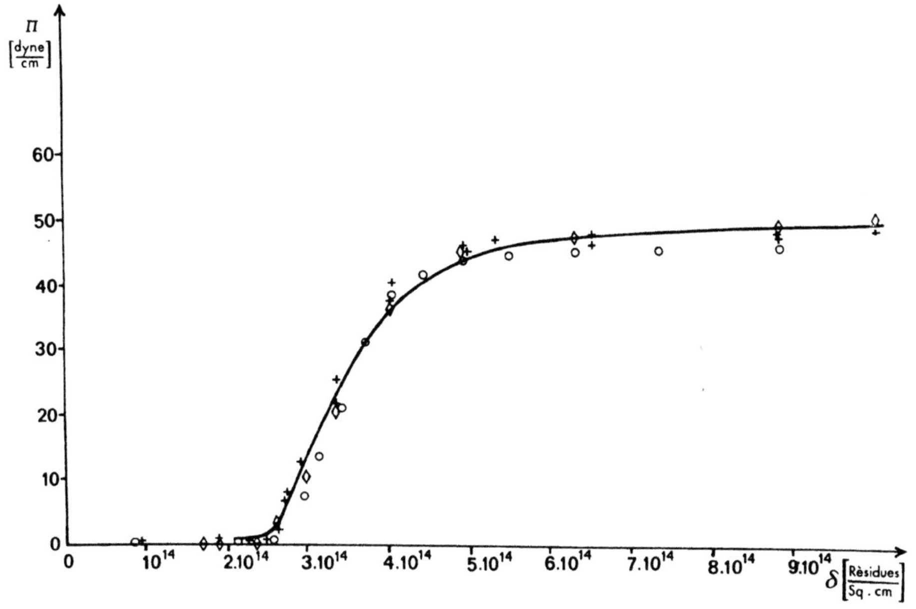

Fig. 2. Surface pressure of HPA spread films at different pH values of the substrate in the absence of Ca^{2+} ions. pH = 3.8 (◇); pH = 6 (+); pH = 8 (○). π: surface pressure; δ: surface density of HPA.

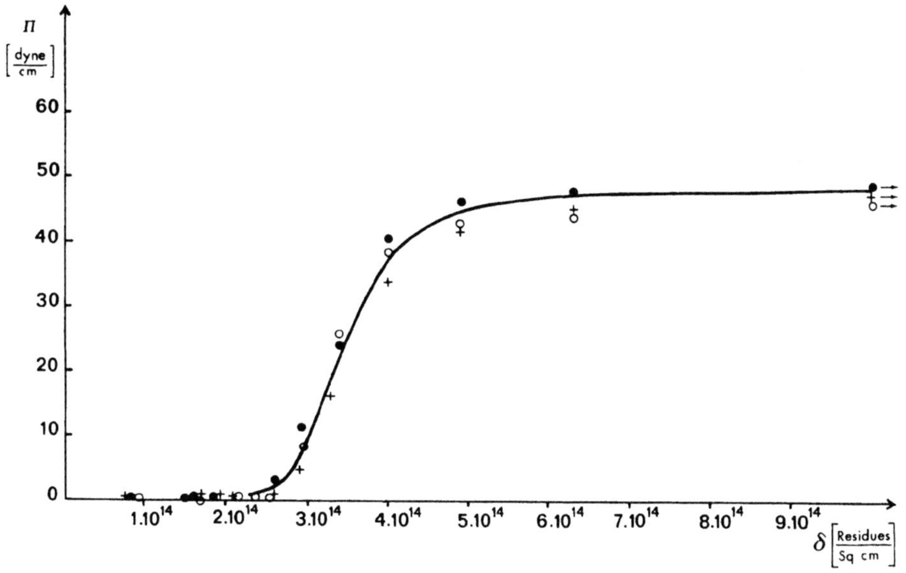

Fig. 3. Surface pressures of HPA spread films in the presence of Ca^{2+} ions at various pH and for various amounts of bound Ca^{2+} ions. π = surface pressure; δ = surface density of HPA. Concentration of $CaCl_2$ in the substrate: 10^{-5} M.

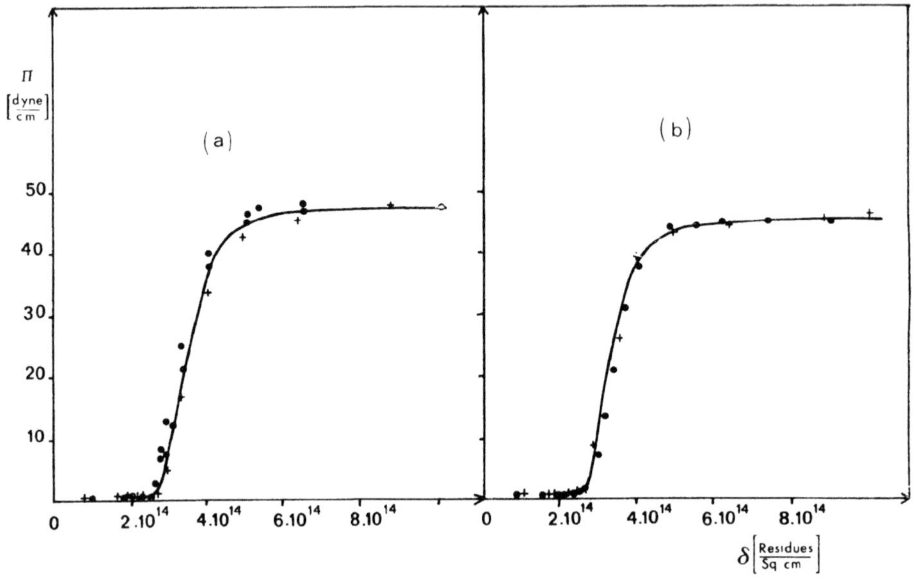

Fig. 4. Effect of bound Ca^{2+} ions on the surface pressure of HPA spread films. (a) pH = 6; (b) pH = 8; for π, r and δ see Figures 2 and 3. Concentrations of $CaCl_2$ in the substrate: zero (●); 10^{-5} M (+).

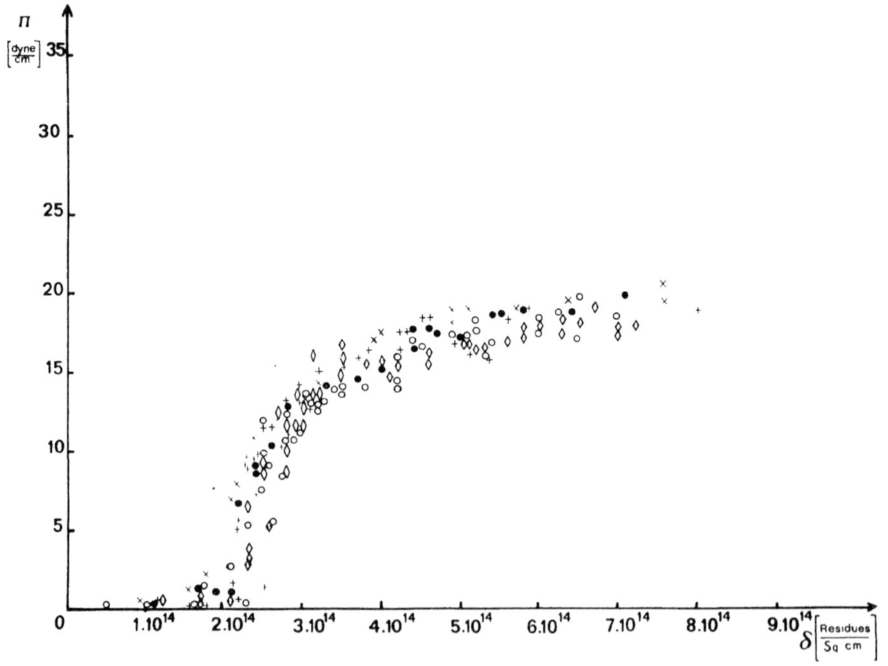

Fig. 5. Surface pressure of HPA films spread on an area equal to 73 cm²; pH = 8. Concentrations of $CaCl_2$ in the substrate: zero (◇); 1.95×10^{-5} M (○); 5.2×10^{-5} M (+); 7.32×10^{-5} M (●); 1.5×10^{-4} M (×). For π and δ see Figures 2 and 3.

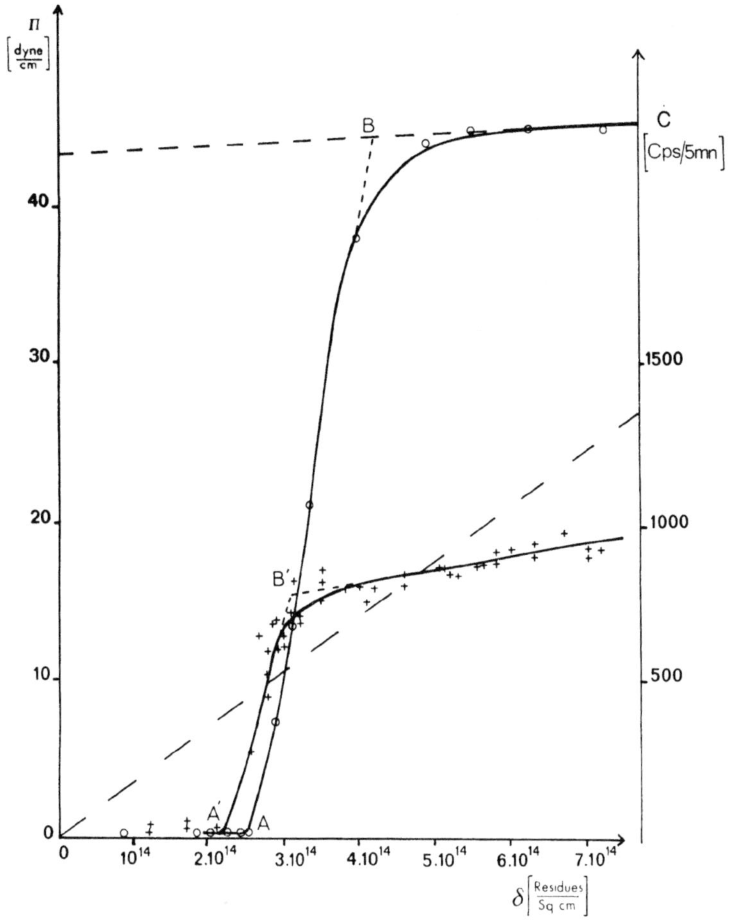

Fig. 6. Comparison of surface pressure measurements of films spread by the two techniques in the absence of calcium ions. pH = 8. Spreading only (+); spreading with subsequent compression (○) Counting rates measured for the spread and uncompressed films (— —), π and δ as in Figures 2 and 3

pressure measurements were reproduced in Figure 6 as a function of the surface density of HPA. The substrate did not contain calcium ions and its pH was equal to 8.

3.3. Binding of calcium ions by the HPA films

In Figure 7 the results of a typical set of experiments were reproduced:

– the counting rate C_0 originating in the β-emission of the dissolved Ca^{2+} ions was proportional to their concentration in the substrate (curve 0);

– the counting rate C_t measured in the presence of the HPA film was higher (curve t);

– the counting rate C_f originating in the β-emission of the bound ^{45}Ca ions was obtained by substracting C_0 from C_t and was plotted in Figure 7b (curve f) as a function of the concentration of calcium ions in the substrate.

Using the calibration method described in reference [12], the surface density of the Ca ions bound to the HPA films was deduced from the values of C_f.

The effect of pH, calcium and salt concentrations in bulk, on the surface density of bound calcium ions by the HPA films has been studied.

The Figures 8 and 9 represent the results obtained at two given pH values, equal to 6 and 7.5, and at two ionic strengths of the substrate. The amount of Ca ions is expressed by r, the ratio of the numbers of Ca ions and of the number of monomers of the spread HPA.

The effect of the pH of the substrate, at a constant Ca^{2+} concentration and ionic

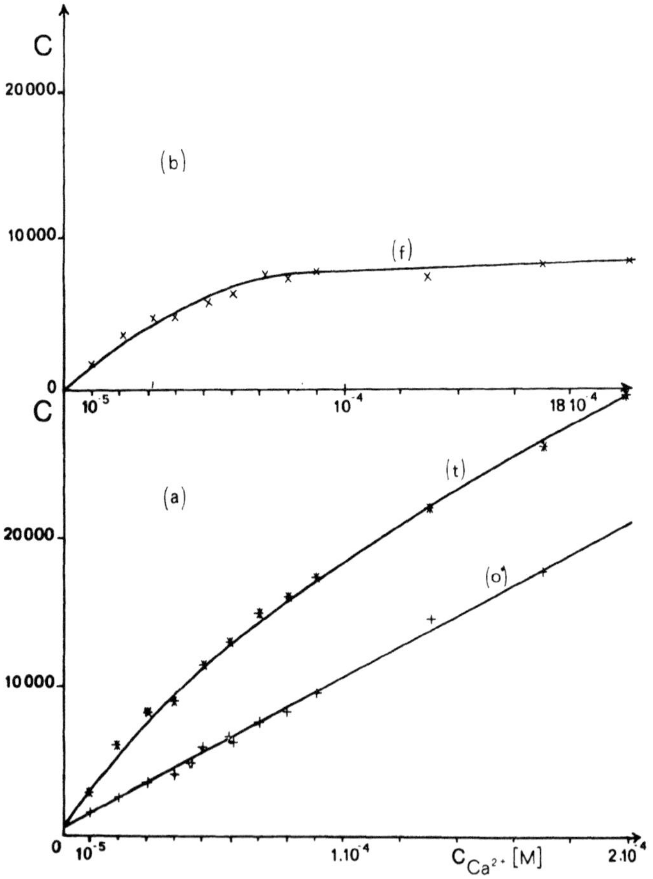

Fig. 7. Effect of Ca^{2+} concentration in bulk on the counting rates measured in the presence and in the absence of HPA films (a) and deduction of the counting rate due to the HPA films (b). (O) = counting rate in the absence of the HPA film. (t) = total counting rate in the presence of the HPA film. (f) = counting rate due to the Ca ions bound by the HPA films. c_{Ca} = concentration of Ca^{2+} ions in the substrate. pH = 8. Phosphate buffer. Na = 6.6×10^{-3} N.

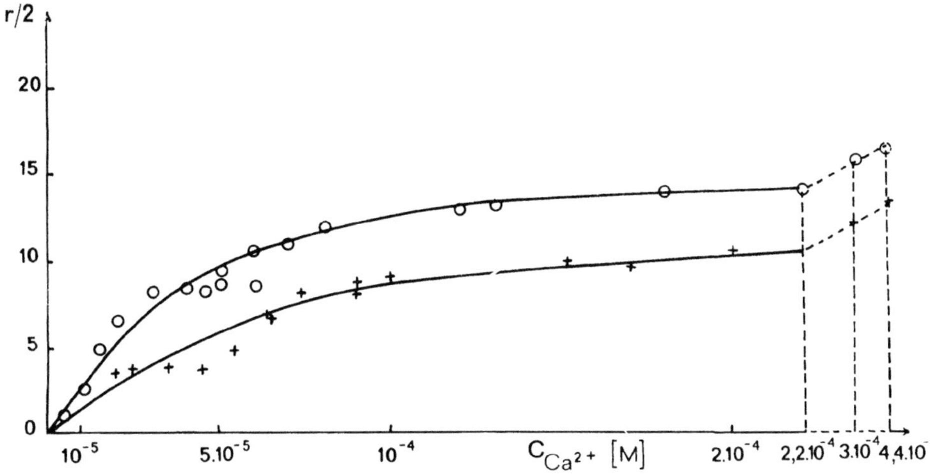

Fig. 8. Variation of bound Ca^{2+} ions with their concentration in the substrate. r = ratio of Ca^{2+} ions/monomer; surface density of carboxyl groups = 1.85 10^{15} groups cm^{-2}. c_{Ca} = concentration of free Ca ions in the substrate. pH = 6. Concentration of Na^+: 6.6×10^{-3} N (○); 3.43×10^{-2} N (+).

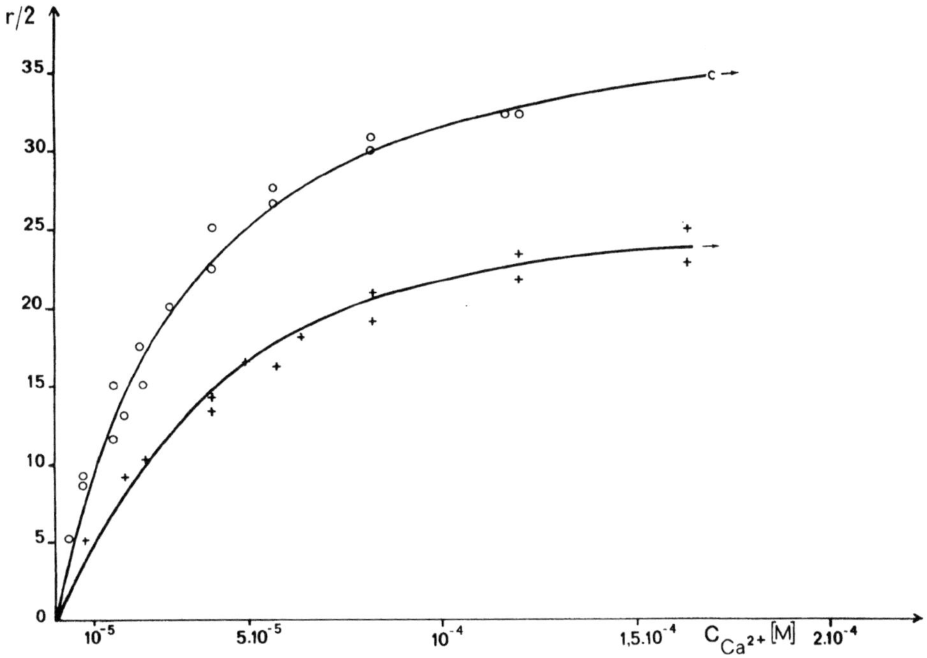

Fig. 9. Effect of Ca^{2+} ion concentration in the substrate on their binding by HPA films. pH = 7.5; for c_{Ca}, r, surface densities of monomers and c_{Na^+} see Figure 8.

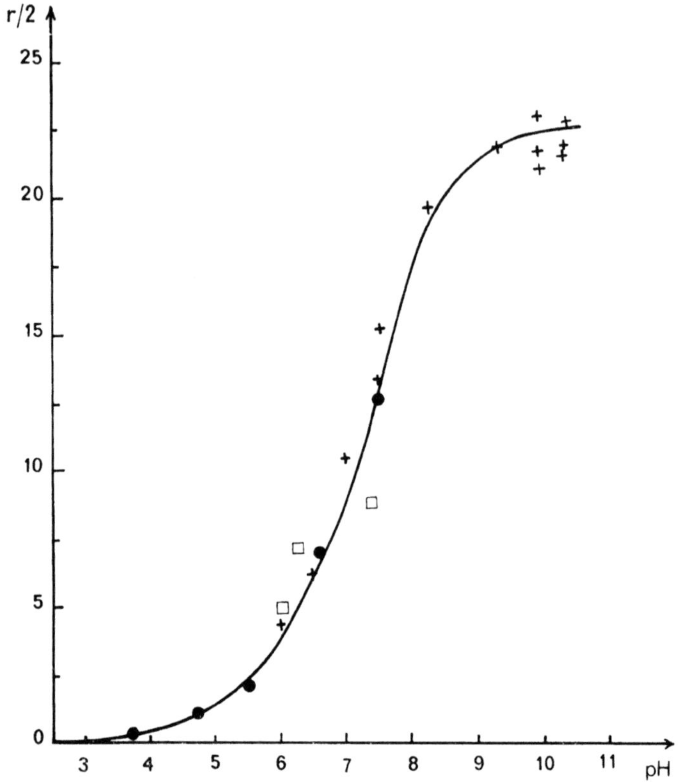

Fig. 10. Variation of the binding of Ca^{2+} ions by the HPA films with the pH of the substrate. The concentration of Cl_2Ca in the substrate is constant and equal to 2×10^{-5} M. r = ratio of Ca^{2+} ions/monomers. The compositions of the buffers were as follows: pH 3.8 (formic acid + NaCl); pH = 6 or 8 (phosphate); pH = 9 ($CO_3NaH + CO_3Na_2$); pH = 10–11 (NaOH + NaCl). The total (Na^+) = 6.6×10^{-3} N. Results obtained in the presence of air (●) or in an atmosphere of N_2 (+).

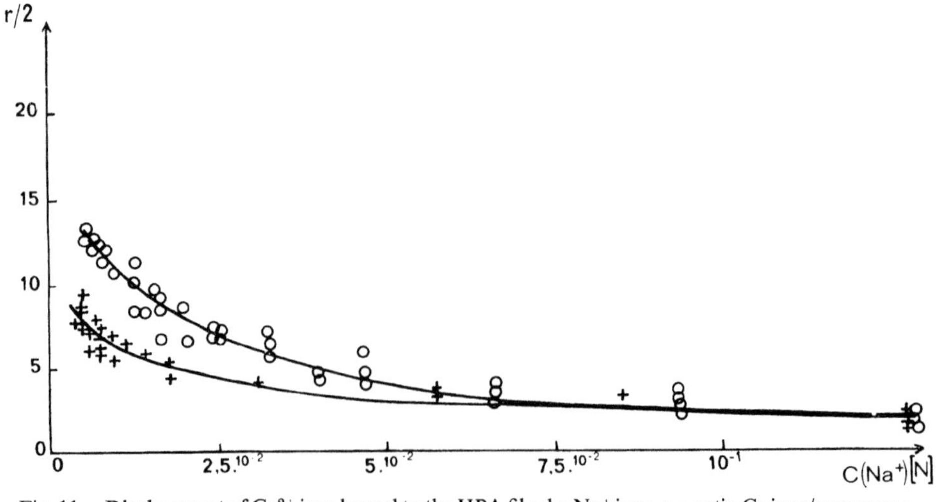

Fig. 11. Displacement of Ca^{2+} ions bound to the HPA film by Na^+ ions. r = ratio Ca ions/monomers. c_{Na^+} = concentration of Na ions in the substrate. Surface density of COOH groups = 1.85×10^{15} groups cm^{-2}. $c_{CaCl_2} = 2 \times 10^{-5}$ M; pH = 7.5 (○); pH = 6 (+).

strength of the substrate, on the binding of Ca^{2+} ions by the HPA spread films is shown in Figure 10. To avoid the possible contamination by CO_2, the experiments done with substrates which pH were higher than 7.5, were performed in a N_2 atmosphere.

The displacement of the bound Ca^{2+} ions by the dissolved Na^+ ions has been studied at constant Ca^{2+} concentration and pH of the substrate. At the following pH values: 6.8; 7.5 and 10.5, and for a given density of the HPA spread film the results were different, but the curves obtained, by plotting r as a function of Na^+ concentration, were analogous. In Figure 11 we have reproduced the results obtained at pH = 7.5.

4. Discussion of the Results and Conclusion

The HPA monolayers compressed after spreading on various aqueous substrates were condensed and solid at all the pH values and concentrations of calcium ions (Figures 2, 3 and 4). At the lowest pressure of the condensed state (point A, Figure 6) the area per residue was equal to 38.5 A^2. At point B, the pressure was equal to 45 dyne cm^{-1} and the area per residue was 24 A^2. When the HPA films were formed by delivering various amounts of the copolymer, the results obtained were qualitatively the same as those reported above but less satisfactory. These films were also condensed (Figures 5 and 6). At the lowest pressure (point A′, Figure 6) the area per residue was equal to 46 A^2. At point B′, the pressure was 15 dynes cm^{-1} and the residue area was 33 A^2. The difference between the results obtained for films spread by the two techniques may have been due to the inclusion of different amounts of solvent. One of us has shown [15] that polar solvents may be retained even by films which were spread on large areas. In the present study the use of such a solvent (methyl alcohol) was essential for dissolving and spreading the HPA.

Furthermore, the physical meaning of the two values of the collapse pressure was not clear. None of them was equal to the spreading pressure of the solid HPA. The last pressure was lower than 1 dyne cm^{-1}. It was concluded that the natures of the phases ejected at the high collapse pressures and of the original, solid, HPA were different. This behaviour is opposite to that of a short chain acid, lauric acid [15].

The results which have been reproduced in Figures 2, 3, 4 and 5 have shown that neither the change of pH nor the presence of calcium ions had any effect on the state of the film. It might have been inferred that the HPA was not ionised or that it did not interact with calcium ions. The results reproduced in Figures 7, 8, 9, 10 and 11 have demonstrated the binding of calcium ions. The surface potential measurements have shown the ionisation of the HPA films (Part II). Therefore it has been concluded from the study of the surface pressure of these films that their state is determined mainly by the Van der Waals interactions between the non-polar parts of the macromolecules. The electrostatic interactions, depending on the ionisation of the HPA, seemed to be negligible.

The results of the study of the Ca^{2+} ions binding, reproduced in Figures 1 and 7, have been obtained under the same conditions as those corresponding to the isotherms shown in Figure 5. The slope of the lines of Figures 1 and 6 is equal to r: the number of bound Ca^{2+} ions per monomer of the HPA film. This number was independent of the pressure of the film. Other authors have found an analogous result for the binding of Ca^{2+} ions by the condensed stearic acid monolayers spread on a large initial surface and then compressed. Their technique of film formation was different from ours. The difference between the two isotherms shown in Figure 6 may not have been relevant for the binding of Ca^{2+} ions.

The effect of the pH and of the ionic strength on this binding (Figures 8, 9, 10, 11) points to a possible relation between the ionisation of the carboxyl groups of the HPA film and their interaction with cations. This relation has been studied and reported in the second part of this paper.

References

1. Segal, J. R.: *Biophys. J.* **8**, 470 (1968).
2. Gilbert, D. L. and Ehrenstein, G.: *Biophys. J.* **9**, 447 (1969).
3. Forstner, J. and Manery, J. F.: *Biochem. J.* **124**, 563 (1971).
4. Forstner, J. and Manery, J. F.: *Biochem. J.* **125**, 343 (1971).
5. Langmuir, I. and Schaeffer, V. J.: *J. Amer. Chem. Soc.* **59**, 2400 (1937).
6. Deamer, D. W. and Cornwell, D. G.: *Biochim. Biophys. Acta* **116**, 555 (1966).
7. Goerke, J., Harper, H. H., and Borowitz, M.: *Surface Chemistry of Biological Systems*, Plenum Press, U.S.A., 1970, p. 23.
8. Schmitt, A., Varoqui, R., and Skoulios, A.: *Compt. Rend. Acad. Sci.* **268**, 1469 (1969).
9. Varoqui, R. and Schmitt, A.: *Compt. Rend. Acad. Sci.* **270**, 788 (1970) C.
10. Varoqui, R. and Strauss, U. P.: *J. Phys. Chem.* **72**, 2507 (1968).
11. Pefferkorn, E., Schmitt, A., and Varoqui, R.: *Compt. Rend. Acad. Sci.* **267**, 349 (1968) C.
12. de Heaulme, M., Hendrikx, Y., Luzzati, A., and Ter-Minassian-Saraga, L.: *J. Chim. Phys.* **64**, 1363 (1967).
13. Caldwell, P. C.: in A. W. Cuthbert (ed.), *Calcium and Cellular Function*, Macmillan, London, 1970, p. 10.
14. *Stability Constants*, The Chemical Society, London, 1964, p. 182.
15. Ter-Minassian-Saraga, L.: *J. Chim. Phys.* **52**, 80 (1955).

POLYSOAPS AT INTERFACES AS MODELS FOR POLYELECTROLYTES

II: *Ion Exchange and Surface Potential*

L. TER-MINASSIAN-SARAGA and C. THOMAS

Laboratoire de Physico-Chimie des Surfaces et des Membranes, C.N.R.S. and U.E.R. Biomédicale, Université René Descartes, 45 rue des Saints Pères – 75270 Paris, Cedex 06

1. Introduction

Since 1957 [1], it is known that the membrane permeability of axons to K^+ ions depends on the concentration of Ca^{2+} in the medium. It is suggested that the membrane bears negatively charged sites liable to bind Ca^{2+} ions [2, 3]. The desorption of Ca^{2+} may produce either a change in the structure of the membrane [2] or a variation in the charge density at the outer surface of the membrane, bringing about an increase of the interfacial potential [3]. It is assumed that either of these effects may explain the change in the membrane permeability to monovalent cations although the role of the 'interfacial potential' has been emphasized. It is found that this potential may vary also with the pH [4].

Gels and concentrated polyelectrolyte solutions have been suggested as models to approximate the properties of biological membranes [5, 6]. Polyions may bind their counter-ions by an ion-pairing mechanism as demonstrated by specific volume measurements [7] and ultrasonic absorption [8]. It is inferred that, when bound, the charged constituents may loose some of their hydration water. The effect of the linear charge density of the polyion and of the resulting shift in potential in its neighbourhood is emphasized [9, 10a].

Monolayers at the surface of aqueous electrolyte solutions may be useful models for polyelectrolytes [11, 14] and for biological membranes [15]. The average distribution of the fixed charges, limited to a plane near the interface, is well defined and the variation of the interfacial potential may be measured. Thus the surface potential-pH plots resemble those of acid-base titrations of polyacids [16, 19]. From these curves the degree of dissociation of the fatty acid in the monolayer may be deduced [17, 18].

The effect of the ionic strength in the substrate and of the soap concentration in the monolayers on the surface potential of the film is usually interpreted by generalizing to these systems the models and theories utilized for polarized solid or liquid electrodes [20].

Recently Jaffé *et al.* [21] study monolayers of ionized polymers and deduce their degree of ionization using a classical Gouy model for polarized systems.

The generalization of this model or of more sophisticated ones [22], set up for polarized electrodes, to unpolarized systems e.g. colloids [23] and monolayers [20], asks for the following comments:

The surface potential is the result of the difference between two outer

Eric Sélégny (ed.). Polyelectrolytes, 219–239.

potentials [24]: one of them corresponds to the free water surface and is constant, the other one is that of aqueous substrate covered by the monolayer. The work corresponding to a unit charge taken from the bulk of the substrate to the interface beyond the region occupied by the permanent dipoles and all the charged sites of the monolayer, is equal to the shift in the potential occurring in the space charge region of the substrate covered by the monolayer. This shift has a first contribution χ from the distribution and orientation of the permanent dipoles of the film and solvent. The second one ϕ_0 comes from the net surface charge density on the plane of the fixed ones and from the counter-ions and co-ions distribution in the field set up by this charge. In the last region the dielectric constant may vary with the distance to the interface [20, 25].

There are two essential differences between the monolayers and the polarized systems:

– the first one is that monolayers are not polarized by an external electric field and have no available electrochemical energy to exchange.

– the second one concerns the non-polarized solid or liquid electrode on one side, and monolayers on the other side. For these, any electrostatic energy can have only the property of an excess thermodynamic function over the value that function would have in an ideal system containing uncharged constituents. This situation is analogous to that of the aqueous electrolyte solutions as it has been inferred by Overbeek [26]. For these homogeneous systems Debye and Guntelberg [27] assume that thermal fluctuations of ionic densities lead to local perturbations of the average electrostatic potential – equal to zero. The perturbations are related to interactions and explain the origin of the activity coefficients of the ions in solution.

The colloids, monolayers and unpolarized electrodes are heterogeneous systems the interfaces of which introduce a major perturbation. They may modify selectively the homogeneous distribution of the different charged constituents by selective adsorption of ions, ionization of surface groups and so on. Thus, charged surfaces and spatial distribution of the counter-ions sets in leading to a shift of the potential in this space charge region. This first order perturbation shift in potential has the character of a Debye or Guntelberg micropotential [23] and the resulting interactions produce excess values for the thermodynamic potential of the ions compared to those of the ideal uncharged ones.

A second perturbation potential – the discrete ion effect – is calculated by Levine *et al.* [20, 23] for the counter-ions. It is related to the finite size of these ions, their high concentration near the interface and the image forces.

However the model used for these last calculations is more realistic for electrodes than for monolayers of polyelectrolytes. For the last ones the penetration of the counterions inside the region occupied by the fixed charges is probable. Haydon *et al.* [28] raised this question for monolayers. For polyelectrolytes this penetrability is sometimes accounted for. Then, the potential shift between a penetrated interface and bulk is called Donnan potential [10] and may be very different from the one calculated for the unpenetrated electrodes.

In the present work, we use the model of the penetrated interface to interpret our results. We study the interfacial potential of ionized films of a hydrophobic polyacid H.P.A.: the copolymer of maleic acid and of hexadecylvinylether.

The amount of bound Ca^{2+} is investigated in Part 1 [29] as a function of pH, ionic strength and Ca^{2+} concentration.

In the absence of Ca^{2+}, the surface potential is practically independent of the ionic strength as for another solid monolayer – stearic acid – at high pH [17]. Then, assuming almost complete ion pairing between the sites and the counterion Na^+, a simple relation is found between the degree of dissociation of the H.P.A. and the surface potential of its monolayers.

In the presence of Ca^{2+} ions, the surface potential increases slightly with the amount of bound Ca^{2+}. By assuming that Ca^{2+} is fixed on one site we verify that the variation of the interfacial potential is related to the displacement of the equilibrium of ion exchange Ca^{2+}/Na^+ with the amount of bound Ca^{2+} studied in Part 1 [29].

The biological implications of our study are shown by comparing our results to those of Hille [30] on the effect of pH and Ca^{2+} concentration on the charge and potential at the nerve surface.

2. The Monolayer and its Surface Potential

At a given pH and ionic composition, the H.P.A. monolayer is constituted by four types of residues: (a), (b), (c) and (d), shown on Figure 1. The residues (a), (b) and (c) are ionized: (a) and (b) are neutralized by the counter-ions $M^{Z_1^+}$ and $M^{Z_2^+}$ (c) is neutralized by the counter-ions of the diffuse layer beyond $x=d$. The ions $M^{Z_1^+}$ and

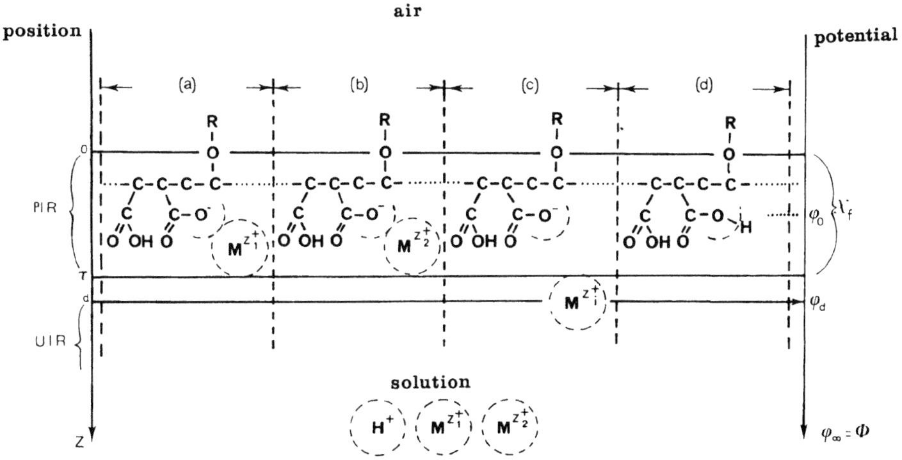

Fig. 1. Model of the HPA monolayer at a given pH and ionic composition of the substrate. The residues (a), (b) and (c) are ionized, (d) is in the acid form. In (a) and (b) the ions M^{Z_1+} and/or M^{Z_2+} form ion pairs with the carboxylates extending to an average depth $x=\tau$; (c) is neutralized by counterions of the diffuse layer beyond $x=d$. The plane at $x=d$ separates the Paired Ion Region (P.I.R.) from the Unpaired Ion Region (U.I.R.). For the definition of ϕ_0, ϕ_d and χ see the text.

M_2^{z+} are bound to the negative sites and form ion-pairs extending to an average depth $x=\tau$. The plane located at $x=d$ separates these paired ions and the corresponding paired ion region 'P.I.R.', from the unpaired ions of the unpaired ion region 'U.I.R.' at $x>d$. The criterium of the choice of the distance d is not unique. For polyelectrolytes it is assumed that d is such that for $x \geqslant d$, $ze\Psi_{(x)} \leqslant kT$, where ze is the charge of the counter-ion [31]. For electrodes it is either assumed that the plane at d separates the fully hydrated counter-ions from the adsorbed partially unhydrated ones [32], or that $x=d$ is the position of nearest approach of the co-ions or of the unadsorbed ones [33].

In our model the finite sizes of the negative sites and of the counterions may hinder complete penetration of the last ones into the plane of the first ones. This penetration decreases the net charge density neutralized by the diffuse layer. But the ion pairs behave as discrete dipoles and contribute to the χ potential produced by the monolayer and the water molecules.

Let δ_1 and δ_2 be the number of ion pairs formed by two monovalent counterions, and $\bar{P}_1$ and $\bar{P}_2$ their effective dipole moments. The contribution $\Delta\eta$ of the ion pairs is equal to:

$$\Delta\eta_{\text{I.P.}} = \eta_{\text{I.P.}} - \eta_{\text{OH}_2} = 4\pi(\delta_1\bar{P}_1 + \delta_2\bar{P}_2). \tag{1}$$

It includes the reorientation effect of the water molecules, as η_{OH_2} is the potential χ for the pure water surface.

The contribution of the undissociated polar groups of the monolayer (see Figure 1) to the potential $\Delta\chi_f$ produced by the film is then equal to:

$$\Delta\chi_f - \Delta\eta_{\text{I.P.}} = 4\,\pi\delta\bar{P}_u + 4\,\pi\delta(1-\alpha)\,\bar{P}_{\text{OH}}, \tag{2}$$

where $\bar{P}_u$ is the effective dipole moment of one undissociated residue minus the effective dipole moment of one hydroxyle $\bar{P}_{\text{OH}}$ and α is the degree of dissociation of the residue.

As the inner potentials are equal in our case, the difference between the outer potential Ψ_f of the substrate covered by the monolayer, and Ψ_{OH_2} of the pure substrate surface, is equal to the surface potential:

$$\Delta V = \psi_f - \psi_{\text{OH}_2} = \Delta\chi_f + \phi_d = \Delta\chi_f + \frac{2kT}{e}\sin h^{-1}\left(\frac{\pi}{2\varepsilon ckT}\right)_\sigma^{1/2}, \tag{3}$$

where ϕ_d varies with the charge density on the plane $x=d$ and with the ionic strength of the substrate [26].

From (1) (2) and (3) it is obtained:

$$\Delta V = 4\pi\delta\bar{P}_u + 4\pi\delta(1-\alpha)\,\bar{P}_{\text{OH}} + 4\pi\delta_1\bar{P}_1 + 4\pi\delta_2\bar{P}_2 + \phi_d. \tag{4}$$

Our monolayer is complex. The conformations of its molecules may be such that some of its dipoles may lie below the plane of the charges at $x=0$, in the region τ, and contribute a $\Delta\eta$ term to the potential ϕ_0 (see Figure 1). Those exposed to the air phase are supposed to remain unionized. Then ϕ_0 is equal to:

$$\phi_0 = 4\pi\delta(1-\alpha)\,\bar{P}_{\text{OH}} + 4\pi\delta_1\bar{P}_1 + 4\pi\delta_2\bar{P}_2 + \phi_d. \tag{5}$$

Using [5], [4] becomes:

$$\Delta V = \text{Constant} + \phi_0 . \tag{6}$$

Furthermore when $\delta_2 = 0$, $\alpha\delta \simeq \delta_1$, and $\phi_d \ll \phi_0$ (4) yields:

$$\Delta V/4\pi\delta = \bar{P}_u + \bar{P}_{\mathrm{OH}} + \alpha(\bar{P}_1 - \bar{P}_{\mathrm{OH}}) . \tag{7}$$

When $\alpha = 0$ or $\alpha = 1$ it is found from (7) that:

$$\Delta V_{\alpha=0}/4\alpha\delta = \bar{P}_u + \bar{P}_{\mathrm{OH}} \text{ and } \Delta V_{\alpha=1}/4\pi\delta = \bar{P}_u + \bar{P}_1 . \tag{8}$$

From (7) and (8) it follows that α is equal to:

$$\alpha = (\Delta V - \Delta V_{\alpha=0})/(\Delta V_{\alpha=1} - \Delta V_{\alpha=0}) . \tag{9}$$

3. Ionic Reactions and Surface Potential in Monolayers

3.1. Titration curves of H.P.A. monolayers*

These curves should conform to the Henderson-Hasselbach equation for two reasons [21]:

(1) Our monolayers are formed of polyacids which show such potentiometric titration curves in solution [34].

(2) Monolayers of fatty acids lead to ΔV vs. pH curves resembling the titration curves of polyacids [17].

For our H.P.A. monolayers one may expect at least a titration curve of the form obtained by Strauss [34] in solution:

$$\mathrm{pH} = pK_1^0 - \log\left\{1/2\,\frac{(1-\alpha)}{\alpha} + 1/2\left[\frac{(1-\alpha)^2}{\alpha} + 4\,\frac{K_2^0}{K_1^0}\,\frac{(2-\alpha)}{\alpha}\right]\right\} + \\ + \frac{1}{2.3\,RT}\,\frac{\partial \bar{G}^e}{\partial \alpha}, \tag{10}$$

where $0 \leqslant \alpha \leqslant 2$.

When $0 \leqslant \alpha \leqslant 1$ the Equation (10) becomes:

$$pK_{\mathrm{app}} = \mathrm{pH} + \log\frac{1-\alpha}{\alpha} = pK_1^0 + \frac{1}{2.3\,RT}\,\frac{\partial \bar{G}^e}{\partial \alpha}, \tag{11}$$

where the barred $\bar{G}^e$ indicates a surface thermodynamic potential. One may expect that if a difference between two titration curves, obtained for a solution or for a monolayer of H.P.A., is observed it could be related to different values of $\bar{G}^e$. For rigid polyions it may be assumed that only $\bar{G}^e_{\mathrm{elec.}}$, the electrostatic contribution to the free enthalpy, varies during the titration.

Writing that the effective (17) or fictive pH at the interface is pH_τ different of pH_∞, where pH_∞ is the bulk pH, it is found that:

$$\mathrm{pH}_\infty = \mathrm{pH}_\tau - \frac{e\phi_0}{2.3\,kT} = \mathrm{pH} . \tag{12}$$

* See also Appendix.

Eliminating pH between (11) and (12) it is obtained:

$$pK_{\text{app}} = pK_1^0 - \frac{e\phi_0}{2.3\,kT}, \tag{13}$$

where

$$pK_1^0 = \text{pH}_\tau + \log\frac{1-\alpha}{\alpha} \quad \text{and} \quad \frac{\partial \bar{G}^e}{\partial \alpha} = -\,Ne\phi_0 \quad (N\text{: Avogadro number}). \tag{14}$$

Eliminating ϕ_0 from (6) and (12) the expression (13) becomes

$$pK_{\text{app}} = pK_1^0 - \frac{e}{2.3\,kT} \quad (\Delta V\text{-constant}). \tag{15}$$

Finally as ΔV is a linear function of α(9) the variation of pK_{app} with α should also be linear.

3.2. Cation exchange by monolayers

This reaction is studied under the conditions that $Z_1 = Z_{\text{Na}} = 1, Z_2 = Z_{\text{Ca}} = 2$ and $\phi_d \ll \phi_0$.

It is assumed that the ions Ca^{2+} are paired with one site, the unpaired charge of Ca^{2+} being neutralized by a coadsorbed anion or by displacement of a second Na^+ ion from the interface.

This reaction is written as follows:

$$\overline{\text{Na}}^0 + \overline{\text{Ca}}^{2+} \rightleftarrows \text{Ca}^+ + \text{Na}^+ . \tag{16}$$

It means that an uncharged $\bar{\text{N}}\text{a}^0$ paired site acquires a positive charge by the reaction (16). The equilibrium constant of the reaction (16) is equal to

$$K_{\overline{\text{Na}}^0}^{\overline{\text{Ca}}^+} = \frac{a_{\overline{\text{Ca}}^+}}{a_{\overline{\text{Na}}^0}} \times \frac{a_{\text{Na}^+}}{a_{\text{Ca}^{2+}}} = \frac{ar/\alpha}{1 - r/\alpha}\,\frac{f_{\overline{\text{Ca}}^+}}{f_{\overline{\text{Na}}^0}} \times \frac{a_{\text{Na}^+}}{a_{\text{Ca}^{2+}}}, \tag{17}$$

where a_i are the activities of the ions in bulk, $f_{\bar{i}}$ their activity coefficients in the region τ, r/α and $1 - r/\alpha$ the fraction of ionized groups occupied by Ca^{2+} and Na^+ respectively. Only $f_{\overline{\text{Ca}}^+}$ is determined by the electrostatic interactions as the Na^+ occupied sites are neutral. According to the model of Guntelberg this excess electrostatic partial free energy and the corresponding activity coefficient of the charged site may be written as follows:

$$\mu_{\overline{\text{Ca}}^+}^{\text{e}} = e\phi^0 = kT \log f_{\overline{\text{Ca}}^+} . \tag{18}$$

Eliminating $f_{\overline{\text{Ca}}^+}$, the relation (17) yields:

$$K_{\overline{\text{Na}}^0}^{\overline{\text{Ca}}^+} = \frac{r/\alpha}{1 - r/\alpha} \times \frac{a_{\text{Na}^+}}{a_{\text{Ca}^{2+}}} \exp\left(e\phi_0/kT\right), \tag{19}$$

where it is supposed that $f_{\overline{\text{Na}}0} = 1$.

From (19) can be deduced an apparent equilibrium constant of ion exchange $K_{\overline{\mathrm{Na}}^0}^{\overline{\mathrm{Ca}}^+}$ app as follows:

$$K_{\overline{\mathrm{Na}}^0}^{\overline{\mathrm{Ca}}^+}\ \mathrm{app} = \frac{r/\alpha}{1 - r/\alpha} \times \frac{a_{\mathrm{Na}^+}}{a_{\mathrm{Ca}^{2+}}}. \tag{20}$$

Using (19) and (20), and using (6) the following variation of pK_{app} with ΔV is found:

$$\log K_{\overline{\mathrm{Na}}^0}^{\overline{\mathrm{Ca}}^+}\ \mathrm{app} = \log K_{\overline{\mathrm{Na}}^0}^{\overline{\mathrm{Ca}}^+} - \frac{e}{kT}\ (\Delta V - \mathrm{constant}). \tag{21}$$

The apparent value of the equilibrium of Ca^{2+}/Na^+ exchange should decrease if the reaction about an increase of the measured surface potential with the amount of bound Ca^{2+} ions. This increase in potential is the consequence of the reversal of the charges of the sites in the region τ and eventually of the sign of the potential ϕ_d in (6). The details concerning the H.P.A. spreading solvent and techniques used in the present study are described in Part 1 [29].

4. Experimental Methods

The surface potential is measured at 24.5°C±0.5 using the device described elsewhere [14].

The accuracy of the results is 1 mV, but their reproducibility is less satisfactory. The errors originating in the inaccuracy of the spread volumes of solution and of the film area are negligible or systematic. The values of the surface potential show a time effect. An aging effect is observed for the spreading solutions of the H.P.A.

Moreover, a Gaussian distribution of the surface potential results is observed for a given system. We ascribe this fact to the polydispersity of the unfractioned H.P.A. and to the small quantities of the spread poly-acid on our small surfaces – 10 cm^2. Thus, considering the results of ten measurements, the standard deviation to the most probable value is ±15 mV and the minimum error is ± 5%.

5. Results

5.1. Titration of H.P.A. Films in the Absence of Calcium

Various surface densities δ of H.P.A. are spread on the substrates containing Na^+ ions at different ionic strengths (10^{-2}–10^{-1} M) and pH.

5.1.1. *Effect of the Film Surface Density*

At constant pH and Na^+ concentration, within the considered range of surface densities (1.61×10^{13}–9.68×10^{14} residues cm^{-2}) the plot of ΔV vs. δ are linear (Figure 2). The upper limit of the linear part corresponds to the inflection point observed on the pressure – area curves reported in Part I (Figures 2–6).

The linear part of ΔV is followed by a plateau, starting at 2.42×10^{14} residues cm^{-2},

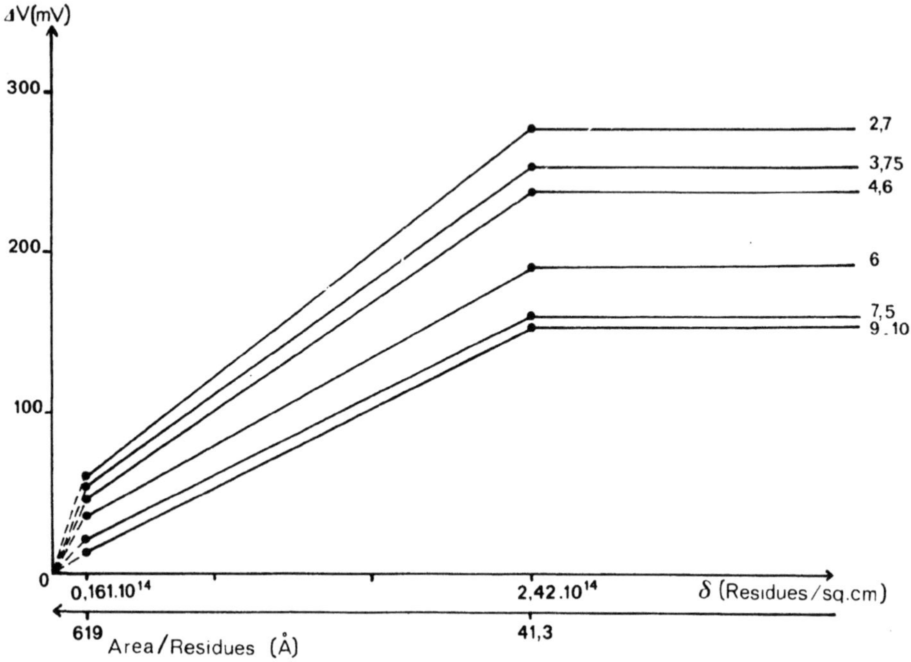

Fig. 2. Surface potential ΔV versus δ monomers/cm^2 at constant pH and $[Na^+]_{bulk} = 6.6 \times 10^{-3}$ M in the absence of Ca^{2+} ions. Curves are labelled with pH values.

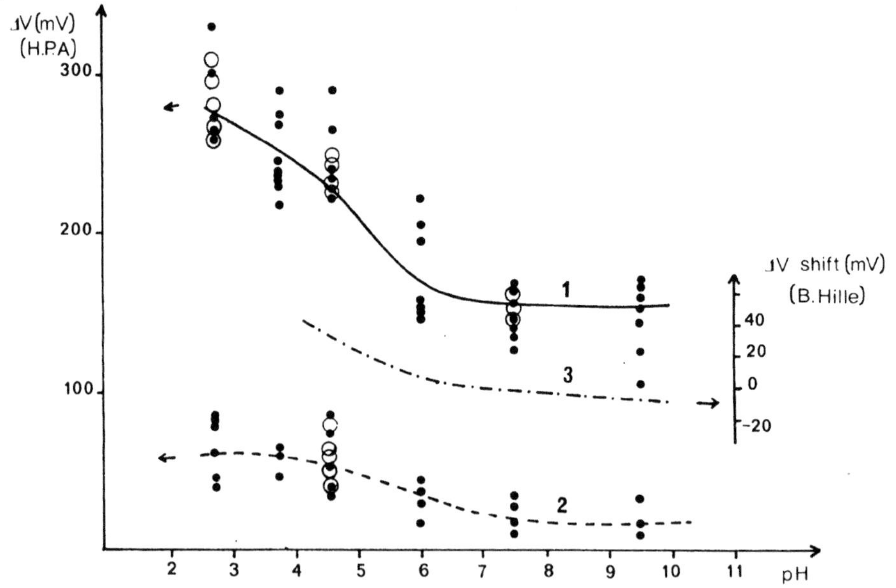

Fig. 3. Surface potential ΔV versus pH at constant δ monomers/cm^2 and $[Na^+]_{bulk}$ in the absence of Ca^{2+} ions.

curve 1: $\delta \geqslant 2.42 \times 10^{14}$ monomers/cm^2.
curve 2: $\delta = 0.32 \times 10^{14}$ monomers/cm^2.
● $[Na^+]_{bulk} = 6.6 \times 10^{-3}$ M (ionic strength 10^{-2} M).
○ $[Na^+]_{bulk} = 1.06 \times 10^{-1}$ M (ionic strength 10^{-1} M).
curve 3 Ranvier node ΔV shift from Hille [30].

corresponding to an area of 41.3 Å^2 residue^{-1}. This value is slightly lower than the projected area – 45 Å^2 – of one stereomodel residue.

5.1.2. *Effect of the pH on the Degree of Ionization of the Interfacial Carboxyl Groups*

Figure 3 shows the plots of ΔV vs pH at constant surface densities of residues δ; as for stearic acid [18] they resemble potentiometric titration curves.

At constant pH there is no apparent salt effect on ΔV in the limit of the experimental accuracy as can be seen from Figure 3. Therefore we infer that ϕ_d may be neglected in the relation (4) and calculate the values of the degree of ionization α utilizing the relation (9).

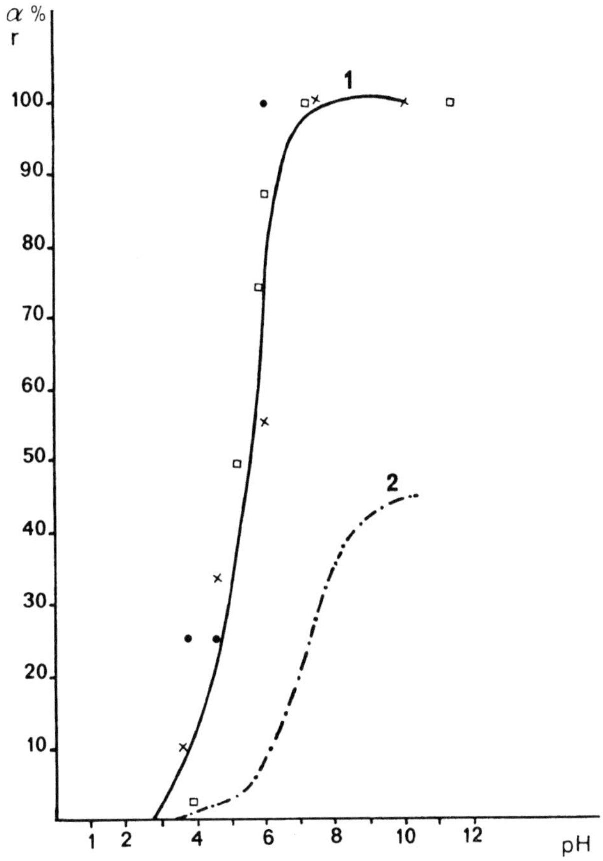

Fig. 4. Curve 1: α = ratio of dissociated COOH/monomers versus pH. α is calculated from (9).
× $\delta \geqslant 2.42 \times 10^{14}$ monomers cm^{-2}.
● $\delta = 0.32 \times 10^{14}$ monomers cm^{-2}.
Curve 2: r = ratio of bound Ca^{2+}/monomers versus pH form Part I Figure 10 [29].
$[Ca^{2+}]_{bulk} = 2 \times 10^{-5}$ M.
$[Na^+]_{bulk} = 6.6 \times 10^{-3}$ M.
Relative increase of membrane maximum conductance $\bar{g}_{Na}$ with pH from Hille [30]: □

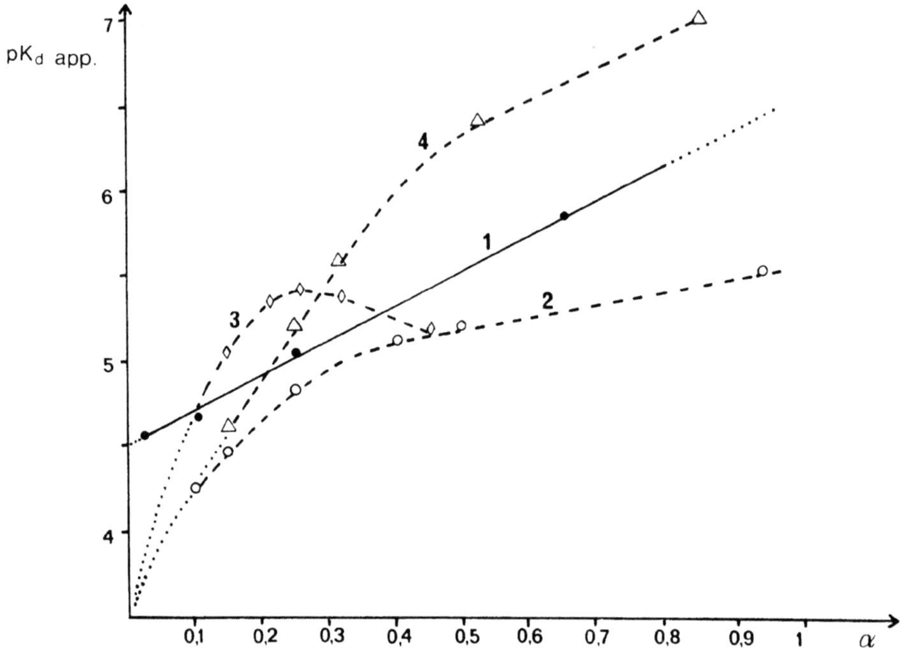

Fig. 5. pK_d of dissociation of HPA vs α in the absence of Ca^{2+} ions.
● from (11) surface film of hexadecyl HPA.
○ ethyl HPA, ◇ butyl HPA, △ hexyl HPA solutions from Strauss [34].

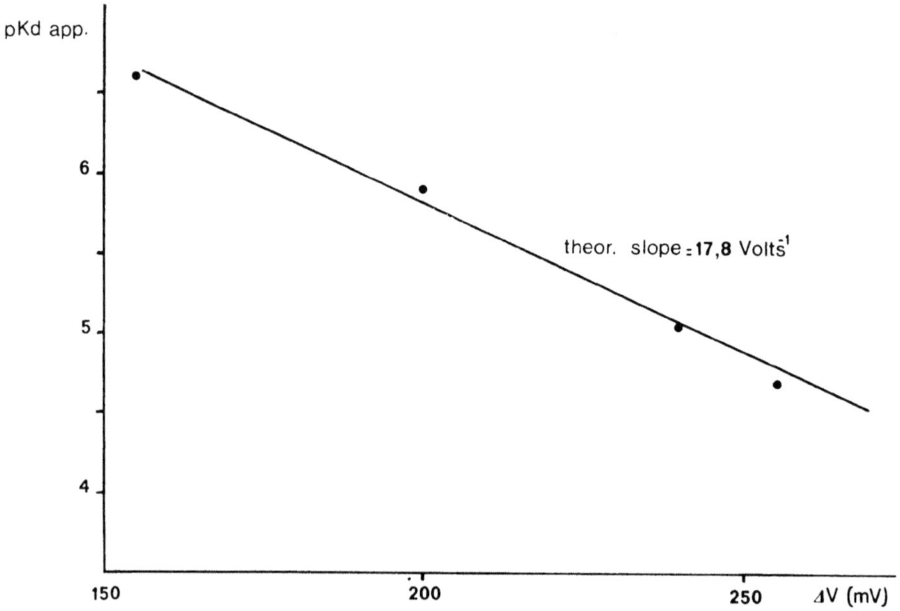

Fig. 6. $pK_{d\,\mathrm{app}}$ of dissociation of surface film of hexadecyl HPA vs ΔV. The straight line has the theoretical slope equal to 17.8 V^{-1} according to (15).

Goddard [18], in his study of stearic acid films, shows that the total drop of the surface potential with pH between 2 and 12 is of the order of 400 mV. This drop consists of two parts. The first one, found between pH 2 and 9, is equal to 200 mV and is ascribed to the formation of a 1/1 acid salt. We obtain this first drop of approximately 200 mV for the H.P.A. films and ascribe it to the dissociation of the first –COOH of the maleic acid residues. The second dissociation, which *pK* is high is not occurring even at the highest pH equal to 10 used. Therefore we assume that at pH=2.7 $\alpha=0$ and at pH$\geqslant$9 $\alpha=1$ (with $0<\alpha<2$).

The values of α obtained from those of the surface potentials at pH=2.7 and pH=9 and the Equation (9) are reproduced on Figure 4. The results obtained for monolayers at two surface densities are comparable. The plot of α vs pH confirms that the degree of dissociation α may be independent of the surface density of the monolayers in the particular case of our system.

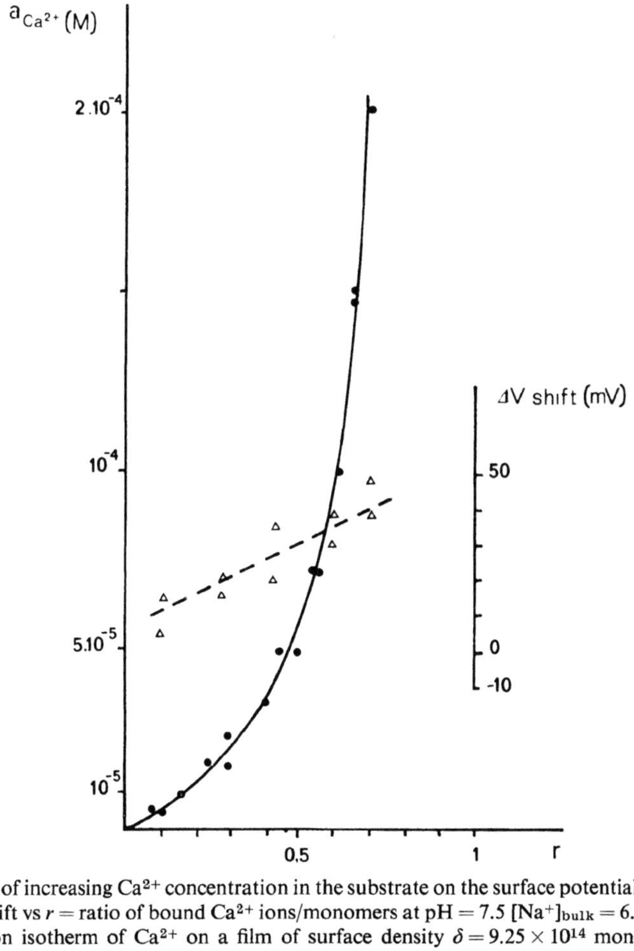

Fig. 7. Effect of increasing Ca^{2+} concentration in the substrate on the surface potential of HPA films. --△--ΔV shift vs r = ratio of bound Ca^{2+} ions/monomers at pH = 7.5 $[Na^+]_{bulk} = 6.6 \times 10^{-3}$ M. ——Adsorption isotherm of Ca^{2+} on a film of surface density $\delta = 9.25 \times 10^{14}$ monomers cm^{-2} is reproduced from Figure 9 Part I [29].

5.1.3. *Relation between the Interfacial Potential and the Apparent Constant of Dissociation*

The values of pK_{app} are calculated for a monolayer at a given surface density and various pH according to the relation (11). They are plotted vs α on Figure 5. On Figure 6 these values of pK_{app} are plotted as a function of the corresponding surface potential according to the Equation (15). Straight lines are obtained in both cases. Their slopes have opposite signs because ΔV decreases when α increases.

The absolute value of the slope of the pK_{app} vs ΔV curve is close to $e/2.3kT = 17.8\ \mathrm{V}^{-1}$. The value of pK_1^0 obtained by extrapolating the pK_{app} vs α straight line to $\alpha = 0$ is equal to 4.5. On Figure 5 our results are compared with those obtained by Strauss [34] for ethyl, butyl, or hexyl H.P.A. titrated in solution.

5.2. Binding of Calcium by H.P.A. Films

Monolayers of a given surface density are spread on the surface of substrates at various pH and concentrations of Ca^{2+} or Na^+ ions.

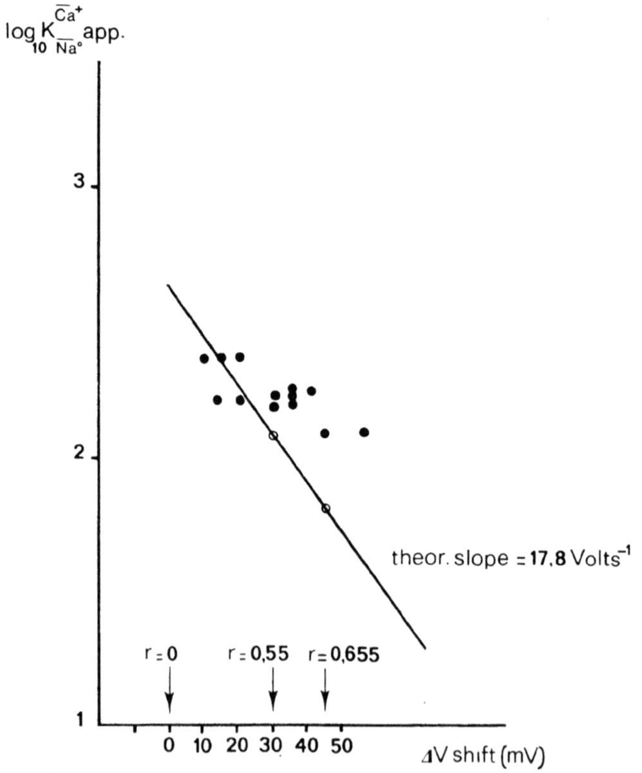

Fig. 8. $\mathrm{Log}K_{\mathrm{Na}0}{}^{\mathrm{Ca}+}$ app vs ΔV shift (or $\Delta(\Delta V)$) when $[Ca^{2+}]_{\mathrm{bulk}}$ varies in the substrate. ΔV shift (or $\Delta(\Delta V)$) is the difference between the surface potential of HPA film in the presence and in the absence of Ca^{2+} ions. The straight line has the theoretical slope 17.8 V^{-1} according (21).

5.2.1. *Effect of the Concentration of Ca^{2+} at Constant δ, pH and Ionic Strength*

The surface potential measured for H.P.A. at pH = 7.5 and $\alpha = 1$ are shown on Figure 7. The adsorption isotherm of Ca^{2+} on a film of the same density is reproduced from Part I [29]. Using these two sets of results the validity of the law (21) is tested. We plot the values of log $K^{\overline{Ca}^{+}}_{\overline{Na}^{0}app}$, as a function of $\Delta(\Delta V)$ on the Figure 8 where $\Delta(\Delta V)$ is the ΔV shift or the differentce between the surface potentials of the H.P.A. film in the presence or in the absence of Ca^{2+} in the system.

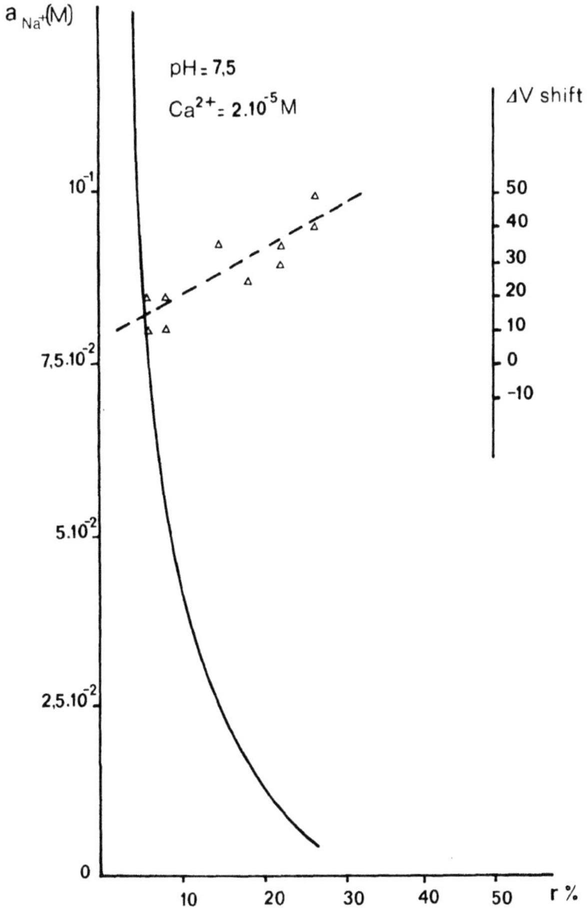

Fig. 9. Effect of the displacement of bound Ca^{2+} ions by Na^{+} ions on the surface potential of HPA films. $[Na^{+}]_{bulk}$ varies in the substrate; pH = 7.5; $\delta = 9.25 \times 10^{14}$ monomers cm^{-2}; $[Ca^{3+}]_{bulk} = 2 \times 10^{-5}$ M. --△-- ΔV shift versus r = ratio of bound Ca^{2+} ions/monomers. —— Adsorption isotherm of Ca^{2+} on a film of the same surface density δ is reproduced from from Part I (29).

5.2.2. *Displacement of Bound Ca^{2+} Ions at Constant δ, pH and Ca^{2+} Concentration in the Substrate*

To increase the ionic strength of the substrate, NaCl is added to it at pH = 7.5 and $\alpha = 1$.

A parallel decrease of the surface potential of the monolayer is observed and corresponds to the displacement of the Ca^{2+} ions related in Part I [29]. The curve of Figure 9 reproduces these results.

If this displacement is a chemical effect, the Equation (21) may describe it. We test this assumption on Figure 10 by plotting log $K_{\bar{Na}^0}^{\bar{Ca}^+}$ app vs $\Delta(\Delta V)$ or surface potential shift.

5.2.3. *Effect of the Degree of Dissociation of the H.P.A. Monolayer on the Amount of Bound Ca^{2+} Ions at Constant Ionic Strength and Ca^{2+} Concentration of the Substrate*

The variation of the amount of Ca^{2+} ions bound by a H.P.A. concentrated film at various pH values of substrate is reproduced on the Figure 4 from Part I [29]. On Figure 11 we plot log $K_{\bar{Na}^0}^{\bar{Ca}^+}$ app vs ΔV according to (20) and (21). The values of ΔV are

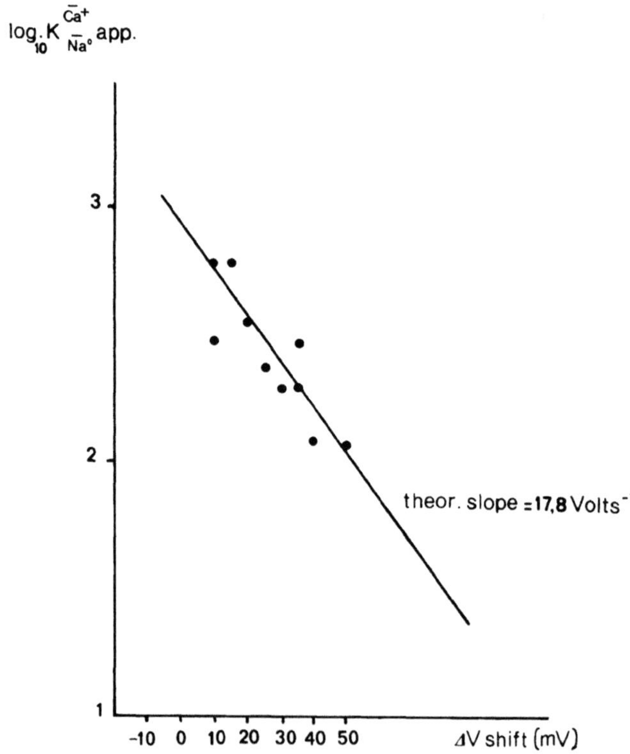

Fig. 10. Log $K_{Na^0}^{Ca^+}$ app vs ΔV shift when $[Na^+]_{bulk}$ varies in the substrate. The straight line has the theoretical slope 17.8 V^{-1} according to (21).

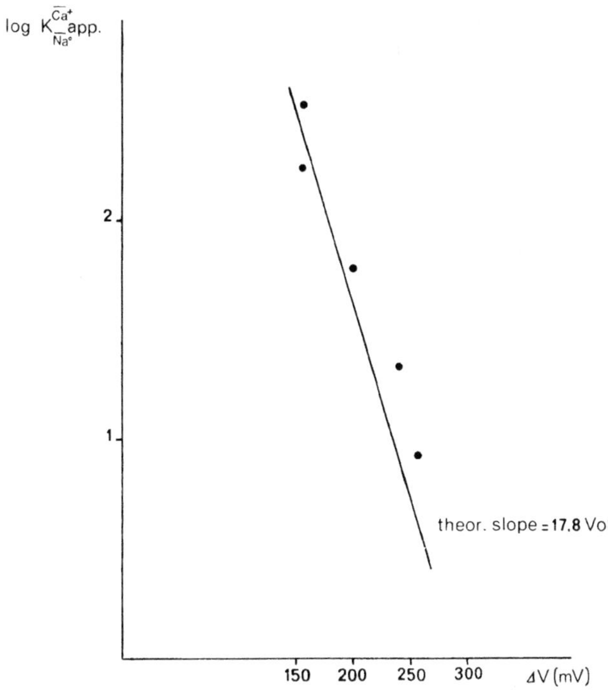

Fig. 11. Log $K_{Na_0}{}^{Ca+}$ app vs ΔV when pH varies in the substrate. $[Na^+]_{bulk} = 6.6 \times 10^{-3}$ M; $[Ca^{2+}]_{bulk} = 2 \times 10^{-5}$ M; $\delta = 9.25 \times 10^{14}$ monomers cm^{-2}. The straight line has the theoretical slope 17.8 V^{-1}. Adsorption isotherm of bound Ca^{2+} is shown on curve 2 Figure 4.

those measured in the absence of Ca^{2+} ions. We have verified at pH = 7.5 that the amount of bound Ca^{2+} ions is too low to change significantly the values of ΔV at constant pH.

6. Discussion of the Results

The studies of the compression isotherms of monolayers of H.P.A. show that these films are solid and insensitive to pH or Ca^{2+} ions (Part I, [29]. Nevertheless their surface potential shift with the ionic strength (present work) is of the order of 20 mV when $\Delta \log c = 1$.

That H.P.A. is ionized is shown by the titration curves of Figure 3. The polyions thus obtained seem to possess such a high charge density in the range of the considered values of α that ion pairing takes place between the negative carboxylates and their counterions as it is shown on the model (Figure 1).

This hypothesis is verified by the very slight variation of the surface potential with regard to the ionic strength (Figure 3). Indeed let pH = 7.5 and $\alpha = 1$, a monolayer of an average surface density $\delta = 2.42 \times 10^4$ residues cm^2, and surface charge density

$\delta = 1.16 \times 10^5$ u.e.s. would lead to a calculated variation of ϕ_d equal to $\phi_d = 57$ mV for a change of the ionic strength from 10^{-2} to 10^{-1} M. The experimental value found for the variation of the surface potential is 20 mV within the limits of the experimental accuracy: 15 mV. The contribution of the permanent dipoles to the surface potential may be taken as independent of the ionic strength of the substrate. Then the observed very slight dependence of the surface potential on the ionic strength can be explained by a small value of ϕ_d in Equation (3) owing to counter-ion pairing with their sites. It is accepted that this process of counter-ion pairing is responsible for the limiting value of the zeta potential [20] of ionized monolayers at oil/water interface, which is smaller than the surface potential of the same films [28]. The difference between these values accounts for the contribution $\Delta\eta_{\text{I.P.}}$ of the ion pairs in the region τ (Figure 1). Therefore in the absence of Ca^{2+} our results for ΔV may be interpreted according to the relation (7) where ϕ_d is neglected. This first order approximation leading to Equation (7) may be justified by the proportionality between the surface potential and density at pH 7.5 and 9 (Figure 2).

This behaviour is different from the theoretical law calculated using the Gouy classical expression of ϕ_d in the absence of counterion pairing. The values of α shown on the Figure 4 may be deduced from ΔV using (9).

In the present work we are not interested to justify or calculate the absolute value of the effective moments P_i of the ion pairs.

Under these conditions we find that the variation of the degree of dissociation with pH is independent of the average surface concentration of the monolayer when δ varies six times. This fact confirms the conclusions of Part I [29] that long range electrostatic intermolecular interactions are absent and also that ϕ_d is negligible. Indeed the real surface charge density is determined only by the linear charge density on the polyions of the H.P.A. which governs the short range interactions and the formation of the ion pairs. Owing to this very high charge density the second dissociation of the maleic acid belonging to the residues of the H.P.A. does not occur as will be shown below.

The titration curve of Figure 4 satisfies the Henderson-Hasselbach equation and our Equation (15). The slopes of the lines representing pK_{app} vs ΔV or vs α are practically equal to the theoretical value of: $e/2.3\,kT = 17.8\ \text{V}^{-1}$ predicted by (9) and (15). The conclusion expressed by the second Equation (14) is verified e.g. the excess electrostatic partial free energy of an anionic site may then be taken as equal to $ze\phi_0$ where ϕ_0 is the potential at the charged sites. It may be related also to the measured surface potential of the monolayer even when the distribution of the counterions is ignored. This is not allowed for polyelectrolytes membranes, solutions or gels for which neither the distribution of the fixed charged sites nor the potential shift ϕ_0 may be measured.

Finally, the values of the pK_{app} we deduce are comparable to those obtained by Strauss [34] in the range of the values of α considered: $0.1 < \alpha < 1$. This author titrates three shorter side chains analogous H.P.A.: the copolymers of maleic acid and ethyl or butyl or hexyl vinyl-ethers. We extrapolate our results shown on the Figure 5 to

$\alpha=0$ and to $\alpha=1$ and find $pK_1^0=4.5$ and $pK_2^0=6.6$ respectively for the first and the second dissociation constants of maleic acid. pK_1^0 is higher than the corresponding bulk value of 3.5. The difference may be ascribed to a difference in the local state of the water near the air water interface.

pK_2^0 verifies the value 6.7 found in the same way by Strauss. This fact may be considered as an a posteriori verification of our method of calculating α neglecting ϕ_d in Equation (3). In this case the state of the water seems to be very similar in the neighbourhood of the polyion anywhere, either in bulk or in the monolayer.

The process of charging a film by dissociation of its carboxylic groups, as it is described in Appendix, is generalized to a monolayer which exchange the ions Ca^{2+} and Na^+. The verification of the relation (21) corresponding to the equilibrium (16) is attempted using three types of experiments – increase of Ca^{2+} binding with Ca^{2+} concentration in bulk or with the decrease of ionic strength, or with pH – respectively on Figures 8, 10 and 11. As long as the amount of bound Ca^{2+}, as well as the surface potential, are not too high the slope of the lines obtained by plotting $\log K_{Na^0}^{Ca^+}$ app vs ΔV shift is close to the theoretical value of: $-e/2.3\,kT=-17.8\ V^{-1}$. Indeed, on Figure 10, although the results are scattered, they show the effect of surface potential on the Ca^{2+}/Na^+ exchange predicted by (21): the displacement of the Ca^{2+} ions by Na^+ ions the concentrations of which in bulk is increased. In the same way the variation of Ca^{2+} binding with the degree of the dissociation of the monolayer is well described by (21) as it is shown on Figure 11. Finally the plot of $\log K_{Na^0}^{Ca^+}$ app vs $\Delta(\Delta V)$ on Figure 8 according to Equation (21) yields a curve which is tangent to the theoretical line for small values of $\Delta(\Delta V)$ and of the amounts of bound Ca^{2+}.

However when ΔV and the amount of bound Ca^{2+} are high the Equation (21) seems to be unsuitable. In this case the positive charge brought by the bound Ca^{2+} ions may lower the pK of the dissociation of the second carboxyl of the maleic acid residue. Therefore, on Figure 8, the points on the theoretical line for $\Delta V=230$ mV and 240 mV corresponding to the amounts r of bound Ca^{2+} equal to 0.55 and 0.665 are obtained by adjusting in equations (20) and (21) the value of α. Then α is equal respectively to 1.15 and 1.3.

In conclusion the relation (21) may not be verified if the possible influence of bound Ca^{2+} on the ionization of the maleic acid residue is not considered.

Our results for Ca^{2+} binding and their interpretation may be compared with those of Goerke *et al.* [19] for stearic or oleic acid monolayers. The affinity constant for one Ca^{2+} ion bound to one lipid site is deduced using a Donnan type of treatment.

Their 'constant' may be deduced from our Equation (20) by dividing the last one by a_{Na}. From our data for H.P.A. we find an affinity constant for one Ca^{2+} ion binding to one carboxylate equal to 1.5×10^4. This value is close to 1×10^4 obtained by Goerke *et al.* [19] for a monolayer of oleic acid at 20 dyne cm^{-1}.

The independence of the surface pressure of the oleic acid monolayers with respect to Ca^{2+} ions found by Goerke *et al.* [19] is a second feature common to oleic acid and H.P.A. monolayers [29]. Finally the decrease of the apparent affinity constant with the degree r of bound Ca^{2+} is stronger for stearic than for oleic acid monolayers. For them at pH 6.8 and pCa = 3 the values of r are 0.5 and 1 respectively for stearic and oleic acid monolayers.

It is suggested that Ca^{2+} binding to stearic acid may occur through coordination [35] or by association of one Ca^{2+} ion to two neighboring carboxylates leading to condensation of the monolayer. The results of Goerke *et al.* [19] for oleic acid and ours on H.P.A. monolayers show that this behaviour may not be general. It seems to be related to the distances between the ionized sites.

For the last two monolayers these distances are larger than for the stearic acid films.

The oleic acid molecules are kept apart by the kink produced in their chain by the double bound. The carboxylates of our ionized H.P.A. are separated by the hexadecyl-vinylether group. These large intersite distances may hinder the simultaneous ion pairing of one Ca^{2+} ion to more than one negatively charged sites.

6.1. Biological implications

These can be seen on Figures 3, 4 and 12. They reproduce the results obtained by Hille [30] who studies the effect of pH and Ca^{2+} on Na^+ currents at the Ranvier nodes during the action potential. It is claimed that these currents and the membrane conductance may depend on the surface potential of the membrane and that pH and Ca^{2+} binding may modify this potential. These modifications are represented as shifts of the plots of potential depending membrane properties. Furthermore it is inferred that the pH modifies the ionization of carboxylic groups in the membrane and its conductance.

On Figure 3 the supposed biological pH dependent shift is compared to the measured surface potential of our H.P.A. monolayer. It may be concluded that the conducting

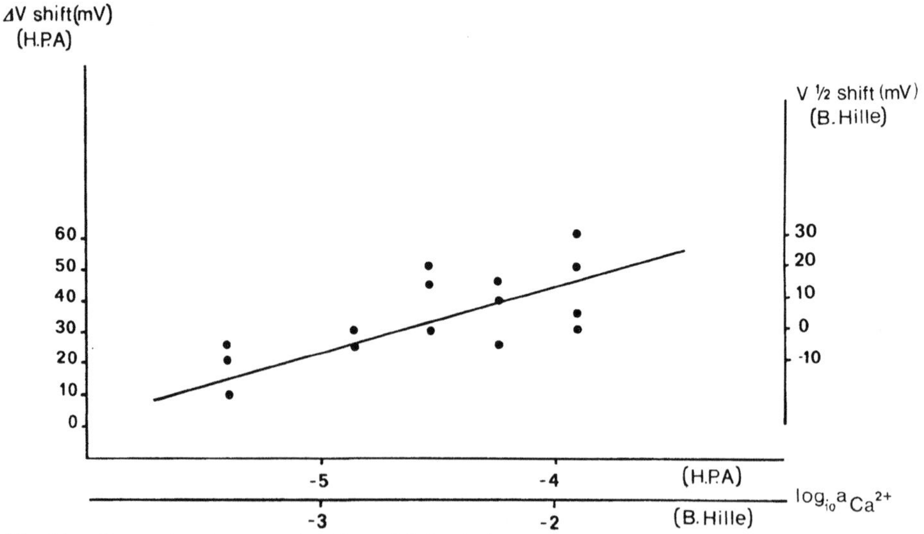

Fig. 12. Comparison between the effect of Ca^{2+} binding on the biological membrane potential and on the surface potential of HPA monolayer. Straight line: biological membrane potential shift calculated by Hille [30]. Points: measured variation of surface potential of HPA monolayer at pH 7.5 $\delta = 2.42 \times 10^{14}$ residues cm^{-2}.

structure of the membrane possesses a high density of carboxylic groups: about 10^{14} groups cm^{-2} which may be possibly maleic acid residues. On Figure 4, the relative increase of the membrane maximum Na^+ conductance, $\bar{g}_{Na}$, with pH is compared to our measured values of the degree of the first ionization of maleic acid. The agreement is surprisingly good. On Figure 12 the supposed biological membrane potential shift by Ca^{2+} binding is compared with the measured shifts of the surface potential of the H.P.A. monolayers at pH=7.5. The ranges of Ca^{2+} concentration in bulk are different as the Na^+ concentration are not equal for the two series of experiments: with biological membranes or with monolayers. But the slopes are qualitatively the same verifying that as for the effect of the pH shown in Figures 3 and 4 the surface density of the COOH groups on the membranes is of the order of 10^{14} groups cm^{-2}.

7. Conclusion

Hydrophobic polyacids spread as monolayers at the surface of aqueous solutions behave as polyelectrolytes solutions with regard to pH variations. When dissociated they seem to form ion-pairs with the monovalent counterions Na^+. The binding of Ca^{2+} ions proceeds through pairing of one ion with one carboxylate leading to an increase of the surface potential or charging of the monolayer. This surface potential may be related to the electrostatic free energy of the charging of one site by Ca^{2+} binding as for COOH dissociation. It governs the variation of the apparent equilibrium of these ionic reactions. The same reactions and variations of the surface potential may occur at the surface of a membrane or at the specialized sites reponsible for membrane conduction. This can be inferred from the comparison of the results obtained for membranes and monolayers of the maleic acid copolymer H.P.A.

Appendix

The Charging of a Surface by the Dissociation of Carboxylic Groups

This charging may be represented by an equation analogous to (16):

$$- \overline{COOH} \rightleftarrows \overline{COO}- + H^+, \quad \text{[A1}$$

where H^+ is located in the bulk of the solution.

The equilibrium constant of the reaction (A1) is equal to:

$$K_0 = \frac{a_{\overline{COO^-}} \times a_{H^+}}{a_{\overline{COOH}}} = \frac{\alpha}{1-\alpha} \frac{f_{\overline{COO^-}}}{f_{\overline{COOH}}} \times a_{H^+}, \quad \text{A2)}$$

where K_0 is the intrinsic dissociation constant of COOH. The activity coefficient $f_{\overline{COO^-}}$ is related to the excess electrostatic free energy of the negatively charged site as assumed by Katchalsky [10b].

$$kT \log f_{\overline{COO^-}} = Z_{(-)} e\phi_0, \quad \text{[A3]}$$

where $Z_{(-)} = -1$.

Eliminating $f_{\overline{COO^-}}$ from (A2) using (A3) it follows for K_0:

$$K_0 = \frac{\alpha}{1-\alpha} \times a_{H^+} \exp\left(-e\phi_0/kT\right) = K_{app} \exp\left(-e\phi_0/kT\right). \quad \text{(A4}$$

This equation is analogous to (11) where

$$\frac{\partial G^e}{\partial \alpha} = -e\phi_0 \quad \text{(A5)}$$

or to the Henderson-Hasselbach Equation (14) which is obtained from 11, 12) and (13).

In our derivation of (A4) we do not allow for the activity of protons in the surface which could be formally defined as:

$$a_{\overline{H}^+} = a_{H^+} \times \exp(-e\phi_0/kT) = a_{H^+} \times f_{\overline{COO^-}}. \quad \text{(A6)}$$

Therefore this activity has no physical meaning.

On the contrary if we keep in mind the origin of the exponential the following trivial result is obtained:

$$\frac{\partial G^e}{\partial \alpha} = -e\phi_0, \quad \text{(A7)}$$

where $\partial G^e \partial \alpha$ is the partial excess electrostatic free energy of the residue charged by ionization. The same conclusion may be reached as for the Ca^{2+}/Na^+ reaction (16) which produces a charged site or for the protonation of an amino group at low pH [10b].

Acknowledgments

The authors wish to express their gratitude to Professor M. Dodé (Laboratoire de Chimie Thermodynamique, Centre d'Orsay Université de Paris-Sud), for his interest in this study.

Thanks are also due to Dr R. Varoqui (C.R.M. Strasbourg) for his kind gift of the pure copolymer of maleic acid and hexadecylvinylether H.P.A.

References

1. Frankenhauser, B. and Hodgkin, A. L.: *J. Physiol. London* **131**, 218 (1957).
2. Tasaki, I., Watanabe, A., and Lerman, L.: *Am. J. Physiol.* **213**, 1465 (1967).
3. Gilbert, D. L. and Ehrenstein, G.: *Biophys. J.* **9**, 447 (1969).
4. Mozhayeva, G. N. and Naumov, A. P.: *Nature* **228**, 164 (1970).
5. Ling, G. N.: in L. M. Gross (ed.), *A Physical Theorie of the Living State*, Blaisdell Publ. Co., New York, Toronto, London, 1962.
6. Morawetz, H.: in *Macromolecules in Solution*, Interscience Publishers, New York, London, Sydney, 1965.
7. Strauss, U. P. and Po Leung, Y.: *J. Am. Chem. Soc.* **87**, 1475 (1965).
8. Zana, R., Tondre, C., Rinaudo, M., and Milas, M.: *J. Chim. Phys.* **9**, 1258 (1971).
9. Rice, S. A. and Nagasawa, M.: in *Polyelectrolyte Solutions*, Academic Press, New York, London, chap. V–IX.

10. in B. E. Conway and R. G. Barradas (eds.), *Chemical Physics of Ionic Solutions,* Electrochemical Society series, 1966.
a. Strauss, U. P. and Gross, L. M.: p. 361.
b. Katchalsky, A., Alexandrowicz, Z., and Kedem, O.: p. 295.
11. De Heaulme, M., Hendrikx, Y., Luzzati, A., and Ter-Minassian-Saraga, L.: *J. Chim. Phys. Physicochim. Biol.* **64**, 1363 (1967).
12. De Heaulme, M.: *J. Chim. Phys. Physicochim. Biol.* **66**, 653 (1969).
13. De Heaulme, M., Hendrikx, Y., Luzzati, A., and Ter-Minassian-Saraga, L.: *Adv. Chem. Ser.* **79**, 23 (1968).
14. Plaisance, M. and Ter-Minassian-Saraga, L.: *J. Colloid. Interf. Sci.* **38**, 489 (1972).
15. in D. Chapman (ed.), *Biological Membranes*, Academic Press, London and New York, 1968.
16. Schulman, J. H. and Hughes, A. H.: *Proc. Roy. Soc. London* **A138**, 430 (1932).
17. Betts, J. J. and Pethica, B. A.: *Trans. Faraday Soc.* **52**, 1581 (1956).
18. Goddard, E. D. and Ackilli, J. A.: *J. Colloid. Interf. Sci.* **18**, 585 (1963).
19. Goerke, J., Harper, A. H., and Borowitz, M.: in *Surf. Chem. of Biol. Systems*, Plenum Press, 1970, p. 23.
20. Levine, S., Mingins, J., and Bell, G. M.: *J. Phys. Chem.* **67**, 2095 (1963).
21. Jaffé, J., Ruysschaert, J. M., and Bricman, G.: *J. Polymer Sci.* (A.2) **8**, 817 (1970).
22. Conway, B. E.: in *Electrode Process*, Ronald Press, New York, 1965.
23. Levine, S. and Bell, G. M.: *J. Phys. Chem.* **67**, 1408 (1963).
24. in *Inform. Bull. of Inter. Union of Pure Appl. Chem.*, No. 3 January, 1970.
25. Sanfeld, A., Hurwitz, H., and Steinchen-Sanfeld, A.: *Electrochemica Acta* **9**, 923 (1964).
26. Overbeek, J. Th. G. and Verwey, E. J. W.: in *Theory of the Stability of Lyophobic Colloids*, Elsevier Publ. Co. Inc., New York, Amsterdam, London, Brussels, 1945.
27. Fowler, R. H. and Guggenheim, E. A.: in *Statistical Thermodynamics*, Cambridge at the University Press, 1939.
28. Haydon, D. A. and Taylor, F. H.: *Phil. Trans. Roy. Soc. London Ser.* **A252**, 225 (1960) and **A253**, 255 (1960).
29. This volume, Part I.
30. Hille, B.: *J. Gen. Physiol.* **51**, 221 (1968).
31. Manning, G. S.: *J. Chem. Phys.* **51**, 924 (1969); **51**, 934 1969); **51**, 3249 (1969).
32. Bell, G. M. and Levine, S.: *J. Coll. Interf. Sci.* **41**, 275 (1972).
33. Levine, S.: *J. Coll. Interf. Sci.* **37**, 619 (1971).
34. Dubin, P. L. and Strauss, U. P.: *J. Phys. Chem.* **74**, 2842 (1970).
35. Deamer, D. W. and Cornwell, D. G.: *B.B.A.* **116**, 555 (1966).

SEDIMENTATION AND DIFFUSION OF POLYELECTROLYTES IN THE PRESENCE OF ADDED SALT

MITSURU NAGASAWA

Department of Synthetic Chemistry, Nagoya University, Chikusa-Ku, Nagoya, Japan

1. Introduction

The purpose of the study on transport phenomena of polyelectrolytes may be to discuss the response of both the polymeric ion and the counter-ions in solutions to externally applied fields: those considered include the electric field, centrifugal field, shearing force and the chemical potential gradient. That is, sedimentation, diffusion, viscosity, electrophoresis, electric conductivity, etc. may be included in the transport phenomena of linear polyelectrolytes. Among them, here we shall discuss sedimentation and diffusion which both are closely related. Just as in all other phenomena, it is convenient to divide our discussion into two problems; the concentration dependence of transport coefficients, which is mainly concerned with the interaction between two polymeric ions, and the limiting magnitudes of sedimentation and diffusion coefficients, which are determined mainly by the radius of gyration of polymeric ion in an infinitely large volume of solvent.

In the discussion on the former, we should take into account both the hydrodynamic interaction between polymeric ions which are more or less randomly coiled and the electrostatic interaction. The hydrodynamic interaction is predominant in the study of sedimentation and diffusion of non-ionic polymers, whereas the electrostatic interaction is also important for polymeric ions.

In the discussion on the latter, that is, on the limiting values, too, the effect of ionic atmosphere should be taken into account in addition to the change of radius of gyration which is only important for the limiting sedimentation and diffusion coefficients of non-ionic polymers.

2. The Concentration Dependences of the Sedimentation and Diffusion Coefficients in the Presence of Added-Salt

If we define the fluxes of solutes, $(J_k)_c$, in relation to the cell, as they are experimentally observed, we have the following phenomenological equation

$$(J_k)_c = \sum_i (L_{ki})_v X_i . \qquad (1)$$

Here, it is assumed that one molecule of polyelectrolyte is dissociated into one polyion of valence $-Z$ and Z cations of valence $+1$, and also that a neutral salt of 1–1 valence type which has a common counter-ion with the polyelectrolyte is added to the solution. The suffixes p, $+$, and $-$ denote polyion, counter-ion, and

Eric Sélégny (ed.), Polyelectrolytes, 241–260.

by-ion, respectively. X_i is the generalized force acting on the component i per unit mass, and the $(L_{ki})_v$ are phenomenological coefficients. Since the solutes have charges, the force X_k in general contains three contributions: the chemical potential gradient, the electric force and the centrifugal force (in sedimentation). Thus,

$$X_k = (1 - \bar{v}_k \varrho_m)\,\omega^2 r - \sum_j \left(\frac{\partial \mu_k}{\partial n_j}\right)_{T,P,n_b} \left(\frac{\partial n_j}{\partial r}\right)_{T,P} + Z_k e E\,, \tag{2}$$

where $\bar{v}_k$, μ_k, Z_k and n_j are the partial specific volume, chemical potential, valence, and concentration of component k or j, respectively. In Equation (2), ω is the angular velocity of rotation, r is the distance from the centre of rotation, ϱ_m is the mass density of the solution, and E is the electric field. In diffusion, the first term of the above equation does not appear. In sedimentation and diffusion, E in the third term of Equation (2) is replaced by $(\partial\psi/\partial r)_{T,P}$, the electrostatic potential gradient set up automatically in the solution. Then, the equation for the flow of solute, $(J_k)_c$, may be expressed in the form

$$(J_k)_c = (S_k)_v\, n_k \omega^2 r - \sum_j (D_{ki})_v \left(\frac{\partial n_j}{\partial r}\right)_{T,P} + (l_k)_v\, n_k \left(\frac{\partial \psi}{\partial r}\right)_{T,P}, \tag{3}$$

where the sedimentation coefficient $(S_k)_v$, the mutual diffusion coefficient $(D_{kj})_v$, and the ionic mobilities $(1_k)_v$ are defined as

$$(S_k)_v = \frac{l}{n_k} \sum_i (L_{ki})_v\,(1 - \bar{v}_i \varrho_m)\,, \tag{4}$$

$$(D_{kj})_v = \sum_i (L_{ki})_v \left(\frac{\partial \mu_i}{\partial n_j}\right)_{T,P,n_b}, \tag{5}$$

$$(l_k)_v = \frac{Z_k e}{n_k} \sum_i (L_{ki})_v \tag{6}$$

and b denotes all components other than solute j.

2.1. In sedimentation velocity experiments [1, 2]

When a steady state is established, no current can flow in the cell, so that

$$\sum_k (Z_k e)\,(J_k)_c = 0\,. \tag{7}$$

Therefore, we have the following equation for the sedimentation potential

$$\left(\frac{\partial \psi}{\partial r}\right)_{T,P} = \frac{\sum_i \sum_j Z_k (D_{ki})_v \left(\frac{\partial n_j}{\partial r}\right)_{T,P}}{\sum_k Z_k (l_k)_v n_k} - \frac{\sum_k Z_k (S_k)_v n_k}{\sum_k Z_k (l_k)_v n_k}\,. \tag{8}$$

Insertion of Equation (8) into Equation (3) gives

$$(J_k)_c = (S_k)_v^{\text{app}} n_k \omega^2 r - \sum_j (D_{kj})_v^{\text{app}} \frac{\partial n_j}{\partial r} \tag{9}$$

in which the sedimentation coefficient observed experimentally, $(S_k)_v^{\text{app}}$, and $(D_{kj})_v^{\text{app}}$ are related to the true sedimentation coefficient and diffusion coefficient by the following equations:

$$(S_k)_v^{\text{app}} = (S_k)_v - (l_k)_v \frac{\sum_k Z_k (S_k)_v n_k}{\sum_k Z_k (l_k)_v n_k}, \tag{10}$$

$$(D_{kj})_v^{\text{app}} = (D_{kj})_v - (l_k)_v \frac{\sum_k Z_k (D_{kj})_v n_k}{\sum_k Z_k (l_k)_v n_k}. \tag{11}$$

If the sedimentation coefficients of the ions constituting the added neutral salt are

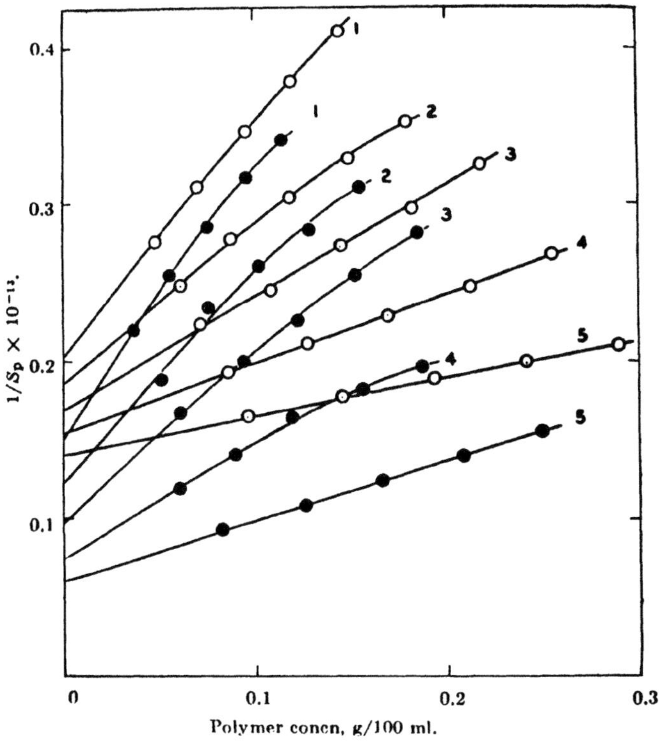

Fig. 1. Examples of the sedimentation coefficient vs. polymer concentration plots. (Reproduced from Reference [2].) Samples, sodium polystyrene sulphonate (Na-PSS). Molecular weight (M_w), open circle 3.9×10^5; filled circles 2.3×10^6, NaCl concentrations, 1, 0.005; 2,0.01; 3, 0.02; 4, 0.05; 5.02 N. Speed of rotation, 59.780 rpm.

equal, that is, if $S_+ = S_-$ as in NaCl, Equation (10) simplifies to

$$\frac{1}{(S_p)_v^{\mathrm{app}}} = \frac{1}{(S_p)_v}(1 + k_s n_p + k_2 n_p^2 + \cdots) \tag{12}$$

with

$$k_s = \frac{Z_p(l_p)_v}{[(l_+)_v - (l_-)_v]\, n_s}, \tag{13}$$

$$k_2 = -k_s^2,$$

where n_p and n_s are the concentrations of polyion and added salt, respectively. Thus, $(S_p)_v$, which is called the limiting sedimentation coefficient, can be obtained as the limit of the observed sedimentation coefficient, $(S_p)_v^{\mathrm{app}}$, at infinite dilution of the marcromolecule.

Examples of the relationship between the observed sedimentation coefficient and the polyion concentration are shown in Figure 1. The limiting sedimentation coefficient, $(S_p)_v$, is observed to depend upon the ionic strength. These differences in $(S_p)_v$ are to be expected because of the expansion of the polyion coil as the ionic strength is decreased, and also because the ionic atmosphere influences the motion of the polyion.

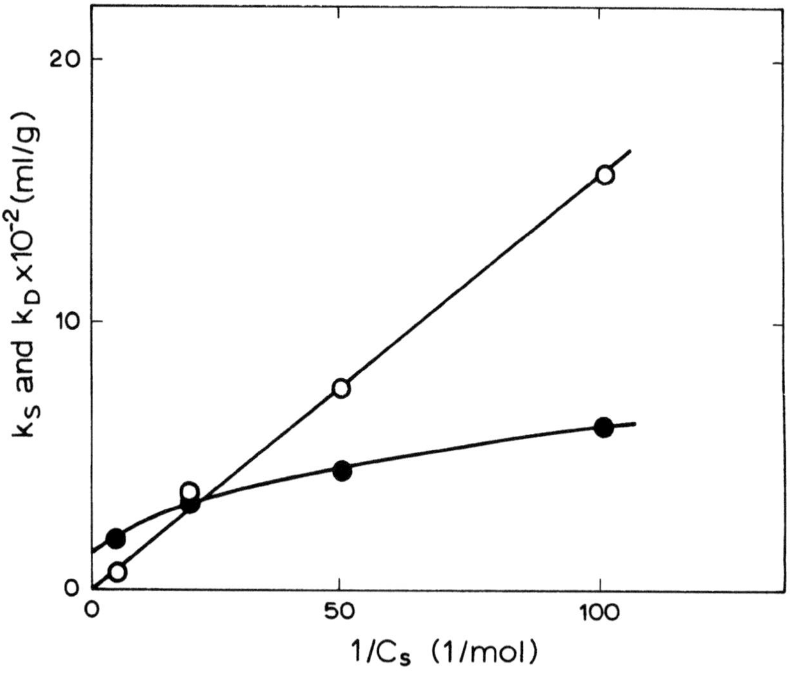

Fig. 2. Relationship between polymer concentration dependence coefficients of sedimentation k_s (filled circle) and diffusion k_D (open circles) and NaCl concentration. (Reproduced from Reference [4].) Calculated from Figures 1 and 4.

There have been some reports that the limiting sedimentation coefficient is independent of ionic strength. These results would be obtained if the molecular weights of the samples used in the studies are low or the backbone of the samples is stiff, so that the polyion coil may be free-draining for solvent.

According to Equation (13), the slope in Figure 1 should be linear with respect to the reciprocal concentration of added salt. This prediction is not verified by the experimental data in Figure 1, as shown in Figure 2. One possible source of disagreement between theory and experiment is the neglect of the hydrodynamic interaction between polyion coils; this point will be discussed again in our analysis of the concentration dependence of the diffusion coefficient. If the molecular weight of the polyion is so low that the hydrodynamic interaction between polyions is negligibly small, the concentration dependence of the sedimentation coefficient should be explicable by taking into account only the electrostatic interaction between the polyions. In this limiting case we should recover the behaviour predicted by Equation (13). An experimental test of this prediction is shown in Figure 3. The observed value of k_s is much smaller than that calculated. Now, in the derivation of Equation (2) it was assumed that all solutes, (counter-ions included), are distributed uni-formly throughout the solution. Of course, in any real solution the counter-ions are attracted to the polyion skeleton and cluster around it. An approximate correction for this

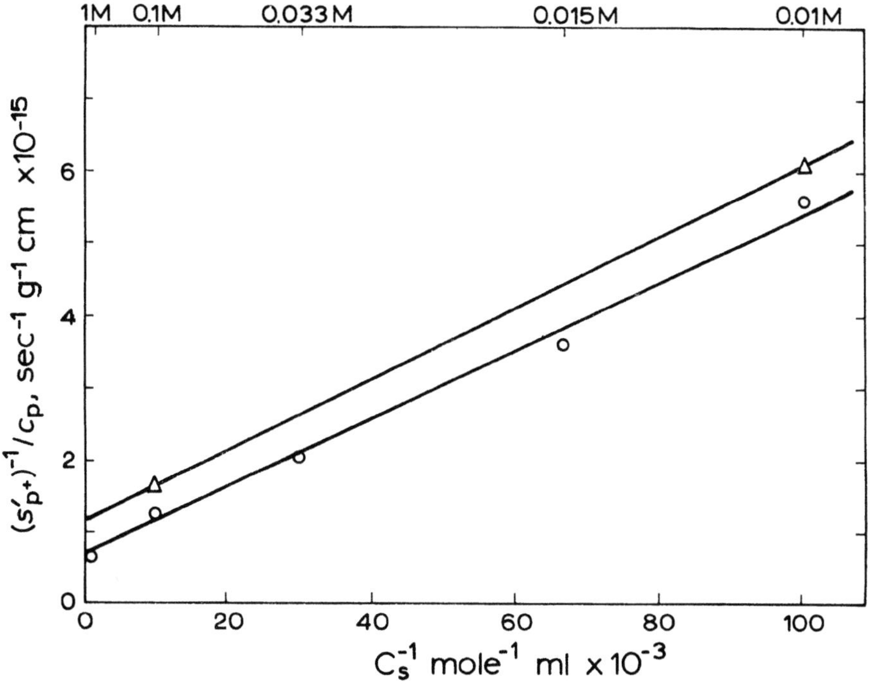

Fig. 3. Relationship between polymer concentration dependence coefficient and NaCl concentration. (Reproduced from Reference [3].) Sample poly L-lysine.

effect involves substituting for the true charge of the polyion an effective charge Z_{eff} [3]. The value of Z_{eff} required compares favourably with the effective charge needed to explain the thermodynamic properties of the polyelectrolyte solution, such as the osmotic pressure. A discussion of the effective charge approximation is found elsewhere.

Equation (12) predicts that the sedimentation potential decreases as the polyion concentration decreases. Therefore, the sedimentation coefficient of the corresponding discharged polymer may be obtained as the limit of the observed sedimentation coefficient at the infinite dilution of polymer. However, this procedure does not completely eliminate the unknown contribution to S_p from the ionic atmosphere, as discussed later. Furthermore, it is to be noted that the proposed extrapolation procedure is valid only when the sedimentation coefficients of both ions constituting the added-salt are equal, that is, if $S_+ = S_-$. If $S_+ \neq S_-$, the difference $S_+ - S_-$ leads to another sedimentation potential which effects the sedimentation coefficient of the polyion even in the limit of infinite dilution of polymer (secondary salt effect) [1].

2.2. In diffusion (4)

It is sufficient to consider one dimensional movement of solutes and, moreover, to assume that the cross coefficients L_{ki} $(k \neq i)$ are negligibly small. In this case, Equation (3) becomes

$$(J_k)_c = -(D_k)_v \left(\frac{\partial n_k}{\partial x}\right) - (l_k)_v n_k \left(\frac{\partial \psi}{\partial x}\right)_{T,P}. \tag{15}$$

The gradient of the diffusion potential, $(\partial\psi/\partial x)_{T,P}$, may be calculated from the condition that, in the steady state, no electric current may pass through the solution, i.e.,

$$\sum_k (Z_k e)(J_k)_c = 0. \tag{16}$$

Thus,

$$\left(\frac{\partial \psi}{\partial x}\right)_{T,P} = \frac{\sum_k (Z_k e)(D_k)_v \left(\frac{\partial n_k}{\partial x}\right)_{T,P}}{\sum_k (Z_k e)(l_k)_v n_k}. \tag{17}$$

The insertion of Equation (17) into Equation (15) yields

$$(J_k)_c = -(D_k)_v \left(\frac{\partial n_k}{\partial x}\right) + (l_k)_v n_k \frac{\sum_k (Z_k e)(D_k)_v \left(\frac{\partial n_k}{\partial x}\right)_{T,P}}{\sum_k (Z_k e)(l_k)_v n_k} =$$

$$= \frac{-\left(\sum_{j\neq k} (Z_j e)(l_j)_v n_j\right)(D_k)_v \left(\frac{\partial n_k}{\partial x}\right) + (l_k)_v n_k \left[\sum_{j\neq k} (Z_j e)(D_k)_v \left(\frac{\partial n_j}{\partial x}\right)\right]}{\sum_k (Z_k e)(l_k)_v n_k}. \tag{18}$$

The flow of polymer, $(J_p)_c$, may be determined independently of the flow of other solutes if a light absorption technique for monitoring the concentration is used. Studies of diffusion in solutions of synthetic polymers have traditionally used the schlieren, the Rayleigh, or the Gouy interference techniques. With none of these can the flow of polymer be observed independently of the flow of small ions. Instead, what is measured is the sum of the flows of all solutes.

If we represent the refractive index of the solution and the specific refractive index of a solute k by α and α_k respectively, the change in the refractive index of the solution at a position x in the cell is given by

$$\frac{\partial \alpha}{\partial t} = -\frac{\mathrm{d}}{\mathrm{d}x}\left[\sum_k \alpha_k (J_k)_c\right]. \tag{19}$$

Substitution of Equation (18) into Equation (19) leads to

$$\frac{\partial \alpha}{\partial t} = \frac{\frac{\mathrm{d}}{\mathrm{d}x}\left[-\sum_k \alpha_k (D_k)_v \left(\frac{\partial n_k}{\partial x}\right)\left(\sum_{j\neq k} Z_j (l_j)_v n_j\right) + \sum_k \alpha_k (l_k)_v n_k \left\{\sum_{j\neq k} Z_j (D_j)_k \left(\frac{\partial n_i}{\partial x}\right)\right\}\right]}{\sum_k Z_k (l_k)_v n_k}. \tag{20}$$

We have assumed that the sample solution has been dialysed against a solvent which contains a neutral salt of 1–1 type having a common counter-ion with the polyelectrolyte solution until the Donnan membrane equilibrium is attained. Moreover, we have also assumed that the diffusion coefficient of the polyion is so small that the chemical potentials of the simple ions are always equal throughout the entire solution at every moment during the diffusion process. That is, the diffusion process proceeds under conditions such that the Donnan equilibrium condition between the polymer phase and the solvent phase is always maintained. Then,

$$n_+ (x) = Z_p n_p (x) + n_- (x) \tag{21}$$

and

$$[Z_p n_p (x) + n_- (x)] n_- (x) = n_s^2. \tag{22}$$

From Equation (21) and (22), the concentrations and concentration gradients of the simple ions can be determined as functions of n_p, n_s, and $\mathrm{d}n_p/\mathrm{d}x$:

$$n_\pm (x) = \tfrac{1}{2}[(Z_p^2 n_p^2 + 4n_s^2)^{1/2} \pm Z_p n_p] \tag{23}$$

$$\frac{\partial n_\pm}{\partial x} = \frac{Z_p}{2}\left[\frac{Z_p n_p}{(Z_p^2 n_p^2 + 4n_s^2)^{1/2}} \pm 1\right]\frac{\mathrm{d}n_p}{\mathrm{d}x}, \tag{24}$$

The gradient of the refractive index of the solution is related to the concentration gradients by

$$\frac{\mathrm{d}\alpha}{\mathrm{d}x} = \sum_k \alpha_k \frac{\mathrm{d}n_k}{\mathrm{d}x} = \alpha_{\mathrm{app}} \frac{\mathrm{d}n_p}{\mathrm{d}x} \tag{25}$$

where

$$\alpha_{\text{app}} = \left[\alpha_{\text{p}} + \tfrac{1}{2}(\alpha_+ + \alpha_-)\frac{Z_{\text{p}}^2 n_{\text{p}}}{(Z_{\text{p}}^2 n_{\text{p}}^2 + 4n_{\text{s}}^2)^{1/2}} + \frac{Z_{\text{p}}}{2}(\alpha_+ - \alpha_-)\right]. \tag{26}$$

Inserting Equations (23)–(26) into Equation (20) and using Equation (22), we obtain an equation for the diffusion coefficient, D_{app}, determined using an optical detection system based on the use of the refractive index increment:

$$\frac{\partial \alpha}{\partial t} = \frac{d}{\mathrm{d}x}\left[D_{\text{app}}\frac{\mathrm{d}\alpha}{\mathrm{d}x}\right], \tag{27}$$

where

$$D_{\text{app}} = \sum_k K_k D_k \left(\sum_k Z_k n_k l_k\right)\alpha_{\text{app}} \tag{28}$$

and

$$\begin{aligned} K_{\text{p}} &= -(\alpha_{\text{p}} + Z_{\text{p}}\alpha_+)\, l_+ n_+ + (\alpha_{\text{p}} - Z_{\text{p}}\alpha_-)\, l_- n_- \\ K_+ &= [(\alpha_{\text{p}} + Z_{\text{p}}\alpha_+)\, l_{\text{p}} n_{\text{p}} + (\alpha_+ \pm \alpha_-)\, l_- n_-]\frac{Z_{\text{p}}}{2}\left(\frac{Z_{\text{p}} n_{\text{p}}}{(Z_{\text{p}}^2 n_{\text{p}}^2 + 1)^{1/2}} - 1\right) \\ K_- &= -[(-Z_{\text{p}}\alpha_- + \alpha_{\text{p}})\, l_{\text{p}} n_{\text{p}} + (\alpha_+ + \alpha_-)\, l_+ n_+]\frac{Z_{\text{p}}}{2}\left(\frac{Z_{\text{p}} n_{\text{p}}}{(Z_{\text{p}}^2 n_{\text{p}}^2 + 1)^{1/2}} - 1\right). \end{aligned} \tag{29}$$

If $(n_{\text{p}}/n_{\text{s}}) \ll 1$, and also if we assume that $\alpha_+ = \alpha_- = \alpha_{\text{s}}$ and $l_+ = l_- = l_\pm$, D_{app} can be approximated by

$$D_{\text{app}} = D_{\text{p}}[1 + (-k_s + 2gA_2M)\, n_{\text{p}} + \cdots], \tag{30}$$

where

$$k_{\text{s}} = \frac{Z_{\text{p}}(l_{\text{p}})_v}{[(l_+)_v - (l_-)_v]\, n_{\text{s}}} \tag{31}$$

$$A_2 = Z_{\text{p}}^2/4n_{\text{s}} \tag{32}$$

$$g = \frac{\alpha_{\text{s}} Z_{\text{p}}}{\alpha_{\text{p}}}\left(2\frac{D_\pm}{D_{\text{p}}} + \frac{l_{\text{p}} D_\pm}{l_\pm D_{\text{p}}}\right) + \frac{1}{Z_{\text{p}}}\frac{D_\pm l_{\text{p}}}{D_{\text{p}} l_\pm}. \tag{33}$$

In Equation (32), A_2 is the second virial coefficient of the polyelectrolyte solution, and k_s is defined in Equation (13). $D_\pm$ is the diffusion coefficient of the neutral salt. The differences between α_+ and α_-, and also between l_+ and l_- can be taken into account easily, and are found to be of secondary importance. Using the appropriate values for the constants in Equation (33), g is found to be close to unity.

Thus, it may be concluded that the diffusion coefficient of a polyion, $(D_{\text{p}})_v$, can be obtained by extrapolation of the observed diffusion coefficient to infinite dilution. The concentration dependence coefficient of D_{app} is given by

$$k_{\text{D}} = -k_s + 2A_2M \tag{34}$$

which is of the same form as the concentration dependence of the diffusion coefficient in a two component system (if $g=1$) [5]. The fact that the diffusion of a polyelectrolyte in neutral salt solutions may be approximately treated as the diffusion of a neutral polymer in a two component system arises from the imposition of the Donnan equilibrium condition. A similar conclusion concerning the analysis of light scattering and sedimentation equilibrium data has been presented by Casassa and Eisenberg [6].

In the above derivation, it is again assumed that the counter-ions as well as the polymeric ions and by-ions distribute uniformly in the solution, and moreover, the hydrodynamic interaction between polyion coils is negligible. As the hydrodynamic interaction is included in both k_s and k_D, however, the effect should be cancelled in Equation (34). Because the counter-ions are strongly attracted around the polymeric ions, the polyelectrolyte solutions are highly non-ideal so that both k_s and A_2 may deviate from the calculated values of Equations (31) and (32). However, it is reasonable to assume that the equality in Equation (34) should hold if we use the experimental values for all k_D, k_s and A_2.

An example of the relationship between the observed diffusion coefficient and the polymer concentration is shown in Figure 4. D_{app} was measured using a Rayleigh

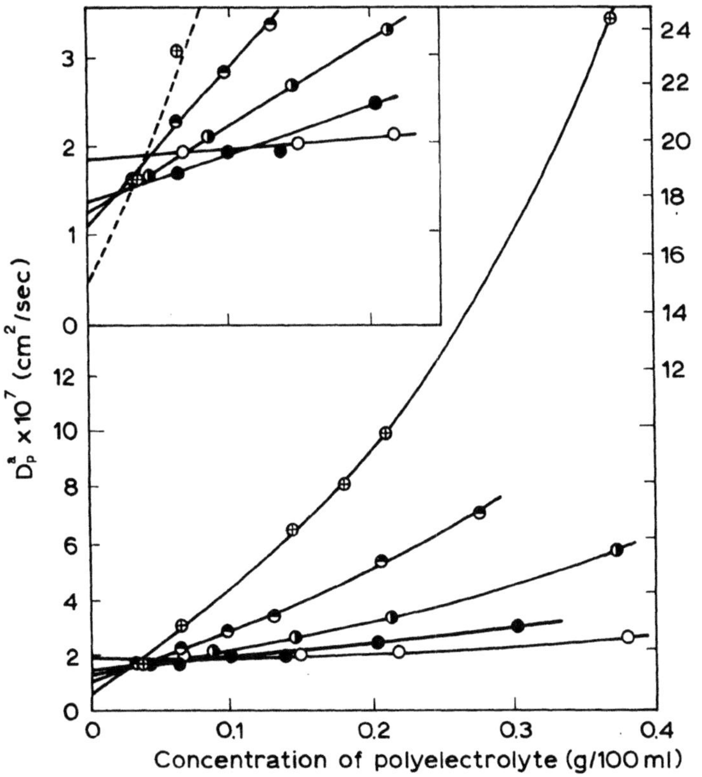

Fig. 4. Polymer concentration dependence of diffusion coefficient. (Reproduced from Reference [4].) Sample, Na-PSS. Mol. Wt. $M_w = 4.9 \times 10^5$, $M_n = 3.2 \times 10^5$.

interference optical system. The changes in D_p with ionic strength arise from the same sources as does the concentration dependence of the sedimentation coefficient. That is, D_p is changed both by the change in the radius of gyration of the polyion with ionic strength and by the effect of ionic atmosphere on the flow of macromolecules.

Equation (34) implies that the slope in Figure 4 should be linear with respect to the reciprocal of the concentration of added salt, since not only k_s but also A_2 is linear in the reciprocal of the concentration of added salt. Figure 2 also shows a replot of the data displayed in Figure 4. In the case of sedimentation, it was pointed out that k_s is not accurately predicted because of the neglect of hydrodynamic interaction between polyions. However, in diffusion k_D depends mainly on the second term of Equation (34) and, hence, the contribution of the hydrodynamic interaction to k_D is small. Thus k_D is found to be proportional to n_s^{-1} with good precision. Moreover, combining k_s and k_D, and also using M_w as determined from light scattering, we can calculate the second virial coefficient, A_2, which can then be compared with the values determined independently from light scattering or osmotic pressure measurements. In Figures 5 and 6, the light scattering and osmotic pressure measurements using the same sample as used in the experiments of Figure 4 are shown [7, 8]. The second virial coefficient A_2 can be estimated from the data taking into account the third virial coefficient. The values of A_2 estimated from light-scattering data and osmotic pressure measurements are found to agree with each other, as shown in Figure 7. The

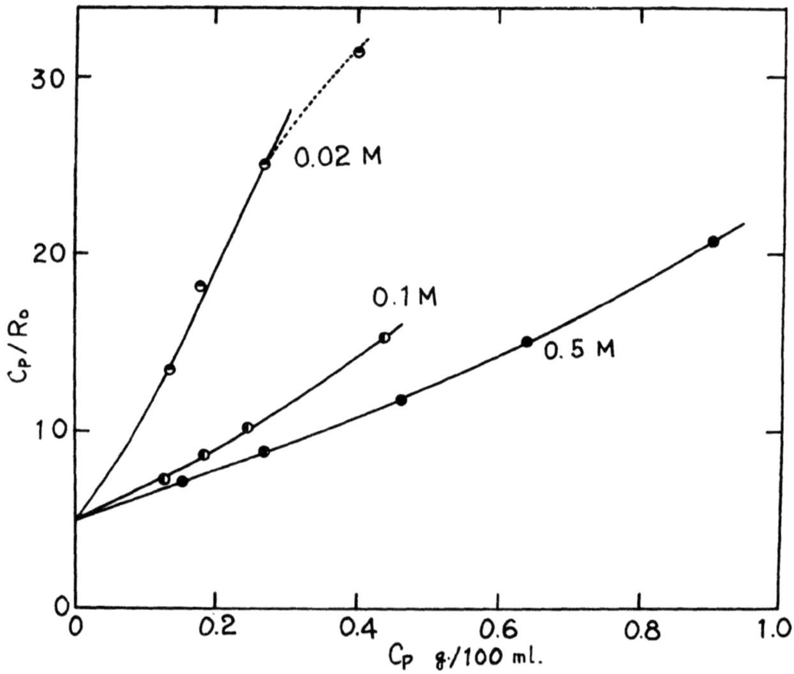

Fig. 5. Light scattering of Na-PSS solution. (Reproduced from Reference [7].) Sample, the same as in Figure 4. NaCl concentrations, denoted in the figure.

TABLE I

C_S M	$D_p^0 \times 10^7$, $cm^2\ s^{-1}$	$S_p^0 \times 10^{13}$, s	$M_{SD} \times 10^{-5}$, g mol^{-1}	$k_D \times 10^{-2}$, ml g^{-1}	$k_S \times 10^2$, ml g^{-1}	$(k_D + k_S)/2\ M_w \times 10^{-e}$, ml mol g^2	$A_2 \times 10^{-4}$, mol mol g^2
0.2	1.83	8.13	2.78	0.680	1.97	2.70	4.1
0.05	1.35	7.25	3.28	3.91	3.25	7.29	8.6
0.02	1.25	6.41	3.20	7.75	4.56	12.5_3	14.8
0.01	1.09	5.88	3.28	16.1_7	6.18	22.7_7	22
0.005	(0.6)	5.13	(4.6)	(64)	7.65	(73)	41

[a] $M_n = 3.2 \times 10^5$; $M_w = 4.9 \times 10^5$.
(Reproduced from Reference [4].)

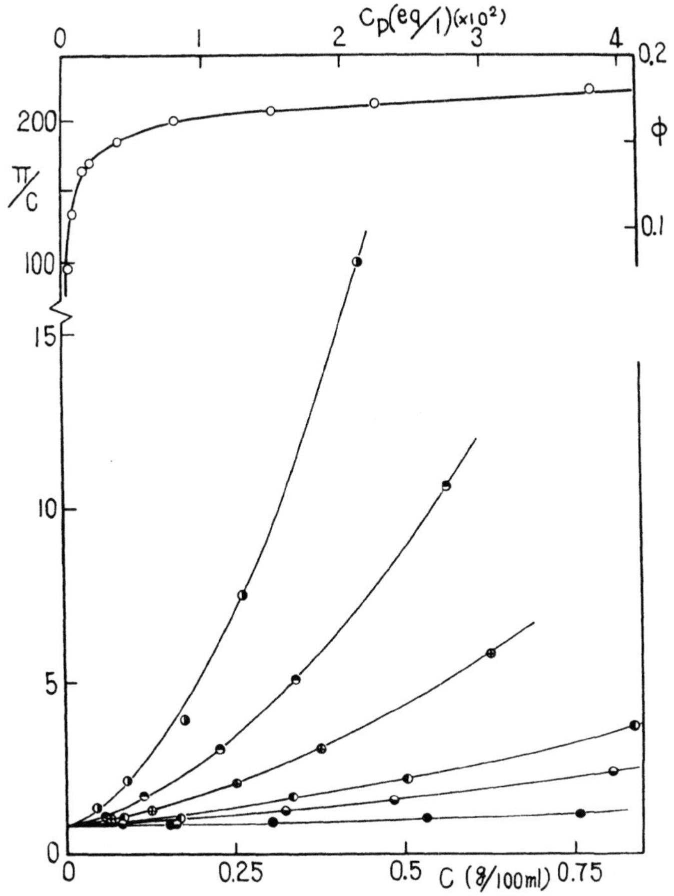

Fig. 6. Osmotic pressure measurements of Na-PSS solutions. (Reproduced from Reference [8].) Samples, the same as in Figure 4. NaCl concentrations, ○, 0; ◑, 0.005; ⊖, 0.01; ⊕, 0.02; ◐, 0.05; ⊖, 0.1; ●, 0.5 N ϕ denotes the osmotic coefficients.

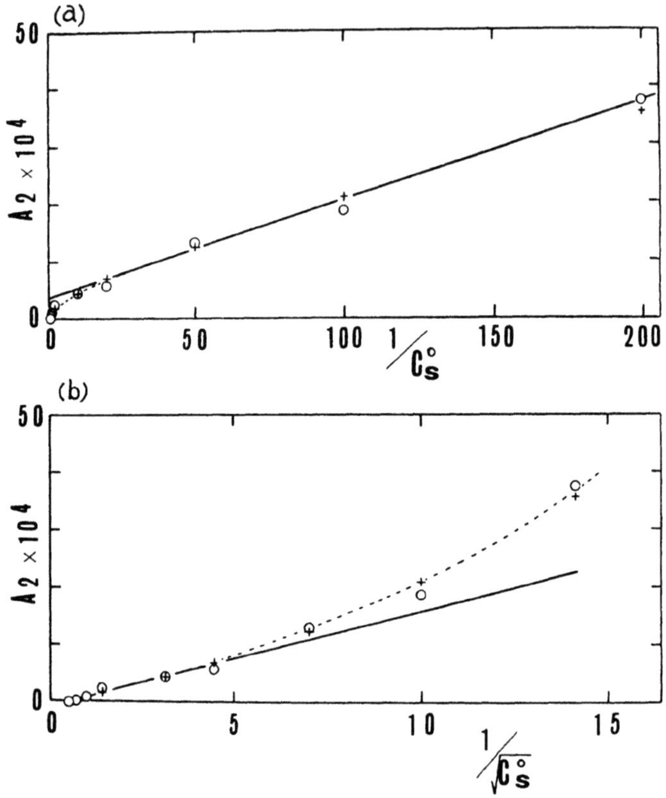

Fig. 7. Plots of second virial coefficient vs. reciprocal ionic strength (*a*) and reciprocal square root of ionic strength (*b*). (Reproduced from Reference [8].) +, A_2 (LS) calculated from Figure 5 and others; ○, A_2 (OS) calculated from Figure 6.

agreement is reasonable because A_2 of polyelectrolyte solutions is independent of molecular weight. One comparison between $(k_s + k_D)/2M_w$ and A_2 is shown in Table I [4]. As seen, the agreement displayed is quite satisfactory.

3. The Effect of the Ionic Atmosphere on the Limiting Sedimentation and Diffusion Coefficients

In general, the diffusion coefficient of an electrolyte is a function of concentration. One of the reasons for the concentration dependence of diffusion coefficient is the electrolyte effect of ionic atmosphere around each ion. A similar effect of ionic atmosphere should be observed for diffusion and sedimentation coefficients of a polyelectrolyte. Moreover, the effect of ionic atmosphere should remain even at the limit of infinite dilution of polyelectrolyte if the concentration of added neutral salt is kept constant, since the ionic atmosphere does not disappear.

The importance of the effect of ionic atmosphere can be easily recognized in electrophoresis since the electric forces on the polyion and on the solvent containing the counter-ions have opposite directions. The solution of the hydrodynamic equation which describes the movement of solvent relative to the polyion has been given by various authors. [9–11]. The theory was well accounted for by experiments. In contrast to the case of electrophoresis, the effect of ionic atmosphere on the limiting sedimentation and diffusion coefficients has been neglected in most of the analyses of data thus far published. Indeed, in all of the studies published the data have been analyzed under the assumption that all ions are distributed uniformly throughout the solution. If this assumption were valid the ionic atmosphere surrounding a polyion would disappear in the limit of infinite dilution of the macromolecule. Unfortunately, the assumption that all of the ions are uniformly distributed is a gross over-simplification of the true ionic distribution. Clearly, the ionic atmosphere surrounding a macro-ion cannot disappear even in the limit of infinite dilution of the macro-ion so long as the concentration of added salt is kept finite. Therefore, it is not to be expected that the sedimentation coefficient of the corresponding un-ionized macromolecule can be obtained by extrapolating the observed sedimentation coefficient of the macro-ion to infinite dilution. Moreover, the effect of the ionic atmosphere of the flow about the macro-ion is expected to increase as the concentration of added salt increases if the macro-ion is a solid sphere having a fixed number of charges on the surface.

If the macro-ion is a linear polyelectrolyte, the sedimentation velocity is changed not only by the effect of the ionic atmosphere on the flow of solvent about the polymer, but also by the change in conformation of the polymer chain as the ionic strength is changed. According to Flory [12], the sedimentation coefficient is related to the expansion factor, α, defined by the ratio of the radius of gyration of the polymer $\langle S^2\rangle^{1/2}$ to its value at the θ point, $\langle S^2\rangle_0{}^{1/2}$, by [12]

$$S_p = \frac{M}{N_A}(1 - \bar{v}\varrho_m)/\Theta\,, \tag{35}$$

$$\Theta = \eta_0 P_0 \langle S^2\rangle_0^{1/2}\,\alpha\,,$$

where

$$\alpha^2 = \langle S^2\rangle/\langle S^2\rangle_0$$

and M is molecular weight, η_0 is the viscosity of solvent, ϱ_m is the density of solution, $\bar{v}$ is the partial specific volume of polymer, P_0 is a constant and N_A is Avogadro's number. The expansion factor α is also related to the intrinsic viscosity $[\eta]$ by the equation

$$[\eta] = \Phi_0 \frac{\langle S_0^2\rangle^{3/2}}{M}\,\alpha^3\,, \tag{36}$$

where Φ_0 is also a constant.

Using Equations (35) and (36), the expansion factor can be eliminated and the following Mandelkem and Flory constant is obtained:

$$\Phi_0^{1/2} P_0^{-1} = \frac{N_A \eta_0}{1 - \bar{v}\varrho_m} S_p [\eta]^{1/3} M^{-2/3}. \tag{37}$$

It is known that $\Phi_0^{1/3} P_0^{-1}$ has the value 2.5×10^6 for non-ionic polymer solutions, independent of solvent, temperature, and molecular weight [13–15]. In polyelectrolyte solutions, however, Equations (37) can be neither constant nor equal to 2.5×10^6 if the ionic atmosphere of the polyion influences significantly the sedimentation velocity of polymer. Thus, one test for the influence is to determine whether or not $\Phi_0^{1/3} P_0^{-1}$ is independent of ionic strength and molecular weight.

The Mandelkern and Flory constant can also be calculated from the limiting diffusion coefficient D_p and $[\eta]$ such as

$$\Phi_0^{1/3} P_0^{-1} = [\eta]^{1/3} M^{1/3} \eta_0 D_p / kT. \tag{38}$$

Using data obtained for sodium polystyrene sulphonate, the Mandelkern-Flory coefficient $\Phi_0^{1/3} P_0^{-1}$ is displayed in Figure 8. As seen, $\Phi_0^{1/3} P_0^{-1}$ is not constant, but changes with ionic strength and also with molecular weight. At a constant molecular weight the uncertainty in $\Phi_0^{1/3} P_0^{-1}$ is determined almost entirely by the uncertainty in S_p, since $[\eta]$ can be determined very accurately. We conclude that the changes in

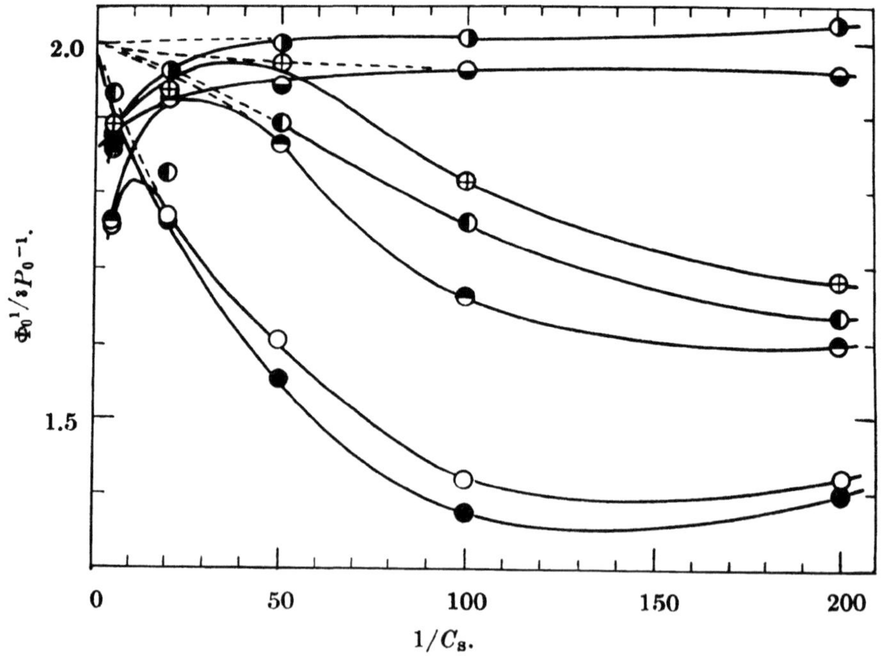

Fig. 8. Ionic strength dependence of the Mandelkern-Flory coefficient. (Reproduced from Reference [2].) Samples, Na-PSS Mol. Wt(M_w), ◑, 3.9×10^4; ⊖, 4.9×10^4; ⊕, 1.00×10^6; ◐, 1.26×10^6 (M_v); ⊖, 1.55×10^6, ○, 2.10×10^6 (M_v); ●, 2.34×10^6.

$\Phi_0^{1/3}P_0^{-1}$ with ionic strength are clearly larger than the combined experimental error. All experimental values in Figure 8, excluding the datum at 0.2 NaCl, tend to the value of 2.0×10^6 as the ionic strength is increased. This value, which is supposed to correspond to that for an un-ionized polymer, is smaller than the commonly accepted value for non-ionic polymers, namely 2.5×10^6. However, the difference is not significant since the meaning of molecular weight in Equation (37) is not clear. If the molecular weight determined from sedimentation and diffusion, M_{SD}, is inserted in Equation (37), we have 2.5×10^6 for $\Phi_0^{1/3}P_0^{-1}$. The sudden decrease of $\Phi_0^{1/3}P_0^{-1}$ at $C_2 = 0.2$ appears unusual, but it is often observed for linear polyelectrolytes if the ionic strength is very high [16, 17]. The physical processes responsible for this behaviour are not known.

The values of $\Phi_0^{1/3}P_0^{-1}$ calculated from S_p and D_p using Equation (37) and (38) are compared in Table II. The values agree with each other again if we employ M_{SD}.

TABLE II

$\Phi_0{}^{1/3}P_0{}^{-1} \times 10^{-6}$ from $[\eta]$, D, W_w	$\Phi_0{}^{1/3}P_0{}^{-1} \times 10^{-6}$ from $[\eta]$, S, M_w	$\Phi^{1/3}P_0{}^{-1} \times 10^{-6}$ from $[\eta]$, D, M_{SD}	$\Phi_0{}^{1/3}P_0{}^{-1} \times 10^{-6}$ from $[\eta]$, S, M_{SD}
2.90	1.88	2.46	2.62
2.86	2.09	2.54	2.63
2.92	1.94	2.50	2.66
2.75	1.96	2.45	2.59

Beside the effect of ionic atmosphere, there are various factors which may affect $\Phi_0^{1/3}P_0^{-1}$. The first may be the effect of drainage. If there is partial free drainage of solvent through the polymer coil, the powers of α in Equations (35) and (36) may be considerably different from 1 and 3 [18]. However, we expect that $\Phi_0^{1/3}P_0^{-1}$ will be constant over a wide range of α, because the effect of partial free drainage on S_p and $[\eta]$ tend to compensate one another (e.g. as in Equation (37)). The second may be that, even in the limit of non-draining sphere, the sedimentation coefficient of Equation (35) and the intrinsic viscosity on Equation (36) may be proportional to lower powers of α than 1 and 3, respectively, as first pointed out by Kurata and Yamakawa [18]. Although there are various experimental works, the dependence of the sedimentation coefficient on α has not yet been completely solved [19, 19a].

In spite of these ambiguities, it is clear from the following estimation that the effect of ionic atmosphere can cause a change in $\Phi_0^{1/3}P_0^{-1}$. First, we shall assume that the electroviscous effect disappears at infinite dilution of the macro-ion, and that the magnitude of the intrinsic viscosity is little influenced by flow perturbations originating from the ionic atmosphere. These assumptions are supported by the experimental observations that: (a) The intrinsic viscosity of globular proteins are independent of the charges on their surface [20] and (b) that the intrinsic viscosities of linear polyions are at least qualitatively accounted for by theories which take into account only the expansion of the polymer coil [21–23]. In contrast to $[\eta]$, the sedimentation coefficient S_p cannot be expressed in terms of α only: If S_p were determined only by α, the following linear relationship should hold from the analogy to $[\eta]$:

$$\left(\frac{M}{S_p}\right)^3 = (\text{const})_1 + (\text{const})_2\, M\,. \tag{39}$$

However, the experimental results do not support this speculation as seen in Figure 9.(2).

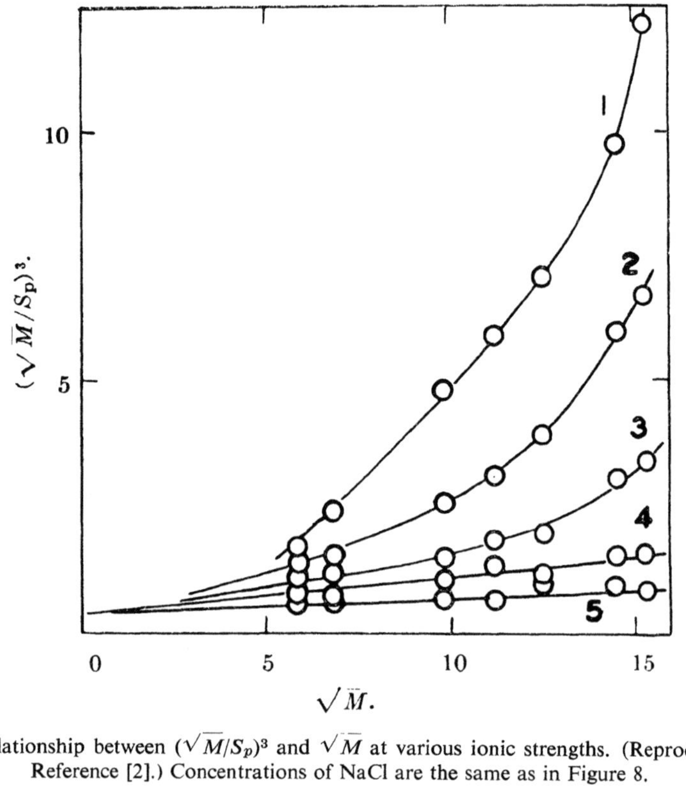

Fig. 9. Relationship between $(\sqrt{M}/S_p)^3$ and $\sqrt{M}$ at various ionic strengths. (Reproduced from Reference [2].) Concentrations of NaCl are the same as in Figure 8.

The electrophoretic effect of ionic atmosphere in diffusion was discussed by Onsager and Fuoss [24, 25]. The effect of ionic atmosphere on the sedimenting velocity of a macro-ion in the presence of an added neutral salt may be discussed in a similar way. The forces k_i acting on the ions must be balanced by other forces k_0 acting on the solvent molecules. Denoting bulk concentrations of the ions by n_i, we have

$$n_0 k_0 = -\,n_+ k_+ - n_- k_- - n_p k_p\,, \tag{40}$$

where suffix denotes the respective ion. A spherical shell between radius r and $r+dr$

from the centre of a macro-ion is subject to a force given by

$$(n'_+k'_+ + n'_-k'_- + n'_pk'_p + n_0k_0)\,4\pi r^2\,\mathrm{d}r, \tag{41}$$

where the variation in n_0 is neglected. If the central ion is a small ion so that the velocities of ions relative to solvent are uniform independent of r, we may assume $k'_i = k_i$. However, if the central ion is a macro-ion, the velocity of solvent should change with r and, hence, $k'_i \neq k_i$ in general.

Eliminating n_0k_0 from Equations (40) and (41), we have for the resultant force acting on the shell

$$[(n'_+k'_+ - n_+k_+) + (n'_-k'_- - n_-k_-) + (n'_pk'_p - n_pk_p)]\,4\pi r^2\,\mathrm{d}r. \tag{42}$$

This force is in the opposite direction to that driving the sedimentation and causes the sphere of r to move with a velocity. At the limit of infinite dilution of polyelectrolyte and at a constant concentration of added salt, the velocity is given by

$$\mathrm{d}v = -\frac{[(n'_+k'_+ - n_+k_+) + (n'_-k'_- - n_-k_-)]\,4\pi r^2\,\mathrm{d}r}{6\pi\eta_0 r}. \tag{43}$$

Since $n'_+k'_+ > n_+k_+$ and $n'_-k'_- < n_-k_-$ for a central macro-anion, Equation (43) may be approximated by

$$\mathrm{d}v = -\frac{4\pi r^2 (n_+ - n_-)\,k_+\,\mathrm{d}r}{6\pi\eta_0 r}, \tag{44}$$

where k_+ is the force exerted on the solvent by a counter-ion. If the counter-ion sediments at a velocity relative to the solvent, $v(r)$, which is a function of the distance r from the centre of the macro-ion, k_+ is given by

$$k_+ = 6\pi\eta_0 a_+ v(r), \tag{45}$$

where a_+ is the effective radius of the counter-ion.

At least, at the limit of non-drainage of solvent through the polyion sphere, the change of the polyion velocity arising from the solvent flow generated by the counter-ions outside the sphere, denoted ΔV, is given by integration of Equation (44) from the sphere surface to infinity.

$$\Delta V = \int_R^\infty \mathrm{d}v. \tag{46}$$

Suppose that a polymeric ion is represented by a hydrodynamically equivalent sphere. That is, suppose that a polymeric ion is represented as a sphere of radius R inside which all fixed charged groups are distributed uniformly. The density of the fixed charges in the sphere, ν_m, is given by

$$\nu_m = Z/(\tfrac{4}{3}\pi R^3), \tag{47}$$

where Z is the number of fixed charges on a polyelectrolyte molecule. If the sphere is in an infinite volume of a neutral salt solution (such as NaCl) of concentration n_s, the Z counter-ions are distributed around the polymeric ion sphere in a manner which can be described by the Poisson-Boltzmann equation. The difference $(n_+ - n_-)$ in Equation (44) is then calculated as

$$n_+ - n_- = - v_m + v_m (1 + \beta)\, e^{-\beta} \frac{\sinh t}{t} \quad (r<R) \tag{48}$$

$$= - \frac{v_m}{e} \beta \left(\cosh \beta - \frac{\sinh \beta}{\beta} \right) \frac{e^{-t}}{t} \quad (r>R) \tag{49}$$

where

$$\beta = \kappa R, \tag{50}$$

$$\mathrm{t} = \kappa r, \tag{51}$$

and

$$\kappa^2 = \frac{8\pi e^2}{DkT} n_s . \tag{52}$$

By assuming that

$$u(r) = U \quad (r > R), \tag{53}$$

where U is the sedimentation velocity of the polyion relative to the cell, from Equations (44), (46) and (48), Nagasawa and Eguchi obtained [2].

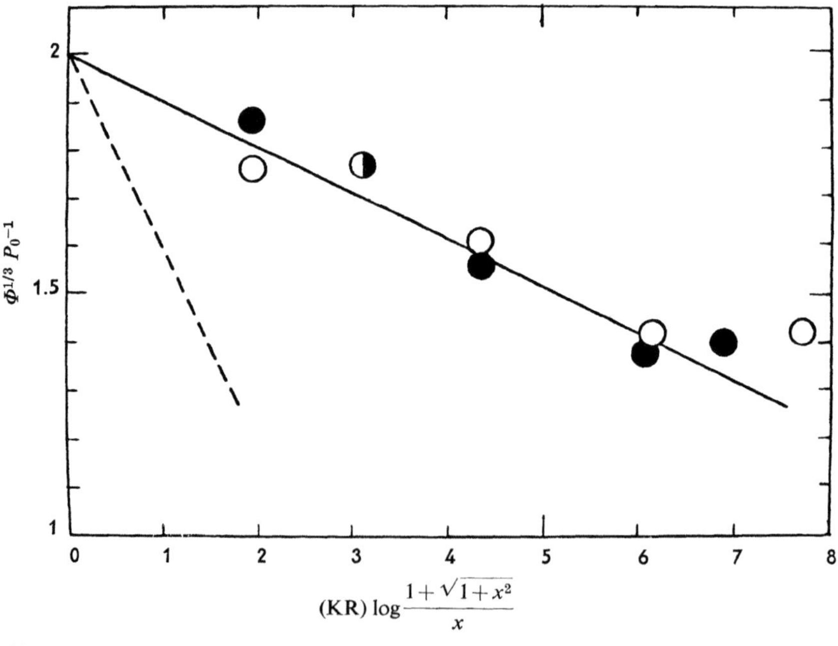

Fig. 10. Ionic strength dependence of the Mandelkern-Flory coefficient. (Reproduced from Reference [2].) Samples, Na-PSS. ●, $M_w = 2.34 \times 10^6$; ○, $M_r = 2.10 \times 10^6$.

$$\Phi_0^{1/3}P_0^{-1} = (\Phi_0^{1/3}P_0^{-1})_0 \left[1 - 2.3\,\frac{DkT}{2e^2}\log\frac{1+(1+x^2)^{1/2}}{x}\right] \tag{54}$$

where $(\Phi_0^{1/3}P_0^{-1})_0$ is the value of $\Phi_0^{1/3}P_0^{-1}$ for the corresponding unionized polyion.

Comparison between the experimental data shown in Figure 8 and Equation (46) shows that the behaviour of two samples with high molecular weights satisfies the linear relationship between $\Phi_0^{1/3}P_0^{-1}$ and $\log[(1+(1+x^2)^{1/2})/x]$ (see Figure 10). The dotted line in Figure 10 was calculated from Equation (54) assuming a=2.5 Å. The disagreement between the experimental slope and the calculated slope is to be expected because the calculated values of $n_+ - n_-$ must be much larger than the true values. Disagreement of the type displayed in Figure 10 (i.e. correct functional form but incorrect magnitude) is often observed in the study of polyelectrolyte solutions, and is usually explained by assuming ion binding. As the molecular weight of the polymer decreases, $\Phi_0^{1/3}P_0^{-1}$ deviates from linearity. Nagasawa and Eguchi [2] have speculated that the deviation arises from an increasing departure of the molecule from the model, more specifically from increased hydrodynamic drainage as M decreases. Although it has often been reported that the limiting sedimentation coefficients of low molecular weight polyelectrolytes are independent of ionic strength, much more work is required to firmly establish this interpretation. In particular, the assumption of Equation (35) should be reconsidered. At least, however, we can see that the effect of ionic atmosphere may cause such a large dependence of the limiting sedimentation coefficient on added salt concentration as observed.

The change in the limiting diffusion coefficient with ionic strength can be explained in the same fashion as can the sedimentation coefficient. That is, there are effects arising from both the change in the radius of gyration of the polyion and from the hydrodynamic flow originating in ionic atmosphere. If the limiting sedimentation and diffusion coefficients are combined, both effects mentioned are self-compensating, and we may calculate the molecular weight of the polyion independent of the ionic strength of the solution. This method has been long used by many investigators. A comparison between the molecular weights thus determined, M_{SD}, and the number and weight average molecular weights determined by osmotic pressure and light scattering M_n and M_w, are shown in Table I. The observed differences between M_{SD} and the other two may arise from the polydispersity of the sample used.

References

1. Pederson, K. O.: *J. Phys. Chem.* **62**, 1282 (1958).
2. Nagasawa, M. and Eguchi, Y.: *J. Phys. Chem.* **71**, 880 (1967).
3. Alexandrowicz, Z. and Daniel, E.: *Biopolymers* **1**, 447, 473 (1963).
4. Suzuki, Y., Noda, I., and Nagasawa, M.: *J. Phys. Chem.* **73**, 797 (1969).
5. Mandelkern, L. and Flory, P. J.: *J. Chem. Phys.* **19**, 984 (1951).
6. Casassa, E. F. and Eisenberg, H.: *J. Phys. Chem.* **65**, 427 (1961).
7. Takahashi, A., Kato, T., and Nagasawa, M.: *J. Phys. Chem.* **71**, 2001 (1967).
8. Takahashi, A., Kato, N., and Nagasawa, M.: *J. Phys. Chem.* **74**, 944 (1970).
9. Hermans, J. J. and Fujita, H.: *Konink1. Ned. Akad. Wetenschap. Proc.* **B58**, 182 (1955).

10. Hermans, J. J.: *J. Polymer Sci.* **18**, 529 (1955).
11. Overbeek, J. Th. and Stigter, D.: *Rec. Trav. Chim.* **75**, 543 (1956).
12. Flory, P. J.: *Principles of Polymer Chemistry*, Cornell University Press, Ithaca, N.Y., 1953.
13. Mandelkern, L. and Flory, P. J.: *J. Chem. Phys.* **20**, 212 (1952).
14. Mandelkern, L., Krigbaum, W. R., Sheraga, H. A., and Flory, P. J.: *J. Chem. Phys.* **20**, 1392 (1952).
15. Fox, Jr., T. G. and Mandelkern, L.: *J. Chem. Phys.* **21**, 187 (1953).
16. Takahashi, A., Soda, A., and Kagawa, I.: *Nippon Kagaku Zasshi* **83**, 873 (1962).
17. Eisenberg, H. and Casassa, E. F.: *J. Polymer Sci.* **47**, 29 (1960).
18. Kurata, M. and Yamakawa, H.: *J. Polymer Sci.* **28**, 785 (1958); **29**, 311 (1958).
19. Noda, I., Saito, N., Fujimoto, T., and Nagasawa, M.: *J. Phys. Chem.* **71**, 4648 (1967).
19a. Sakato, K. and Kurata, M.: *Polymer J.* **1**, 260 (1970).
20. Townend, R., Weisenberger, L., and Timasheff, S. N.: *J. Am. Chem. Soc.* **82**, 3175 (1960).
21. Eisenberg, H. and Woodside, D.: *J. Chem. Phys.* **36**, 1844 (1962).
22. Takahashi, A. and Nagasawa, M.: *J. Am. Chem. Soc.* **86**, 543 (1964).
23. Lapanje, S. and Kovac, S.: preprint of IUPAC Symposium on 'Macromolecular Chemistry', Prague, 1965.
24. Onsager, L. and Fuoss, R. M.: *J. Phys. Chem.* **36**, 2689 (1932).
25. Robinson, R. A. and Stokes, R. H.: *Electrolyte Solutions*, Butterworths, London, 1959, p. 134.

TRANSPORT PHENOMENA IN POLYELECTROLYTE-SALT SOLUTIONS

R. VAROQUI and A. SCHMITT

Center of Macromolecular Research (C.N.R.S.) 6, rue Boussingault, 67083 Strasbourg-Cedex, France

1. Introduction

The interpretation of sedimentation and diffusion velocity data for non-electrolytes in binary macromolecular component systems as well as in solutions of more than two components is based on well-tried and familiar equations. Whether the same principles can be extended to multicomponent polyelectrolyte solutions has been the subject of several controversies.

In fact complications due to the introduction of charges cannot be avoided. Moreover macro-ions are ordinarily studied in the presence of a low molecular weight electrolyte and the driving force which acts on the polyion and the small ions gives rise to different particle velocities; as a result, an electrical potential gradient is set up and the motion of the ions arises as a balance of both applied external forces and electrical coupling effects.

A moderately high concentration of salt has been assumed to prevent the appearance of a macroscopic electrical potential difference and it was customary to assume that the polyion motion then proceeds identically to that of the uncharged particle of the same shape and dimensions.

Although the presence of an excess of salt swamps out the appearance of any electrical potential difference – the so-called primary charge effect – it must still be kept in mind that in polyelectrolyte solutions the small ions of opposite charge to those of the polyion – the counterions – are not randomly distributed through the solution, a fraction of them being localized in the close vicinity of the polyion. Whatever the driving forces involved, the macro-ion steadily undergoes electrostatic retention from his non-electroneutral environment and a priori the motion of the polyion cannot be viewed as proceeding identical to that of an uncharged particle for which only hydrodynamical interactions are occurring.

The validity of some of the assumptions made in the interpretation of transport data for poly-ionic solutions has been questioned [1, 2]. However, no real quantitative estimate of the errors has as yet been established and the question of which molecular parameters are actually attainable by transport measurements remains to be answered.

We shall make an attempt here to present a coordinated and unified picture for both mass transport (diffusion and sedimentation) and electrical transport (electrophoresis). The general approach of the thermodynamics of irreversible processes which yields correct expressions for the flux-force equations for electrically neutral components will be used in the derivation of transport parameters.

Eric Sélégny (ed.), Polyelectrolytes, 261–276.

In this text emphasize is laid on the general formulation by using the concept of macroscopic friction coefficients.

The generalized friction model which has been widely used in the interpretation of transport and electrokinetic processes in ion exchange membranes may be extended in due form to polyelectrolyte solutions. The significance of the different particle friction coefficients and how they may be evaluated in terms of molecular parameters is explained. The usefulness of the formulation will be substantiated by computing the electrical interaction parameters explicitely for some polyelectrolytes of different shape and charge; however, no undue stress will be laid on the application to numerous models. Only limiting laws with respect to the salt-to-polymer concentration ratio are considered, namely the case of the most practical importance, when an excess of salt is present, is discussed. The case of salt-free-polyelectrolyte solutions which has been recently developed by the authors along the same lines [3, 4] is not treated in this paper.

2. Theory

2.1. Specification of the Model

We consider a polyelectrolyte of algebraic valence Z_3 which dissociates in an aqueous medium into a polyion of total electrical charge eZ_3 ($e=1.6\times10^{-19}$ coulomb) and ν_3 oppositely charged counter-ions of valence Z_1*.

A low molecular weight electrolyte which we shall denominate salt in the text is added; we assume that one mole of salt gives rise to ν_1 ions identical to the counter-ions and ν_2 co-ions of valence Z_2 carrying charges of the same sign as those of the macro-ion.

The electroneutral component, the solvent, is denoted by subscript w, the counter- co-and poly-ions by subscripts 1, 2 and 3 respectively. The ionic concentration of the ith species will be denoted c_i and the concentration of the electroneutral components by c_w, c_s and c_p for the solvent, the salt and the polyelectrolyte components respectively.

2.2. The Phenomenological Equations

The phenomenological particle flow equations are given by system (1). Flows are connected linearly to generalize forces, coupling effects are expressed by way of the cross coefficients L_{ij} and use of the Onsager reciprocal relationship is made:

$$J_i = \sum_{j=1}^{3} L_{ij}\nabla(-\mu_j) \quad (i=1,2,3) \tag{1}$$

$$L_{ij} = L_{ij} \tag{2}$$

* Z_3 may be different from the stoichiometric valence Z_{3s} if ion binding or ion condensation as shown by Manning occurs. A fraction of the counter-ions is then associated with the polyion and the effective degree of ionization is depressed. This point and its incidence in transport processes has been discussed more fully in a previous paper [4].

$$J_i = J_i^r - \frac{c_i}{c_w} J_w^r. \tag{3}$$

The flows J_i expressed in moles per unit area and unit time are relative to that of the solvent J_w^r, r means that J^r is referred to an arbitrary frame of reference. Practically, one is more often concerned with dilute solutions and the difference between the exchange fluxes J_i and the absolute fluxes J^r is then immaterial, J_i being very nearly the flux measured in the laboratory using the measuring cell as frame of reference. The forces $\nabla(-\mu_j)$ are related to the gradients of the partial molal Gibbs free energy associated to one mole of i, i.e. the gradients of the generalized electrochemical potential.

The set of linear Equations (1) which implies that only three fluxes and forces are linearly independent supposes the system to be submitted to specified constraints which are summarized below:

$$\sum_{i=1}^{3,w} c_i \nabla(-\mu_i) = 0, \tag{4}$$

$$\sum_{i=1}^{3,w} J_i^r \bar{V}_i = 0, \tag{5}$$

$$\nabla T = 0. \tag{6}$$

Equation (4) expresses that the sum of the forces acting on one volume element of the fluid is zero. This is the important condition of mechanical equilibrium or momentum conservation which implies that the external forces are opposed by pressure gradients.

Equation (5) signifies that the solution is restrained from bulk flow, a situation which always holds when the fluid is confined to a fixed closed vessel. Only isothermal conditions are considered – Equation (6) – and furthermore we suppose the absence of chemical reactions. A more complex situation would for instance be encountered if bound or condensed counter-ions were exchanging at a finite rate with atmospheric isotopic counter-ions [6].

It is worthwhile to emphasize the constraints (4) which indeed underlies the application of Einstein's generalized friction model. The conditions (4) to (6), which have been thoroughly discussed [7, 9] for unidimensional transport processes in multicomponent systems of uncharged and charged particles, are mostly fulfilled experimentally for the transport processes we shall discuss here (for a more precise account concerning this point the reader is referred to the pertaining literature [10, 11]).

We note that the linear form of Equation (1) makes it possible to write alternatively a set of Equations in which forces are presented by linear functions of the flows by introducing resistance coefficients rather than conductivity coefficients.

For the present purpose it is most suitable to use the formulation of the binary friction model, expressing the total force by:

$$\nabla(-\mu_j) = \sum_{i=1}^{3,w} f_{ji}(v_j - v_i) \quad (j = 1, 2, 3), \tag{7}$$

v_i is the mean (time average) velocity of particle i and f_{ij} is the neutral binary friction coefficient. Equation (7) states that the total force acting on one mole of particle j is opposed by frictional forces between j and all particles i. Relationships between the L_{ij} and the f_{ij} and concentrations are established at once from (1) and (7) and the additional relation (8).

$$J_i = c_i\, v_i. \tag{8}$$

These relations are given in the Appendix (For a detailed report on the friction model the reader is also referred to the work of Laity [12], Spiegler [13], Bearman [14] and other reviews [10]). Formulation (7) is self-consistent as long as the condition (4) of momentum conservation holds, then symmetrical relations-equivalent to Equations (2) – can be written:

$$c_i f_{ij} = c_j f_{ji}. \tag{9}$$

Because linear flux equations are used in Equation (1) the range over which linear behaviour holds must be checked experimentally; it supposes the phenomenological coefficients to be independent on the magnitude as well as on the nature of the applied forces. We shall reconsider this important point later on in the text.

2.3. Transport parameters

From the general flow equations transport parameters can be derived in terms of the phenomenological coefficients if a proper choice of forces appertaining to a specified experimental situation is made. We shall distinguish between transport of ions in an applied external electrical field and mass transport in diffusion and sedimentation at zero electrical current. As usual, transport parameters will be defined as the rate of motion of ions or electroneutral components per unit applied force.

2.3.1. *Electrophoretic mobility*

Relation (12) which gives the electrophoretic mobility μ_i – the velocity per unit electrical field in a chemically homogeneous medium – is obtained at once noting that the force acting on the ion is proportional to the electrical field E:

$$\nabla(-\mu_j) = Z_j \mathscr{F} E, \tag{10}$$

$$J_i = \sum_j Z_j L_{ij} \mathscr{F} E \quad (i, j = 1, 2, 3), \tag{11}$$

$$u_i = \frac{J_i}{c_i E} = \frac{\mathscr{F}}{c_i} \sum_j Z_j L_{ij}. \tag{12}$$

Conductivity

The total electrical current I is given by:

$$I = \mathscr{F} \sum_i Z_i J_i \quad (i = 1, 2, 3) \tag{13}$$

and the specific conductance K of the solution is defined by:

$$K = \frac{I}{E} = \mathscr{F}^2 \sum_{i,j} Z_i Z_j L_{ij} \quad (i, j = 1, 2, 3). \tag{14}$$

2.3.2. *Mass transport*

In diffusion and sedimentation experiments the electrical current is usually zero and it is therefore suitable to define flows of electroneutral components.

Defining a set of new fluxes I, J_s, J_p such that:

$$\begin{aligned} I &= \sum_{i,j} Z_i L_{ij} \nabla(-\mu_j) \quad (j, i = 1, 2, 3) \\ J_s &= \frac{J_2}{\nu_2} \\ J_p &= J_3 \end{aligned} \tag{15}$$

it has been shown [3] that the new forces X_e, X_s and X_p associated respectively to I, J_s and J_p are given by

$$\begin{aligned} X_e &= \frac{\nabla(-\mu_1)}{Z_1} \\ X_s &= \nu_1 \nabla(-\mu_1) + \nu_2 \nabla(-\mu_2) = \nabla(-\mu_s) \\ X_p &= \nu_3 \nabla(-\mu_1) + \nabla(-\mu_3) = \nabla(-\mu_p). \end{aligned} \tag{16}$$

A new matrix $(L_{\alpha\beta})$ of phenomenological coefficients connecting the new fluxes to the new forces is now defined:

$$\begin{pmatrix} I \\ J_s \\ J_p \end{pmatrix} = (L_{\alpha\beta}) \begin{pmatrix} X_e \\ X_s \\ X_p \end{pmatrix}. \tag{17}$$

Putting $I=0$ and substituting X_e in J_s and J_p in terms of the X_s, X_p and $L_{\alpha\beta}$ coefficients, a final system of phenomenological equations for the transport of salt and polyelectrolyte components is obtained.

$$\begin{aligned} J_s &= L_{ss} \nabla(-\mu_s) + L_{sp} \nabla(-\mu_p), \\ J_p &= L_{ps} \nabla(-\mu_s) + L_{pp} \nabla(-\mu_p). \end{aligned} \tag{18}$$

The L_{ps} and L_{pp} coefficients are related to the former L_{ij} by the following equations:

$$\begin{aligned} L_{ps} &= \frac{1}{\nu_2} \left[L_{23} - \left(\sum_{ij} Z_i Z_j L_{i3} L_{j2} \right) \left(\sum_{i,j} Z_i Z_j L_{ij} \right)^{-1} \right] \\ L_{pp} &= L_{33} - \left(\sum_{i,j} Z_i L_{i3} \right)^2 \left(\sum_{i,j} Z_j Z_i L_{ij} \right)^{-1}. \end{aligned} \tag{19}$$

(a) *Sedimentation*

Expressing the forces in terms of the gravitational field $\omega^2 r$ and defining the sedimentation constant S_p as the velocity of the polyelectrolyte per unit field in an homogeneous solution, S_p is obtained in rel. (21):

$$\begin{aligned} \nabla(-\mu_p) &= M_p(1 - v_p\varrho)\,\omega^2 r \\ \nabla(-\mu_s) &= M_s(1 - \bar{V}_s\varrho)\,\omega^2 r \end{aligned} \tag{20}$$

$$S_p = c_p^{-1}\left(\frac{J_p}{\omega^2 r}\right)_{c_p,\,c_s} = c_p^{-1}\left[L_{ps}M_s(1 - \bar{V}_s\varrho) + L_{pp}M_p(1 - \bar{V}_p\varrho)\right] \tag{21}$$

$$\begin{aligned} M_p\bar{V}_p &= \nu_3 M_1\bar{V}_1 + M_3\bar{V}_3 \\ M_s\bar{V}_s &= \nu_1 M_1\bar{V}_1 + \nu_2 M_2\bar{V}_2 \end{aligned} \tag{22}$$

M_s, M_p, $\bar{V}_s$ and $\bar{V}_p$ are the molecular weights and the partial specifique volumes of the salt and polyelectrolyte components, ϱ is the solution density.

(b) *Isothermal Diffusion in a Concentration Gradient*

The polymer flow depends on two independent forces $\nabla(-\mu_p^c)$ and $\nabla(-\mu_s^c)$, superscript c designates the chemical part of the general potential μ_i in rel. (16):

$$J_p = L_{pp}\,\nabla(-\mu_p^c) + L_{ps}\,\nabla(-\mu_s^c) \tag{23}$$

J_p cannot as a general rule be expressed in terms of a unique diffusion coefficient. However, we are concerned with dilute polymeric solutions and we shall limit considerations to the most practical case of solutions containing an excess of salt with respect to the polymer. Moreover the experimental conditions can be chosen so as to have $\nabla\mu_s^c \simeq 0$ when before starting the diffusion run the interdiffusing polymeric solution is dialyzed against the pure salt solution. At constant μ_s^c the diffusion coefficient D_p is then defined by:

$$D_p = -\left(\frac{J_p}{\nabla c_p}\right)_{\mu_s{}^c} = RTL_{pp}c_p^{-1}(1 + 2Bc_p + \cdots) \tag{24}$$

in which B is the second virial coefficient. It has been shown that for polyelectrolytes the virial expansion (24) is correct if a third component which provides a high concentration of ions is present [15].

We shall now express the transport parameters in terms of friction coefficients by making use of the relations given in Appendix I. A close physical insight into the signification of the transport parameters will be gained from this formulation.

$$u_3 = \frac{\mathscr{F}}{f_3}\left[Z_3 + \frac{f_{31}u_1^0}{\mathscr{F}} + \frac{f_{32}u_2^0}{\mathscr{F}}\right] \tag{25}$$

$$S_p = \frac{M_p(1 - \bar{V}_p\varrho)}{f_3}\left[1 + \frac{M_s(1 - \bar{V}_s\varrho)}{M_p(1 - V_p\varrho)}\cdot\frac{f_{31} + f_{32} - \nu_3 f_{1w}}{\nu_1 f_{1w} + \nu_2 f_{2w}}\right] \tag{26}$$

$$D_p = \frac{RT}{f_3} \tag{27}$$

$$f_3 = f_{3w} + f_{31} + f_{32} \tag{28}$$

$$u_1^0 = Z_1 \mathscr{F} f_{2w} d^{-1} \tag{29}$$

$$u_2^0 = Z_2 \mathscr{F} f_{1w} d^{-1} \tag{30}$$

$$d = f_{1w} f_{2w} + f_{2w} + f_{21} f_{1w}. \tag{31}$$

Relations (25) to (31) which were derived for the limiting situation $c_p \ll c_s$ are limiting laws with respect to the salt-to-polymer concentration ratio. In practice they apply as long as experimental results are extrapolated to infinite polyelectrolyte dilution and the ionic strength is kept at finite values. u_1^0 and u_2^0 are the electrophoretic mobilities, see Equation (12), of co- and counter-ions in a polyelectrolyte free salt solution at identical salt composition.

3. Discussion

Equations (25) to (27) are cast into a form which allows the definition of a complex friction coefficient f_3 explicited by rel. (28). The f_3 parameter may be viewed as representing the total force the macro-ion experiences from its whole environment when the relative velocity is unity. The term $RT f_3^{-1}$ should therefore be identical to the limiting polyion self-diffusion coefficient in an homogeneous medium at the same salt composition. This important conclusion may be set up using a more rigorous approach.

Let us imagine an experiment for which $v_1 = v_2 = 0$; $v_3 \neq 0$; such an experimental situation is easily devised using for instance labelled polyions and measuring isotopic fluxes in a self-diffusion process, from (7):

$$J_3 = D_3^a \nabla(-c_3) = c_3 v_3 \tag{32}$$

$$\nabla(-\mu_3) = RT c_3^{-1} \nabla(-c_3) = (f_{31} + f_{32} + f_{3w})\, v_3 \tag{33}$$

and the self diffusion coefficient D_3^a of the polyion is found to be*

$$D_3^a = \frac{RT}{f_3}. \tag{34}$$

Actually, if we devise a transport process in a three component polyelectrolyte system at infinite dilution in which the salt is maintained steadily at constant chemical potential throughout the measuring cell, the dissipative process occurring in the system can be completely described by the frictional properties of the polyion. This point becomes apparent if we focus attention on the significance of the diffusion Equation (27). In the limiting case of excess salt and the condition $\nabla \mu_s^c = 0$, no driving force acts on the small ions; the polyion in its own concentration gradient acquires with

* This result is strictly valid in the limit of infinite polyion dilutions, a situation for which polyion-polyion interactions are vanishing. For more concentrated solutions one has to take into account isotopic interactions f_{33} when flux equations for both labelled and unlabelled particles are developed. For this point see References [3] and [4].

respect to its environment a net velocity and the f_3 term gives the total local interactions the polyion has to 'overcome' during its stochastic motion. (This picture is self-consistent from a macroscopic point of view but it cannot yet of course be transposed to the microscopic level).

In the sedimentation expression (26) the coupling term in parentheses would be negligibly small if $M_p \gg M_s$ or if for instance the salt is already in equilibrium in the measuring cell, then S_p should have been defined at constant μ_s rather than at constant c_s and formulae (26) and (27) would have obtained analogue closed forms.

The additional terms appearing in the brackets of rel. (25) and (26) are obviously relative to coupling effects between flows of ions in electrophoresis and flows of salt and polyelectrolyte in sedimentation.

Significance and evaluation of the friction coefficients. The f_{iw} are familiar hydrodynamic parameters expressing viscous drag effects exerted by the ions on the solvent; f_{3w} can be computed according to current hydrodynamic theories of non charged macromolecules and f_{1w}, f_{2w} may be approximated by their values in ordinary electrolyte solutions.

On the other hand the f_{3i} parameters express interactions between the polyion and the ionic atmosphere. It has been convincingly demonstrated that ionic atmospheres around charged polymers are not suppressed even in a high salt medium, rather the double layer around the polyion is compressed as the ionic strength is increased. Therefore the f_{3i} parameters are correlated to long range electrostatic forces occurring between small ions and the polyion. This charge effect which does not vanish at infinite dilution should be clearly distinguished from the so-called primary charge effect which occurs in the more general situation when the salt is not present at excess and leads to a measurable electrical potential in the cell [16–17].

The evaluation of the f_{3i} parameters can be performed by relating these coefficients to the self-diffusion coefficient of co- and counter-ions in a polyelectrolyte solution.

In fact, instead of looking at the polyion motion referred to its environment it is possible to imagine the converse situation of self-diffusion of co- and counter-ions where the ionic atmosphere is at motion and the polyion is kept stationary. The electrical forces experienced in the two processes are identical. This argument is easily confirmed making use of Equation (9) and writing f_3 into the alternative form

$$f_3 = f_{3w} + (f_{13} + f_{32}) \cdot C_s / C_p. \tag{35}$$

The f_{i3} terms being now related to the ratio D_i^0 / D_i of the self-diffusion coefficient of co- and counter-ions in absence and in presence of the polyelectrolyte:

$$D_i = \frac{RT}{f_{ii} + f_{ij} + f_{i3} + f_{iw}} \quad (i, j = 1, 2)$$

$$D_i^0 = \frac{RT}{f_{ij} + f_{i3} + f_{iw}} \quad (i, j = 1, 2)$$

$$f_{13} = \frac{RT}{D_i^0}\left[\frac{D_1^0}{D_1} - 1\right] \tag{36}$$

$$f_{23} = \frac{RT}{D_2^0}\left[\frac{D_2^0}{D_2} - 1\right].$$

The derivation of D_i, D_i^0 according to (36) proceeds identically as for rel. (34); however, in the present case the derivation rests on the argument that $f_{1w} + f_{12} + f_{11}$ and $f_{2w} + f_{21} + f_{22}$ have identical values in absence and in presence of the polyion.

Formulation (36) is most suitable since self-diffusion coefficients of small ions may be directly evaluated in terms of molecular parameters, for many polyelectrolyte models. Before this point is illustrated some attention should be drawn to the criterion which permits an assessment of the validity of the friction model in the present case.

We have implicitly allowed the friction coefficients to be independent of the magnitude and the nature of applied forces, that is to say these coefficients are completely defined by the equilibrium properties of the solution as shown for example by Bearman for self-diffusion processes in binary liquid solutions [14]. Nevertheless, for ionic solutions polarization effects resulting from the application of an external field of forces may give rise to distorted ionic atmospheres and the identification of a unique interaction parameter f_{ij} in electrical and self-diffusion processes becomes questionable. However, it has been proved that as far as polyelectrolytes are concerned, the perturbation of the counter-ion distribution with respect to the equilibrium situation is fairly small despite the high polarizability of polyelectrolyte solutions [18]. Moreover, linear forces – fluxes relations have usually been reported from experimental investigations and for both polyelectrolyte and pure salt solutions electrical and self-diffusion determinations have led to nearly identical frictional parameters [19–20]. The friction model might therefore be used with confidence as long as systems not too far from equilibrium are concerned.

4. Application to Polyelectrolyte Models

We wish to relate transport parameters for two stiff symmetrical polyelectrolyte models by evaluating separately for each model the friction coefficients in terms of the molecular parameters.

A number of important works have been devoted to the computation of self-diffusion rates of small ions in polyelectrolyte solutions. Since this question extends beyond the present necessarily limited discussion we shall refer the reader for this question to the original memoirs [21–25]. Let us just mention that the effect of the electrostatic field of the macro-ion on the macroscopic self-diffusion constant of small ions has been computed analytically or numerically for polyelectrolytes of different shape and charge. The theory of self-diffusion starts usually with a cell model, the polyion being held fixed in space and the rate of motion of labelled ions in the spatially periodic electrostatic field set up by the polyion is obtained by solving the modified diffusion equation.

4.1. Polyelelectrolyte models

Particle denoted (PS) is a porous sphere, the N monomers are evenly distributed over the sphere of radius R_0. The total charge is smeared out over the whole sphere and a volume charge density σ_v is defined:

$$\sigma v = v_3 e V_p^{-1}; \quad V_p = 4/3\pi R_0^3.$$

An analytical expression for D_i relative to this model is found in a paper of Bell [23] and the electrical potential distribution $\phi(R)$ is approximated by a step function according to a relation given by Liffson [26]:

$$\begin{aligned} \frac{D_i}{D_i^0} &= \frac{1 + (1 + 2\varrho_v)\,\chi_i/3}{(1 + \varrho_v \chi_i)\,[1 + (1 - \varrho_v)\,\chi_i/3]} \\ \varrho_v &= c_p \mathcal{N} v_p \\ \chi_i &= \exp(-\phi_i) - 1 \\ |\phi_i| &= \sin h^{-1}\left(\frac{c_f}{2c_s}\right)\left[1 - \frac{XR_0}{1 + (X/\mu)\tanh \mu R_0 \cosh \mu R_0}\right]. \end{aligned} \tag{37}$$

The limiting friction parameter is derived as:

$$\lim f_{3i} = f_{iw} V_p c_s \mathcal{N}\left[1 - \frac{3}{\chi_i + 3}\right] \quad (i = 1, 2) \tag{38}$$

and the hydrodynamic theory of Debye-Bueche [27] has been used for the evaluation of f_{3w}:

$$f_{3w} = 6\pi\eta R_0\, \Psi(\sigma). \tag{39}$$

The parameters appearing in (37) and the volume factor $\Psi(\sigma)$ are defined in the original papers. Formula (39) holds good for an uncharged polymer. For a charged polymer the solvent velocity induced by the polyion motion at any location is necessarily perturbed by the presence of the stationary cloud of counterions. The estimation of the error made by using (39) is a difficult problem which we will not undertake in this article. Refinement relative to this point should demonstrate whether the error in doing so is larger than is at present believed.

The sphere model which has a rather historical interest could nevertheless perhaps provide an approximation for some stiff biocolloids or approach the behaviour of low charged linear polyelectrolytes despite the crude choice of a uniform charge density inside the sphere.

Particle denoted (R) is the classical rod model of total length L [24]; A charging parameter ξ is assigned a key role in this model:

$$\xi = \frac{e^2}{\varepsilon k T b} \quad b = L/v_3 .$$

An analytical expression for D_i and ϕ_i given by Manning [24] has been used to evaluate f_{3i}:

$$\lim_{c_p \to 0} f_{3i} = \frac{f_{iw} \nu_3 \xi}{6}.$$

The f_{3w} parameter is computed according to the relation of Kirkwood and Riseman [28] with the same restrictions already outlined for Equation (39):

$$f_{3w} = N f_m (L_n N)^{-1}. \tag{42}$$

From the above, for the sphere model the charge effect depends on the ionic strengths since f_{3i} is proportional to c_s and depends also on c_s by the χ_i parameter. On the other hand for the rod-like model it is surprisingly found that in the context of $c_p \ll c_s$, f_{3i} is only related to the linear charge on the polyion whatever the amount of added salt.

4.2. The Reduced Friction Coefficient

The data which have been calculated for a mono-mono valent salt (sodium chloride) are given on Figures 1 and 2. On both graphs f_3/f_{3w} is plotted against the degree of ionization $\alpha_s = \nu_{3s}/N$ – the stoechiometric charge per monomer unit.

For the sphere the charge effect is almost completely suppressed at 1 molar salt

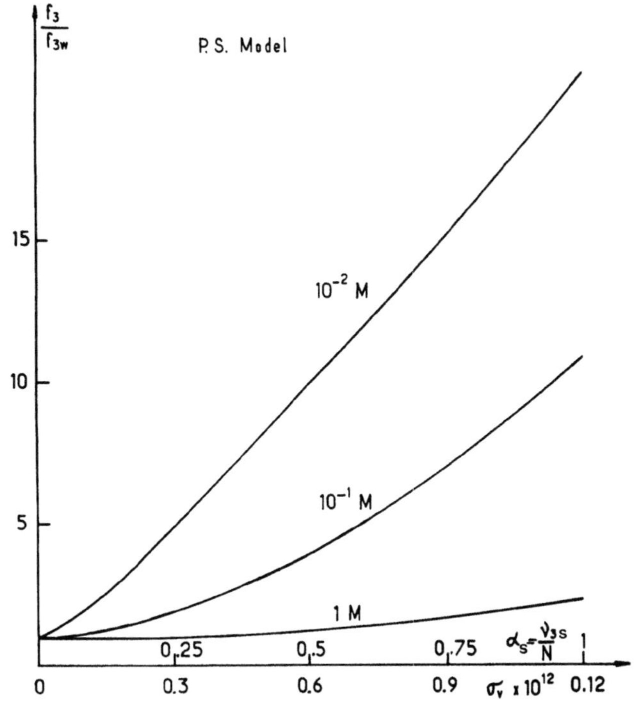

Fig. 1. Reduced friction coefficient vs. α_s for the (PS) model at different salt composition.

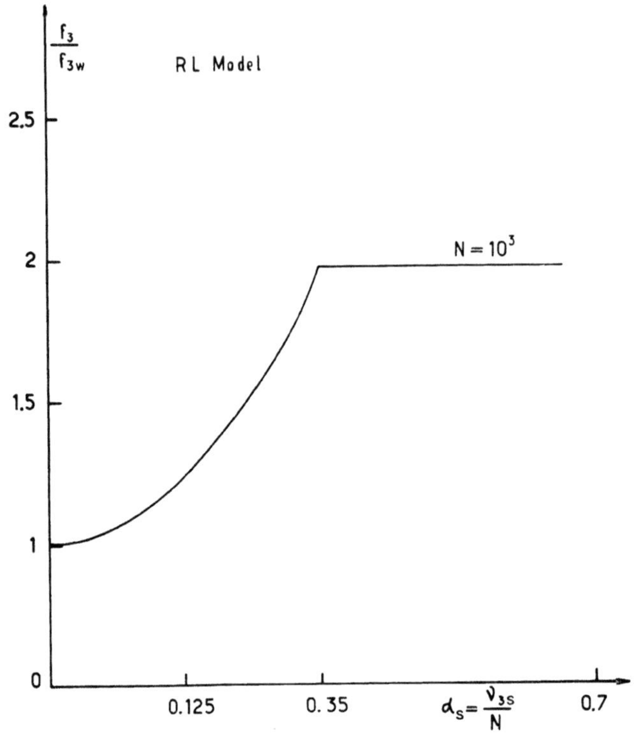

Fig. 2. Reduced friction coefficient vs. α_s for the (R) model.

concentration, but increases considerably at lower ionic strengths reaching for example for 0.1 M a value close to 10 when each of the thousand monomers carries one elementary charge.

For the rod the charge effect is less enhanced at higher charge densities which is mainly a consequence of the peculiar aspects of ionic interactions occurring between a charged rod and small ions. In Manning's theory [24] it is shown that at infinite dilutions as ξ (stoichiometric) > 1 sufficient counterions will 'condense' on the polyion to lower ξ (effective) to a value one. For $\xi > 1$ the effective ionization ν_3 is related to the total number ν_{3s} of ionizable groups on the polyion by:

$$\nu_3 = \nu_{3s}\xi^{-1}. \tag{43}$$

In computing the data for the rod-like model the $(\nu_{3s} - \nu_3)$ condensed ions have been considered as rigidly attached to the polyion defining one kinetic unit with it and therefore, as a result of condensation, the effective degree of ionization cannot exceed a critical value and the charge effect reaches an upper limiting value for $\xi = 1$.

4.3. Sedimentation of charged polymers

Sedimentation of polyelectrolytes appears to be a more involved process than diffusion since generally the salt is not brought to chemical equilibrium in the ultracentrifuge cell. $\nabla(-\mu_s)$ could be zero for example if low molecular weight polyelectrolyte is centrifuged and the salt has already reached its equilibrium distribution. For this situation S_p cannot be defined at constant composition but rather the force acting on the polyelectrolyte should then include the chemical term

$$\left(\frac{\partial \mu_p}{\partial c_s}\right)_{c_p} \left(\frac{\partial c_s}{\partial r}\right)$$

and rel. (26) would take a different form.

The coupling term L_{ps} is generally negative; namely for the two models the following relation has been verified for a number of ordinary salts:

$$\nu_3 f_{1w} > f_{31} + f_{32}. \tag{44}$$

Rel. (44) is verified at once for the rod model noting that for most common mono-mono valent salts,

$$f_{1w} > \frac{\xi}{6}(f_{1w} + f_{2w}) \quad \xi \leqslant 1. \tag{45}$$

The variation of S_p in NaCl according to the polyion charge and ionic strengths is given on Figure 3.

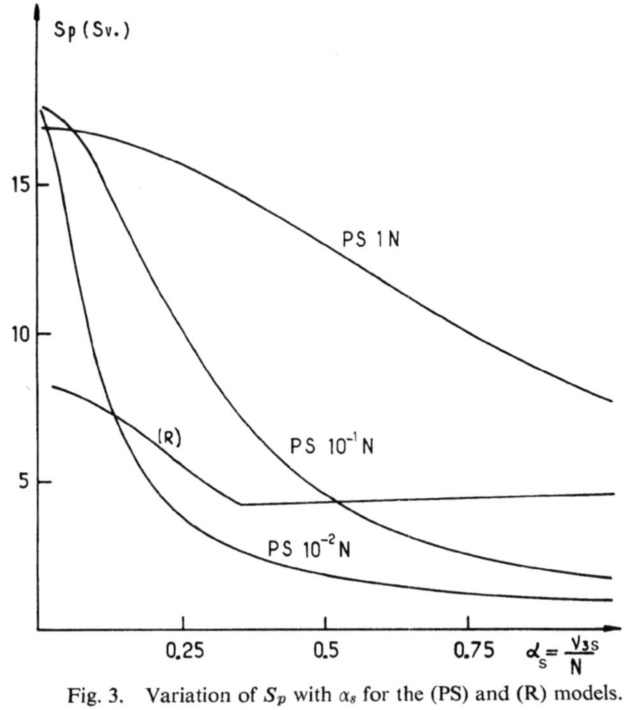

Fig. 3. Variation of S_p with α_s for the (PS) and (R) models.

For the porous sphere S_p decreases rapidly with increasing α especially at low salt concentrations. For the rod S_p decreases up to the critical charge density $\xi = 1$, reaches a minimum and increases slowly almost linearly beyond this point.

It must also be noted that the Svedberg relationship

$$\frac{RTS_p}{D_p(1 - V_p\varrho)} = M_p(1 + r_M) \tag{46}$$

in which r_M is given by the right-hand side parenthesis in rel. (26), leads to an apparent molecular weight smaller than the true value. As reported on Figure 4, r_M

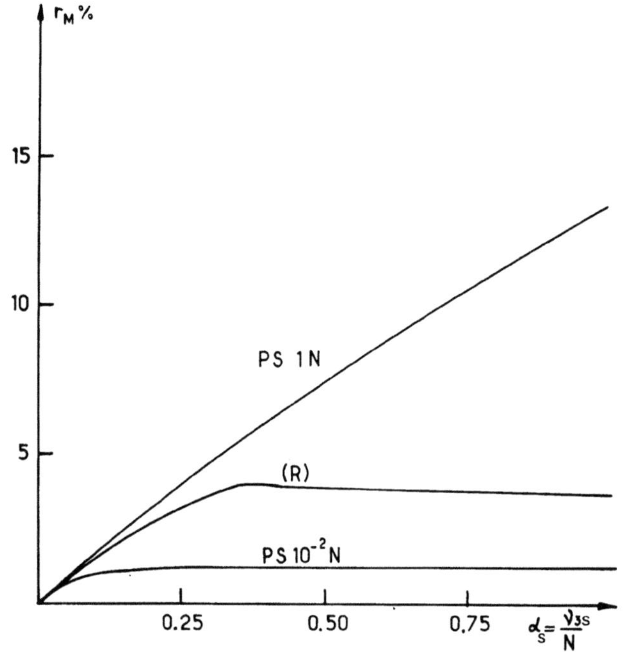

Fig. 4. Variation of r_M with α_s for the (PS) and (R) models.

can reach quite important values. This correction term which for the rod is inversely proportional to the monomer weight has been computed chosing $M_m = 340$. For this model r_M does not depend on the molecular weight and moreover f_3 varies almost proportionally to N:

$$f_3 = f_m N (\log N)^{-1} + \frac{\alpha \xi N}{6} (f_{1w} + f_{2w}) \tag{47}$$

so that in complete analogy with uncharged rod polymers S_p depends only slightly on the molecular weight through the logarithmic term. In the classical representation

$$S_p = KN^a$$

the exponent a goes approximately from 0.1 to zero as N goes from 10^2 to infinity.

4.4. Numerical Values

In deriving the data given in Figure 1 to Figure 4 a mono-monovalent salt has been chosen and the numerical values are relative to the sodium-chloride salt. The following additional relations and numerical data have been used:

Sedimentation:

$$
\begin{aligned}
M_p &= NM_m + v_3 (M_1 - 1)\\
M_s &= M_1 + M_2\\
\bar{V}_p &= \frac{\bar{V}_3 M_m + \alpha_s \bar{V}_1}{M_m + \alpha_s (M_1 - 1)}\\
\bar{V}_s &= \frac{V_1 M_1 + V_2 M_2}{M_1 + M_2}\\
M_m &= 340 \text{ g/mole}^{-1}; \quad \bar{V}_3 = 0.55 \text{ cm}^3 \text{ g}^{-1}.
\end{aligned}
\tag{49}
$$

Numerical values for $\bar{V}_1$ and $\bar{V}_2$ are found in a paper of Pederson [16] and the solution density ϱ is taken from standard data [25].

Polyelectrolyte model:

$$
\begin{aligned}
N &= 10^3\\
R_0 &= 100 \text{ Å}\\
1 &= 2.5 \text{ Å (distance between the center of two successive monomers).}
\end{aligned}
\tag{50}
$$

Frictional parameters:

$$
\begin{aligned}
f_{1w} &= 1.86 \times 10^{15} \text{ cgs} \qquad \text{(from Reference [16])}\\
f_{2w} &= 1.22 \times 10^{15} \text{ cgs}\\
\sigma &= 7.5\\
\Psi(\sigma) &= 0.847\\
f_{3w} &= 0.866 \times 10^{17} \text{ cgs} \qquad \text{(sphere model)}\\
f_m &= 3\pi\eta\, 1\\
f_{3w} &= 1.85 \times 10^{17} \text{ cgs} \qquad \text{(rod model).}
\end{aligned}
\tag{51}
$$

Physical constants:

$$
\begin{aligned}
\eta &= 9 \times 10^3 \text{ cgs} \qquad \text{(solvent viscosity)}\\
T &= 298 \text{ K}\\
\varepsilon &= 78.5 \qquad \text{(dielectric constant)}\\
\mathcal{F} &= 96493 \text{ coulomb}\\
\mathcal{N} &= 6.02 \times 10^{23}.
\end{aligned}
\tag{52}
$$

Appendix

The phenomenological coefficients L_{ij} are related to the friction coefficients by following the procedure described in the text and expressing the fluxes with regard to the solvent $v_w = 0$. Following relation are given in Reference [13]:

$$M_{ij} = c_i f_{ij} = M_{ji}$$

$$M_{ii} = \sum_{\substack{j \\ i \neq j}} M_{ij}$$

$$L_{11} = \frac{c_1^2}{d'}(M_{22}M_{33} - M_{23}^2) \quad L_{12} = \frac{c_1 c_2}{d'}(M_{13}M_{23} + M_{12}M_{33})$$

$$L_{22} = \frac{c_2^2}{d'}(M_{11}M_{33} - M_{13}^2) \quad L_{23} = \frac{c_2 c_3}{d'}(M_{12}M_{13} + M_{11}M_{23})$$

$$L_{33} = \frac{c_3^2}{d'}(M_{11}M_{22} - M_{22}^2) \quad L_{13} = \frac{c_1 c_3}{d'}(M_{12}M_{23} + M_{13}M_{22})$$

$$d' = M_{11}M_{22}M_{33} - M_{11}M_{23}^2 - M_{22}M_{13}^2 - 2M_{12}M_{23}M_{13} - M_{33}M_{12}^2 .$$

References

1. Gosting, L. J.: *Adv. Protein Chem.* **11**, 429 (1956).
2. Nagasawa, M. and Eguchi, Y.: *J. Phys. Chem.* **71**, 880 (1967).
3. Varoqui, R. and Schmitt, A.: *Biopolymers* **11**, 119 (1972).
4. Schmitt, A. and Varoqui, R.: *J. Chem. Soc., Faraday Trans. II* **69**, 1087 (1973).
5. Manning, G. S.: *J. Chem. Phys.* **46**, 2324 (1967); *J. Chem. Phys.* **51**, 934 (1969).
6. Schmitt, A.: Thesis, Strasbourg, 1972.
7. Hooyman, G. J., Holtan, H., Mazur, P., and De Groot, S. R.: *Physica* **19**, 1095 (1953).
8. De Groot, S. R., Mazur, P., and Overbeek, J. T. G.: *J. Chem. Phys.* **20**, 1825 (1952).
9. Mijnlieff, P. F. and Overbeek, J. T. G.: *Koninkl. Ned., Akad. Wetensch. Proc. Ser. B* **65**, 221 (1962).
10. Katchalsky, A. and Curran, P. F.: *Non-Equilibrium Thermodynamics in Biophysics*, Harvard University Press, Cambridge, 1967.
11. Fitts, D. O.: *Non-Equilibrium Thermodynamics*, McGraw-Hill, Inc. New York, 1962.
12. Laity, J.: *J. Phys. Chem.* **63**, 80 (1959); *J. Chem. Phys.* **30**, 682 (1959).
13. Spiegler, K. S.: *Trans. Faraday Soc.* **54**, 1408 (1958).
14. Bearman, R. J. and Kirkwood, J. G.: *J. Chem. Phys.* **28**, 136 (1958).
    Bearman, R. J.: *J. Chem. Phys.* **31**, 751 (1959); *J. Phys. Chem.* **65**, 1961 (1961).
15. Hill, T. L.: *Introduction to Statistical Thermodynamics*, Addison-Wesley, Mass., 1960, Chap. 19.1 and 19.2.
16. Pedersen, K. O.: *J. Phys. Chem.* **62**, 1282 (1958).
17. Alexandrowicz, Z. and Daniel, E.: *Biopolymers* **1**, 447 (1963); **6**, 1500 (1968).
18. Oosawa, F.: *Polyelectrolytes*, Marcel Dekker, Inc., 1971, Chap. 5.
19. Wall, F. T., Terayama, H., and Techakumpuch, S.: *J. Polym. Sci.* **20**, 477 (1956).
20. Huizenga, J. R., Grieger, P. F., and Wall, F. T.: *J. Am. Chem. Soc.* **72**, 2636 (1950); *J. Am. Chem. Soc.* **72**, 4228 (1950).
21. Lifson, S. and Jackson, J. L.: *J. Chem. Phys.* **36**, 2410 (1962).
22. Jackson, J. L. and Coriell1, S. R.: *J. Chem. Phys.* **38**, 959 (1963); *J. Chem. Phys.* **39**, 2418 (1963).
23. Bell, G. M.: *Trans. Faraday Soc.* **60**, 1752 (1965).
24. Manning, G. S.: *J. Chem. Phys.* **46**, 2324 (1967); *J. Chem. Phys.* **51**, 934 (1969).
25. Pefferkorn, E. and Varoqui, R.: *European Polymer J.* **6**, 663 (1970).
26. Lifson, S.: *J. Chem. Phys.* **27**, 700 (1957).
27. Debye, P. and Bueche, A. M.: *J. Chem. Phys.* **16**, 573 (1948).
28. Riseman, J. and Kirkwood, J. G.: in F. R. Eirich (ed.), *Rheology*, vol. 1, Acad. Press Inc., New York, 1956.
29. *Handbook of Chemistry and Physics* (44th ed.).

ELECTRIC POLARISABILITY OF RIGID POLYELECTROLYTES

G. WEILL and C. HORNICK
Centre de Recherches sur les Macromolécules C.N.R.S., 67083 Strasbourg, Cedex, France

Aqueous solutions of polyelectrolytes present a very large dielectric increment which has been recognized as originating from the polarisation of the counter-ion atmosphere. By definition, the polarizability α is linked to the partition function Q of the system through [1].

$$\alpha = \frac{kT}{E}\frac{\partial \ln Q}{\partial E}. \tag{1}$$

This expression is easily transformed into [2]

$$\alpha = \frac{\langle \mu^2 \rangle_{E=0}}{kT}, \tag{2}$$

where $\langle \mu^2 \rangle_{E=0}$ is the square average dipole moment in the absence of an electrical field. For a molecule with no permanent dipole moment $(\langle \mu \rangle_{E=0}=0)$ the square average dipole moment measures the fluctuations in the distribution of the counter-ions around the uniform most probable distribution. One can therefore expect that the study of the polarisability give interesting information on the counter-ion atmosphere and particularly on the repulsion between counter-ions trapped in the electric potential of the polyion, which will limit the build up of large fluctuations.

Several theoretical models have been developed since the initial calculation of Mandel [3] which was based on a linear polyelectrolyte with discrete binding sites and no counter-ion repulsion. They all retain the rigid character of the linear polyelectrolytes. Most of the measurements have, however, been performed on solutions of flexible polyelectrolytes which adopt a gaussian conformation at high ionic strengths. In order to perform a quantitative comparison between theory and experiment measurements on rigid polyelectrolytes of well known molecular structure were required. Such a polyelectrolyte can be prepared by ultrasonication of solutions of DNA [4]. The anti-parallel double helical structure warrants the absence of permanent electric moment as it can indeed be checked by the measurement itself if one deduces the polarisability from a transient electro-optical effect resulting from the orientation of the rigid molecules in the electric field [5]. This method, while avoiding some of the difficulties linked to the relation between dielectric constant and polarisability due to internal field effects, gives an internal check of rigidity of the molecule with varying ionic conditions. It lacks, however, the interesting information on the dynamic behaviour of the counter-ion atmosphere that one can deduce from the dispersion of the dielectric constant. In this respect both methods are complemetary [6].

Eric Sélégny (ed.), Polyelectrolytes, 277–284.

1. Short Survey of the Methods

Several optical phenomena can be used to detect the degree of orientation attained under the effect of an electric field:

- light scattering,
- linear dichroïsm,
- fluorescence polarisation,
- birefringence (Kerr effect).

All of them have been used in our study of DNA fragments. [7] The Kerr effect measurements are however the most sensitive and have been used to study the variation of polarisability with the nature and concentration of counter-ions. If a saturation of orientation can be approached, the degree of orientation for an arbitrary electric field can be directly derived from any electro-optical measurement, the amplitude of the optical phenomenon at complete orientation being easily extrapolated. If low degrees of orientation only are accessible (as it will be the case at higher ionic strengths) a second measurement (flow birefringence, angular dependence of the light scattering) or an hypothesis (direction of the transition moment) is needed to separate the electric and optical factor in the electro-optical factor. As an example we describe how Kerr effect and flow birefringence have been coupled in our experiments [8].

For a rigid particle with an optical anisotropy per unit volume $(g_1 - g_2)$ and an electrical anisotropy $\alpha_{\parallel} - \alpha_{\perp}$ the limiting birefringence induced by an hydrodynamic field and an electric field are respectively given by:

$$\lim \left[\frac{\Delta n}{G\eta_0 c}\right]_{\substack{c\to 0\\ G\eta_0\to 0}} = \frac{2\pi}{15 n_0 \eta_0 D}\,\bar{v}(g_1 - g_2), \tag{3}$$

$$\lim \left[\frac{\Delta n}{cE^2}\right]_{E=0} = \frac{2\pi}{15 n_0}(g_1 - g_2)\frac{(\alpha_{\parallel} - \alpha_{\perp})}{kT}, \tag{4}$$

where G is the shear gradient, η_0 the viscosity of the solvent, n_0 the index of refraction of the solvent, c the concentration of the polyelectrolyte, $\bar{v}$ its partial specific volume, D its rotatory diffusion constant.

D can be deduced from the variation of the extinction angle in the flow birefringence experiment:

$$\left[\frac{\mathrm{d}x}{\mathrm{d}G}\right]_{\substack{c\to 0\\ G\to 0}} = \lim \left[\frac{\frac{\pi}{4} - x}{G}\right]_{\substack{c\to 0\\ G\to 0}} = \frac{1}{12D} \tag{5}$$

or from the decay of the electric birefringence at the end of application of the field [5]:

$$\Delta n(t) = \Delta n(0)e^{-6Dt}. \tag{6}$$

Both values must be consistent and the length L of the particle derived from the relation [9]:

$$D = \frac{3kT}{\pi \eta_0 L^3}\left[\ln \frac{2L}{d} - 0.8\right]. \tag{7}$$

where d is the particle diameter, in agreement with the length derived from the angular dependence of the light scattering, assuming a rodlike particle.

If so, relations (3) and (4) can be combined to give $(g_1 - g_2)$ and $\alpha_{\|} - \alpha_{\perp}$.

All DNA fragments used had molecular dimensions close to the typical results given below. More details on the sonication and experimental conditions will be found in References [7] and [8].

-- Molecular weight: 3.7×10^5

-- Mass per unit length : 200

(from the asymptotical behaviour of the angular dependence of light scattering)

-- Overall length : 1850 Å

(ratio of the two preceding lines)

- $\varrho\sqrt{12}$: 1770 Å

(length deduced from the radius of gyration ϱ assuming a fully rigid rod)

-- Rotatory diffusion constant and corresponding length

– from Kerr effect $1660 < D < 4160$ $1500 < L < 2100$ Å

– from Flow birefringence $L = 1650$ Å

- $g_1 - g_2$: 7×10^{-3}

- $\alpha_{\|} - \alpha_{\perp}$ at $c = 10^{-4} g$/cc Na-salt in the absence of added NaCl: 2×10^{-15} cm^3

Considering the different averages arising from the polydispersity the agreement is very good and the measured polarisability can be assumed to be that of a rodlike fragment of DNA with the regular Crick and Watson structure.

2. Variation of the Polarisability with the Nature and Concentration of Counter-ions

The results are summarized on Figure 1, where the curves are average over all the measurements. Three interesting features appear clearly:

(1) There is no significant differences between the Li, Na and K salts of DNA, their polarisability of $2 \mp 1 \times 10^{-15}$ cm^3 (for a molecular weight of 3.7×10^5 dalton) in the absence of added salt decreasing to $0.8 \mp 0.3 \times 10^{-15}$ at 8×10^{-3} M.

(2) The polarisability of the Mg^{++} salt is much bigger in the absence of added salt ($\mp 2 \times 10^{-15}$ cm^3) but drops more rapidly with the concentration of added salt ($6 \mp 2 \times 10^{-15}$ cm^3 in 2×10^{-3} M Cl_2 Mg).

(3) The TMA salt has a low polarisability which doesn't vary appreciably with the ionic strength.

An acceptable theory should account for these quantitative and qualitative features.

3. A Comparison with the Theories

The theory of Alexandrowicz and Katchalsky [10] and Lifson [11] gives a satisfying picture of the distribution of counter-ions around the polyelectrolyte, and of the fraction of counter-ions trapped in the electrostatic potential of the polyion. The

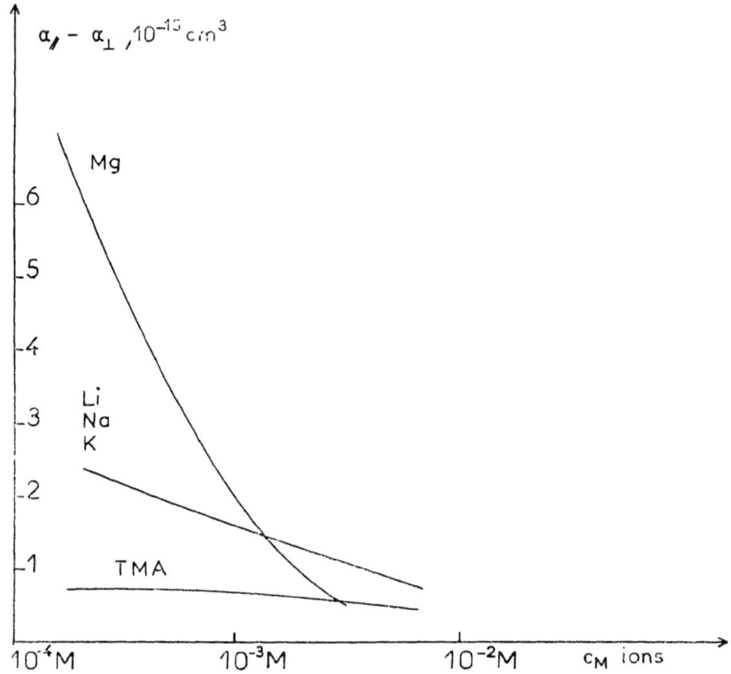

Fig. 1. Variation of the anisotropy of polarisability $\alpha_{\parallel} - \alpha_{\perp}$ of a DNA rod-like fragment with the nature and concentration of added salt.

addition of salt reduced the thickness of the layer of trapped counter-ions without changing appreciably the fraction of free counter-ions in the range of ionic strengths used in this work. This suggest that the decrease of the polarisability has its origin in the increase of counter-ion repulsion resulting of their higher concentration in the polyion atmosphere. We shall therefore discuss our results in the framework of those of the theories where description of the counter-ion distribution is possible, discarding O'Konski [12] theory in which an arbitrary conductivity is attributed to the surface layer as well as Schwartz [13] or Takashima's [14] theories in which the inner properties of the counter-ion layer are replaced by a set of boundary conditions valid for an unrealistic negligible thickness. One should however note that their result has a correct functional dependence (see Reference [7]).

Starting from Equation (2) and a linear polyelectrolyte of length L one can express the square average dipole moment as:

$$\langle \mu^2 \rangle_{E=0} = n_+ Z e_0^2 \frac{\int_{-L/2}^{+L/2} x^2 p(x)\, \mathrm{d}x}{\int_{-L/2}^{+L/2} p(x)\, \mathrm{d}x}, \tag{8}$$

where n_+ is the number of trapped counter-ions, Z their valency and $p(x)$ their probability to be at an abcissa x from its center along the linear polyion.

If any repulsion between counter-ion is neglected, $p(x)$ is a constant and (2) becomes:

$$\alpha = \frac{n_+ Z^2 e_0^2 L^2}{12kT} \tag{9}$$

a quite general result identical with the original Mandel's result [6] for a discrete site model. Introducing in (9) the molecular parameters of our DNA sample, assuming a fraction of free counter-ions of .2 [15] one finds:

$$\alpha = 130 \times 10^{-15}\ \mathrm{cm}^3$$

a result nearly two order of magnitude bigger than the experimental result, therefore emphasizing the role of counter-ion repulsion.

The first calculations taking these repulsions in account have been performed by McTague and Gibbs [16]. Their matrix method was solved numerically after the choice of an average repulsion between discrete sites. A similar matrix method soluble in closed form has been devised by Oosawa *et al.* [17] for the case of mixtures of mono and divalent counterions. It starts from rewriting Equation (2) as:

$$\langle \mu^2 \rangle_{E=0} = \tfrac{1}{12} n_+ L^2 \left[\Delta m^2 + 2\Sigma_k\, \Delta_k m \right],$$

where

$$\begin{aligned} \Delta m^2 &= \langle m_i^2 \rangle - \langle m \rangle^2 , \\ \Delta_k m &= \langle m_i m_{i+k} \rangle - \langle m \rangle^2 , \end{aligned} \tag{10}$$

where m_i is the charge of the ion occupying the ith site. The presence of an interaction W_{ik} between counterions (repulsive as well as attractive) will make $\Delta_k m$ different from zero. But an evaluation of W_{ik} and of its change with the thickness of the layer would be rather arbitrary.

The overall energy of repulsion between counter-ions can however be obtained from a consideration of the free energy of the charging process. It has a rather simple expression if one considers the simplified model of Oosawa [2] where the layer of trapped ion is considered as a phase of radius a. Oosawa [2] and later on Schurr [19] developed the fluctuations in Fourier components of wave length L/k. The counterion repulsion potential ϕ_k associated with the kth component is:

$$\phi_k = \frac{2}{L} \int_0^L \phi(r) \cos \frac{2\pi k r}{L}\, \mathrm{d}r , \tag{11}$$

where $\phi(r)$ is the interaction potential between two ions at distance r, and the polarisability becomes:

$$\alpha = \frac{n_+ Z^2 e_0^2 L^2}{12kT} \sum \frac{1}{k^2} \frac{1}{1 + \dfrac{n_+ \phi_k}{2kT}} . \tag{12}$$

A slightly different evaluation of the sum in (12) by Oosawa and Schurr (the latter correcting an assumption of rapid convergence in (11)) leads to two different but numerically very close results, for

$$\alpha = \frac{n_+ Z^2 e_0^2 L^2}{12kT} \frac{1}{1 + \dfrac{2n_+ Z^2 e_0^2}{\varepsilon kTL} \ln \dfrac{R}{a}}, \qquad \text{Oosawa} \qquad (13)$$

where R is the radius of the cylindrical cell of length L corresponding to the average volume offered to one molecule in the solution and ε the dielectric constant;

$$\alpha = \frac{n_+ Z^2 e_0^2 L^2}{12kT} \sum \frac{1}{k^2} \frac{1}{1 + \dfrac{2n_+ Z^2 e_0^2}{kT} \left(\ln \dfrac{L}{\sigma} - \ln 2\pi k \right)}, \qquad \text{Schurr} \qquad (13')$$

where σ is the parameter of the radial gaussian distribution of counter-ions around the polyion.

Since our results show the predominance of the repulsion term one can rewrite (13) and (13′) neglecting one in the denominators:

$$\alpha = \frac{L^3 \varepsilon}{24 \ln \dfrac{R}{a}}, \qquad \text{Oosawa} \qquad (14)$$

$$\alpha = \frac{L^3 \varepsilon}{24 \ln \dfrac{L}{\sigma}}. \qquad \text{Schurr} \qquad (14')$$

For our concentrations and according to the semi quantitative significance of a or σ one can adopt for R/a (or L/σ) in the absence of added salt $R/a \sim 1000$ Å/100 Å ~ 10 ($L/\sigma \sim 1800/100$); $\ln R/a \sim 2-3$. Therefore:

$$\alpha \sim \frac{5.8}{24} \cdot \frac{\varepsilon}{3} 10^{-15} \text{ cm}^3$$

which has the right order of magnitude, ε being probably smaller in average than its value of 80 in a region of high counterion concentration [20].

The expression (14) and (14′) lead to a right prediction for the decrease of the polarisability with the layer of bound counter-ion, the thickness a (or σ) decreasing rapidly with the addition of salt. However it doesn't explain the higher value for the Mg^{++} salt in the absence of added salt, the effect of the valency disappearing in the final result when counter ion repulsion dominates the phenomenon. It may however be the result of a cooperative binding of Mg^{++} on DNA which would counterbalance the general effect of repulsion (see Equation (10)). The rapid drop would however result from the higher repulsion between divalent counterion for a given thickness a. The lower value of α for the TMA salt is consistent with what is known of its small binding which at the same time reduces n_+ and makes the repulsion very small.

4. Conclusion

The semi-quantitative agreement between the theory and experiments, if gratifying, call for some less optimistic remarks. In fact expression (14) shows that for highly charged polyelectrolytes, the polarisability should not give more than an estimation of the variation of thickness of the layer of trapped counter-ions. The lack of difference between the three alcaline ions seems to deny to measurements of static polarisability the possibility to use it for a detailed knowledge of the distribution of the counter-ions. More differences should however appear when using mixtures of monovalent and divalent counter-ions. In this case the increased binding energy of the divalent counter-ions will be partly overcome by the decreased counterion repulsion for monovalent counter-ion [17]. The maximum of dielectric constant observed should be studied with rigid polyelectrolytes. It arises from larger possible fluctuations in the distribution of counter-ions.

The demonstration here given of the predominant role of counter-ion repulsion in the limitation of the fluctuations in the distribution of counterion have also interesting implication for the interpretation of the dispersion of the dielectric constant. In Oosawa's theory, the relaxation time τ_k associated with the kth fluctuation is:

$$\tau_k = \frac{L^2}{2\pi k}\frac{kT}{\xi}\left(1 + \frac{n + \phi_k}{kT}\right)^{-1}, \tag{15}$$

where ξ is the friction constant of the ion. The second term in the parenthesis can be interpreted as a reduction of the translatory diffusion constant of the counter-ion due to the field of the other ions. Using values of the translatory diffusion constant typical for a free ion ($\sim 10^{+5}$) and a 'repulsion factor' of the order of 50 as observed in the static experiment one should obtain a longest $\tau(k=1)$ of the order of 10^{-7} which seems much smaller that what has been observed [6]. This problem remains unsolved. Some ways for direct observation of the mobility of the counter-ions along a rigid polyelectrolyte may perhaps now be thought either from inelastic light scattering or pulsed gradient spin echo techniques.

A last point should not be forget in the comparison of dielectric and electrooptic measurements of the polarisability. The latter one measures in fact the anisotropy of polarisability $\alpha_{\parallel} - \alpha_{\perp}$. We have considered our result as giving $\alpha_{\parallel}$ and compared it to theories which neglect any transversal displacement of the counter-ions. While this is reasonable it doesn't means that transversal movements or end effects do not play a role in the value of the dielectric constant and its dispersion.

References

1. Van Vleck, J. H.: *Electric and Magnetic Susceptibilities,* Oxford University Press, 1932.
2. Oosawa, F.: *Biopolymers* **9**, 677 (1970).
3. Mandel, M.: *Mol. Phys.* **4**, 489 (1961).
4. Litzler, R., Cerf, R., and Sadron, C.: *Compt. Rend Acad. Sci.* **259**, 473 (1964).
5. Benoit, H.: *Ann. Phys.* **6**, 561 (1951).

6. see Mandel, M.: this volume, p. 285.
7. Weill, G., Hornick, C., and Stoylov, S.: *J. Chim. Phys.* **65**, 182 (1963).
8. Hornick, C. and Weill, G.: *Biopolymers* **10**, 2345 (1970).
9. Ullmann, R.: *Macromolecules* **2**, 27 (1969).
10. Alexandrowicz, Z. and Katchalsky, A.: *J. Polymer Sci.* **A1** 3231 (1963).
11. Lifson, S. and Katchalsky, A.: *J. Polymer Sci.* **13**, 43 (1964).
12. O'Konski, C. T.: *J. Phys. Chem.* **64**, 605 (1960).
13. Schwartz, G.: *Z. Physik. Chem.* **19**, 286 (1959).
14. Takashima, S.: *Advances in Chem. Series* **63**, 232 (1967).
15. Daune, M.: *Biopolymers* **7**, 659 (1969).
16. McTague, J. and Gibbs, J. H.: *J. Chem. Phys.* **44**, 4295 (1966).
17. Minakata, A., Imai, N., and Oosawa, F.: *Biopolymers* **11**, 347 (1972).
18. Oosawa, F.: *J. Polymer Sci.* **23**, 421 (1957).
19. Schurr, M.: *Biopolymers* **10**, 1371 (1971).
20. Padova, J.: *J. Chem. Phys.* **39**, 1552 (1963).

DIELECTRIC PROPERTIES OF POLYELECTROLYTES IN SOLUTION

M. MANDEL and F. VAN DER TOUW

Gorlaeus Laboratoria, Afdeling Fysische Chemie III, Rijksuniversiteit Leiden, The Netherlands

Dielectric measurements belong to the oldest techniques used by physical chemists to gain information about molecular properties and molecular behaviour. Dielectric investigations are generally focused on two main aspects: (1) the equilibrium value of the electric permittivity (or dielectric constant) ε which is related to the equilibrium polarization of the system under the influence of an external field; (2) the dispersion or frequency dependence of ε which is related to the change in time of the polarization when the external field is established or switched off. Both the equilibrium or static electric permittivity and the dielectric dispersion of polyelectrolyte solutions have been subjects of experimental investigations of which Oncley's pioneering work on protein solutions should be mentioned [1]. Owing to experimental difficulties measurements in the past have often been limited to a small frequency range but more recently experiments covering a much broader domain have become available. (In our own laboratory experiments are performed between 2.5 kHz and 100 MHz; Minakata and Imai [2] claim to have measured between 30 Hz and 6 MHz). Although it should be ideal to measure dielectric properties at frequencies as low as possible, determinations of ε below 50 kHz become increasingly difficult for two different reasons: (1) the conductivity contribution of the solution admittance becomes predominant; (2) the effects of electrode polarization which has to be corrected for increase very rapidly with decreasing frequency. Special experimental equipment is necessary and correction methods must be devised to take care of the influence of the electrode effects.

In our laboratory much attention has been given to the experimental aspects of determining the dielectric properties of polyelectrolyte solutions. Especially the correction method for electrode polarization was investigated very extensively. It was found [3] that many of the assumptions which are quite generally admitted are not justified (e.g. the assumption that the electrode effects only depend on the conductivity of the solution, the assumption that for a system between flat, parallel electrodes, the contribution of the electrode polarization to the measured capacity is proportional to the square of the distance between the electrodes, etc.) Although these investigations were long and tedious they made it possible to acquire results as well in the region between 2.5 kHz–700 kHz as at higher frequencies 1 MHz–100 MHz (the latter especially due to the efforts of van Beek who developed the experimental method [4]) in which we are now fully confident but many of which are not yet published (contributions by Briedé, Muller, Vreugdenhil and Zwolle are to be acknowledged).

In the present paper we shall present a summary of some of these results obtained with aqueous solutions of the following polyelectrolytes:

Eric Sélégny (ed.), Polyelectrolytes, 285–300.

poly-(styrene-sulphonic acid), PSS
poly-(acrylic acid), PAA
poly-(methacrylic acid), PMA
poly-(glutamic acid), PGA
calf thymus DNA, DNA.

Most experiments were performed on salt-free solutions and, if possible with polyions of different molecular weight. For weak polyelectrolytes the influence of the degree of ionization was investigated. The following general features were observed:

(1) there are two separated dispersion regions, one with a lower critical frequency $f_{c,1}$ in the domain 10^4–10^5 Hz, the higher with a critical frequency $f_{c,2}$ in the region 10^6–10^7 Hz (Figures 1, 2 and 3). The lower dispersion region corresponds to the one which we observed previously when our measurements were limited to frequencies up to 10^6 Hz (Mandel and Jenard [5], van der Touw and Mandel [6]) and which was also found by Takashima [7, 8] for DNA and PGA. The higher dispersion region corresponds to the one observed by Sachs *et al.* [9] for PAA and PSS. It was already pointed out by the latter that the critical frequency of the high frequency dispersion region is *molecular weight independent* in contrast to the lower critical frequency, an observation which is fully confirmed by all our results, as can be seen from Table I and Figure 4. (Recently Minakata and Imai [2] have published results on PAA and PSS and also found both dispersion regions; they failed however to notice this striking difference in molecular weight dependence of the critical frequency for both regions). It should also be noted that for all different polyelectrolyte solutions investigated

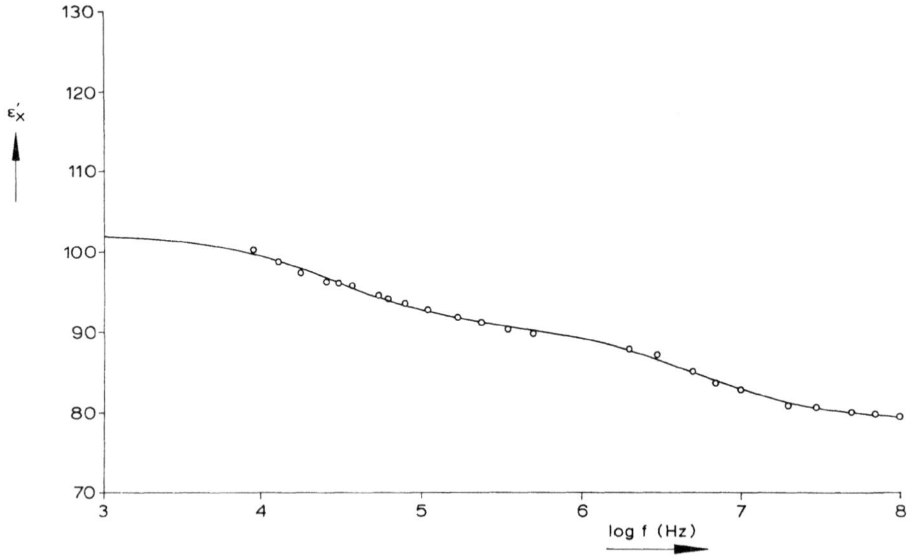

Fig. 1. Electric permittivity ε'_x as function of the frequency f for PAA-Na$^+$ (mol. weight 10^5), $C_p = 5 \times 10^{-3}$ monomol l^{-1}, $\alpha = 0.40$. (The drawn curve is the calculated dispersion curve according to the Cole-Cole formula with the help of the least-squares procedure.)

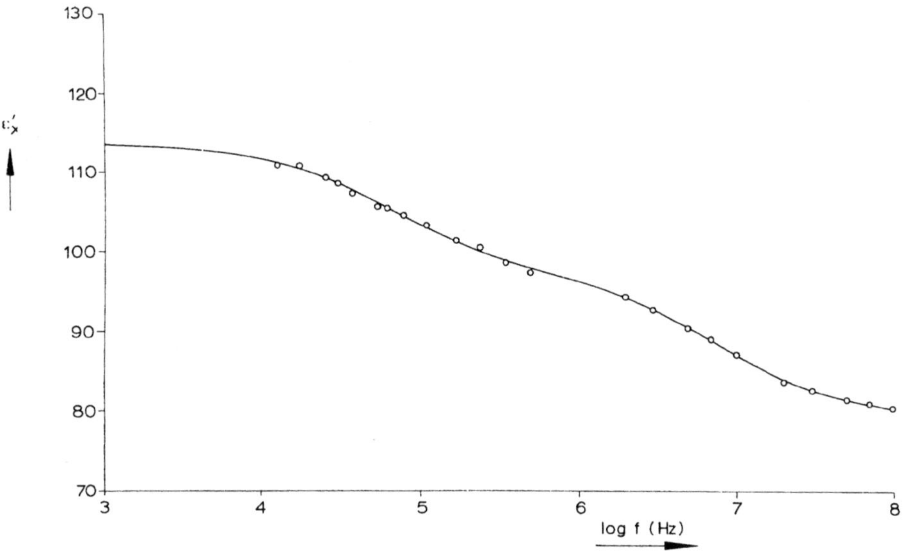

Fig. 2. Electric permittivity ε'_x as function of the frequency f for PSS-Na$^+$ (mol. weight 10^5), $C_p = 4.7 \times 10^{-3}$ eq. l^{-1}, $\alpha = 1.0$ (drawn curve as in Figure 1).

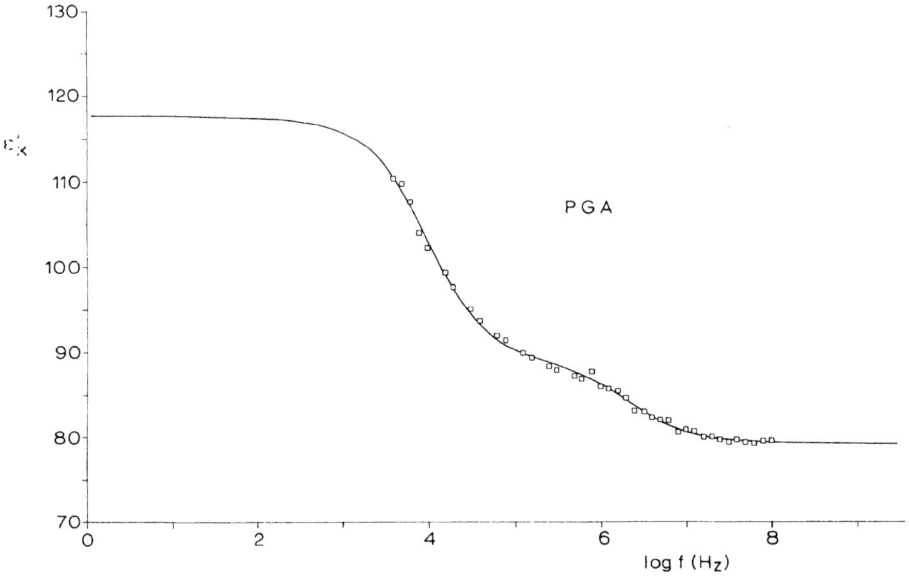

Fig. 3. Electric permittivity ε'_x as function of the frequency f for PGA-Na$^+$ (mol. weight 8×10^4), $C_p = 10^{-3}$ monomol l^{-1}, $\alpha = 0.80$ (drawn curve as in Figure 1).

TABLE I

Polymer	Mol. weight.	C_p(eq/l)	α	$f_{c,1}$(Hz)	$\Delta\varepsilon_s$	$f_{c,2}$(Hz)	$\Delta\varepsilon_2$
PSS-Na^{+}[a]	9×10^6	1.1×10^{-3}	1	(4.2×10^2)	(4×10^3)	1.8×10^6	14.5
PSS-Na^{+}	1×10^5	1.5×10^{-3}	1	2.3×10^4	37.5	2.4×10^6	14.3
PSS-Na^{+}	1×10^5	3×10^{-3}	1	3.8×10^4	42.5	4.1×10^6	17.4
PSS-Na^{+}	1×10^5	4.7×10^{-3}	1	7.9×10^4	37.4	7.5×10^6	18.9
PAA-Na^{+}	1×10^5	5×10^{-3}	0.4	2.2×10^4	26.8	4.2×10^6	13.7
PMA-Na^{+}	0.9×10^5	5×10^{-3}	0.4	2×10^4	35.1	4.2×10^6	13.3
PGA-Na^{+}	0.8×10^5	10^{-3}	0.8	9.8×10^3	38.5	2.3×10^6	9.5
DNA-Na^{+}	<400000	10^{-3}	~ 1	7.1×10^3	40.6	1.2×10^6	13.6

[a] For this high molecular weight PSS the extrapolation to zero frequencies is hazardous because of the low value of $f_{c,1}$.

$f_{c,2}$ is of the same order of magnitude regardless of the nature of the macromolecule or its degree of dissociation. Perhaps a slight correlation between $f_{c,2}$ and the macromolecular concentration C_p can be noticed.

(2) Both dispersion regions *cannot* be discribed by a simple Debye-curve. This indicates the existence, for both regions, of a distribution of relaxation times. It has been found that the experimental points can be fitted to a Cole-Cole function [10], in which such a distribution of relaxation times is accounted for by a factor β, $0 < \beta < 1$, ($\beta = 1$ corresponding to a Debye curve), with β always approximately equal to 0.7.

(3) Extrapolation of the experimental results for a given system at the low frequency

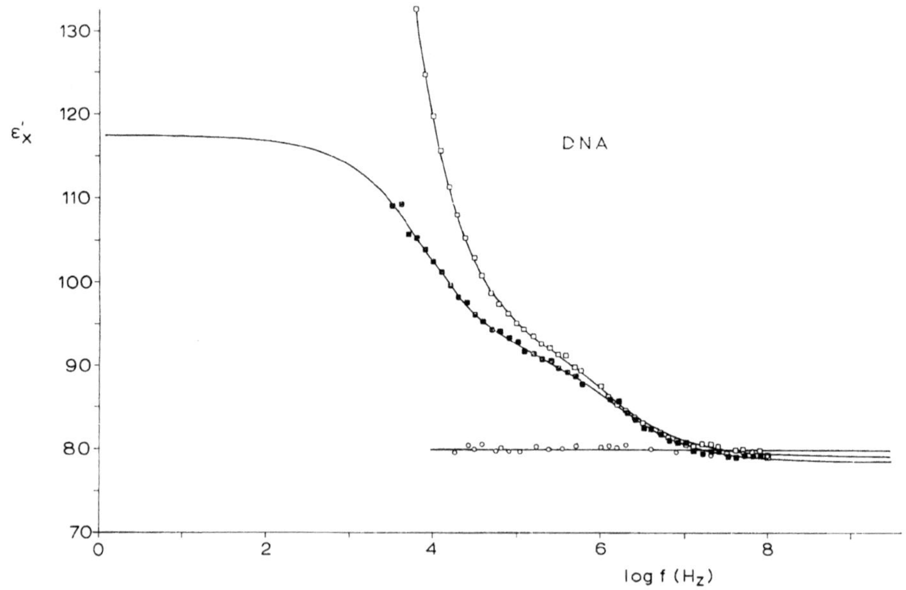

Fig. 4. Electric permittivity ε'_x as function of the frequency f for DNA-Na^{+} for two different degrees of polymerization P, $C_p = 10^{-3}$ eq. l^{-1}, $\alpha \sim 1$; ■: $P = 1200$; □: $P = 18\,000$.

side to $f=0$ (with the help of a computerized fitting procedure) yields the static electric permittivity ε_s of the solution or the dielectric increment $\Delta\varepsilon_s=\varepsilon_s-\varepsilon_w$, where ε_w is the electric permittivity of water. Extrapolation on the high frequency side to $f=\infty$ yields $\varepsilon_{s,\infty}$. This value was found in all cases to be close to ε_w indicating that no important dispersion involving the solute particles is to be expected at higher frequencies than those investigated.

(4) The total dielectric increment can be split up into a contribution $\Delta\varepsilon_1$ disappearing after the low frequency dispersion region and $\Delta\varepsilon_2$ characteristic for the contribution disappearing during the second high frequency dispersion region. It is found that $\Delta\varepsilon_2$ is generally smaller than $\Delta\varepsilon_1$ (except for very low molecular weights) the latter increasing with the molecular weight in contrast to the former (Figure 5). In general also $\Delta\varepsilon_1$ and $\Delta\varepsilon_2$ increases with the degree of ionization α (Figure 6). Both $\Delta\varepsilon_s/C_p$ and $\Delta\varepsilon_2/C_p$ are functions of C_p.

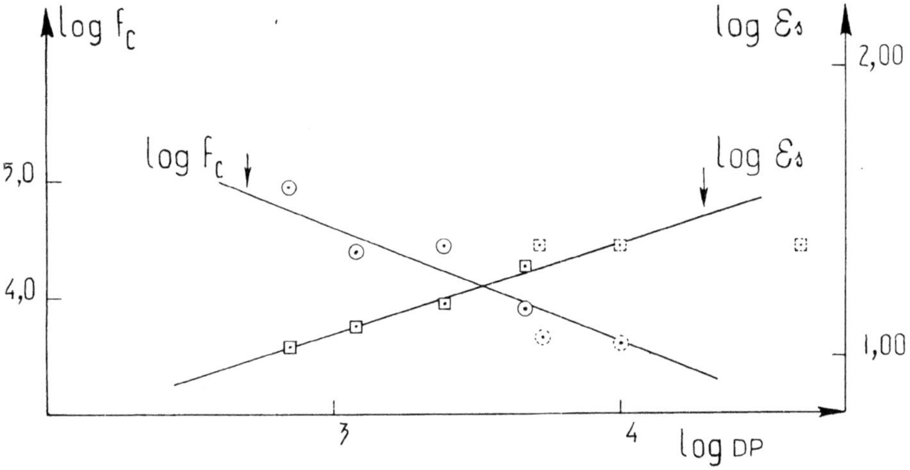

Fig. 5. Variation of $f_{c,1}$(○) and ε_s(□) with degree of polymerization (D.P) for PMA-Na$^+$, $C_p=$ $=5\times10^{-3}$ eq. l^{-1}, $\alpha=1$. (Dotted points correspond to high molecular weights for which the extrapolation is hazardous because of the low value for $f_{c,1}$.)

(5) Although there seems to be a small but definite influence of the nature of the monovalent counterion on ε_s and $f_{c,1}$ (Figure 7), a much more spectacular effect is observed if the monovalent ion is partially or totally replaced by a bivalent alkaline earth counterion (Figure 8). In general ε_s rises with increasing M^{2+}/M^+ ratio up to a maximum and then decreases again in a somewhat parallel way as observed in the decrease of the solution viscosity.

We shall now present a theoretical approach both to the value of ε_s and to the mechanism underlying the two dispersion regions. Although the solutions which were investigated were not infinitely diluted with respect to the macromolecular component ($C_p\sim10^{-3}$—5×10^{-3} eq. l^{-1}) it will assumed in the theoretical interpretation of the

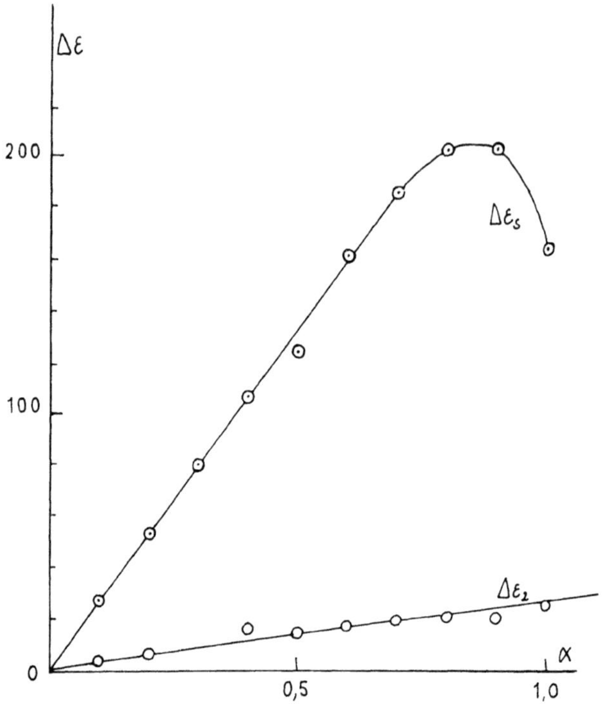

Fig. 6. Variation of $\Delta\varepsilon_s$ and $\Delta\varepsilon_2$ with α for PMA-Na$^+$ ($P = 5200$), $C_p = 5 \times 10^{-3}$ monomol l^{-1}.

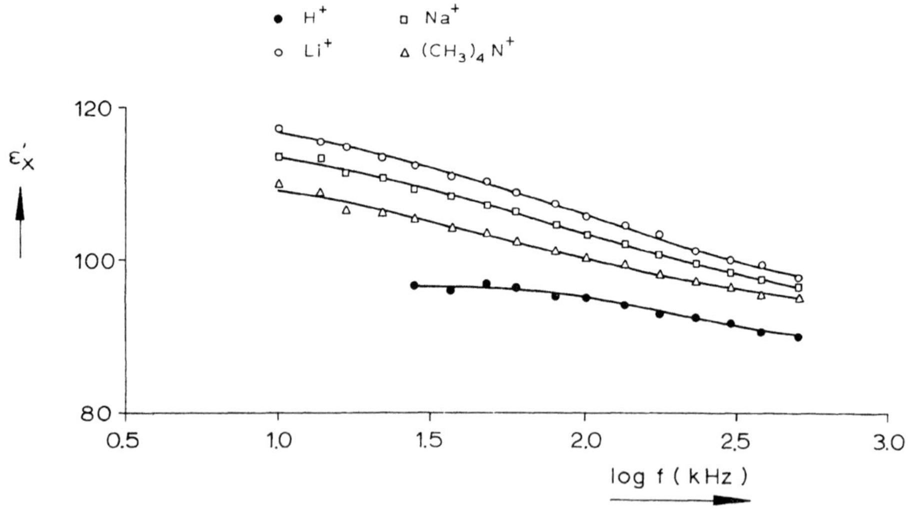

Fig. 7. Low frequency dispersion of the electric permittivity ε'_x for PSS (mol. weight 5000), $C_p = 4.2 \times 10^{-3}$ eq. l^{-1}, $\alpha = 1$, and different monovalent counter-ions (●: H^+; ○: Li; □: Na; △ $(CH_3)_4N^+$).

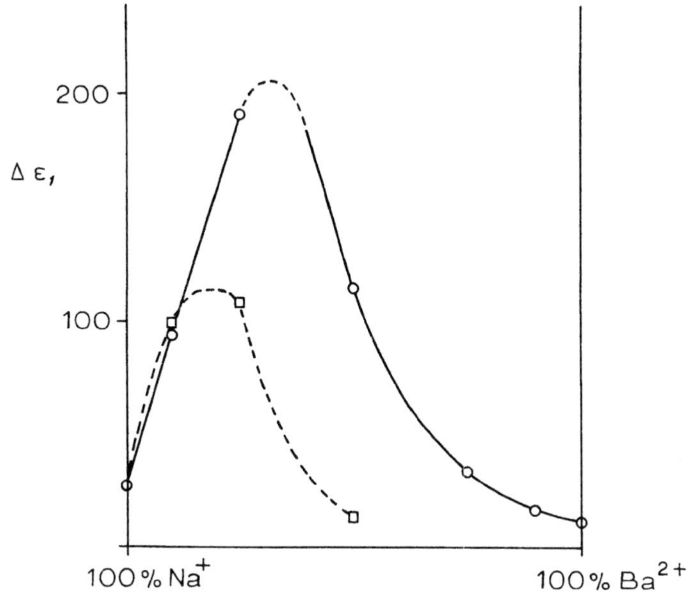

Fig. 8. Change in $\Delta\varepsilon_1$ with increasing ratio Ba^{2+}/Na^+ for PSS. (mol. weight 5000), $C_p = 4.2 \times 10^{-3}$ eq. l^{-1}, $\alpha = 1$; ○:PSS neutralized by NaOH-Ba$(OH)_2$ mixtures; □:PSS neutralized by NaOH in the presence of $BaCl_2$ in various concentrations.

results that interactions between the polyions can, in a first approximation, be neglected. (This assumption is not fully justified however as for most flexible polyelectrolytes which have been studied the specific dielectric increment $\Delta\varepsilon_s/C_p$ is not constant over the concentration range investigated.) In the past, different interpretations for the origin of the polarization due to polyelectrolytes in aqueous solutions have been proposed: (a) orientational polarization due to the permanent dipole moment of the polyelectrolyte (e.g. Oncley [1], Takashima [8], Hanss [11]); (b) Maxwell-Wagner effects only or combined with surface conductivity on the polyelectrolyte (Allgen [12], Dintzis *et al.* [13], O'Konski [14]); (c) deformation of the ionic atmosphere particularly that part containing the counterions which are closely associated to the polyion (Mandel [15], MacTague and Gibbs [16], Schwarz [17], Oosawa [18]). In as far as the two first mechanisms are concerned neither one can account for all the experimental facts consistently, particularly in as far as $\Delta\varepsilon_s$ and the first dispersion region is concerned and without introducing extremely improbable values for some molecular parameters. On the other hand it has been shown previously [5, 6] that the value of $\Delta\varepsilon_s$ could be understood, at least qualitatively, in terms of a model calculation in which a large dipole moment is induced by the electric field in the macromolecular region by a perturbation of the average distribution of the associated counterions [15, 19]. Although the model is rather crude (rigid rod-like polyion, no repulsion forces between associated ions, neglect of the finite size of the polyelectrolyte in as far as the interactions between counterions and the charged sites on the polyelectrolyte are concerned, etc.) it has been possible to explain most of the striking features of ε_s, such as its molecular weight dependence, the influence of the degree of dissociation and of the valency of the counter-ions. Also the right order of magnitude of $f_{c,1}$ could be predicted and the origin of the distribution of relaxation times in the low frequency dispersion region could be accounted for.

In its simple form only one single relaxation mechanism can be predicted. The theory in itself is thus unable to explain the second high frequency dispersion region, which was attributed previously to a Maxwell-Wagner dispersion (due to interfacial space charge between the macromolecular regions and the solvent). The extensive investigation of the second dispersion region has made clear that it cannot be accounted for by this mechanism, neither in as far as $\Delta\varepsilon_2$ is concerned nor with respect to the value of the critical frequency $f_{c,2}$ which should be lower according to the Maxwell-Wagner theory (as was also pointed out by Minakata and Imai [2]). Furthermore a Maxwell-Wagner type mechanism would lead to values of $\Delta\varepsilon_2$ and $f_{c,2}$ depending markedly on the size, shape and the exact nature of the polyion contrary to what is observed. We now wish to propose that the high frequency dispersion region can also be explained in terms of ionic atmosphere deformations as will be presented succinctly below and discussed extensively elsewhere [19].

We shall use the following model in discussing the dielectric properties of the polyelectrolyte solution. Assume a linear distribution of N identically charged $(-e)$ sites equally spaced and situated symmetrically with respect to the center of mass (origin of the internal coordinate x) at positions $x_1, x_2, \ldots\ x_k, \ldots\ x_N$. (The linear distribution of these sites not necessarily implies a rod-like polyion; it may also correspond to the projection of the real sites on a principal axis). The distance between the extreme sites will be L. Assume further that $n\,(n<N)$ identical ions of charge $+ze$ are distributed along the sites of the polyion. These associated ions cannot leave the radially directed potential well nor the polyelectrolyte at its ends. If the polyelectrolyte, would be infinitely large the potential energy of interaction of a counter-ion and the polyelectrolyte is to be identical at any site of the latter. It will be assumed that for a macromolecule of finite but sufficiently large dimensions this is still to be true, thus neglecting end-effects on the interaction between the ions and the charged sites and interactions between the ions themselves. (Although this assumption may be considered to be rather crude, the simultaneous neglect of the end-effects and of ion interactions may be partially compensating each other as the former will tend to drive the ions towards the center of mass whereas the other will have an opposite effect). Then the probability $P_i^0(x_k)$ to find (in the absence of an external field) ion i at site x_k for a given orientation of the principle axis of the polyelectrolyte (characterized by the orientation of the unit vector $\mathbf{1}_L$ in the direction of this axis) will be the same at all sites, $P_i^0(x_k)=N^{-1}$. It is also reasonable to assume that $P_i^0(x_k)$ is independent of the exact orientation of the polyion with respect to the external coordinate system (which means that the distribution of the ions and the orientation of the macromolecule are not correlated) and that, for sufficiently diluted solutions the orientation of $\mathbf{1}_L$ is random.

Under these conditions the dipole moment of the polyion for any given orientation and due to the distribution of the n ions along the N sites (neglecting the contribution of the permanent dipole moments in the polyelectrolyte), to be represented by $\langle \mathbf{m}_L \rangle_{0,L}$, is given by the following expression.

$$\langle \mathbf{m}_L \rangle_{0,L} = \left[\sum_{(k)} (-e)\, x_k + \sum_{i=1}^{n} \sum_{k=1}^{N} (ze)\, x_k P_i^0 (x_k) \right] \mathbf{1}_L . \tag{1}$$

This expression is zero for any value of $\mathbf{1}_L$ so that further averaging over all possible orientations does not give rise to an average dipole moment contribution from these associated ions, $\langle \mathbf{m}_L \rangle_0 = 0$. The fact that $\langle \mathbf{m}_L \rangle_{0,L}$ as given by (1) is zero follows from the symmetric distribution of the sites with respect to the center of mass which is also the origin with respect to which the dipole moment is defined. Note that for a distribution of charges with net charge different from zero, as is presently the case ($eN \neq zen$), the value of the dipole moment is not uniquely defined but depends on the *arbitrary* choice of the origin. There is however the extra condition to be considered that, in the absence of an external field, all average dipole contributions should vanish (as usually there is no remanent polarization). Under the assumptions that no correlation exists between the orientation of the polyion and the distribution of the ions on the one hand and that all orientations of the polyelectrolyte are equally probable on the other hand, $\langle \mathbf{m}_L \rangle_0$ can only vanish if the origin for the dipole moment coincides with the center of mass where the principal axis of rotation meet. Finally it should be observed that in the absence of an external field the average *position* $x^{(i)}$ of any ion i is just zero.

$$\langle x^{(i)} \rangle_{0,L} = \sum_{k=1}^{N} p_i^0 (x_k)\, x_k = P_i^0 \sum_k x_k = 0 . \tag{2}$$

Suppose now that an external field is applied which establishes an average field $\mathbf{E}$ inside the system. The field acting on a molecule will be an internal field, different form $\mathbf{E}$ and function in principle of the position and orientation of all molecules. An explicit expression for this internal field is difficult to derive; we shall assume that, up to a first approximation, it may be represented by $\sigma\mathbf{E}$, where σ is a numerical constant not differing too much from unity.

The internal field modifies the potential energy of an ion at a given site by the quantity $-\sigma\mathbf{E} \cdot x_k ze \mathbf{1}_L$. This yields the following expression for the probability $P_i(x_k)$ for finding ion i at site x_k.

$$P_i(x_k) = \frac{e^{\sigma x_k ze \mathbf{E} \cdot \mathbf{1}_L / kT}}{\sum_k e^{\sigma x_k ze \mathbf{E} \cdot \mathbf{1}_L / kT}} = P_i^0 (x_k) \left[1 + \frac{\sigma x_k ze}{kT} (\mathbf{E} \cdot \mathbf{1}_L) + \cdots \right] . \tag{3}$$

In the last member of Equation (3) an expansion in powers of E up to linear terms only has been used, non-linear effects not being taken into account. If (3) is substituted into the expression for the average moment of the molecule at constant orientation of the polyion in the presence of the field, defined analogously to (1) with $P_i^0 (x_k)$ replaced by $P_i (x_k)$, the following result is obtained.

$$\langle \mathbf{m} \rangle_{E,L} = \frac{n(ze)^2 \sigma \mathbf{E}}{kT} \cdot \left\langle \sum_{k=1}^{N} x_k^2 \right\rangle_{0,L} \mathbf{1}_L \mathbf{1}_L = \frac{n(ze)^2 \sigma}{12\, kT} L^2 (\mathbf{1}_L \mathbf{1}_L \cdot \mathbf{E}). \tag{4}$$

Here use has been made of the fact that for a symmetric distribution of equally spaced sites $x_k = kd$ (where k is an integer ranging from $-(N-1)/2$ to $(N-1)/2$ and $d \sim L/N$). If all orientations of the polyion are equally probable the average value of the dyade $\mathbf{1}_L \mathbf{1}_L$ over all orientations is just $U/3$, where U is the unit tensor of second rank.

$$\langle \mathbf{m} \rangle_E = \tfrac{1}{3} \left[\frac{n(ze)^2 \sigma L^2}{12\, kT} \right]. \tag{5}$$

This result, which was already derived previously [15] predicts that the dielectric increment should be proportional to C_p, z^2, n and L^2,

$$\Delta\varepsilon_s = \frac{4\pi}{36} C_p \frac{n(ze)^2 \sigma L^2}{kT} \tag{6}$$

in accordance with the experimental facts. It also explains the maximum observed in the change of $\Delta\varepsilon_s$ with increasing M^{2+}/M^+ ratio for mixed counter-ions. In this case nz^2 in (6) has to be replaced by $(n_{M^+} + 4n_{M^{2+}})$ which increases with increasing M^{2+}/M^+ whereas for rather flexible polyelectrolytes L^2 decreases with the same ratio as is known from e.g. viscosity measurements. The calculation of $\Delta\varepsilon_s$ according to (6) necessitates knowledge of n and L, even if σ is assumed to be close to unity.

It is interesting to note that the result given by (4) can also be derived in a slightly different way by introducing the position of two arbitrary ions $x^{(i)}$ and $x^{(j)}$ as shown in the appendix.

$$\langle \mathbf{m} \rangle_{E,L} = \sum_{i=1}^{n} \sum_{j=1}^{n} \langle x^{(i)} x^{(j)} \rangle_{0,L} \frac{(ze)^2 \sigma}{kT} (\mathbf{E} \cdot \mathbf{1}_L) \mathbf{1}_L. \tag{7}$$

If interactions between ions are neglected there will be no correlation between the position of both ions and (7) simplifies into (4).

The frequency dependence of $\Delta\varepsilon_s$ can be expressed with the help of the Kubo formalism [20]. It can be shown that quite generally the following relation holds

$$\Delta\varepsilon(\omega) = \Delta\varepsilon(0) \int_0^\infty e^{i\omega\theta} \frac{\partial}{\partial\theta} \Gamma(\theta)\, \mathrm{d}\theta, \tag{8}$$

where $\Delta\varepsilon(\omega)$ is the frequency dependent increment and $\Delta\varepsilon(0)$ the value of the static increment; $\Gamma(\theta)$ is the normalized decay function for the electric polarization which is defined by

$$\Gamma(\theta) = \frac{\langle \mathbf{m}(0) \cdot \mathbf{m}(\theta) \rangle_0}{\langle \mathbf{m}(0) \cdot \mathbf{m}(0) \rangle_0} = \frac{\sum_{k=1}^{n} \langle x^{(k)}(0)\, \mathbf{1}_L(0)\, x^{(k)}(\theta)\, \mathbf{1}_L(\theta) \rangle_0}{\sum_{k=1} \langle x^{(k)}(0)\, \mathbf{1}_L(0)\, x^{(k)}(0)\, \mathbf{1}_L(0) \rangle_0} : \mathbf{EE}, \tag{9}$$

where (7) has been used. Here $\mathbf{m}(0)$ and $\mathbf{m}(\theta)$ represent the instantaneous dipole moment at a given time $t=0$ and after a time interval θ respectively; $\mathbf{1}_L(0)$ and $\mathbf{1}_L(\theta)$ define the orientation of the macromolecule and $x^{(k)}(0)$ and $x^{(k)}(\theta)$ the position of ion k in an analogous way. If it is assumed that no correlation exists between the fluctuations in the orientation of the polyion and the fluctuations in the distribution of the small ions, then $\Gamma(\theta)$ can be written as follows.

$$\Gamma(\theta)=\frac{\langle \mathbf{E}\cdot\mathbf{1}_L(0)\,\mathbf{1}_L(\theta)\cdot\mathbf{E}\rangle_0\,\langle x^{(k)}(0)\,x^{(k)}(\theta)\rangle_{0,L}}{\langle \mathbf{E}\cdot\mathbf{1}_L(0)\,\mathbf{1}_L(0)\cdot\mathbf{E}\rangle_0\,\langle x^{(k)}(0)\,x^{(k)}(0)\rangle_{0,L}}=\Gamma_L(\theta)\,\Gamma_d(\theta). \tag{10}$$

Here $\Gamma_L(\theta)$ and $\Gamma_d(\theta)$ stand for the decay function of orientation and ionic distribution respectively. If all decay functions are assumed to be exponential with respect to θ, the following relation will exist between the overall relaxation time τ and the relaxation time of orientation and ion-distribution, τ_L and τ_d respectively.

$$\frac{1}{\tau}=\frac{1}{\tau_L}+\frac{1}{\tau_d}. \tag{11}$$

This relation shows that if τ_L and τ_d are widely differing the overall relaxation time will be determined by the smallest of the two. In the opposite case both decay mechanisms will influence τ. For flexible uncharged macromolecules in solution no relaxation time of the same order of magnitude as the largest relaxation time τ observed here with charged polyelectrolytes is found [21] whereas for more rigid uncharged macromolecules there is no conclusive evidence that a τ_i exists with a value approximately equal to this τ. The assumption is thus justified that the observed values of τ correspond to the relaxation time of the ion-distribution τ_d. This is also confirmed by dielectric measurements on aqueous solutions of rigid, ellipsoidal charged particles, such as the alfalfa mosaic cirus, of which the orientational relaxation time can be calculated from the molecular geometry as revealed by electron microscopy ($\tau_i \geqslant 3\times10^{-5}$ s). In this case the ion distribution relaxation time was found to determine the observed dielectric dispersion and its order of magnitude was comparable to the values of τ for the polyelectrolytes which are discussed here [22].

Mandel [17] has estimated that τ_d should be proportional to L^2/ukT in the case where interactions are to be neglected (u standing for the mobility of the associated ions in their motion from site to site along the macromolecular chain). Oosawa [19] derived an expression for τ_d by solving the diffusion equation for the associated ions taking into account repulsion between these ions but neglecting the attraction of the counterions to the charges on the macromolecule. If only the first term in the Fourier expansion of τ_d determines its mean value dominantly, as can be shown to be generally the case, the following expression is obtained.

$$\tau_d=\frac{L^2}{(2\pi)^2\,ukT\,(1+Q)}. \tag{12}$$

Here Q is a factor which takes into account the repulsion between counter-ions. If this term can be neglected with respect to unity the Oosawa's expression becomes identical to the result found by Mandel (except in as far as the proportionality factor is concerned). For the static dielectric increment of the solution Oosawa found an expression which is equal to the r.h.s. of (6) multiplied by the same factor $(1+Q)^{-1}$. From experimental evidence it is clear that Q cannot be much larger than unity. As pointed out already by Hornick and Weyl [23] this would lead to dielectric increments which do not depend on the valency of the associated counter-ions, contrary to experimental findings. Also in the case of the rigid alfalfa mosaic virus, for which L can be estimated from electron microscopy, the observed relaxation time indicates that Q should not have a value much different from unity [22]. Under these circumstances and in view of the uncertainty on σ and u, it will be assumed that Q may be neglected in a first approximation. This is further justified by the fact that in Oosawa's model the repulsion effects between the associated ions are probably overestimated by the neglect of the interactions between these ions and the charged sites along the macromolecular chain of finite length.

So far, neither theory accounts for the two relaxation domains that are observed experimentally. We still feel that the low-frequency dispersion and the total dielectric increment is correctly described by the mechanism discussed above in detail. In as far as the high-frequency dispersion region is concerned, for which Maxwell-Wagner effects may be ruled out, our present ideas go towards an explanation which is analogous to the one applied for the low-frequency dispersion, i.e. a mechanism essentially based upon the perturbation of the ion distribution by the electric field. Whereas the dispersion observed at the lower frequency range is essentially determined by fluctuations in the ion-distribution corresponding to displacements of the ions from their average position which are large (of the same order of magnitude as the overall dimensions of the macromolecular particle), the fluctuations which are responsible for the dispersion at the high frequency end correspond to much smaller ionic displacements. Let us assume that in a given fluctuation from the average distribution an ion moves from one site to another with a mobility u' but finds after a given distance b its motion opposed by a potential barrier higher than kT. Such fluctuations of relatively short duration will give rise to an increment $\Delta\varepsilon_2$ of a value defined analogously to (6) but in which L^2 is to be replaced by b^2. On the other hand over a longer time interval fluctuations will appear in which the potential barrier is overcome and which leads to the total increment as given by (6). Whereas the former fluctuations determine a relaxation time proportional to b^2/u' the latter are characterized by a relaxation time proportional to L^2/u, where u will be smaller than u'. It then follows that

$$\Delta\varepsilon_s/\Delta\varepsilon_2 = L^2/b^2 \tag{13}$$

$$\tau_1/\tau_2 = L^2u'/b^2u. \tag{14}$$

Such an *ad hoc* explanation based upon the existence of potential barriers which oppose the motion of the associated ions along the chain seems to be highly speculative. Some support for it is however found in recent results [22] concerning the dielectric properties of the alfalfa virus particle of rigid geometry in aqueous solutions for which only one frequently dispersion region was observed. For this particular charged particle a single mean relaxation time $\tau = \tau_1 = 5 \times 10^{-7}$ s was found which could be interpreted in terms of fluctuations over the entire length of the particle as for the low frequency dispersion of non flexible polyelectrolytes. Apparently due to its rigid conformation and elongated shape the alfalfa virus particle does not present potential barriers which oppose the motion of associated counter-ions over its total length. For the more flexible charged macromolecules considered here the average conformation may be considered to consist of several more or less linear parts of an average length b. In the region where two of these linear parts meet local deviations in the average potential of interaction between the counter-ions and the charged sites in the axial direction are to be expected which may give rise to the potential barriers introduced above. Within the framework of such a model the value of b (and therefore τ_2) would depend on the nature of the macromolecule but, for polyelectrolytes of sufficiently large dimensions, not on the molecular weight, in accordance with experimental results. Also, in as far as the average conformation of the polyelectrolyte depends through intermolecular interactions on C_p, b is expected to be a function of the macromolecular concentration too, also in agreement with experimental results.

The order of magnitude of b may be estimated from the experimental values of τ_2 if the assumption is made that u' does not differ too much from the mobility u_0 of the counter-ions in ordinary dilute electrolyte solutions. Values of b calculated with the assumption $u' \sim u_0$ and using (12) (in which Q is put equal to zero and L is replaced by b) are collected in Table II. It can be seen that b increases with increasing stiffness of the macromolecular chain. With the value of b the average length L for each macromolecule may be calculated using (13). The order of magnitude of these average dimensions is not unreasonable as can be seen from Table II. Finally, using again the assumption that $u' \sim u_0$ and subsequently that $\sigma = 1$, the value of n may be estimated from the ratio $\Delta\varepsilon_2/\tau_2$. Values of the fraction x of associated counter-ions ($x = n/\alpha P$, where P is the degree of polymerization) calculated this way are collected also in Table II. It may be observed that these values are rather low compared to results

TABLE II

Polymer (conc. (eq./l))	P	α	$\Delta\varepsilon_s$	$\Delta\varepsilon_2$	$\tau_2(s)$	b(Å)	x	L(Å)
PGA-Na^+ (10^{-3})	600	0.8	38.5	9.5	7.3×10^{-8}	600	0.3	1200
PSS-Na^+ (4.7×15^3)	500	1	17.4	18.9	2.1×10^{-8}	320	0.3	450
PAA-Na^+ (5.2×10^{-3})	1500	0.4	26.8	13.7	3.8×10^{-8}	430	0.3	600
DNA-Na^+ (10^{-3})	1200	1	40.6	13.6	1.3×10^{-7}	810	0.2	1400
DNA-Na^+[a] (10^{-3})	18000	1	(300)	12.1	1.1×10^{-7}	720	0.2	(3600)

[a] For this high molecular weight DNA the extrapolation to zero frequencies is hazardous because of the low value of $f_{c,1}$.

obtained by other methods [24]. It should be kept in mind however that the assumption $\sigma u' \sim u_0$ is rather crude and that x will increase with decreasing ratio $\sigma u'/u_0$. The complete neglect of interaction effects may also have some bearing on the low values of x and it is also not to be excluded that only part of the total fraction of associated counter-ions contribute to the deformation of the ionic atmosphere mechanism responsible for the dielectric properties. It is remarkable however that such a simple model calculation yields values for the different molecular quantities discussed here which are of a correct order of magnitude.

It has been pointed out above that the experimental results show that for both dispersion regions a distribution of relaxation times rather than one single relaxation time is necessary to describe the frequency dependence of the dielectric increment. This could be due to the fact that both u' and u are not single valued but may have different values for different fractions of the associated counterions participating in the mechanism giving rise to the polarization which is detected. Also it may be argued that for different orientations of the macromolecule with respect to the field E the effective lengths L and b over which the ions can move will be different.

Summarizing we may state that all the experimental facts concerning the dielectric properties of polyelectrolyte solutions discussed in this paper may be understood and even interpreted in a semi-quantitative way using the theory based on the perturbation of the average distribution of associated counter-ions by the electric field. For polyelectrolytes of non-rigid geometry the assumption must be introduced that, at least from the dielectric point of view, the average conformation may be represented as a collection of more or less linear parts which is as far as the motion of the ions along the chain is concerned are separated by potential barriers higher than kT. Fluctuations in the ion distribution within such linear parts give rise to the dispersion mechanism observed at the high frequency region and determines $\Delta\varepsilon_2$. Fluctuations in the overall distribution over the average dimensions of the polyion determine the total static dielectric increment $\Delta\varepsilon_s$ and are responsible for the low-frequency relaxation.

Appendix

Let us represent a certain configuration β of the associated counter-ions by their internal coordinates $x_\beta^{(i)}$ $(i=1, 2 \dots n)$, each $x_\beta^{(i)}$ corresponding to the coordinate of a given charged site on the polyion. Such a configuration will be characterized by a potential energy $V_\beta = V_\beta(x_\beta^{(1)}, x_\beta^{(2)}, \dots x_\beta^{(n)})$. In the absence of an external field $V_\beta = V_\beta^0$ In the presence of the external field V may be represented by the following expression.

$$V_\beta = V_\beta^0 - \sum_{j=1} z_j e x_\beta^{(j)} \sigma \mathbf{E} \cdot \mathbf{1}_L . \tag{A.1}$$

The normalized probability P_β of this configuration is given by

$$P_\beta = \frac{e^{-V_\beta/kT}}{\sum\limits_{(\beta)} e^{-V_\beta/kT}}, \tag{A.2}$$

where $\Sigma_{(\beta)}$ stands for a summation over all possible configurations. In the absence of an external field the normalized probability for this configuration P_β^0 will be defined analogously to (A.2) with V_β replaced by V_β^0. The expression for P_β may be expanded in a power series in E up to linear terms only for the problem in which we are concerned. This yields

$$P_\beta = P_\beta^0 \left[1 + \frac{e\sigma}{kT} (\mathbf{E} \cdot \mathbf{1}_L) \sum_{j=1}^{n} z_j \cdot (x_\beta^{(j)} - \langle x^{(j)} \rangle_{0,L}) + \cdots \right], \tag{A.3}$$

where $\langle x^{(j)} \rangle_{0,L}$ stands for the average position of ion j in the absence of the field and is defined by

$$\langle x^{(j)} \rangle_{0,L} = \sum_{(\beta)} x_\beta^{(j)} P_\beta^0 = \frac{\sum_{(\beta)} x_\beta^{(j)} e^{-V_\beta{}^0/kT}}{\sum_{(\beta)} e^{-V_\beta{}^0/kT}} . \tag{A.4}$$

For a symmetric distribution of charged sites the moment of the polyelectrolyte due to a given distribution of associated counter-ions may be defined as follows

$$\mathbf{m}_\beta = \sum_{(i)} z_i e x_\beta^{(i)} \, \mathbf{1}_L . \tag{A.5}$$

It follows quite generally that the average dipole moment, for a given orientation, will be represented by

$$\begin{aligned} \langle \mathbf{m} \rangle_{E,L} &= \sum_{(\beta)} \mathbf{m}_\beta P_\beta = \\ &= \sum_{(i)} z_i e \left[\langle x^{(i)} \rangle_{0,L} + \frac{e\sigma}{kT} (\mathbf{E} \cdot \mathbf{1}_L) \times \right. \\ &\left. \times \sum_{j=1}^{n} \langle z_j x^{(i)} (x^{(j)} - \langle x^{(j)} \rangle_{0,L}) \rangle_{0,L} \right] \mathbf{1}_L . \end{aligned} \tag{A.6}$$

If it is assumed that in the absence of the electric field the average position of any ion is $\langle x^{(i)} \rangle_{0,L} = 0$, then the first term on the r.h.s. of (A.6) disappears and the expression simplifies into

$$\langle \mathbf{m} \rangle_{E,L} = \sum_{(i)} \sum_{(j)} z_i z_j e^2 \frac{(\mathbf{E} \cdot \mathbf{1}_L)}{kT} \langle x^{(i)} x^{(j)} \rangle_{0,L} \, \sigma \, \mathbf{1}_L , \tag{A.7}$$

which is exactly equal to (7). In the absence of correlations between the position of any two different ions i and j this further simplifies into (4)

$$\langle m \rangle_{E,L} = \sum_{(i)} z_i^2 e^2 \frac{(\mathbf{E} \cdot \mathbf{1}_L)}{kT} \langle x^{(i)} x^{(i)} \rangle_{0,L} \, \sigma \, \mathbf{1}_L . \tag{A.8}$$

The equivalence between (A.8) and (4) follows from the fact that $\langle x^{(i)} x^{(i)} \rangle_{0,L}$ may also be expressed with the help of the reduced probability $P_i^0(x_k)$ to find ion i at a given site x_k in the absence of the field.

$$\langle x^{(i)} x^{(i)} \rangle_{0,L} = \langle (x^{(i)})^2 \rangle_{0,L} = \sum_{k=1}^{N} x_k^2 P_i^0(x_k) = P_i^0(x_k) \sum_{k=1}^{N} x_k^2 , \tag{A.9}$$

References

1. Oncley, J. L.: *J. Am. Chem. Soc.*, **60**, 1115 (1938); *Chem. Rev.* **30**, 433 (1942).
2. Minakata, A. and Imai, N.: *Biopolymers* **11**, 329 (1972).
   Minakata, A.: *Biolpoymers* **11**, 1567 (1972).
3. van der Touw, F. and Mandel, M.: *Trans. Faraday Soc.* **67**, 1336 and 1343 (1971).
4. van Beek, W. M., van der Touw, F., and Mandel, M.: (in preparation).
5. Mandel, M. and Jenard, A.: *Trans. Faraday Soc.* **59** 2158 and 2270 (1963).
6. van der Touw, F. and Mandel, M.: Book of Abstracts of the IUPAC symposium on Macromolecules, Leiden 1970, I, 375.
7. Takashima,S.: *J. Mol. Biol.* **7**, 445, (1963); *Biopolymers* **5**, 889 (1967).
8. Takashima, S.: *Arch. Bioch. Biophys.* **77**, 454 (1958).
9. Sachs, S. B., Raziel, A., Eisenberg, H., and Katchalsky, A.: *Trans. Faraday Soc.* **65**, 77 (1969).
10. Cole, K. S. and Cole, R. H.: *J. Chem. Phys.* **9**, 341 (1941).
11. Hanss, M.: *Biopolymers* 4, 1035 (1966).
12. Allgèn, L. G.: *Biochim. Biophys. Acta* **13**, 446 (1954).
    Allgèn, L. G. and Roswall, S.: *J. Polymer Sci.* **23**, 635 (1957).
13. Dintzis, H. M., Oncley, J. L., and Fuoss, R. M.: *Proc. Natl. Acad. Sci.* **40**, 62 (1954).
14. O'Konski, C. T.: *J. Phys. Chem.* **64**, 605 (1960).
15. Mandel, M.: *Mol. Phys.* **4**, 489 (1961).
16. McTague, J. P. and Gibbs, J. H.: *J. Chem. Phys.* **44**, 4245 (1966).
17. Schwarz, G.: *Z. Physik* **145**, 563 (1956); *Z. Physik. Chem.* **19**, 286 (1959); *J. Phys. Chem.* **66**, 2636 (1962).
18. Oosawa, F.: *Biopolymers* **9**, 677 (1970).
19. van der Touw, F. and Mandel, M.: (to be published).
20. Kubo, R.: *Lectures Theoret. Phys. (Boulder)* **1**, 120 (1959).
21. de Brouckère, L. and Mandel, M.: *Adv. Chem. Phys.* **1**, 77 (1958).
22. Van der Touw, F., Briedé, J. W. H., and Mandel, M.: *Biopolymers* **12**, 111 (1973).
23. Hormick, C. and Weill, G.: *Biopolymers* **10**, 2345 (1971).
24. Katchalsky, A.: *Pure Appl. Chem.* **26**, 327 (1971).

RADIOFREQUENCY PROPERTIES OF POLYELECTROLYTE SYSTEMS

K. ARULANANDAN*, S. S. SMITH**, and K. S. SPIEGLER***

University of California, U.S.A.

Abstract. Electrical measurements of heterogeneous media composed of solid polyelectrolytes and dilute aqueous solutions (or pure water) are interpreted in terms of a simple electrical network. It is demonstrated that this model network represents an extension of Maxwell's equation for the conductance of dilute suspensions of spheres to condensed systems. Discussing past work on cation exchange resinsolution systems, it is shown that the three empirical geometrical parameters of the model explain quantitatively the change of the low-frequency ($\leqslant$ 1000 Hz) conductivity of the heterogeneous mixture with the conductivity of the interstitial solution, and that the same parameters also explain (a) the potential differences between two NaCl solutions of different concentration separated by the mixture, and (b) the frequency-dispersion of the dielectric constant and of the conductivity of these mixtures in the radiofrequency range (20–90 MHz).

Application of the model to clay-solution aggregates showed that the radiofrequency dispersions of these systems at 10–50 MHz are fairly well described by the model. These dispersions were found to depend on the clay type (kaolinite, illite, montmorillonite), degree of consolidation, and soil fabric; these properties are reflected in the values of the geometrical parameters and of the dielectric constant of the solid which fit the dispersion curves best.

The influence of the counterion type on the dispersion curves is not well understood.

Frequency-dispersion curves of a commercial anion-exchange membrane in the range 5–50 MHz are shown and discussed.

It is concluded that the simple network used here represents a simple, first approach for the characterization of mixtures of solid polyelectrolytes and electrolyte solutions, and for other similar heterogeneous media.

1. Introduction

The interpretation, from first principles, of the electrochemical properties of heterogeneous media composed of solid polyelectrolytes and aqueous solutions of small-molecular electrolytes or of polyelectrolytes presents many hurdles, since these systems are even more complex than colloidal, quasi-homogeneous, solutions of polyelectrolytes. On the other hand, many heterogeneous media of this kind whose exact geometrical structure is not always regular are of such outstanding practical importance that their characterization by electrical measurements, followed by a less fundamental correlation of their properties, is well worthwhile. Examples of such media are clay-water-electrolyte systems ('clay-water aggregates'), various soils, certain porous water – or oil – bearing rocks, ion-exchange columns whose interstitial pore space is filled with aqueous electrolyte solutions, and even certain molecular-filter membranes.

It will first be demonstrated that Maxwell's formula for the electrical conductivity of a dilute suspension of solid conducting spheres in a conducting medium in terms

* Department of Civil Engineering, University of California, Davis.

** Present address: Converse, Davis & Associates, P.O. Box 2268D, Pasadena, Calif.

*** Sea Water Conversion Laboratory, University of California, Berkeley; to whom correspondence should be addressed.

Eric Sélégny (ed.), Polyelectrolytes, 301–321.

of the conductivities of the components and the volume percentage of the disperse phase is compatible with a simple network consisting of (a) two series elements representing conduction through solid and liquid respectively and (b) an element representing conduction through the liquid only. To extend this model to aggregates of solids and liquids in which the volume percentage of the solid is substantial, an additional parallel element representing conduction through contacting solids is added.

Past work on electrical measurements of columns consisting of spheres of ion-exchange resins and electrolyte solutions, and on the interpretation of the results by the simple equivalent-network model described above is then reviewed. The experimental results covered are conductivity measurements at low frequencies (60 and 1000 Hz), electrical potential differences between two different NaCl solutions separated by such columns, and the variation of the dielectric constant and conductivity of the columns above 20 MHz, all over a range of sulution concentrations. It will be shown that not only does the proposed network describe the variation of the measured parameters with solution concentration and/or frequency, but that the geometrical parameters of the model are roughly the same for all these measurements.

The application of the model to electrical measurements of systems consisting of clays and electrolyte solutions is then discussed. Dispersions (at 10–50 MHz) of consolidated specimens prepared by equilibrating different clays are reported. Interpretation of these measurements in terms of the model parameters exhibits meaningful variations of the values of the dielectric constants and conductivity of the solid clay phase with the nature of the clay, the electrolyte solution, the degree of consolidation of the aggregate and with the soil fabric.

2. The Maxwell Formula

Maxwell [1, vol. I, p. 440, Equation (17)] derived the following formula for the electrical conductivity, k (Ω^{-1} cm^{-1}) of a dilute suspension of spheres of conductivity, k_r, in a solution (or other homogeneous conductor) of conductivity, k_w, in terms of the conductivity of the two phases and of the volume fraction, f, occupied by the spheres:*

$$\frac{k}{k_w} = \frac{2 + (k_r/k_w) - 2f[1 - (k_r/k_w)]}{2 + (k_r/k_w) + f[1 - (k_r/k_w)]} . \tag{1}$$

The application of this formula, and subsequent studies on the extension of Maxwell's work to more condensed media (i.e. media with large f) in which the shape of the solid is not necessarily spherical has been discussed in detail by Meredith and Tobias [2]. If one tries to represent the Maxwell formula by an array of prisms as shown in Figure 1, in which the electrical current passes vertically through a cube of width and height 1 cm, application of Ohm's law leads indeed to an expression for the conductivity, k, of the suspension in terms of the conductivities k_r and k_w of the

* Note that in Maxwell's original the symbol 'k' stands for resistivity $=$ (conductivity)$^{-1}$.

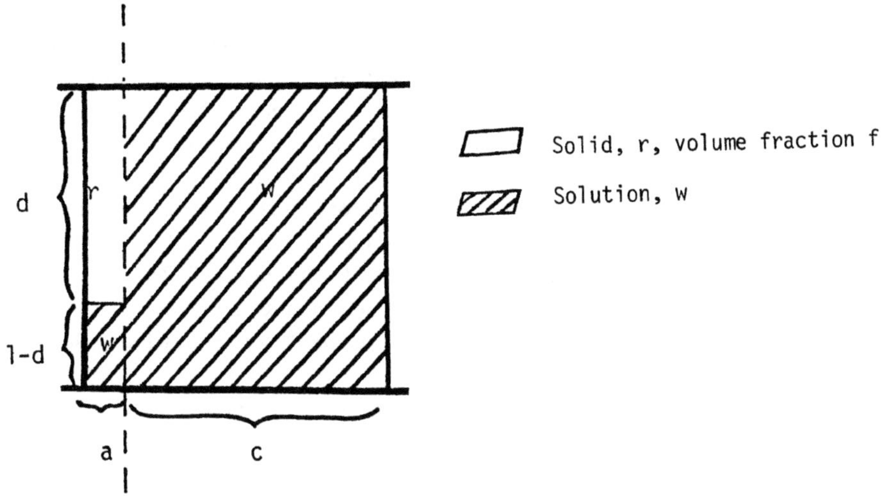

Fig. 1. Network for Maxwell equation. – Model represents 1 cm³ of suspension. Open area represents solid, r, hatched area solution, w. a and c are geometrical parameters of model; $c = 1 - a$. Electric current passes vertically through model.

two separate phases, but instead of the single geometrical parameter, f, two parameters, a $(= 1 - c)$ and d are necessary:

$$k = \left[\left(\frac{1-d}{a}\right)\Big/ k_w + \frac{d}{a}\Big/ k_r\right]^{-1} + ck_w = \frac{(1-a)\,\mathrm{d}k_w^2 + (1-d+ad)\,k_r k_w}{\mathrm{d}k_w + (1-d)\,k_r}. \tag{2}$$

It is of interest that if one identifies the volume of the solid in 1 cm^3 of the suspension f, with the volume of the model, $a \times d$, and if one arbitrarily takes the length, d, in the model equal to $(2+f)/3 \cong 2/3$, the model Equation (2) is indeed equal to the Maxwell Equation (1). Thus the rather complicated true current distribution can be approximated in a thought experiment by the sum of two currents, viz. one which passes alternately through a solution element of volume $a \times (1-d)$ and a solid element, $a \times d$, and another, in parallel, which follows a somewhat tortuous path through the remainder of the solution, characterized by the volume element $c \times 1$.

3. The 'Three-Element Model'

It is of interest to examine whether this model can be extended to conductive porous media of arbitrary geometrical shape in which the solid particles are in contact, i.e. media consisting of two continuous conductive phases. While the a priori prediction of the conductivity of such media is not feasible, because their geometrical shape is not known, the prediction of electrical properties of all two-phase media with intrinsic conductivities k_r, k_w of the two phases, from measurements with systems of the same geometry, but different intrinsic conductivities should be possible provided all these systems of differing k_r and k_w are represented by the same model with the same geometrical parameters (similar to the geometrical parameters 'a' and 'd' in Figure 1 and Equation (2)). For instance, if the geometrical model parameters for a system consisting of a cation-exchange resin column with interstitial solution are known, it should be possible to determine the conductivity of similar systems containing other solutions; the electrical potential differences between two solutions of the same electrolyte at different concentration separated by a 'plug' of this kind; and the frequency-dependence of both the dielectric constant and of the electrical conductivity of the plug. From past measurements of these properties one concludes that this is indeed possible.

The model [3–6] is shown in Figure 2. (a), (b) and (c) represent the electric current

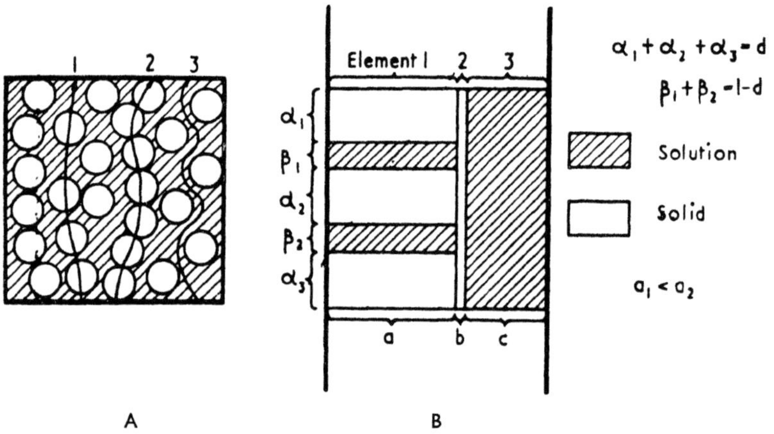

Fig. 2. Electrochemical model of porous plug composed of conducting spheres and solution [5]. – (A) Schematic representation of current path through plug. (1) represents current through solution and spheres in series, (2) through spheres in contact with each other, (3) current through solution. (B) Simplified model representing situation shown in (A). $a + b + c = 1$ cm.

through solution and solid spheres in series, through spheres in contact with each other, and through the solution respectively. The electric current is thought to pass between an upper and a lower electrode in contact with the heterogeneous medium. Applying Ohm's law to this model we obtain for the conductivity of the plug:

$$k = \underbrace{\frac{k_r k_w}{[k_r(1-d)/a] + [k_w d/a]}}_{\text{element 1}} + \underbrace{bk_r}_{2} + \underbrace{ck_w}_{3} \tag{3}$$

4. Conductivity of Ion-Exchange Resin Columns at Low Frequencies ($\leq$1000 Hz)

Figures 3–5 show the conductance, k, of columns composed of various spherical ion-exchange resin particles equilibrated with different interstitial solutions, as functions of the solution conductivity, k_w. The curves drawn are calculated from Equation (3) by choice of the best fitting parameters a, b and d. The experimental points fit relationships of type Equation (3), $k=f(k_w)$ quite well. It might be argued that it is often not hard to fit a smooth curve if one has free choice of three parameters, but it should be noted that the parameters are obviously of geometrical significance, as shown by the fact (Table I) that columns of different particle size and different

TABLE I

Geometrical parameters of 3-element model [4–6]

Solution	Cation-Exchange Resin	Mesh (U.S. standard) dry	Size (mm)	Resin conductivity 26.5 ± 1.5°C Ω^{-1} cm^{-1}	a	b	d
NaCl	Amberlite IR-120[a]	35–40	0.42–0.50	0.029	0.63	0.01	0.95
KCl	Amberlite IR-120	35–40	0.42–0.50	0.048	0.60	0.01	0.93
$CaCl_2$	Amberlite IR-120	35–40	0.42–0.50	0.0075	0.64	0.03	0.95
$AgNO_3$	Amberlite IR-120	35–40	0.42–0.50	0.026	0.58	0.04	0.95
NaCl	Dowex 50[b]	100–200	0.074–0.149	0.030	0.63	0.01	0.95
$CaCl_2$	Amberlite IR-120	20–25	0.71–0.84	0.0075	0.65	0.02	0.90

[a] Product of Rohm and Haas Co., Philadelphia, Pa., U.S.A.
[b] Product of Dow Chemical Co., Midland, Mich., U.S.A.

sphere conductivity, k_r, which have only one common characteristic (viz. their spherical geometry) all yield very similar a, b and d values.* Also, it should be considered that, except at very low solution conductivities, k_w, when the current is carried primarily by element '2' (Figure 2), the curve depends primarily on the values of 'a' and 'd' (since $b \leqslant 0.04$), so that one really deals only with a two-parameter fitting problem.

Moreover, the same model with identical geometrical parameters also gives a fair quantitative prediction of 'plug potentials' (i.e. electrical potential differences between two calomel electrodes in NaCl-solutions of different concentrations separated by the plug) and of the variation of the apparent dielectric constant and the conductivity of the heterogeneous columns at radiofrequencies, as discussed in the following.

* The measurements described in Figures 3 to 5 were performed at 25–28°C. No attempt at temperature control was made, since both abscissa and ordinate values are believed to vary by about the same percentage with the temperature over this small temperature range.

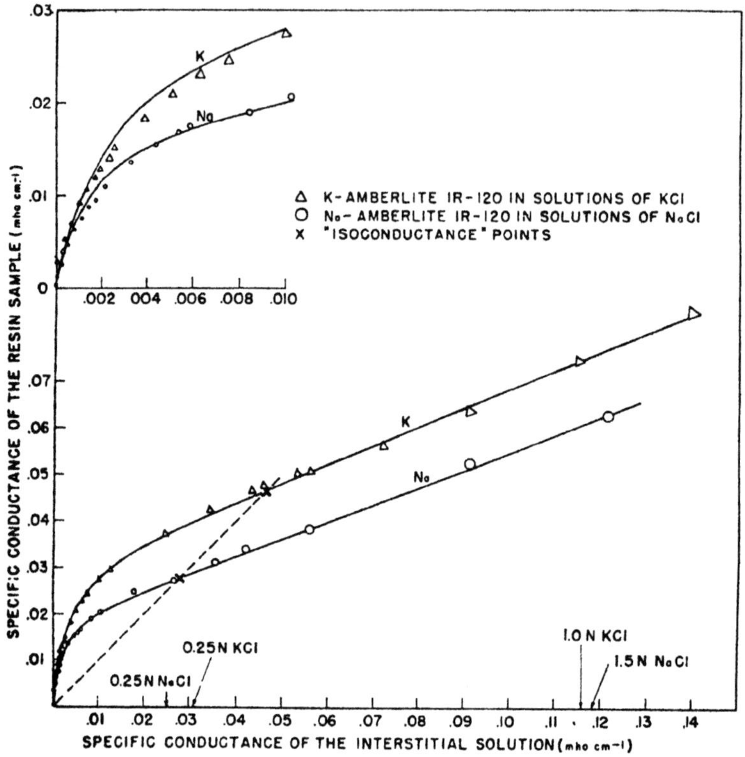

Fig. 3. Specific conductance of resin plug saturated with solution vs. specific conductance of interstitial solution (1000 Hz) [4]. Conductivity values for salt solutions marked on abscissa refer to 25°C.

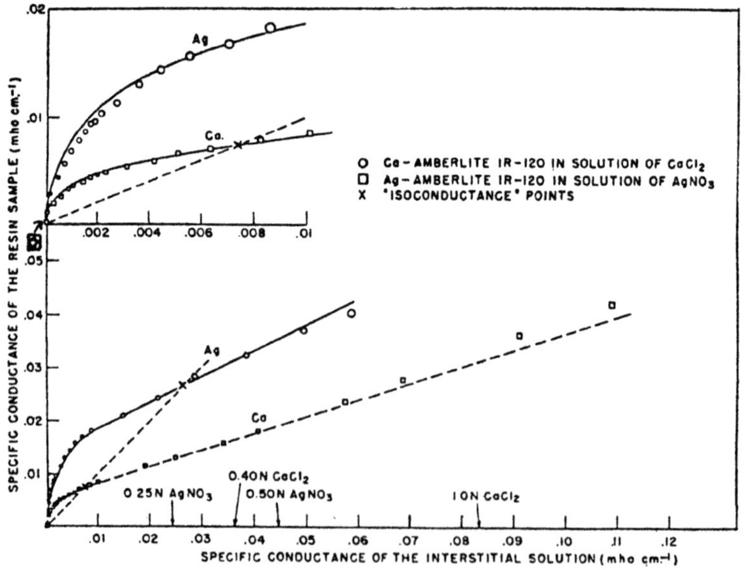

Fig. 4. Specific conductance of resin plug saturated with solution vs. specific conductance of interstititial solution (1000 Hz) [4]. – Conductivity values for salt solutions marked on abscissa refer to 25°C.

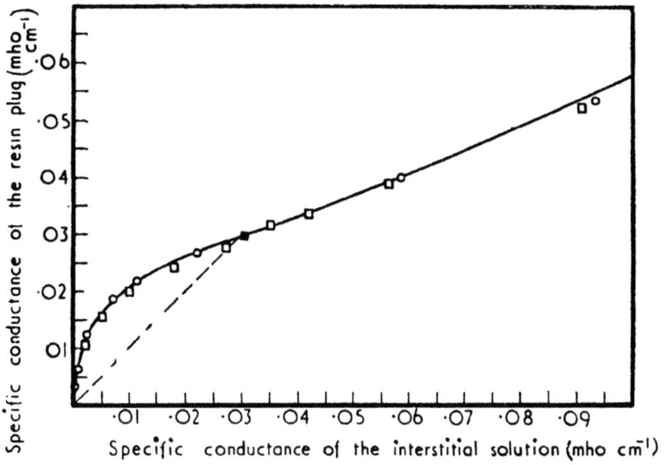

Fig. 5. Specific conductance of resin plug saturated with solution against specific conductance of solution [5]. – Circles represent experimental data with small-size Dowex-50 cation-exchange resin (100–200 mesh, U.S. Standard Screen). Squares represent data from a previous investigation [4] for coarse Amberlite IR-120 (35–40 mesh). Solid line is calculated from theory based on model (Figure 2); cross mark shows isoconductance point; 60 Hz, temp. 25–28 °C.

5. Electrical Potential Differences Across an Ion-Exchanger Column

The model of Figure 2 has also been used to predict the potential difference, E', between Ag/AgCl electrodes inserted in NaCl solutions at the top and the bottom respectively of a column consisting of a cation-exchange resin (Dowex-50, 100–200 mesh) and interstitial NaCl solution [5]. Solutions of different concentrations were considered, but the ratio between the mean ionic activities of the NaCl solutions separated by the plug was always 3:1, with the more concentrated solution at the bottom. For details of the experimental procedure and of the interpretation method, the reader is referred to the original publication [5]. Briefly, the 'plug potential' (i.e. the electrical potential difference across the column) of the model is determined by the potential differences across the three elements (*a*), (*b*) and (*c*) shown in Figure 2. If elements (*a*) or (*b*) alone were present, the potential difference across the model would be that of an ideally cation-selective membrane viz. 28.3 mV for the 3:1 activity ratio of the NaCl solutions [5]. On the other hand, if (*c*) alone were present, the potential difference across the model would be the diffusion potential between the two NaCl solutions, viz. about − 6.1 mV. In reality, all three elements are present, and, by Kirchhoff's law, each contributes to the total potential difference in accordance with its conductance. Thus, at very low NaCl-concentrations, when the conductance of element (*c*) is relatively low, the plug potential approaches that of an ideal cation-selective membrane. As the solution concentrations increase, the contribution of element (*c*) increases also. This contribution is negative; the membrane potential decreases with increasing solution concentration, i.e. the column which had acted

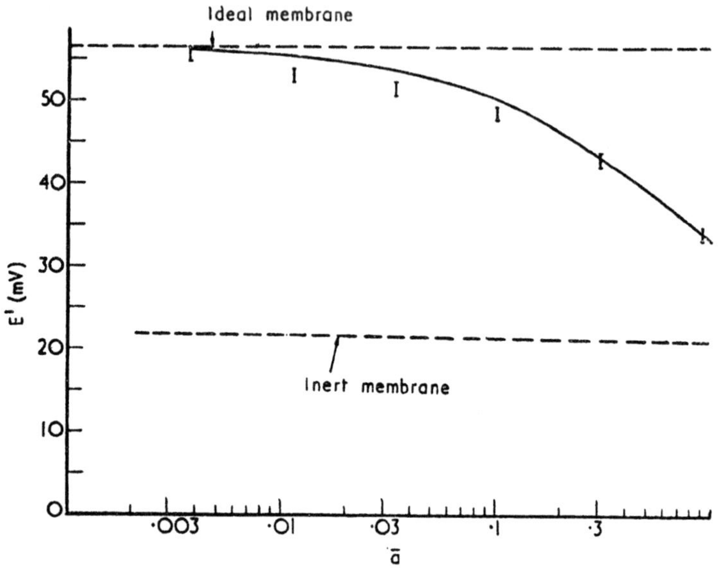

Fig. 6. Potential difference, E', between upper and lower Ag/AgCl electrode against average activity a of sodium chloride in solutions separated by cation-exchange resin plug (Dowex 50, 100–200 mesh) [5]. – Solution activity ratio 3:1. Vertical lines represent potential difference measured between Ag/AgCl electrodes at 25°C, solid line calculated from theory.

like a near-ideal cation-selective membrane in the presence of very dilute solutions, becomes electrochemically less cation-selective as the solution concentration increases.

This behavior is shown in Figure 6 in which the potential difference between the upper and lower Ag/AgCl electrode inserted in NaCl solutions separated by a cation-exchange resin is plotted against the average ion activity, a, in the solutions. ('Ion activity' is taken to be the mean ionic activity in the aqueous NaCl solutions.) It is important that the *potential differences* predicted from the model by use of the geometrical parameters $a=0.63$, $b=0.01$, $d=0.05$, and of the solid conductivity, $k_r=0.030\ \Omega^{-1}\ \mathrm{cm}^{-1}$ – all determined from *conductivity* plots similar to Figures 3–5 – approach the measured values fairly closely.* In other words, the geometrical parameters of the model, as determined from conductance measurements, seemed to be valid for the prediction of 'plug potentials' also. It would be of interest to perform measurements in systems composed of other electrolytes and polyelectrolytes, in order to determine if this correlation [proven in Reference (5) for one solution-resin system only] is of general validity.

* Note that the difference of the electrode potentials of the two Ag/AgCl electrodes, viz. 28.3 mV was added to the calculated membrane potential in order to obtain the total potential drop across the system Ag|AgCl|NaCl, a'|Plug|NaCl, a''|AgCl|Ag.

6. Radiofrequency Measurements of Ion-Exchange Resin Columns

In view of the success of the 3-element model for the quantitative description of the variation of the low-frequency conductivity of resin-solution columns with the solution concentration, it is logical to inquire whether the same model can explain the electrical behavior of these or similar systems at high frequencies, at which the admittance of the system as a whole and/or of the model elements is appreciable. In other words, the question is asked: Can the 3-element model describe the change of the complex impedance of these systems with the frequency of the alternating current used to measure the impedance?

Such measurements have indeed been performed and interpreted in terms of the model [6]. In order to correct for the impedances of the line and the electrodes, which are both in series with the system to be measured, the complex impedance of the specimen was determined by vectorial subtraction of the total impedances measured for columns of different height [6–8].

To interpret the results of these measurements in terms of the model (Figure 2), each component of the model was represented as a complex impedance consisting of a resistor and a capacitor in parallel as shown in Figure 7. Inductive elements are thought to be small and are therefore not included in the model.

The admittance, Y, of the model can be calculated from elementary network analysis [9]:

$$\left.\begin{aligned} &\text{Element '}a\text{': } Y_a = \left[\frac{1-d}{ak_w + j\omega\alpha a\varepsilon_w} + \frac{d}{(ak_r + j\omega\alpha a\varepsilon_r)}\right]^{-1} \\ &\text{Element '}b\text{': } Y_b = bk_r + j\omega\alpha b\varepsilon_r \\ &\text{Element '}c\text{': } Y_c = ck_w + j\omega\alpha c\varepsilon_w \\ &\qquad\qquad Y = Y_a + Y_b + Y_c \end{aligned}\right\}, \tag{4}$$

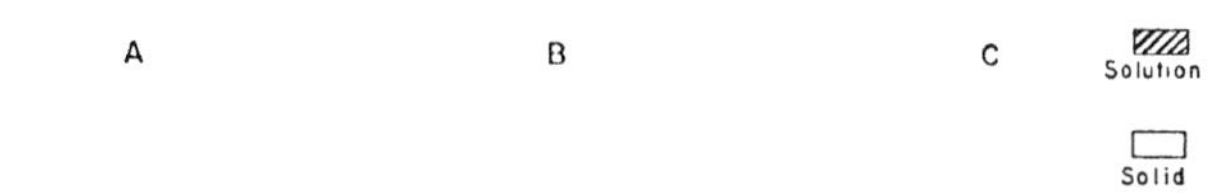

Fig. 7. Model of porous plug [6]. – (A) schematic representation of current path through plug. (1) represents current through solution and spheres in series, (2) through spheres in contact with each other, (3) current through solution. (B) simplified model for d.c. conductance, representing situation shown in (A). Different zones represent separate resistors of dimensions shown. $a+b+c=1$ cm. (C) Extension of same model to a.c. properties. Each zone is represented by a parallel circuit of resistor and capacitor. Impedance of each zone is determined by its dimension (equal to those shown in (B)) and conductivity and dielectric constant of material forming the zone. Size of cube shown is $1 \times 1 \times 1$ cm.

where α is the capacitance of the unit capacitor in the vacuum, i.e., $0.0885\,10^{-12} f$ farad, and Y and k are given in $\Omega^{-1}\,\text{cm}^{-1}$, ω is the angular frequency (s^{-1}), and $j \equiv \sqrt{-1}$. Separating imaginary and real parts, the following expressions for the apparent dielectric constant, $\varepsilon'_{\text{theor}}$, and the apparent specific resistance, ϱ'_{theor}, of the model are obtained:

$$\varepsilon'_{\text{theor}} = \frac{a}{d(1-d)\,S}\left[\frac{\varepsilon_r k_w^2}{1-d} + \frac{\varepsilon_w k_r^2}{d} + \omega^2\alpha^2\left(\frac{\varepsilon_r \varepsilon_w^2}{1-d} + \frac{\varepsilon_r^2 \varepsilon_w}{d}\right)\right] + b\varepsilon_r + c\varepsilon_w \tag{5}$$

$$k'_{\text{theor}} = \left(\varrho'_{\text{theor}}\right)^{-1} = \frac{a}{d(1-d)\,S} \times \left[\frac{k_r k_w^2}{1-d} + \frac{k_r^2 k_w}{d} + \omega^2\alpha^2\left(\frac{\varepsilon_w^2 k_r}{1-d} + \frac{\varepsilon_r^2 k_w}{d}\right)\right] + bk_r + ck_w, \tag{6}$$

where

$$S \equiv \left(\frac{k_w}{1-d} + \frac{k_r}{d}\right)^2 + \omega^2\alpha^2\left(\frac{\varepsilon_w}{1-d} + \frac{\varepsilon_r}{d}\right)^2. \tag{7}$$

These values are shown in Figures 8 and 9 along with the corresponding experimental values, ϱ' and ε', determined from the measured resistance and capacitances after correction for line effects. In these calculations the values of the specific conductances of the interstitial fluids were determined in the same cell and by the same method

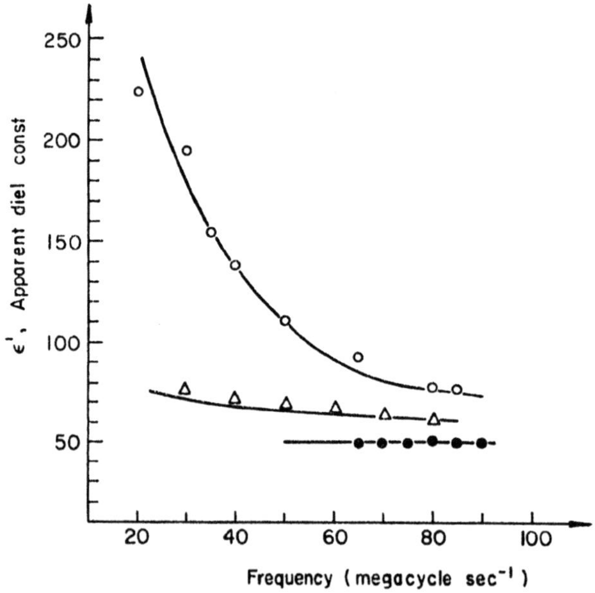

Fig. 8. Apparent dielectric constants of ion-exchange resin columns saturated with different interstitial solutions [6]. – Resin: Amberlite IR-120, Ca form, 25°C. ○ Water; △ 0.0116 M $CaCl_2$; ● 0.065 M $CaCl_2$ —— Calculated: $a = 0.65$, $b = 0.02$, $d = 0.90$, $k_r = 0.075\ \Omega^{-1}\,\text{cm}^{-1}$, $\varepsilon_r = 38$, $\varepsilon_w = 75$.

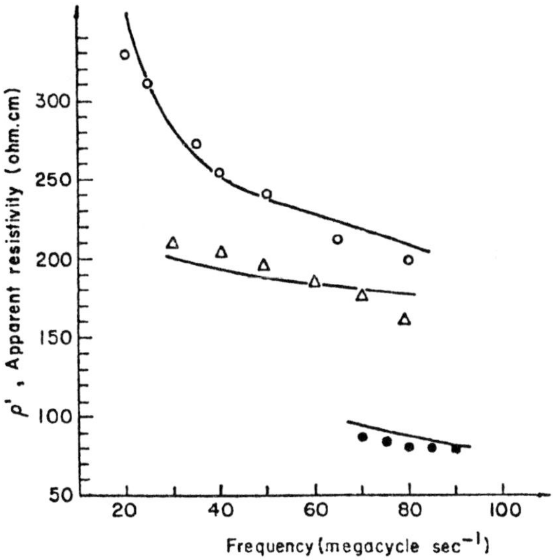

Fig. 9. Apparent specific resistances of ion-exchange resin columns saturated with different interstitial solutions [6]. – Resin: Amberlite IR-120, Ca form, 25°C. ○ Water; Δ 0.0116 M $CaCl_2$; ● 0.065 M $CaCl_2$; —— Calculated: $a=0.65, b=0.02, d=0.90, k_r=0.075\,\Omega^{-1}\text{cm}^{-1}$, $\varepsilon_r=38, \varepsilon_w=75$.

as the measurements of the resin-solution mixtures, and were found to be $k_w=1.0\times\times10^{-6}\,\Omega^{-1}\,\text{cm}^{-1}$ for distilled water and 0.00262 and $0.0125\,\Omega^{-1}\,\text{cm}^{-1}$ for the 0.0116 and 0.065 M $CaCl_2$ solutions, respectively. The dielectric constant for water was taken from the work of Hasted *et al.* [10], as $\varepsilon_w=78.0$ and the dielectric constant of the dilute $CaCl_2$ solution, which is very similar, was estimated from the work of the same authors as 77.7, equal to an equimolar $MgCl_2$ solution. The dielectric constant of the more concentrated $CaCl_2$ solution was determined, using the same procedure as described for the resin-solution mixtures, $\varepsilon_w=75.0$. The specific conductance of the resin was taken as $k_r=0.0075$ mho cm^{-1}, as determined at 1000 Hz, and the dielectric constant of the resin, ε_r, as 38 ± 1, as determined from Figure 8 by extrapolation of the results to high frequencies [6]. All these values were assumed to hold over the whole frequency range of the measurements.

The three geometrical parameters used for calculating the apparent dielectric constant, $\varepsilon'_{\text{theor}}$, and the apparent conductivity, $k'_{\text{theor}}=1/\varrho'_{\text{theor}}$ from Equations (5) and (6) respectively were $a=0.65$, $b=0.02$ and $d=0.90$ ($c\equiv1-a-b=0.33$). It is seen that the model (Figure 7) using geometrical parameters practically identical with those valid for the interpretation of the low-frequency conductivity variation with the conductivity of the interstitial solution (Table I) reproduces reasonably well the sharp decreases with frequency of the apparent dielectric constant and specific resistance of the resin columns in distilled water, as well as the variation of these frequency-dispersion curves with the concentration of the interstitial solution. It is of particular interest that the model explains the high values of the apparent dielectric constants found for the resin-solution systems, which are much higher than the dielectric constants of the two separate phases. The apparent specific resistances predicted by the theory are also fairly close to the measured values.

The fact that this interpretation yields almost the same geometrical parameters as the measurement of other properties of the columns shows that the dispersion above 20 MHz is indeed due to the heterogeneity of the mixture, i.e. a 'Maxwell-Wagner effect.' Moreover, this analysis of the dispersion curves yields the dielectric constant of the solid. The latter provides a rough measure of the water content of the solid phase.

7. Radiofrequency Measurements of Clay-Solution Aggregates

In view of the successful application of the equivalent model (Figure 7) to the interpretation of the variation of the apparent dielectric constant and apparent conductivity of resin-solution systems, it seemed of interest to study clay-water and clay-solution aggregates, in an effort to determine whether the geometrical model parameters computed from these measurements reflect on the state of these aggregates [11,12].

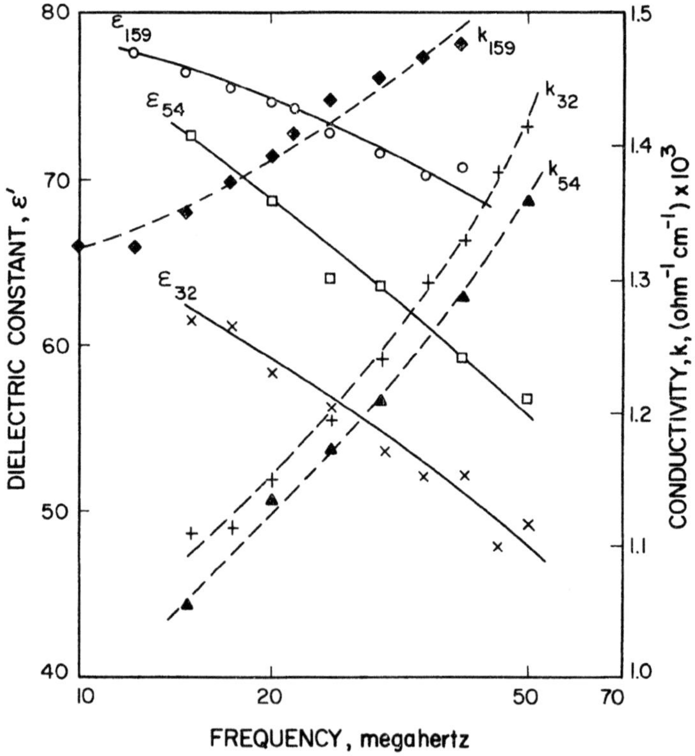

Fig. 10. Change of dielectric constant, ε, and conductivity, k, of illite-water aggregates with compaction [12] – ○, □, × dielectric constants, ◆, ▲, + conductivities at 159, 54 and 32 weight percent water, repectively. Additional experimental details are listed in Table II.

While the results reported in this section deal with dispersions in the frequency range 3–60 MHz, which seem indeed to be interpretable in terms of heterogeneity effects ('Maxwell-Wagner effects') [12], it should be noted that dispersions in polyelectrolyte-solution systems have also been observed at lower frequencies [13–16], and that these dispersions have been interpreted in a different manner. In one case [15] the frequency range of the observed dispersions overlaps somewhat with the Maxwell-Wagner dispersion range of the clay-solution aggregates described here.

The homoionic clay samples in various states described in the following were prepared by equilibration with appropriate chloride solutions, centrifuged and then consolidated. Except for the aggregates whose properties are described in Figure 10 and Table II, all samples were consolidated by increasing consolidation pressure up to 1 kg cm^{-2}. This was followed by leaching with water or dilute solutions in some cases. The electrical measurements were performed at 22 ± 0.5°C.

TABLE II
Consolidation of illite-water aggregates.
Pore fluid electrolyte: NaCl. Dielectric constant of fluid taken as $\varepsilon_W = 79$.

Consolidation pressure (kg cm^{-2})	Water content (weight % based on dry solid)	Pore fluid extract conductivity (Ω^{-1} cm^{-1}) (measured)	Optimized model parameters							
			a	b	c	d	$(a \times d)$	ε_r	k_r	k_W
Paste	159	0.0010	0.53	0.009	0.46	0.74	0.39	44	0.0022	0.00092
1.5	54	0.0010	0.65	0.007	0.34	0.62	0.40	23	0.0016	0.00072
15.0	32	0.0016	0.71	0.005	0.28	0.58	0.41	13	0.0014	0.00083

Figure 10 shows the variation of the dielectric constant and of the conductivity of *illite-NaCl solution aggregates at different degrees of compression*, as indicated by the water content (w/c, weight percent water based on dry solid, as determined by drying at 105°C). The clay used was Grundite (Illinois Clay Products, Morris, Ill.), washed through a 50 μ sieve. Table II lists the geometrical parameters, a, b, c $(\equiv 1-a-b)$, d and the dielectric constant, ε_r, and conductivity of the solid, k_r, as well as the conductivity of the interstitial solution, which gave the best fit of the measured points with Equations (5) and (6). [The dielectric constant of the interstitial solution was taken as a constant value, viz. 79, which is equal to the value of ε_w in a free solution of identical concentration.] These 'optimized' parameters were determined by means of a computer program. In general the model equation described the changes of the dielectric constant and conductivity of the aggregate quite well, as shown, for one example, in Figure 11.

Consideration of the computed geometrical parameters of Table II show that 'b' is very small, i.e. the model is quite close to the one underlying the Maxwell equation (Figure 1), except that in the condensed clay aggregate the numerical values of 'a' are different from those for Maxwell's dilute suspension for which 'a' is quite small.

In terms of the model shown in Figure 2, one can imagine the transport of electric current to take place via two main pathways, viz. solid and solution in series '*a*' and through a tortuous path in the interstitial solution ('*c*'). As compaction increases, the relative contribution of the latter path becomes less. It is of interest that the product $(a \times d)$ which represents the fraction of solid in the model increases only very slightly with consolidation although the percentage of solid in the aggregate increases with increasing consolidation pressure (and hence with decreasing water content). This fact shows that one cannot very strictly quantitatively identify the volume of the model components with the volume of the components in the sample. The trend for decreasing pathways through free solution with increasing consolidation shows up clearly, however. It is also of interest that the conductivity of the interstitial solution computed from the curves of Figure 11 is quite similar to the conductivity of the pore solution measured in the pore fluid extract. Thus the parameters computed by the curve-fitting ('optimization') procedure are not devoid of physical meaning.

The geometrical significance of parameter '*d*' is not quite as clear-cut for clays as it is for ion-exchange resin-solution systems (Table 1). There is a general increasing trend for '*d*' in the series kaolinite → illite → montmorillonite, but in a plot of '*d*' vs. the ratio between computed k_r and the specific k_s values in the experiments reported here, all points fall roughly on a single line, irrespective of clay type.

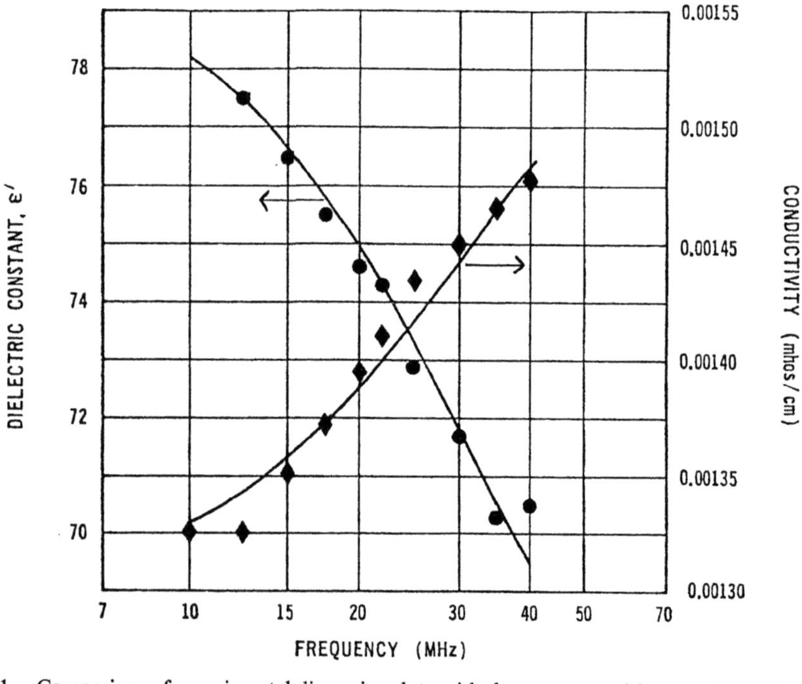

Fig. 11. Comparison of experimental dispersion data with those computed from model [12] – Illite-NaCl solution aggregate. Water content 159% (weight water/dry weight of clay). Pore fluid: aqueous NaCl solution, conductivity: $10^{-3}\ \Omega^{-1}\ cm^{-1}$. ● Dielectric constant. ◆ Conductivity. Solid lines are calculated from optimized (bestfitting) model parameters.

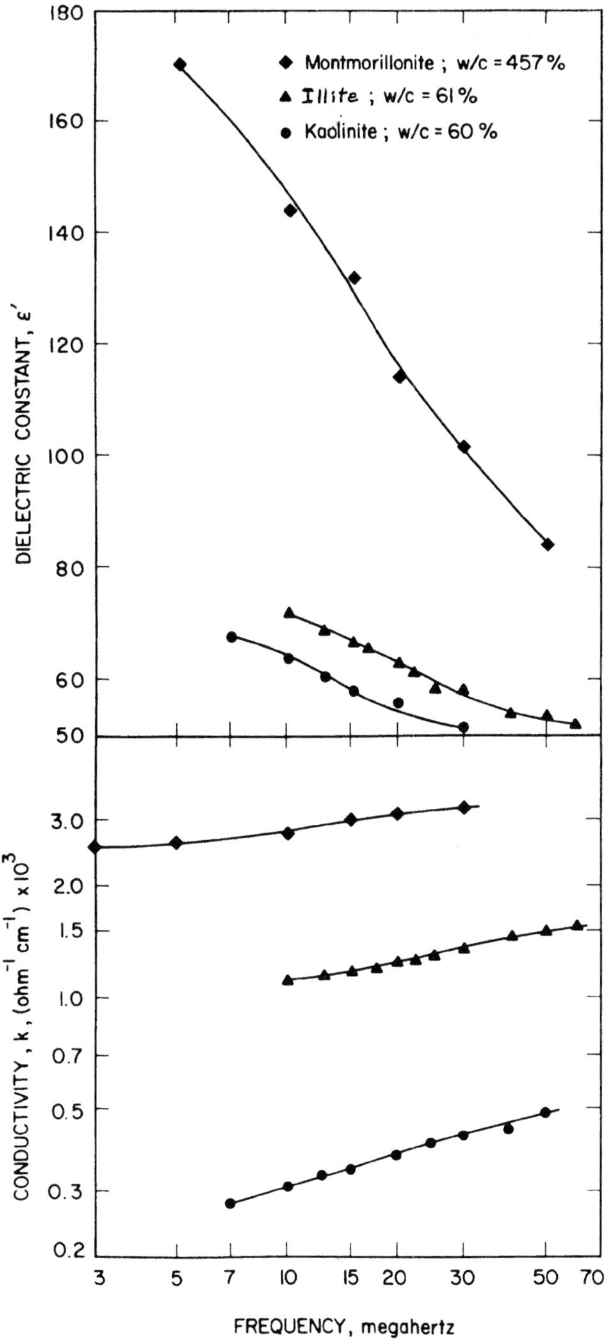

F g. 12. Frequency-dispersion curves of different clay-water aggregates [12]. – Experimental details are listed in Table 3.

Figure 12 shows the dispersion curves of aggregates composed of *different sodium clays* and very dilute NaCl solutions. All aggregates were gradually consolidated to a final load of 1 kg cm^{-2}. It is known that the cation-exchange capacity ('base exchange capacity') of these clays increases in the sequence kaolinite < illite < montmorillonite. This sequence is also very clearly seen in the dispersion curves of Figure 12. The ion-exchange capacity is the main factor determining the high-frequency electric properties, as indeed it strongly influences many other properties of the clay-solution aggregates. It is seen from Table III that while the total water content of the

TABLE III

Model parameters for different clay-water aggregates [12]
Pore fluid electrolyte: NaCl. ε_W taken as 79.

Soil type	Water content (%)	Pore fluid extract conductivity (mho cm^{-1})	Optimized model parameters							
			a	b	c	d	$(a \times d)$	ε_r	k_r	k_W
Kaolinite	60	0.0002	0.56	0.012	0.43	0.43	0.24	10	0.00037	0.0002
Illite	61	0.0014	0.66	0.007	0.33	0.66	0.44	20	0.0017	0.0008
Montmorillonite	457	0.0010	0.82	0.020	0.16	0.93	0.76	63	0.0043	0.0005

aggregates increases with increasing base-exchange capacity, the fraction of the current carried through the liquid phase alone, as characterized by the model parameter 'c' (Figure 2) actually decreases with increasing base-exchange capacity. On the other hand the product $(a \times d)$ which is related to the amount of the solid phase in the aggregate (except for the relative small amount characterized by 'b') increases with increasing base-exchange capacity. This fact, and also the increase of the dielectric constant of the solid phase, ε_r, indicate that the water uptake correlated to the higher base-exchange capacity is due to increased swelling of the solid phase, thus confirming relevant information from other measurements, e.g. X-ray diffraction patterns of clays in varying states of hydration.

The interpretation of the influence of the *soil fabric* on the dispersion is based on similar considerations. Figure 13 shows the dispersion curves for two illite-solution

TABLE IV

Model parameters for different soil fabric [12]

Fabric	Soil type	Water content (%)	Pore fluid extract conductivity (mho cm^{-1})	Optimized model parameters[a]							
				a	b	c	d	$(a \times d)$	ε_r	k_r	k_W
'Flocculated'	Illite	61	0.0014	0.66	0.007	0.33	0.66	0.44	20.	0.0017	0.000[illegible]
'Dispersed'	Illite	56	0.0019	0.76	0.010	0.23	0.81	0.62	29.	0.0019	0.000[illegible]

[a] ε_W was taken as 79.

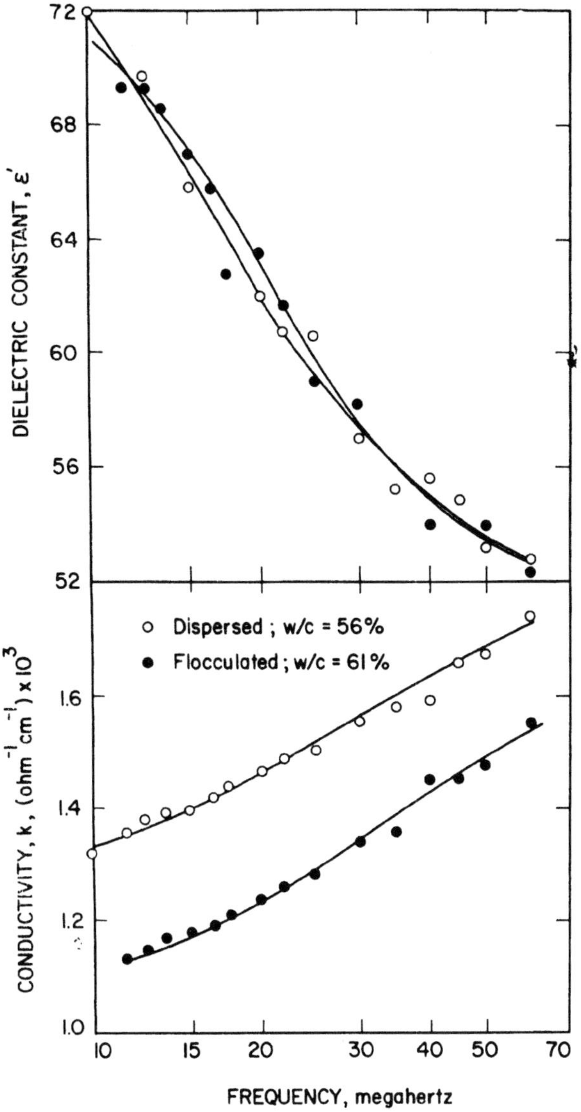

Fig. 13. Frequency-dispersion of dispersed and flocculated illite [12]. – Experimental details are listed in Table IV.

aggregates. In the preparation of one sample ('flocculated soil'), the clay was initially treated with 0.1 N NaCl. This relatively high solution concentration causes flocculation (aggregation) of the solid particles. After consolidation to 1 kg cm^{-2}, the sample was leached with 0.01 N NaCl solution. The second illite sample was treated with 0.01 N sodium oxalate and then consolidated in similar manner. It is known that treatment of clays with sodium oxalate causes a disaggregated ('dispersed'*) soil fabric. Computation of the model parameters (Table IV) indicates clearly that less water is associated with the solid in the 'flocculated' than in the 'dispersed' illite, because the dielectric constant, ε_r, of the latter is higher. Perhaps the higher value of the product $(a \times d)$ for the 'dispersed' sample is also an indication of the same phenomenon. As for the lower value of parameter 'c' in the 'dispersed' sample, it is possible [12] that this may be correlated with the tendency of the particles to parallel

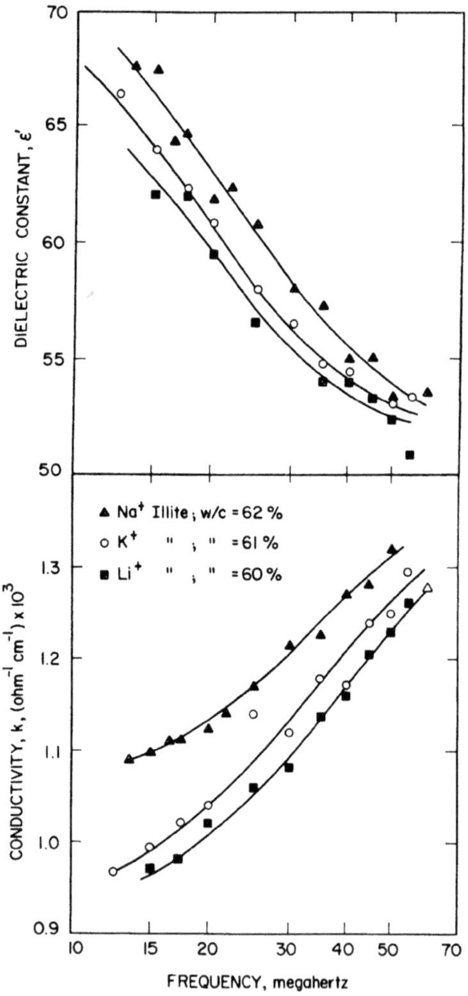

Fig. 14. Frequency-dispersion curves of illite with different counter-ions [12]. – Experimental details are listed in Table V.

* The term 'dispersed' is used here in its accepted meaning in soil mechanics, whereas in the rest of the article, the term 'dispersion' refers to the change of dielectric constant and/or conductivity of a sample with electrical frequency, as is common in electrophysics and electrical engineering.

orientation in this sample, which, in turn causes an increase in the tortuosity of the path of the electric current through the solution (Figure 7A, path 3).

The electrical dispersion of clays depends also on the nature of the type of counter-ion, as shown in Figure 14 and Table V. It is seen that the curves and parameters follow the sequence $Na^+ \rightarrow K^+ \rightarrow Li^+$ rather than the atomic-weight sequence

TABLE V

Model parameters for illite-water aggregates with different cations [12]

Cation	Soil type	Water content (%)	Pore fluid extract conductivity (mho cm^{-1})	Optimized model parameters [a]						
				a	b	c	d	ε_r	k_r	k_W
Na^+	Illite	62	0.0014	0.64	0.002	0.36	0.65	21.	0.0015	0.00080
K^+	Illite	61	0.0011	0.62	0.015	0.36	0.52	13.	0.0012	0.00080
Li^+	Illite	60	0.0012	0.64	0.010	0.35	0.49	14.	0.0011	0.00084

[a] ε_W was taken as 79.

$K^+ \rightarrow Na^+ \rightarrow Li^+$. Tentative explanations for this behavior have been given [12] but this aspect of the subject remains to be further investigated in the future.

8. Radiofrequency Measurements of a Synthetic Ion-Exchange Membrane

Synthetic ion-exchange membranes also exhibit strong variations of the dielectric constant and electrical conductivity in the radiofrequency range [17] and at least one other dispersion range was found at lower frequencies. While it is not the purpose of this communication to present a full report on membrane measurements, one typical result is shown in Figure 15. Since ion-exchange membranes are solid polyelectrolytes which look like single-phase media to the naked eye, but have been shown [18] to be heterogeneous in colloid dimensions (characteristic 'pore' dimension about 10^{-5} cm), they represent a borderline case between systems of macroscopic heterogeneity such as columns of ion-exchange resins, whose radiofrequency properties could be explained by the model of Figure 7, and solutions of electrolytes or polyelectrolytes whose properties can be interpreted by microscopic methods of molecular polarization [15]. It is of interest that when the dispersion curves are approximated by the model Equations (5) and (6) with the best-fitting parameters, the values obtained for the conductivity of the 'interstitial' water ($k_w = 10^{-6}\ \Omega\ \mathrm{cm}^{-1}$), dielectric constant of the solid [$\varepsilon_r = 32$; water content: 40% – dry basis (19)] are physically reasonable. In view of the close proximity of the polymer segments in cross-linked membranes, it is not surprising the direct 'particle-to-particle conduction' parameter, 'b', is quite high, [compared to granular ion-exchange resin columns (Table I)]. It should be noted that this membrane is stated by the manufacturer to be permselective for

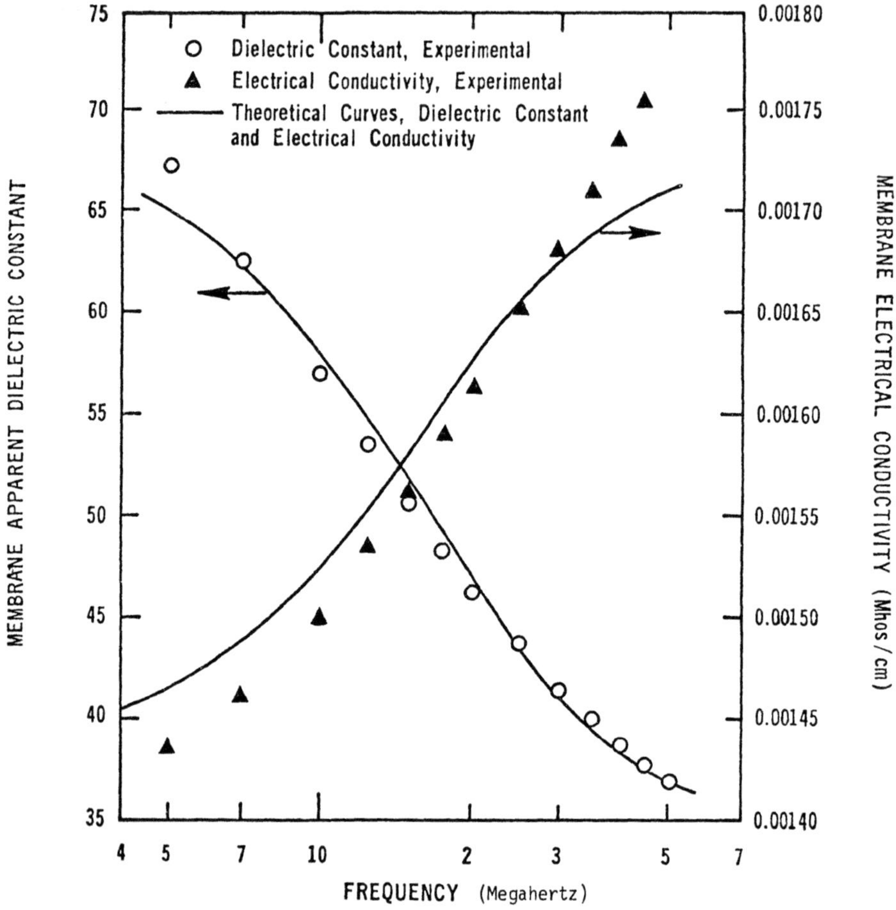

Fig. 15. Frequency-dispersion curves of anion-exchange membrane [16]. – Membrane AMT (Asahi Glass Co., Tokyo) equilibrated with distilled water. Optimized parameters: $a = 0.168$; $b = 0.827$; $c = 1\text{–}a\text{–}b = 0.005$; $d = 0.678$; $\varepsilon_r = 32$; $k_r = 0.001\,74\ \Omega^{-1}\ \mathrm{cm}^{-1}$; $k_w = 10^{-6}\ \Omega^{-1}\ \mathrm{cm}^{-1}$.

monovalent cations [20]; in addition to the heterogeneities mentioned before in this paragraph, the membrane might have a surface film of somewhat different composition than the bulk.

Acknowledgments

The authors thank the National Science Foundation, Grant No. GK-20372, the Office of Saline Water, U.S. Department of the Interior and the Sea Water Conversion Project of the State of California for their support of this work, and Mrs C. Tung for her competent editorial help in the production of this paper.

References

1. Maxwell, J. C.: *A Treatise on Electricity and Magnetism*, 3rd ed. (1891), reprinted by Dover Publications, New York, 1954.
2. Meredith, R. E. and Tobias, C. W.: in C. W. Tobias (ed.), *Advances in Electrochemical Engineering*, vol. 2, Interscience, New York, 1962, Chapter 2.
3. Wyllie, M. R. J. and Southwick, P. F.: *J. Petroleum Technol.* **6**, 44 (1954).
4. Sauer, M. C., Southwick, P. F., Spiegler, K. S., and Wyllie, M. R. J.: *Ind. Eng. Chem.* **47**, 2187 (1955).
5. Spiegler, K. S., Yoest, R. L., and Wyllie, M. R. J.: *Disc. Farad. Soc.* **21**, 174 (1956).
6. Sachs, S. B. and Spiegler, K. S.: *J. Phys. Chem.* **68**, 1214 (1964).
7. Sachs, S. B., Katchalsky, A., and Spiegler, K. S.: *Electrochim. Acta* **15**, 693 (1970).
8. Mandel, M.: *Bull. Soc. Chim. Belg.* **64**, 442 (1955).
9. LePage, W. R.: *Analysis of Alternating-Current Circuits*, McGraw-Hill, New York, 1952.
10. Hasted, J. B., Ritson, D. M., and Collie, C. H.: *J. Chem. Phys.* **16**, 1 (1948).
11. Arulanandan, K. and Mitra, S.: *Proceedings of the 4th Asilomar Conference on Circuits and Systems*, 1970, p. 480.
12. Smith, S. S.: Ph.D. thesis, Department of Civil Engineering, University of California, Davis, Calif., 1971.
13. Mandel, M. and Jenard, A.: *Trans. Faraday Soc.* **59**, 2158 (1963).
14. O'Konski, C. T. and Shirai, M.: in Conway B. E. and Barradas B. G. (eds.), *Chemical Physics of Solid Solutions*, J. Wiley, New York, 1966, p. 391.
15. Sachs, S. B., Raziel, A., Eisenberg, H., and Katchalsky, A.: *Trans. Faraday Soc.* **65**, 77 (1969).
16. Arulanandan, K., Linkart, T. A. *et al.*, unpublished results.
17. Spiegler, K. S. and coll.: 'Study of Membrane-Solution Interfaces by Electrochemical Methods', Research and Development Progress Report No. 353, Office of Saline Water, U.S. Department of the Interior, U.S. Government Printing Office, Washington, D.C. (1968), pp. 72–85 describing work by K. Arulanandan, A. C. Eisenberg and D. Q. Fletcher.
18. Block, M.: 'Applications of Ion-Exchange Membranes', Ph.D. Thesis, Imperial College of Science and Technology, London, 1964.
19. Shaffer, L. H. and Mintz, M. S.: in K. S. Spiegler (ed.), *Principles of Desalination*, Academic Press, New York, 1966, Chapter 6, p. 210.
20. *Selemion Ion Exchange Membranes*, Asahi Glass Co., Ltd., Tokyo, 1965.

ULTRASONIC ABSORPTION AND DENSITY STUDIES OF COUNTER-ION SITE BINDING IN AQUEOUS SOLUTIONS OF POLYELECTROLYTES

C. TONDRE and R. ZANA

Centre de Recherches sur les Macromolécules, C.N.R.S., 6, rue Boussingault, 67083, Strasbourg, Cedex, France

Abstract. The interest of ultrasonic absorption methods for the study of problems specific to polyelectrolyte solutions is pointed out. It is shown that the conjunction of ultrasonic absorption and density measurements should provide quantitative informations (1) on the distribution of site bound counter-ions between those bound without and with dehydration and (2) on the exchange rates between the various types of ions. On the basis of previous ultrasonic absorption measurements a model for counter-ion site binding is presented. Ultrasonic relaxation spectra (absorption vs frequency curves) of a series of polyphosphate solutions are then given. They indicate that two relaxation processes are present. The low frequency relaxation process seems to be essentially dependent on the counter-ion while the high frequency relaxation process seems to be associated with the polyion. In view of these results a model is presented for the kinetics of counter-ion site binding. This model rests primarily (1) on the similarity between counter-ion site binding by polyions and ion-pair formation in simple electrolytic solutions and (2) on Manning's theory for counter-ion condensation in polyelectrolyte solutions. It assigns the ultrasonic absorption found in polysalt solutions to the perturbation by the ultrasonic waves of the equilibria between various states of hydration of the complex formed by the bound ion and that part of the polyion involved in the binding. From this model are derived equations relating the relaxation times and relaxation amplitudes to the rate constants of the equilibria between site bound counter-ions, to the volume changes associated with these equilibria and to the concentrations of the different types of counter-ions. These equations can be shown to account qualitatively for all of the experimental results. However, their use for the determination of the various parameters characterizing site binding equilibria in cobalt polyphosphate solutions appears to be limited by the large number of these parameters and the limited ultrasonic frequency range investigated in this work. To overcome the first limitation extensive density measurements have been performed on polyelectrolyte solutions. Such measurements yield the total volume change associated with site binding which is used in the analysis of the ultrasonic results.

1. Introduction

It is now well established and accepted that in polyelectrolyte solutions part of the counter-ions are bound by the polyions. One usually distinguishes between 'ionic atmosphere binding' of counter-ions and 'site' or 'specific binding' of counter-ions [1]. The former is a loose type of binding; the counter-ions are trapped in the strong electrostatic field in the vicinity of the polyions and their mobility is decreased; the hydration shells of the counterions and of the charged sites on the polyion are not in contact and, therefore, remain unaffected. On the contrary, in the latter type of binding the hydration shells of the site bound counter-ion and of the charged site are in contact or overlapp. In this last case, a positive volume change occurs upon site binding, owing to the release of electrostricted water molecules from the hydration shells. The very existence of site binding has long been challenged although indirect evidences have been accumulated [1]. More recently, direct evidences have been obtained by

Eric Sélégny (ed.), Polyelectrolytes, 323–338. *All Rights Reserved.*

means of dilatometric [2] and refractometric [3] techniques, which are sensitive to the volume change associated with site binding.

From what has been said above it would appear as if counter-ion site binding were a process similar to ion-pair formation in simple ionic solutions. In this case also, ion-pairs include both 'solvent-separated' ion-pairs and 'tight' or 'contact' ion-pairs. However, in contradistinction to the situation which prevails for simple ionic solutions a survey of the literature [1] shows that relatively little is known about site binding in polyelectrolyte solutions and in particular about the repartition of site bound counter-ions between those bound with and without dehydration. The reason for this is that most physicochemical methods (diffusion, osmometry, conductivity...) cannot discriminate very well, if not at all, even between atmospheric binding and site binding. Another quite important problem yet unsolved concerns the rate of exchange between site bound counter-ions and 'free' counter-ions. This question has some implications in regard to problems involving ion exchange as well as transport of ions across membranes, as pointed out by Eigen *et al.* [4a, 4c]. The attempts previously made [5] to study this problem used diffusion techniques. Bound counter-ions were found to have lifetimes of the order of several minutes [5]. These results however have been shown to be incorrect both on theoretical [6] and experimental grounds [7].

This state of the question led us to undertake an extensive study of the ultrasonic absorption of polyelectrolyte solutions in the hope of providing an answer to the questions raised above. Ultrasonic absorption methods have proved to be most useful for the study of the kinetics of ion-pair formation in simple ionic solutions [4b, 8] and should also provide informations on counter-ion site binding, given the similarity between these two phenomena. In the next part the theoretical basis of ultrasonic absorption are presented with special emphasis on the use of this method for the study of site binding in polyelectrolyte solutions. Previously published ultrasonic studies of polyelectrolyte solutions are briefly reviewed in Section 3. Section 4 contains recent results obtained on polyphosphates. These results are discussed and a model is presented for the kinetics of site binding. The interest of the determination of the apparent molal volumes of polyelectrolytes for the quantitative analysis of the ultrasonic results (which has been pointed out in Section 2) is again shown. The results of measurements of apparent molal volumes of polyelectrolytes are presented and briefly discussed in Section 5. In the last paragraph are given the expressions of the relaxation times and relaxation amplitudes derived from the model presented in Section 4. These equations can be shown to account qualitatively for the observed results. They also reveal that both the accuracy on the ultrasonic results as well as the frequency range investigated in this work are not sufficient to permit a quantitative determination of all of the unknown parameters characterizing site binding.

2. Theoretical Basis for the Study of Site Binding by Means of Ultrasonic Absorption Techniques

A bound counter-ion and the part of the polyion on which it is bound may be looked at as contituting a 'complex ion'. On the other hand the different states of hydration of this complex ion may be considered as in chemical equilibrium. When ultrasonic waves are propagated through such a system, the chemical equilibria are perturbed by the pressure changes due to the wave, provided that volume changes are associated with these equilibria [4b, 8]. Energy transfers occur between the system and the waves. These transfers will follow the pressure changes with a certain time lag related to the rate constants of the equilibria. This results in the loss of part of the ultrasonic energy, i.e., in an absorption of the sound wave. If, for the sake of simplicity, one assumes to have only one equilibrium, the ultrasonic absorption coefficient α for this process is then given [4b, 8] at the frequency N by:

$$\frac{\alpha}{N^2} = \frac{A}{1 + 4\pi^2 N^2 \tau^2}, \tag{1}$$

where $\tau = (2\pi N_R)^{-1}$ is the relaxation time characterizing the equilibrium and N_R the relaxation frequency. The relaxation time characterizes the ability of the system to follow the perturbation. The constant A is proportional to $\tau \cdot \Delta V_0^2$ where ΔV_0 is the volume change associated with the reaction under study. The dependence of the ultrasonic absorption on ΔV_0^2 makes this technique much more sensitive to site binding than refractometry [3] or dilatometry [2] which depend only on ΔV_0. Moreover, it is clear that for 'ionic atmosphere binding' no absorption should occur since $\Delta V_0 = 0$. Therefore ultrasonic absorption will be sensitive only to site binding and thus appears as a very selective technique.

On the other hand τ and N_R contain the rate constants of the equilibrium under study and, for bimolecular reactions, the concentrations of the species involved in the reaction [4b, 8]. Had counter-ion site binding been a one step process the determination of τ would have provided a direct estimation of the lifetime of the bound counter-ions and, therefore, of the exchange rate between bound and free counter-ions. On the other hand the study of A as a function of concentration would have permitted to obtain informations on the distribution of bound counterions between those bound with and without dehydration. Unfortunately, as will be shown in Section 4, site binding is a multistep process involving at least two equilibria. All of the unknown quantities involved in such a process (four rate constants, two volume changes and the concentrations of the species) cannot be obtained from ultrasonic absorption data alone. Independent measurements become necessary. For this purpose we have measured the density d of the polyelectrolyte solutions from which can be obtained the apparent molal volume $\bar{V}_{CP}$ of the polyelectrolyte CP (C counterion, P polyion) according to:

$$\bar{V}_{CP} = \frac{M_{CP}}{d_0} - \frac{d - d_0}{d_0 c}, \tag{2}$$

where M_{CP} is the molecular weight per equivalent of monomer, c the concentration in equivalent of monomer per cm^3 and d_0 the density of water. From the extrapolation of the plots of $\bar{V}_{CP}$ vs. c one can obtain the apparent molal volume at zero concentration, $\bar{V}^0_{CP}$. As will be seen below the values of $\bar{V}^0_{CP}$ permit to obtain the total volume change associated with the binding.

3. Brief Review of Previous Ultrasonic Absorption Studies in Relation to Counter-ion Site Binding by Polyions

The first experiments which gave a direct ultrasonic evidence of the contribution of site binding to the absorption of polyelectrolyte solutions have been performed on carboxymethylcellulose [9] (CMC) and polyethylenesulfonic acid (PESA) [10a]. Additions of alkali chlorides to a solution of tetramethylammonium salt of CMC (TMA-CMC) were found to give rise to large increases of ultrasonic absorption while additions of TMA-Cl brought about only negligible changes [9]. On the other hand, the neutralization of PESA by NaOH was found to result in a large increase of absorption while the neutralization by TMA-OH had no effect on the absorption [10a].

These results led us to use TMA^+ ion as a reference ion, i.e. an ion for which there is no ultrasonic absorption caused by site binding [9, 10] or by complex formation [4c]. In the following the contribution of site binding to the absorption of polyelectrolyte solutions has always been obtained by taking the difference between the absorptions of two equimolecular solutions of the same polyelectrolyte: one with the counter-ion C, the other with the counterion TMA^+. It may be safely assumed that this procedure eliminates all of the contributions to the ultrasonic absorption other than that due to site binding [9, 10]. As will be seen in Section 4, it is very difficult to draw any conclusion about the kinetics of site binding if the absorption due to this process is not separated from that due to other equilibria.

Following the above experiments, extensive measurements were undertaken on a number of polyacids fully neutralized by alkali metal and TMA hydroxydes [9, 10]. These studies showed that the magnitude of the absorption due to site binding is strongly dependent on the nature of both the counter-ion and the polyion. Ionic sequences were obtained by writing the alkali metal ions in the order of increasing ultrasonic effect. It was thus found that the ion giving rise to the largest ultrasonic effect depends on the polyion [10b]. The charge parameter ζ [11] which permits to explain certain properties of polyelectrolyte solutions cannot be used to explain these results because polyelectrolytes characterized by the same value of ζ have been found to give rise to site binding absorptions differing by a factor as large as 10 [10b]. All of these results have been explained in terms of a model which postulates two types of site binding according to the respective values of the actual distance d between charged sites and of the diameter D of the counterion [10b]. When D is either too small or too large compared to d ion-pairs are formed. This process is accompanied by a relatively small volume change giving rise to rather small ultrasonic effects. On the other hand, when D is comparable to d, the ion fits into the cavity between two adjacent sites. The binding is then more comparable to a chelation. It is accompanied by a large volume change from which arise large ultrasonic effects.

The results relative to an alternate copolymer of maleic acid and methylvinylether

(P(MA-MVE)) give evidence of the two types of site binding postulated above. This copolymer is a polydicarboxylic acid characterized by two well separated pK_a's [12]. At neutralization degree θ below 0.5 only one carboxylic group per monomer is ionized and d is large. At $\theta > 0.5$ the number of fully ionized monomeric units increases linearly with θ; for these units d is small and chelation may occur with small ions. The excess absorption of P(MA-MVE) neutralized by LiOH or NaOH was found to be very close to zero up to $\theta = 0.5$ and to increase almost linearly with $\theta > 0.5$ [10b].

The above results show the kind of qualitative informations that ultrasonic absorption can provide on site binding. The next paragraph deals with the effect of frequency on the excess absorption of polyelectrolyte solutions in order to obtain informations on the kinetics of counter-ion site binding.

4. Ultrasonic Relaxation Spectra of Polyelectrolyte Solutions

Figures 1, 2 and 3 show the relaxation spectra of a series of polyphosphate (PP) solutions. Our choice in studying this polyelectrolyte was guided by the fact that preliminary studies [10b] showed that it presents large ultrasonic effects and also because polyphosphates have been extensively studied by Strauss *et al.* [1]. All of the curves on Figures 1 to 3 could be fitted by the equation

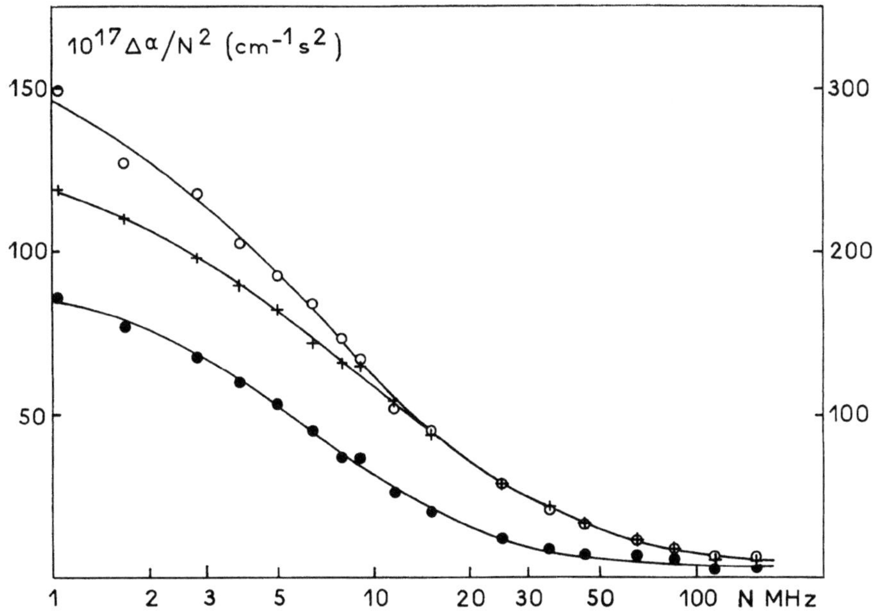

Fig. 1. Ultrasonic relaxation spectra of sodium Polyphosphate at 25°. $c_{Na\text{-}PP} = 0.033N$ (●), 0.0617N (○) and 0.107N (+). The solid lines represent the curves obeying Equation (3) and fitting the best with the experimental results (●, ○ and +).

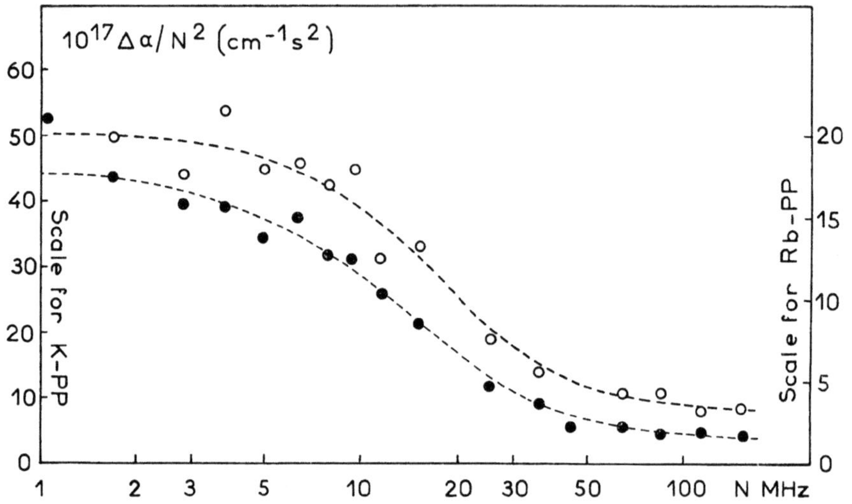

Fig. 2. Ultrasonic relaxation spectra of potassium (●) and rubidium (○) polyphosphates at 25°. The dotted lines represent the theoretical curves fitting the best with the experimental values. $c_{\text{K-PP}} = 0.117\,\text{N}$ $c_{\text{Rb-PP}} = 0.11\,\text{N}$.

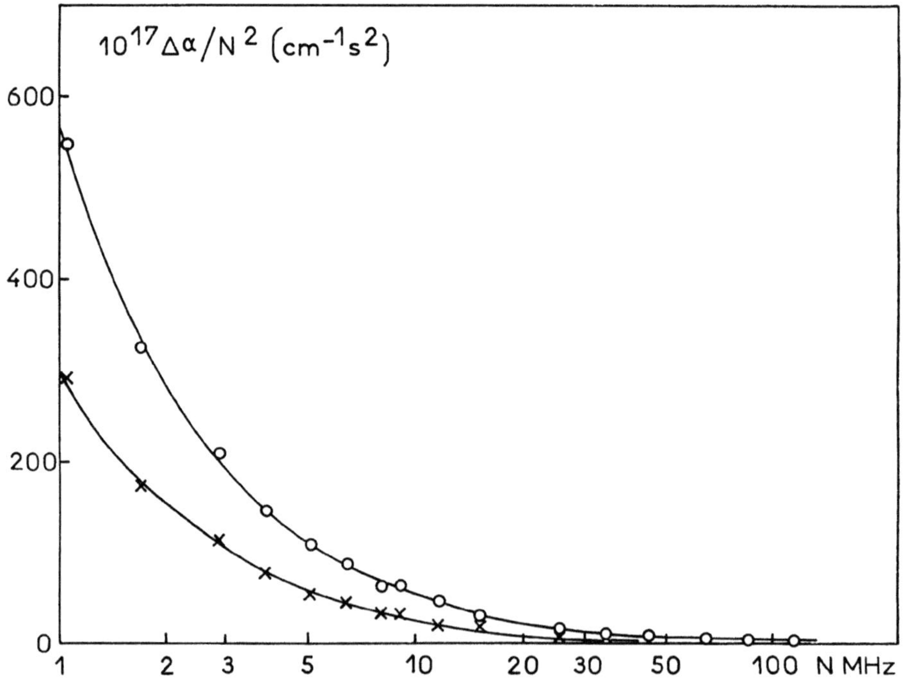

Fig. 3. Ultrasonic relaxation spectra of cobalt polyphosphate at 25°. $c_{\text{Co-PP}} = 0.125\,\text{N}$ (○) and 0.068 N (×). The solid lines represent the curves obeying Equation (3) and fitting the best with the experimental results (○ and ×).

$$\frac{\Delta\alpha}{N^2} = \frac{A_1}{1 + N^2/N_1^2} + \frac{A_2}{1 + N^2/N_2^2} + B, \tag{3}$$

where

$$\frac{\Delta\alpha}{N^2} = \left(\frac{\alpha}{N^2}\right)_{\text{C-PP}} - \left(\frac{\alpha}{N^2}\right)_{\text{TMA-PP}}. \tag{4}$$

The values of the relaxation amplitudes A_1 and A_2 and of the relaxation frequencies N_1 and N_2 are given in Table I together with the values of B. Also given in Table I are some previously reported [10a] results on sodium polyelethylenesulfonate (Na-PESA). For Co-PP 0.125 N a series of values are given which all fit the results within the experimental accuracy, thereby indicating the precision on the relaxation parameters A_1, A_2, N_1 and N_2.

TABLE I

Values of the relaxation parameters

Polysalt	Concentration equiv. l^{-1}	10^{17} A_2 cm^{-1} s^2	N_2 MHz	10^{17} A_1 cm^{-1} s^2	N_1 MHz	10^{17} B cm^{-1} s^2
Li-PP	0.115	10	5	30	18	8
NA-PP	0.033	53	4	32	14	3
,,	0.062	87	3.7	60	19	5.5
,,	0.107	120	4.5	105	22	10
K-PP	0.117	10	6.5	30	16	4
Rb-PP	0.110	–	–	17	17	3
Mn-PP	0.07	620	1.4	30	12	6
Co-PP	0.068	500	1.05	44	11	1
Co-PP	0.125	700	1.4	70	10	2
	,,	950	1.05	90	10	2
	,,	1500	0.72	100	10	2
	,,	1000	1.05	75	12	2
Na-PESA	0.095	22	5.5	12	45	2.5
,,	0.19	40	6	18	40	6
,,	0.36	75	6	42	50	9

Table I indicates that for Na-PP and Co-PP both N_1 and N_2 are, within the experimental accuracy, independent of the polyelectrolyte concentration c in the limited range of concentration investigated in this work. The relaxation amplitudes A_1 and A_2 appear to be almost proportional to c. The results for Na-PESA given in Table I show exactly the same features: relaxation frequencies independent of c and relaxation amplitudes practically proportional to c. It must be remembered that for CMC solutions the absorption due to site binding has also been found to vary almost linearly with concentration in the range 0.03–0.2 eq./1 [9].

The above results are to be compared with those of Atkinson *et al.* [13] on polyacrylic acid and CMC: the excess ultrasonic absorption was also found to be proportional to c, but, in contradistinction to our results, a distribution of relaxation times had to be used in order to fit the absorption data for monovalent and divalent counterions [13]. However in this work [13] the excess absorption was taken as $(\alpha/N^2)_{\text{polyelectrolyte}}$ minus $(\alpha/N^2)_{H_2O}$. Therefore this excess absorption includes that due to site binding as well as that associated to all of the other processes occurring in polycarboxylic acid solutions [14]. The relaxation times for these processes are generally different from those relative to site binding. This is believed to be the reason for the need of a distribution of relaxation times in order to fit the data. A confirmation of this reasoning is found in the results relative to divalent ions. There the ultrasonic effects due to site binding are much larger than the effects due to the other processes [14]; in this case the data of Atkinson *et al.* [13] show two bands of relaxation times and are therefore comparable to ours.

The results of Table I show that for polyphosphates the low frequency relaxation process has an amplitude very sensitive to the nature of the counterion (~ 0 for Rb^+ and 950×10^{-17} cm^{-1} s^2 for Co^{++}). Also, Table I shows that at a given concentration the ratio of the values of A_2 for Na-PP and Na-PESA is larger than 5 while the ratio of the values of N_2 is only 0.7. As will be seen in Section 5, the difference between the value of A_2 is due to a difference between the volume changes associated with the binding of Na^+ to PP and PESA. On the other hand, Table I shows that for polyphosphates A_1 depends only little on the nature of the counterion as compared to A_2. Moreover, the extreme values of N_1 are within a factor 2 for the series of counterions listed in Table I. Finally, the comparison of Na-PP and Na-PESA shows that N_1 depends on the polyion. All of these results are quite similar to those reported for the ultrasonic absorption associated with the formation of ion-pairs in divalent sulfate solutions [15]. For polyelectrolytes, however, the dependence of N_2 on the nature of the counter-ion is not as pronounced as for divalent sulfates. Table I shows that N_2 is equal to 1.05 and 1.4 MHz for CoPP and MnPP respectively while the values 0.56 and 4.6 MHz have been found for $CoSO_4$ and $MnSO_4$ respectively [15]. Our results are to be compared with those of Atkinson *et al.* [13]. These authors reported that in calcium and magnesium polyacrylates the low frequency relaxation process is characterized by values of N_2 practically equal while calcium and magnesium acetates show very different low relaxation frequencies. At the present time no explanation can be given to these results which appear to be specific to polyelectrolytes in the presence of divalent counter-ions and which may indicate that as far as kinetics go, counter-ion site binding by polyions is not completely identical to ion-pair formation in electrolytic solutions . This possibility is examined in Section 6.

The above results permit to present a model for the kinetics of counter-ion site binding. This model rests primarily (1) on the similarity which has been found for the frequency dependence of the absorptions due to site binding and to ion-pair formation and (2) on Manning's theory [16] for counter-ion condensation in polyelectrolyte solutions. In this model, site binding between a counter-ion C and a part P of the polyion writes:

$$\mathrm{P} + \mathrm{C} \rightarrow \mathrm{PC}_1 \underset{k_2}{\overset{k_1}{\rightleftharpoons}} \mathrm{PC}_2 \underset{k_4}{\overset{k_3}{\rightleftharpoons}} \mathrm{PC}_3 . \qquad (5)$$

As for ion pair formation [15], PC_1, PC_2 and PC_3 represent three states of hydration of the counterion-polyion complex with the volume changes ΔV_{21} and ΔV_{32} associated respectively to reaction $PC_1 \rightarrow PC_2$ and $PC_2 \rightarrow PC_3$. However, in this model, the formation of PC_1 from P and C is not considered as an equilibrium obeying the mass action law because Manning [16] has shown that even at infinite dilution a certain number of charged sites are neutralized by condensed counter-ions. It will be assumed that only those condensed counter-ions are active as far as ultrasonic absorption is concerned, while the free counter-ions do not take part to the relaxation processes. Free and bound counter-ions may however exchange very rapidly. As is shown in Section 6 the model represented by Equation (5) is characterized by two concentration independent relaxation frequencies and two relaxation amplitudes proportional to c.

As said in the introduction and in Section 2 ultrasonic data alone (N_1, N_2, A_1 and A_2) do not permit to obtain all of the unknown quantities associated with the process represented by Equation (5) (the rate constants k_1, k_2 and k_4, the volume changes ΔV_{21} and ΔV_{32} and the concentration $[PC_1]$, $[PC_2]$ and $[PC_3]$). Independent measurements become necessary. For this purpose apparent molal volumes of polyelectrolytes have been determined.

5. Apparent Molal Volumes of Polyelectrolytes and Their Use for the Determination of the Volume Change Upon Counter-ion Site Binding

Table II gives the values of the apparent molal volumes at infinite dilution $\bar{V}^0_{CP}$ (see Section 2) for a series of polyelectrolytes. Using the values of the apparent molal volumes of the counter-ions [17] $\bar{V}^0_C$, which are also listed in Table II, one can obtain the apparent molal volume of the polyion, $(\bar{V}^0_P)_C$, by making use of the equation

$$\bar{V}^0_{CP} = \bar{V}^0_C + (\bar{V}^0_P)_C . \tag{6}$$

Equation (6) has been verified to a high degree of accuracy for simple electrolytes. It assumes a complete ionization of the salt at infinite dilution. An examination of Table II reveals that for each of the polyelectrolytes studied in this work the values of $(\bar{V}^0_P)_C$ depend on the counter-ion. This result means that the additivity law does not hold for polyelectrolyte solutions. A simple explanation for this behaviour is provided by Manning's theory which states that even at infinite dilution part of the counter-ions remain condensed on the polyions. On the other hand, dilatometry [2], refractometry [3] and ultrasonic absorption [9, 10] have shown that part of the condensed counter-ions and of the sites on which they are bound are dehydrated. Therefore, the value $(\bar{V}^0_P)_C$ of the apparent molal volume of the polyion P, as obtained from Equation (6) will include the volume change δV_{CP} associated with the binding of C. This is why the subscript C was placed on the apparent molal volume of the polyion P in Equation (6). If we call $(\bar{V}^0_P)_{true}$, the true apparent molal volume of the polyion P it may be assumed that $(\bar{V}^0_P)_{true} \# (\bar{V}^0_P)_{TMA}$ since there are numerous evidences [9, 10, 18, 4c] that a negligible volume change is associated with the condensation of TMA^+ ion (this ion

TABLE II

Apparent molal volumes of polyelectrolytes and volume changes upon counter-ion binding

Polyelectrolyte	Alternate copolymer maleic acid-methyl vinyl ether			Polyacrylic acid			Polyethylene-sulfonic acid			Polyphosphate			
	$\bar{V}^{\circ}_{CP}$	$(\bar{V}^{\circ}_{P})_{C}$	δV_{CP}	$\bar{V}^{\circ}_{CP}$	$(\bar{V}^{\circ}_{P})_{C}$	δV_{CP}	$\bar{V}_{CP}$	$(\bar{V}^{\circ}_{P})_{C}$	δV_{CP}	$\bar{V}_{CP}$	$(\bar{V}^{\circ}_{P})_{C}$	δV_{CP}	$\bar{V}^{\circ}_{C}$
Acid form	56.3	61.7	25.3	46.7	52.1	23	44.9	50.3	2.2	–	–	–	– 5.4
Li salt	36.8	43.1	6.7	28.1	34.4	5.3	43.8	50.1	2.0	17.6	23.9	8.0	– 6.3
Na salt	36.3	42.9	6.5	28.8	35.4	6.3	45.4	52	3.9	18.8	25.4	9.5	– 6.6
K salt	47.1	43.5	7.1	38.2	34.6	5.5	54.6	51	2.9	28.4	24.8	8.9	3.6
Rb salt	49.5	40.8	4.4	41.4	32.7	3.6	59.3	50.6	2.5	30.3	21.6	5.7	8.7
Cs salt	54.5	38.5	2.1	47.6	31.6	2.5	65.1	49.1	1.0	–	–	–	15.9
TMA salt	120.6	36.4	0	113.3	29.1	0	132.3	48.1	0	100.1	15.9	0	84.2
Co salt	–	–	–	–	–	–	–	–	–	19.3	36.7	20.8	–17.4

(1) $\bar{V}^{\circ}_{CP}$, $(\bar{V}^{\circ}_{P})_{C}$, $\bar{V}^{\circ}_{C}$ and δV_{CP} are expressed in cm^3 $equiv.^{-1}$.
(2) $(\bar{V}^{\circ}_{P})_{C} = \bar{V}^{\circ}_{PC} - \bar{V}^{\circ}_{C}$.
(3) $\delta V_{CP} = (\bar{V}^{\circ}_{P})_{C} - (\bar{V}^{\circ}_{P})_{TMA}$.

is known to have a negligible electrostrictive action on the surrounding water molecules [18, 4c]). Therefore, Equation (6) yields

$$(\bar{V}_P^0)_C - (\bar{V}_P^0)_{TMA} \# \delta V_{CP}. \tag{7}$$

The values of δV_{CP} are listed in Table II. The comparison of the results relative to Na-PP and Na-PESA shows that the values of δV_{CP} are in the ratio 2.4. To a first approximation, $\Delta\alpha/N^2$ may be taken as proportional to δV_{CP}^2, even for the multistep process found for site binding. One should therefore expect to find with Na-PP an absorption 6 times larger than with Na-PESA, in good agreement with the experimental results which yield a factor 5 (see Table I).

The relationship between δV_{CP}, ΔV_{21} and ΔV_{32} writes

$$\delta V_{CP} = \Delta V_{21} \frac{[PC_2] + [PC_3]}{c} + \Delta V_{32} \frac{[PC_3]}{c}. \tag{8}$$

Measurements of apparent molal volumes have also been performed on polymethacrylic acid, polystyrenesulfonic acid and carboxymethylcellulose. The results are reported elsewhere [19] together with a method for the evaluation of the contribution of electrostriction to the apparent molal volumes of polyions. Nuclear magnetic resonance measurements together with density measurements have permitted us to show [20] that for CoPP the electrostriction per site of polyion is equal to that of a Co^{++} equivalent and is about 20 cm^3 equiv.$^{-1}$.

6. Basis for the Quantitative Analysis of the Ultrasonic Relaxation Spectra of Polyphosphates

The model on which this analysis is based is represented by Equation (5). The fraction ϱ of condensed counter-ions can be calculated from Manning's work [16]. If c is the total concentration of sites (and also of counter-ions since we used fully neutralized polyphosphates) we then have

$$\varrho c = [PC_1] + [PC_2] + [PC_3]. \tag{9}$$

From Equation (9) the following expressions of the concentrations $[PC_1]$, $[PC_2]$ and $[PC_3]$ can be derived:

$$[PC_1] = \varrho c \frac{k_2 k_4}{T}; \quad [PC_2] = \varrho c \frac{k_1 k_4}{T}; \quad [PC_3] = \varrho c \frac{k_1 k_3}{T} \tag{10}$$

with

$$T = k_1 (k_3 + k_4) + k_2 k_4. \tag{11}$$

On the other hand, within the framework developped by Eigen [4b] and others [21] one can derive the expressions for the relaxation times τ_1 and τ_2 and for the relaxation amplitudes A_1 and A_2 for the two coupled equilibria given by Equation (5).

We thus found:

$$\tau_{1,2}^{-1} = \frac{S}{2}\left[1 \pm \left(1 - \frac{4T}{S^2}\right)^{1/2}\right] = \frac{1}{2\pi N_{R1,2}}, \tag{12}$$

where the signs + and − refer to subscripts 1 and 2 respectively; S, sum of the reciprocals of the relaxation times, is given by

$$S = k_1 + k_2 + k_3 + k_4 = \frac{1}{\tau_1} + \frac{1}{\tau_2} \tag{13}$$

T, given by Equation (11), can be shown to be equal to the product of the reciprocals of the relaxation times.

The relaxation amplitudes can be expressed as follows, using Equations (8) and (10) to (13):

$$A_1 = 1.18 \times 10^{-7}\, \varrho c \left(\frac{1}{\tau_2}\right) \frac{\left(\dfrac{\delta V_{CP}}{\varrho} - \Delta V_{21} \dfrac{k_1}{1/\tau_2}\right)^2 [T - k_1(S - \Sigma)]}{k_1\left[\dfrac{\Sigma}{\tau_1^2} + T\left(\dfrac{1}{\tau_2} - \dfrac{1}{\tau_1} - \Sigma\right)\right]} \tag{14}$$

$$A_2 = 1.18 \times 10^{-7}\, \varrho c \left(\frac{1}{\tau_1}\right) \frac{\left(\dfrac{\delta V_{CP}}{\varrho} - \Delta V_{21} \dfrac{k_1}{1/\tau_1}\right)^2 [T - k_1(S - \Sigma)]}{k_1\left[\dfrac{\Sigma}{\tau_2^2} + T\left(\dfrac{1}{\tau_1} - \dfrac{1}{\tau_2} - \Sigma\right)\right]} \tag{15}$$

with

$$\Sigma = k_1 + k_2. \tag{16}$$

In Equations (14) to (16), the rate constants are in s^{-1}, the volume changes in cm^3 $mole^{-1}$, c in equiv. l^{-1}, the A's in $cm^{-1}\ s^2$ and the constant has been calculated for the temperature 25°.

In Equations (14) and (15) three quantities are unknown: k_1, Σ and ΔV_{21}. A third equation, which would correspond to an additional result on the system under study, is therefore required to solve Equations (14) and (15). This additional result may be either a rate constant, or a fraction of any type of bound counterions, or any volume change.

By using results obtained from NMR studies, it is possible to determine all of the unknown quantities associated with the binding of Co^{2+} by polyphosphate (0.125 N Co-PP solution). Cobalt polyphosphate was selected because of the large excess ultrasonic absorption found with this polyelectrolyte (see Table I) and also because NMR data are available both on Co^{2+} [22] and on cobalt polyphosphate [23]. There are two possible methods of calculation. In both methods counter-ion site binding is assumed to be completely analogous to ion-pair formation. In state PC_1 the hydration shells of P and C are in contact but do not overlap. In state PC_2 the polyion site has lost part of its hydration shell while that of the counterion remains unaffected. In state PC_3 the hydration shells of both C and P are modified. The removal of water

TABLE III

Results of the calculations

	$10^{17}A_1$ (cm^{-1} s^2)	$10^{17}A_2$	N_1 (MHz)	N_2	10^7k_1 (s^{-1})	10^7k_2	10^7k_3	10^7k_4	ΔV_{21} (cm^3 mole^{-1})	ΔV_{32}	$\frac{[PC_1]}{\varrho c}$	$\frac{[PC_2]}{\varrho c}$	$\frac{[PC_3]}{\varrho c}$
First method	70	700	10	1.4	6.2	0.05_5	0.24	0.64	23	4.6	0.006	0.72	0.27
	90	950	10	1.05	6.2	0.10	0.24	0.42	22.9	4.4	0.01	0.63	0.36
	100	1500	10	0.72	6.1	0.14_5	0.24	0.21	22.3	4.2	0.01	0.47	0.52
	75	1000	12	1.05	7.4	0.11	0.24	0.42	22.8	4.3	0.01	0.63	0.36
Second method	70	700	10	1.4	6.2	0.1	0.52	0.37	22.3	3.4	0.006	0.41	0.58
	90	950	10	1.05	6.1	0.17	0.39	0.28	21.9	4.3	0.01	0.41	0.58
	100	1500	10	0.72	6.1	0.18	0.27	0.19	21.8	4.4	0.01	0.41	0.58
	75	1000	12	1.05	7.4	0.15	0.39	0.27	22.1	3.6	0.01	0.41	0.58

molecules from the hydration shell of the counterions is assumed to be responsible of the low frequency relaxation process, as indicated by the experimental results described in Section 4.

First method. In the case of Co^{2+}, NMR measurements [22] have permitted to obtain the value of the rate constant k_{H_2O} for the exchange of a water molecule between the inner hydration sheath of this ion and the bulk water. The most reliable value of k_{H_2O} is $2.4 \times 10^6 s^{-1}$ [22]. Eigen *et al.* [4a, 4c] have shown that k_{H_2O} depends only little on the nature of the ligand which replaces the water molecule in the inner hydration sheath, in the case of small ligands. This result will be assumed to hold true when the ligand is part of a polyion, as in the present case. Also it is generally assumed [15] that for a system such as that represented by Equation (5) one has $k_{H_2O} \# k_3$.

On the other hand by combining Equations (11), (13) and (16) it can be readily shown that:

$$k_1 = [\Sigma(\Sigma + k_3 - S) + T]/k_3. \tag{17}$$

Equations (14), (15) and (17) have been solved with the aid of a minicomputer and the solutions are given in Table III, for the four sets of values of relaxation parameters fitting the results relative to Co-PP 0.125 N. The ΔV's appear to be relatively insensitive to the values of the relaxation parameters.

Second method. NMR measurements of water proton chemical shifts upon addition of Co^{2+} to TMA-PP solutions have been recently reported [23]. In these experiments it was observed that up to a value of about 0.5 for the ratio $r = [Co^{2+}]/[\text{TMA-PP}]$, all of the Co^{2+} ions introduced into the polyelectrolyte solution lose all of their hydration water. On the contrary, at $r > 0.5$, the added cobalt ions loose only little of their hydration water. To solve Equations (14) and (15) within the framework of the kinetic model represented by Equation (5) we have assumed that the cobalt ions bound with complete dehydration correspond to those in state PC_3 in Equation (5). Therefore, using Equation (10), one can write:

$$r = \frac{[PC_3]}{c} = \varrho \frac{k_1 k_3}{T}. \tag{18}$$

Combining Equations (11), (13), (16) and (18) one obtains:

$$k_1 = T\Sigma r / \{[\Sigma(S - \Sigma) - T]\,\varrho + \text{Tr}\}. \tag{19}$$

Equations (14), (15) and (19) have been solved using a minicomputer, in the same manner as above.

The results of the calculations are given in Table III. They do not differ much from those obtained using the first method of calculation, mainly for the values of ΔV_{32} and ΔV_{21}. It must be pointed out that the third set of relaxation parameters yield values of the rate constants, volume changes and fractions of bound counter-ions which are practically identical with both methods of calculation. This is thought to give a strong support to the kinetic model presented in this work.

The values of ΔV_{32} and ΔV_{21} demand some additional remarks. In the model represented by Equation (5), ΔV_{32} and ΔV_{21} must be considered as the volume changes upon inner-sphere and outer-sphere complex formation between Co^{2+} and the part of the polyphosphate chain involved in the binding. In a recent ultrasonic absorption study [24] on ion-pair formation in solutions of divalent sulfates values of ΔV_{21} five to ten times larger than ΔV_{32} were reported, in agreement with the results of Table II. Also, theoretical considerations [25] support the experimentally observed larger values of ΔV_{21} as compared to ΔV_{32}. For Co-PP both ΔV_{21} and ΔV_{32} are 4 to 5 times larger than for divalent sulfates because the electrostriction of polyphosphate is much larger than that of $SO_4^{=}$.

In conclusion, ultrasonic absorption and density measurements in addition to NMR results, did allow us to answer some of the questions raised in the introduction, at least to a certain extent. However, the results obtained in this work have led to quantitative developments only for Co-PP and this report should be taken as a very preliminary one in this field of investigation. More experimental results appear to be needed before definite conclusions can be drawn concerning the detailed nature of the equilibria responsible of the site binding absorption measured for polyelectrolyte solutions. New experiments should be performed in a larger concentration range in order to investigate more thoroughly the effect of the nature of the divalent counterion. Also, the frequency range investigated should be extended to frequencies below 1 MHz, in order to permit a more accurate determination of A_2 and N_2. In this respect a newly developped ultrasonic technique [26] as well as relaxation techniques based on rapid pressure changes [4b] may prove extremely useful.

References

1. Armstrong, R. W. and Strauss, U. P.: *Encyclopedia of Polymer Science and Technology*, Vol. 10, p. 781, Interscience Publ. Inc., New York, 1969.
2. Strauss, U. P. *et al.*: *J. Am. Chem. Soc.* **87**, 1476 (1965); *J. Phys. Chem.* **76**, 254 (1972).
3. Ikegami, A.: *J. Polymer Sci.*, part A, **2**, 907 (1964); *Biopolymers* **6**, 431 (1968).
4. (a) Diebler, H., Eigen, M., Ilgenfritz, G., Maas, G., and Winkler, R.: *Pure Appl. Chem.* **20**, 93 (1969); (b) Eigen, M. and de Maeyer, L.: *Techniques of Organic Chemistry*, Vol. VIII, Part 2, Interscience Publ. Inc., New York, 1963; (c) Eigen, M.: *Pure Appl. Chem.* **6**, 97 (1963).
5. Wall, F. T. *et al.*: *J. Am. Chem. Soc.* **76**, 868 (1954); *J. Chem. Phys.* **20**, 1200 and 1207 (1952).
6. Lifson, S. and Jackson, J.: *J. Chem. Phys.* **36**, 2410 (1962).
7. Gottlieb, M.: *J. Phys. Chem.* **75**, 1981, 1985 and 1990 (1971).
8. Stuehr, J. and Yeager, E.: *Physical Acoustics*, Vol. IIA (ed. by W. P. Mason), Academic Press, New York, 1965.
9. Zana, R., Tondre, C., Milas, M., and Rinaudo, M.: *J. Chim. Phys., Physicochim. Biol.* **68**, 1258 (1971).
10. (a) Tondre, C. and Zana, R.: *IUPAC Symposium on Macromolecules*, Leiden 1970, Book of Abstracts, Vol. I, p. 387; (b) Tondre, C. and Zana, R.: *J. Phys. Chem.* **75**, 3367 (1971).
11. Fuoss, R., Katchalsky, A., and Lifson, S.: *Proc. Natl. Acad. Sci. U.S.* **37**, 579 (1951).
12. Dubin, P. and Strauss, U. P.: *J. Phys. Chem.* **74**, 2842 (1970).
13. Atkinson, G., Baumgartner, E., and Fernandez-Prini, R.: *J. Am. Chem. Soc.* **93**, 6436 (1971).
14. Michels, B. and Zana, R.: *Koll. Zeit.* **234**, 1008 (1968); Zana, R. and Michels, B.: *Proceedings of*

the 7th Int. Cong. Acoustics, Vol. II, p. 41, paper 20 M6, Akademiai Kiado, Budapest, 1971.
15. Bechtler, A., Breitschwerdt, K., and Tamm, K.: *J. Chem. Phys.* **52**, 2975 (1970).
16. Manning, G.: *J. Chem. Phys.* **51**, 924, 933 and 3249 (1969): *Biopolymers* **9**, 1543 (1970).
17. Zana, R. and Yeager, E.: *J. Phys. Chem.* **71**, 521 and 4241 (1967); Millero, F.: *Chem. Rev.* **71**, 147 (1971).
18. Conway, B. *et al.*: *J. Phys. Chem.* **75**, 3031 (1971) and *J. Chem. Phys.* **43**, 243 (1965).
19. Tondre, C. and Zana, R.: *J. Phys. Chem.* **76**, 3451 (1972).
20. Tondre, C., Spegt, P., Zana, R., and Weill, G.: *Biophys. Chem.* **1** (1973, in press).
21. Hammes, G. and Park, A.: *J. Am. Chem. Soc.* **90**, 4151 (1968).
22. Chmelnik, A. and Fiat, D.: *J. Chem. Phys.* **47**, 3986 (1967).
23. Spegt, P. and Weill, G.: *Compt. Rend. Acad. Sci Paris*, Series C, **274**, 587 (1972).
24. Jackopin, L. G. and Yeager, E. B.: *J. Phys. Chem.* **74**, 3766 (1970).
25. Hemmes, P.: *J. Phys. Chem.* **76**, 895 (1972).
26. Eggers, F.: *Acustica* **19**, 323 (1967).

SOME APPLICATIONS OF INFRARED SPECTROSCOPY TO THE STUDY OF POLYELECTROLYTE SYSTEMS

J. C. LEYTE

Gorlaeus Laboratoria, Afdeling Fysische Chemie III, Rijksuniversiteit, Leiden, The Netherlands

In concordance with the title the author wishes to emphasize that this is not a review The examples discussed below were chosen with the problem of polyion counter-ion interaction in mind and they are not intended to form a representation for all the possible work on polyelectrolyte systems using IR.

1. General Remarks

Typically IR studies on polyelectrolyte molecules are concerned with specific chemical groups present in the polymer molecule or its solvent: $-SO_3^-$ groups in polystyrene-sulfonate and $-COOH$ groups in polyacrylic acid may be cited as examples of groups studied by this method.

It is usually assumed that the IR radiation is absorbed by highly localized oscillators giving rise to the well known group frequencies. Although long range order has been shown to introduce interesting features in polymer spectra [1, 2] it is generally safe to restrict the discussion of the factors that influence the vibration frequencies to the immediant neighbourhood of the chemical group studied if such a group is not part of the polymer chain skeleton. Still, the group frequency concept must not be taken literally as the normal coordinate associated with a given frequency seldom consists exclusively of internal coordinates (stretches, bendings) within the group for which it is supposed to be characteristic.

This is immediately apparent from the eigenvalue problem defining L, the matrix connecting the internal coordinates S and the normal coordinates Q [3]

$$S = LQ \qquad GFL = L\Lambda .$$

Even keeping the force constant matrix F invariant, changes in the geometry or the mass distribution of the system affect L through the corresponding changes in the Wilson G matrix. This is confirmed by the results of many vibration analyses in the literature from which it is apparent that it is quite often wrong to state that a certain band has shifted on isotopic substitution: the new normal coordinate may be a significantly different combination of internal coordinates.

Barring vibrational analysis the application of IR to the complicated systems discussed here may consist of

(a) analytical application of the frequency spectrum proving the presence or absence of certain groups;

Eric Sélégny (ed.), Polyelectrolytes, 339–346.

(b) physical chemical interpretation of frequency shifts, band splitting etc.; and
(c) quantitative measurement and interpretation of band intensity.

A disadvantage of the method lies in the high concentrations that are usually necessary to obtain high quality spectra: the lowest concentrations that can be studied in aqueous solutions are roughly in the 10^{-2}–10^{-1} molar range. The time scale on the other hand is clearly advantageous: species existing for more than about 10^{-12} s can be detected if showing characteristic absorption features.

2. Some Results

In the study of polyelectrolytes the IR absorption technique has been used to obtain information on proton dissociation from polyacids, ionbinding, the structure of the water of hydration and conformational changes.

In connection with the general problem of polyion counter-ion interaction some very important results have been published by Zundel [4]. An example is his work on the SO_3^- group in salts of polystyrene sulfonate. The unperturbed $-SO_3^-$ group is expected to have C_{3v} symmetry and the SO stretching modes are therefore classified as A (symmetrical) and E (asymmetrical, degenerate). A non-symmetrical perturbation will be expected to split the degenerate E mode as may be seen from a correlation of the irreducible representations of C_{3v} and C_s

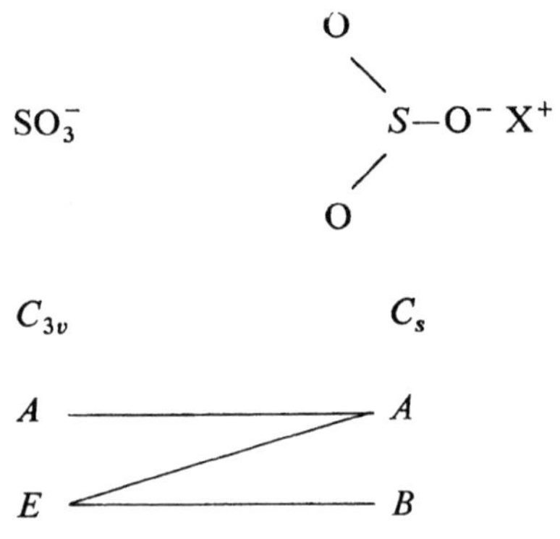

From the absorbtions of the SO_3^- group in the 1100–1250 cm^{-1} region it was found [4] that in the dried salts the splitting depends strongly on the cation. The effect increases with decreasing ionic radius and with increasing charge. For the alkali ions the splitting decreases from 69 cm^{-1} for Li^+ to 25 cm^{-1} for Cs^+ while for Be^{2+} it is as large as 109 cm^{-1}.

From these results it may be concluded that in these salts the ions are placed asymmetrically with respect to the C_{3v} axis of the unperturbed SO_3^- groups. Now, by measuring the intensity of the bending mode of H_2O it is possible to study the E-mode splitting as a function of the degree of hydration of the salts. A continuous decrease of the effect is observed and the amount of splitting is still changing when roughly 5 H_2O molecules per cation are present in the salt according to Zundel.

The decrease of the interaction of the charged groups is therefore a gradual proces in this case.

A detailed picture of the important role of the water of hydration in polyelectrolyte systems is furnished by a study of the dissociation of the – SO_3H groups in polystyrenesulfonate membranes. Using the 'S–OH' vibration at 905 cm^{-1} to calculate the degree of dissociation it was shown [4] that it takes 4 molecules of H_2O per acid group to achieve essentially complete dissociation.

Similar work has been done [5] in an investigation of counter-ion binding by poly-(methacrylic acid), PMA, and poly(acrylic acid), PAA.

In this work the carbonyl stretching mode $\nu(C=O)$ at 1700 cm, and the symmetric $\nu_s(CO_2^-)$, (1415 cm^{-1}), and asymmetric $\nu_a(CO_2^-)$, (1550 cm^{-1}), stretching modes of the carboxylate group were studied. In Tables I and II it is seen that the dry salts show specific effects indeed while in the D_2O solutions (~0.5 eq/l) this specificity has disappeared. The small difference in ν_a between the bivalent and univalent ions is only indirectly related to the charge difference: as the Earth alkali salts were insoluble they were measured as swollen gels and these results can therefore not be compared with the results on the alkali salts.

TABLE I

Polymethacrylates

	Dry		D_2O solution	
	ν_a	ν_s	ν_a	ν_s
Li	1553	1401	1549	1416
Na	1561	1396	1549	1413
K	1554	1392	1550	1415
Cs	1546	1389	1550	1417
Mg	1557	1401	1546	1416
Ca	1540	1400	1546	1417
Sr	1538	1397	1546	1415
Ba	1529	1391	1547	1416

TABLE II

Polyacrylates

	Dry		D_2O solution	
	ν_a	ν_s	ν_a	ν_s
Li	1571	1422	1570	1413
Na	1570	1406	1571	1412
K	1570	1399	1569	1412
Cs	1565	1400	1571	1412
Mg	1582	1431	1565	1413
Ca	1562	1417	1563	1415
Sr	1550	1414	1563	1413
Ba	1553	1405	1564	1414

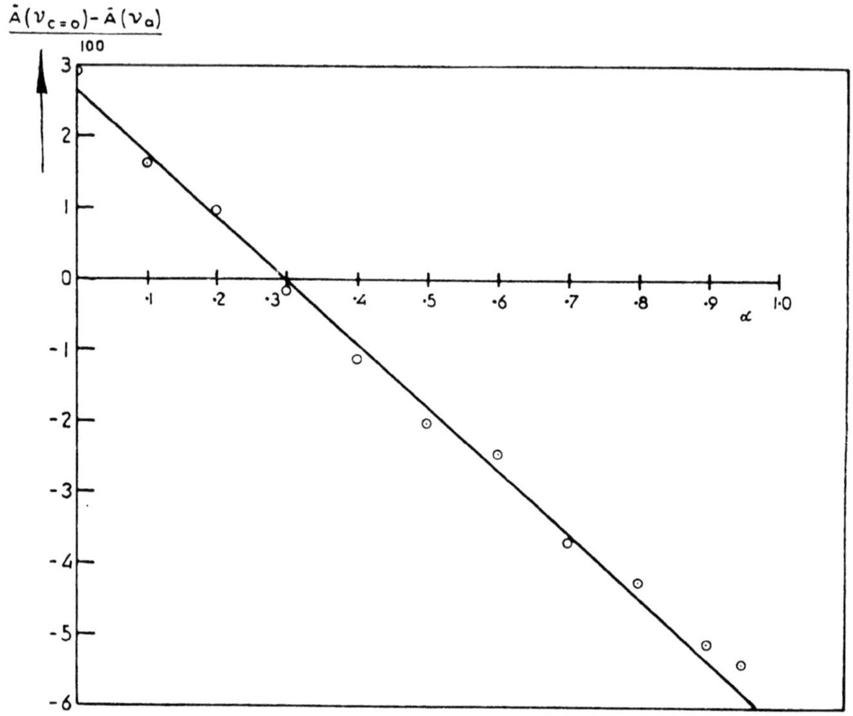

Fig. 1. Absorbtion difference of $\nu_{c=0}$ and ν_a in polyacrylic acid as a function of the degree of ionization α.

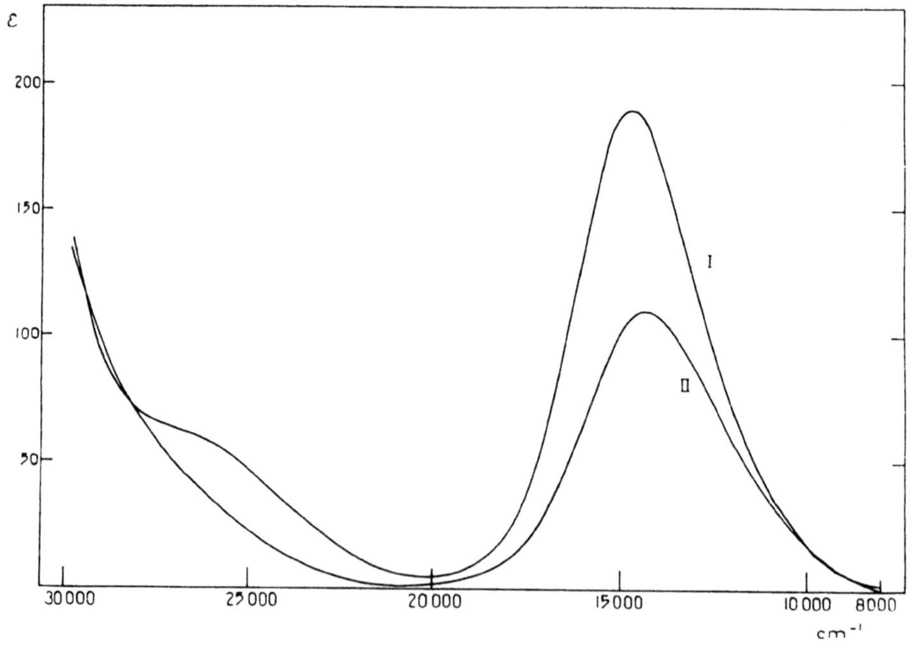

Fig. 2. Visible and near UV spectrum of the PMA/Cu complex at $f = 1.3$ (I) and $f = 3.3$ (II). PMA: 3.00×10^{-2} equiv./l.; $Cu(ClO_4)_2$: 9.00×10^{-2} equiv./l.

From these results it may be noted that the carboxylate ions do not differentiate within the groups of alkali or Earth alkali ions and the hydration of these ions can therefore not be seriously affected.

In Figure 1 the difference in the absorbtion due to the undissociated and dissociated groups is shown to be a linear function of neutralization: no effects connected with site binding can be found though it is well known from thermodynamic methods that at higher charge densities serious depression of the activity coefficients occurs [6].

For aqueous solutions, much more detailed information is gained by the use of IR on strongly interacting systems as for example the Cu-PMA complex [7]. The binding of Cu ions by PMA is most conveniently discussed as a function of the ratio f=(equivalent concentration of neutralized PMA/equiv. concentration of Cu(II)).

Using several experimental methods it has been shown that for $f \leqslant 1.2$ each Cu ion binds two carboxylate ions [8]. At higher charges densities of the polyelectrolyte ($f > 1.2$) a decrease is observed in the intensities of the visible and UV absorbtion

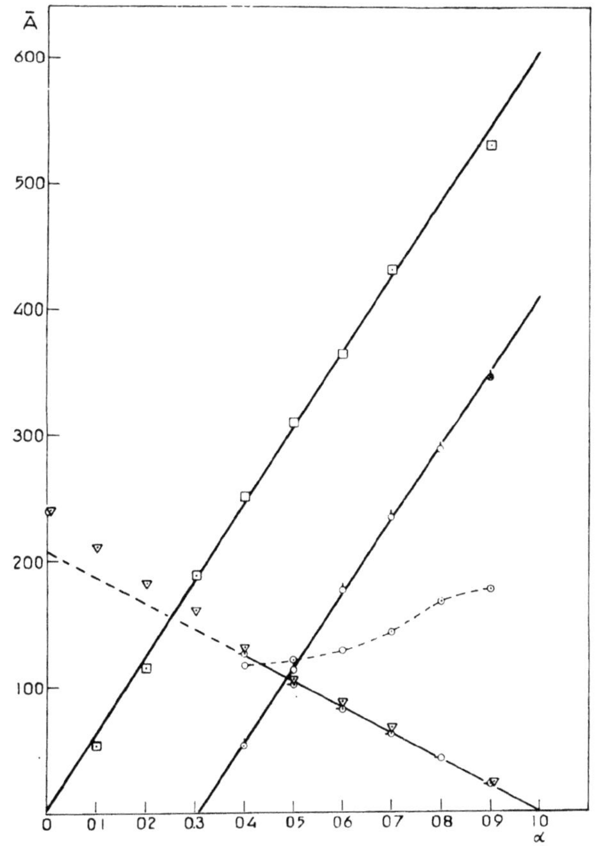

Fig. 3. Absorbance $\bar{A}$ from IR spectrofotometric titrations. PMA: ν_1 (⊡), ν_2 (▽); PMA + 30.0% equiv. Cu(II): ν_1 (⫯), ν_2 (-○), ν_3 (○).

characteristic of the complex (Figure 2). This was seen first for Cu-PAA [9] and it was given an extremely interesting interpretation: partial disintegration of the complex due to expansion of the polymer chain and competition between site binding and interaction with the overall electrostatic field of the polyion; an important polyelectrolyte effect in other words.

The validity of this interpretation can be checked accurately as the absorbtion bands due to the undissociated COOH groups ($\nu(C=O)$) and the chelated and free-carboxylate groups ($\nu_{as}(CO_2^-)$) can be seperately observed in the IR spectrum of the D_2O solutions.

In Figure 3 the absorbtion intensity of these modes is displayed as a function of the degree of neutralization α of PMA in the presence of 30% Cu(II) and for a copper free solution. Here $\nu_1 = \nu_{as}(CO_2^-)$, $\nu_2 = \nu(C=O)$ and $\nu_3 = \nu_{as}(CO_2^-)$ for the chelated carboxylic groups. The absorbtion of the free CO_2^- groups shows conclusively that no carboxylate groups are liberated above $f=1$ ($\alpha=0.3$). Per Cu ion two CO_2^- ions are bound and stay bound as neither the free CO_2^- concentration nor the COOH concentration shows any deviation from the normal change with α. Most interesting, however, is the absorbtion at ν_3, due to the bound CO_2^- ions: far from decreasing above $f=1$ it actually increases.

In summary the results are therefore:

(a) The number of CO_2^- groups bound per Cu(II) ion is constant and equal to 2.

(b) The UV and visible extinction coefficients decrease for $f>1$.

(c) The IR extinction coefficient for $\nu_{as}(CO_2^-)$ of the chelated carboxylic ions increases for $f>1$.

A satisfactory explanation of these results can be given in terms of the following model.

For $f<1$ the chain bears effectively no charge and in the coiled polyelectrolyte a binuclear Cu complex is formed involving two Cu ions and four carboxylate groups. For the acetate ion this complex has been shown to exist in the solid state and in CH_3OH solution but in H_2O it disintegrates into mononuclear complexes. Possibly, within the uncharged coil water is sufficiently scarce to allow the formation of the binuclear complex. Now, in this complex the asymmetrical ligand field gives rise to

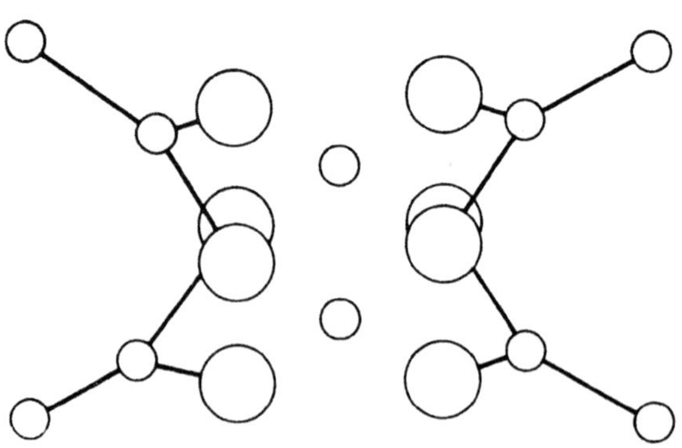

Fig. 4. Binuclear copper carboxylate complex.

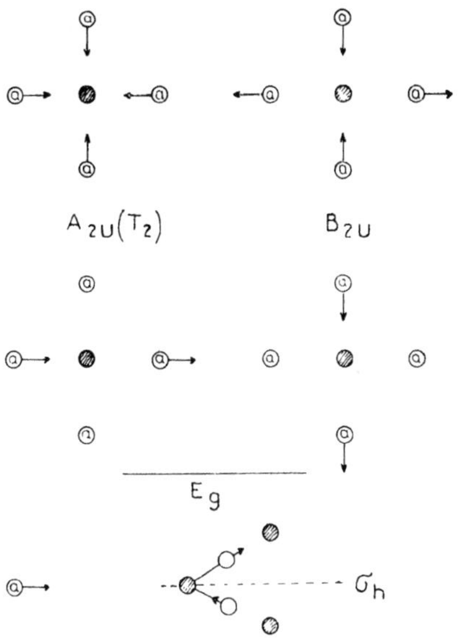

Fig. 5. Schematic representation of the symmetry species of the combined antisymmetric carboxylate vibrations in the binuclear complex.

intense absorbtion in the UV/visible region. In connection with this it is important to note that the absorbtion shoulder at 25000 cm^{-1} characteristic for this type of binuclear complex occurs for $f<1$.

For $f>1$ the chain acquirs net charge, it expands and water molecules break down the binuclear complexes. This results in complexes in which each Cu ion is surrounded by two carboxylate groups and additional H_2O molecules. In this, more symmetrical ligand field the transitions leading to the observed UV/visible absorbtion are less intense: the selection rule forbidding these transitions is less strongly perturbed.

The consequences for the IR absorbtion intensities may now be estimated.

Assuming D_{4h} symmetry for the binuclear complex the asymmetric stretching mode of the CO_2^- groups is involved in vibrations of symmetry species A_{2u}, E_g and B_{2u} of which only A_{2u} is IR active. If no large change is induced in the transition moment of the $\nu_{as}(CO_2^-)$ by the proximity of the Cu(II) ion one would expect the integrated absorbtion per carboxylate group in this complex to be some 25% of the free ion value. At $f=1.3$ the ratio is 33% and as at this f-value part of the complex is already dissociated (as evidenced by the UV and visible regions) this result must be considered satisfactory.

Now, if D_{2h} symmetry is assumed for the mononuclear complex, placing Cu and both CO_2^- groups in σ_h, we find only one of both combinations of $\nu_{as}(CO_2^-)$ to be IR active. At $f=3.3$ the integrated absorbtion per bound carboxylate group is found to be 47% of the free ion value while 50% is expected on the basis of this model. Other configurations for the mononuclear complex would, because of lower symmetry lead to even higher expected absorbtion per carboxylate group.

It is therefore concluded that the dissociation of binuclear complexes existing at $f<1.2$ into mononuclear complexes for $f>1.2$ is responsible for the changes in the UV, visible and IR spectra of the solutions.

References

1. Zerbi, G.: *International Symposium on Macromolecules*, Leiden 1970; Butterworths, London, 1971; p. 499 and references cited there.
2. Zbinden, R., *Infrared Spectroscopy of High Polymers*, Academic Press, New York and London, 1964.
3. Wilson, E. B., Decius, J. C., and Cross, P. C.: *Molecular Vibrations*, McGraw-Hill 1955.
4. Zundel, G.: *Hydration and Intermolecular Interaction*, Academic Press, New York and London, 1969.
5. Leyte, J. C., Zuiderweg, L. H., and Vledder, H. J.: *Spectrochim. Acta* **23A** 1397 (1967).
6. Katchalsky, A.: *International Symposium on Macromolecules*, Leiden 1970; Butterworths, London, 1971; p. 327 and references cited there.
7. Leyte, J. C., Zuiderweg, L. H., and van Reisen, M.: *J. Phys. Chem.* **72**, 1127 (1968).
8. Mandel, M. and Leyte, J. C.: *J. Polymer Sci.* **A2**, 3771 (1964).
9. Wall, F. T. and Gill, S. J.: *J. Phys. Chem.* **58**, 1128 (1954).

OPTICAL ROTATORY DISPERSION AND CIRCULAR DICHROISM OF SYNTHETIC POLYELECTROLYTES AND OF THEIR COMPLEXES

MICHEL VERT

Laboratoire de Chimie macromoléculaire, UER des Sciences Exactes et Naturelles, E.R.A. 471, Université de Rouen, 76130-Mont-Saint-Aignan, France

1. Introduction

The study of the properties of polymers carrying reactive groups, and consequently, of polyelectrolytes, has been undertaken by a wide variety of techniques (potentiometry, conductimetry, viscometry, light-scattering, NMR, ultrasonic absorption, etc...) which can all be considered as classical in the sense that they can be used in a general manner.

There are other techniques which need the use of molecules particularly adapted to the circumstances: isotopically or radioactively marked ones for instances. Optical Rotatory Dispersion (ORD) and Circular Dichroism (CD) are classed amongst these because of the necessity of having asymmetric centres along the macromolecule chains.

For a long time, optical activity was the attribute of only natural macromolecules such as proteins, polynucleotides or polysaccharides.

With the advances in the macromolecule synthesis, a great number of optically active polymers has been successfully prepared. However, their optical activity, has only been studied very recently, and this mainly in order to clarify stereochemical and conformational problems [1].

It was for similar reasons that we first carried out investigations in the field of optically active synthetic polymers [2, 3].

We very soon realized that in the majority of cases their optical activity depended on the chemical reactivity and not on conformational changes. At present we employ optical rotary power, ORD and CD for studying the chemical behaviour of macromolecules.

In the present paper we hope on the one hand to present an original conception of interpreting the optical activity of synthetic polymers, and on the other one to show that optical activity is not only a means to study conformational phenomena but can also prove an excellent method of detection, and even of study, of the ionisation of polyelectrolytes and of the complexation of macromolecules with ions or small molecules. In order to show both the advantages as well as the disadvantages of such a technique, we shall first of all give a brief reminder of the origins of optical rotation and circular dichroism and of their sensibility to secondary structures and chemical modifications.

Eric Sélégny (ed.), Polyelectrolytes, 347–370.

2. Reminder of ORD and CD Phenomena

Optically active substances are distinguished from inactive ones by their action on polarized light in a plane, that is to say by their action on an electromagnetic wave the electrical field vector E of which vibrates in a plane. Polarized light can be decomposed into two circular polarized lights E_L and E_R turning in phase in opposite directions.

Optical activity phenomena result from the respective existence of two refraction indices n_L and n_R (circular birefringence) and of two absorption coefficients k_L and k_R (circular absorption) corresponding to each of the circular components.

The effect of circular birefringence is a rotation of the plane of polarization of the light and is usually expressed in angle of rotation or rotatory power (α). ORD and CD are respectively the variations of rotatory power and of circular differential absorption with wavelength.

Detailed discussion of the theories, apparatus and applications in organic chemistry or for biopolymers can be found in the books of the following authors: Caldwell and

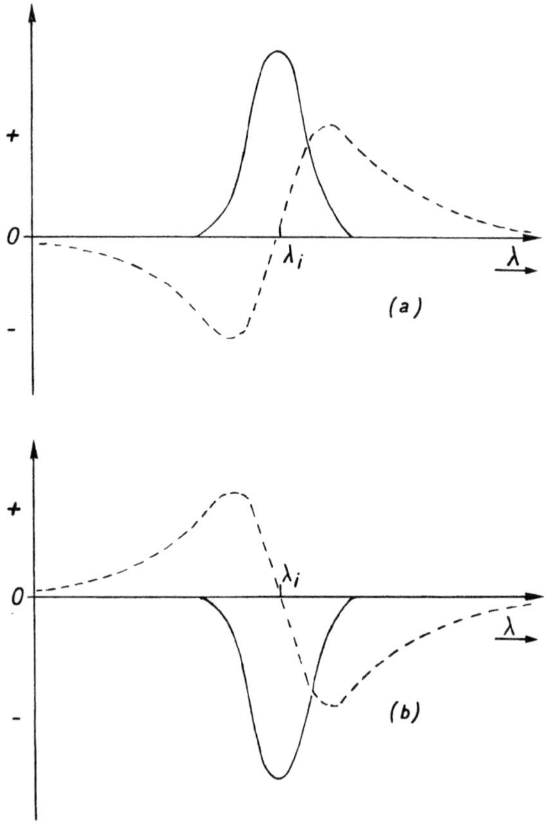

Fig. 1. ORD ---- and CD —— of a positive COTTON effect (a) and of a negative one (b) corresponding to an isolated, optically active, electronic transition λ_i.

Eyring [4], Crabbe [5], Djerassi [6], Fasman [7], Jirgensons [8], Snatzke [9] and Velluz *et al.* [10].

To summarize it can be said that ORD and CD have the same origins as refraction and absorption of natural light in the UV or in the visible region; that is to say the electronic transitions that can be undergone by the electrons of certain atoms or of certain groups of atoms, called chromophores, through the action of electromagnetic waves.

To each optically active electronic transition i characterized by the wavelength λ_i, there corresponds a COTTON effect which is shown in ORD by a characteristic S-shaped curve and in CD by a bell-shaped curve; both these curves may be positive or negative (Figure 1).

The CD, as is UV absorption, is localised around the λ_i whereas the ORD covers the whole of the spectrum because of the dispersion effect. When $\lambda \gg \lambda_i$ it obeys an equation of the form:

$$[\alpha] = \frac{a_i \lambda_i^2}{\lambda^2 - \lambda_i^2}$$

in which a_i is a parameter which is proportional to the rotational force R_i which characterized each COTTON effect:

$$R_i = \mathrm{Im}\,\{\mu_e^i \cdot \mu_m^i\}$$

μ_e^i and μ_m^i have the dimensions of the electric and magnetic dipole moments induced by the ith transition.

In fact organic compounds often possess several accessible electronic transitions, and consequently ORD and CD curves resulting from the addition of these partial contributions are observed (Figure 2).

Outside the zone of COTTON effects the ORD curve follows Drude's equation

$$[\alpha] = \sum_{i=1}^{n} \frac{a_i \lambda_i^2}{\lambda^2 - \lambda_i^2}.$$

**Definitions*:

Specific optical rotation:

$$[\alpha]_\lambda^\tau = \frac{(\alpha)_\lambda{}^\tau \times 100}{l \times c}$$

(α) optical rotation; l pathlength in dm c concentration in g 100 cm^{-3}.

molar specific optical rotation:

$$[\Phi]_\lambda^\tau = \frac{[\alpha]_\lambda{}^\tau \times M}{100}$$

M = Molecular weight

molar specific ellipticity:

$$[\theta]_\lambda^\tau = \frac{(\theta)_\lambda{}^\tau \times M}{l \times c} = 3300 \quad (\varepsilon_l - \varepsilon_r)$$

for macromolecules, M is replaced by M_r mean constitutional repeating unit weight.

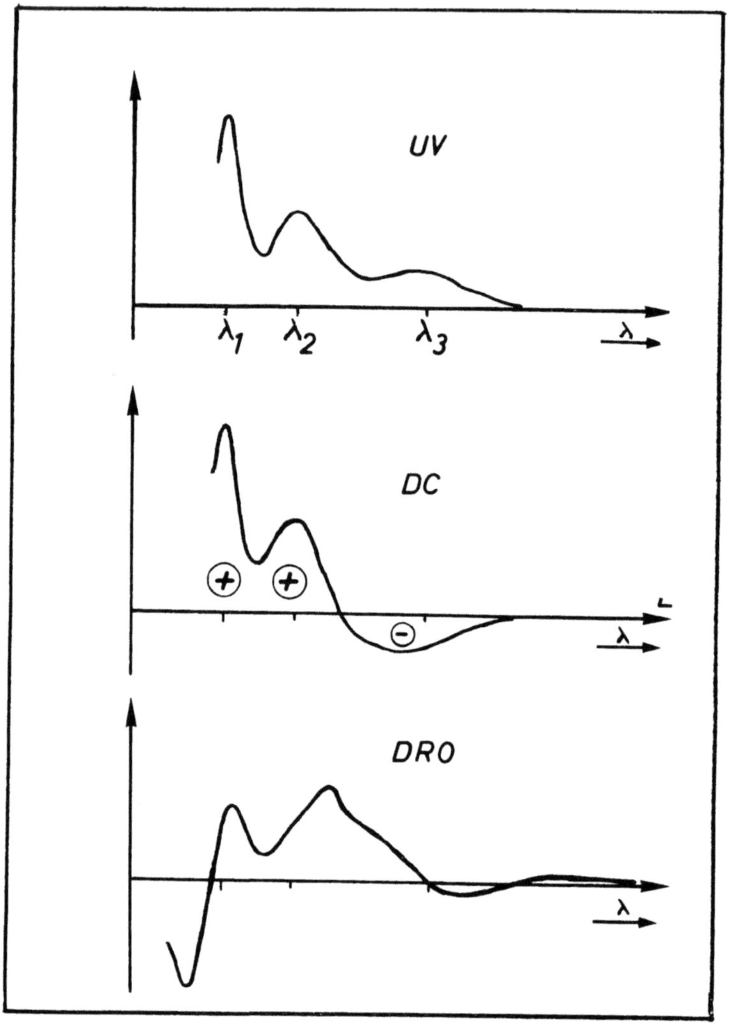

Fig. 2. UV, CD and ORD curves resulting from several optically active electronic transitions λ_1, λ_2, λ_3.

This general equation can often be simplified to the one-term Drude equation

$$[\alpha] = \frac{A}{\lambda^2 - \lambda_c^2}$$

in which A and λ_c are constants having no direct relation with any particular electronic transition. The ORD is then simple. If it does not satisfy the one-term Drude equation, then it is said to be complex and needs either a multiple-term Drude equation with a maximum number of electronic transitions or phenomenological equations such as the Moffitt one:

$$[\alpha] = \frac{a_0\lambda_0^2}{\lambda^2 - \lambda_0^2} + \frac{b_0\lambda_0^4}{(\lambda^2 - \lambda_0^2)^2}.$$

It must be noted that the dispersive effect that complicates the ORD curves presents nevertheless an advantage. Measurements of optical rotary power or ORD far from the COTTON effects can give valuable information about the electronic transitions situated in the UV region and, consequently, about the corresponding chromophores. Moreover, an analytical comparison between ORD and CD will provide, by difference, information on the inaccessible COTTON effects. Finally since the COTTON effect can be positive or negative, transitions that are so frequently overlapped that they become almost indistinguishable on the absorption spectra can often be discerned – even only partially – on the CD curves.

3. Influences on the Optical Activity

3.1. Case of Small Molecules

Apart from intrinsically asymmetric chromophores, it is the existence of a dissymmetry in the space around the chromophores that makes their electronic transitions optically active. This dissymmetry is the result of the presence of one or more asymmetric centres (usually carbon atoms) and thus depends on structural factors (configuration, conformation). If the experimental or external conditions are modified (solvent, temperature, solution-composition), then there is a change in the conformational equilibrium which has some effect on the optical activity, thus justifying the interest of ORD and CD in the structural analysis of optically active compounds and particularly of helical conformation or helix-coil transitions of biopolymers.

However an important fact that should be noted but is often neglected in structural studies in which the external conditions are modified is that *the optical activity depends not only on the chromophores and the molecular structure, but also on the chemical reactivity of the molecules.* If a chemical reaction (solvatation [11], ionisation, complexation [12]) implicates a chromophore submitted to the asymmetry of a chiral or asymmetric centre, then the partial COTTON effects corresponding to this chromophore are modified and this modification affects the ORD and CD curves of the initial molecule.

This remark is all the more justifiable since the chromophore groups generally contain π bonds and non-bonding electrons that favour many chemical reactions or interactions.

The difficulties in interpreting modifications in the optical activity that are consecutive to changes in the external conditions come from the sensitivity of the optical activity to two different types of factors: *structural factors* and *chemical factors*; both give rise to the same effects: displacements, variation in amplitude or sign, inversions in the COTTON effects.

It is materially very difficult to dissociate the influence of chemical reactivity from that of the structure because they are intimately linked together through the intermediary of conformational equilibria. At present except in rare cases, it is impossible to reasonably calculate COTTON effects from molecular parameters.

However, the problem may be simplified in two extreme cases in which the chemical reactivity can be considered as the principal factor:

(a) *The Molecule Cannot Give Rise to Any Conformational Change.* – It is the case for example of rigid bicyclic compounds such as camphor (I) [13] or isofenchone (II) [14].

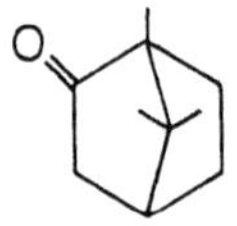

I

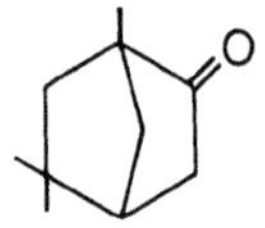

II

In a methanol-ethanol mixture, isofenchone presents at least two COTTON effects of opposite signs within the transition $n-\pi^*$ of $C{=}O$. The different bands could have been attributed to a dimerization but in view of the solvent's aptitude to give strong hydrogen bonds they can reasonably be attributed to an equilibrium in the solvatation which leaves solvated and non-solvated molecules together having different COTTON effects.

(b) *The Molecule Can Take on a Large Number of Statistically Distributed Conformations* – This is the case of most of the optically active compounds in which the atoms that constitute the molecules can be considered to be in a state of free rotation. Experience shows that the effect of a variation in the statistic conformational equilibrium consecutive to a chemical reaction can be considered as secondary to the influence of the modification in the chromophore. Under these conditions any large variation in the COTTON effects linked to a chromophore can be attributed to the chemical modification of this chromophore. This modification can be due to reactions (addition, substitution, ionisation) or to interactions between:

- identical optically active molecules (dimerization, formation of aggregates)
- optically active molecules and solvent (solvatation)
- optically active molecules and molecules or ions in the solution (complexation).

N-tosyl L-tyrosine (III) for example has three ionisable functions the apparent pK_a's of which are respectively about 3.5 (COOH), 10.4 (ϕ—OH) and 12 ($-SO_2NH-$). Figure 3 shows the evolution of the ORD curve in water when these different functions are ionised by addition of an alkaline reagent.

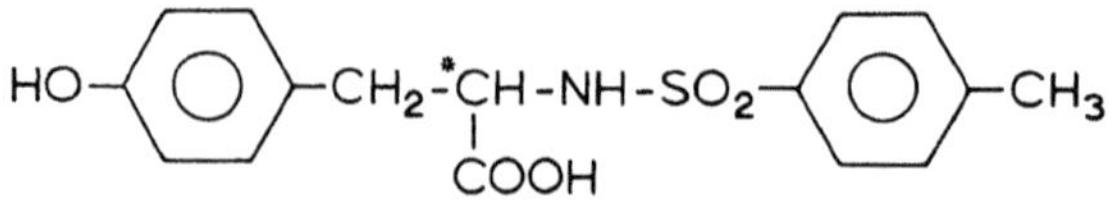

III

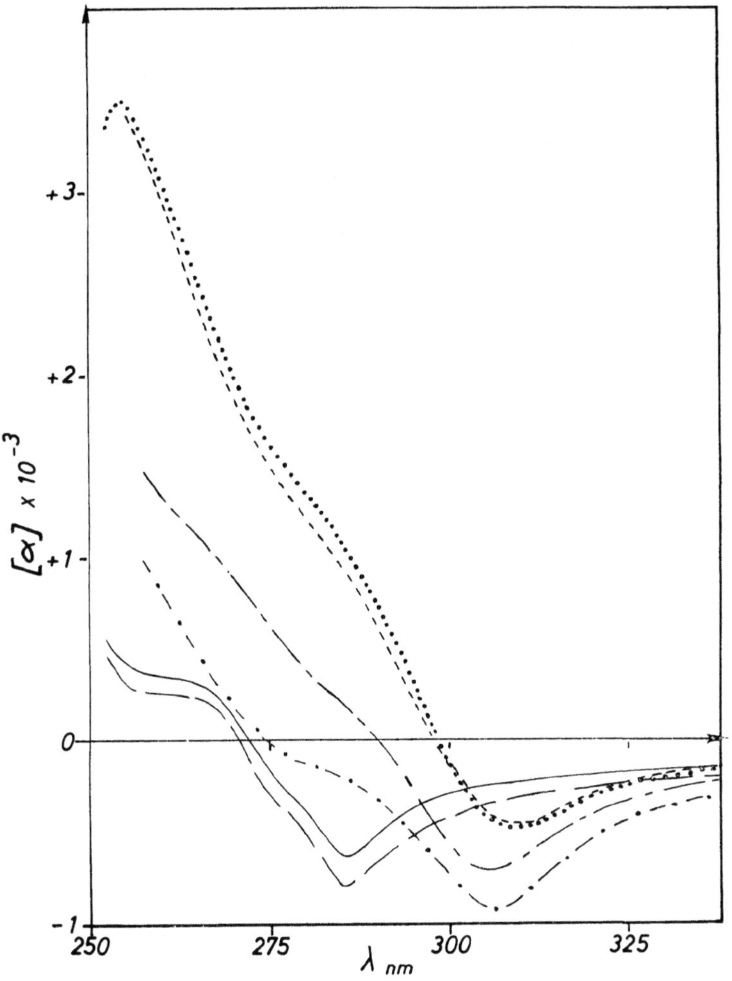

Fig. 3. N-tosyl L-tyrosine (III) ORD curves in function of the pH and of the excess potassium hydroxide concentration: pH = 2.60 ——; pH = 4.80 —— ——; pH = 10.70 . . . . . pH = 11.80 ---; KOH = M/5 —— — ——; KOH = M —·—·—

The molecule has several chromophores able to reflect the modifications of ionisable groups: the $C{=}O$ of COOH, the aromatic phenolic ring and the sulphonamide one. At pH 2.6 the ORD curve corresponds to the non-ionised molecule. One can distinguish a COTTON effect at 280 nm that is associated with one of the aromatic bands in this region of the spectrum and a more important positive one at 225 nm. Up to pH 4.8 only a slight displacement of the curve from 250 to 350 nm without any apparent modification of the aromatic COTTON effects is observed. This displacement is due to a modification in the $C{=}O$ the absorption band of which is at 200 nm. At pH 10.70 ionisation of the phenol causes a bathochrome displacement of the COTTON effects at 274 and 225 nm, these passing at 293 and 232 nm respectively. The peak of the latter is attained at 260 nm. If more alkaline reagent is then added, a negative contribution appears around 275 nm which can be attributed to ionisation of the sulfonamide functions since this is a weak acid needing a great excess of potassium hydroxide in order to become completely ionised.

Simultaneous measurements of optical rotary power at 436 nm, of the pH and of the phenate-ion absorption at 293 in function of τ* show that there is a linear relation between variations in the optical activity and the degree of neutralisation (Figure 4)

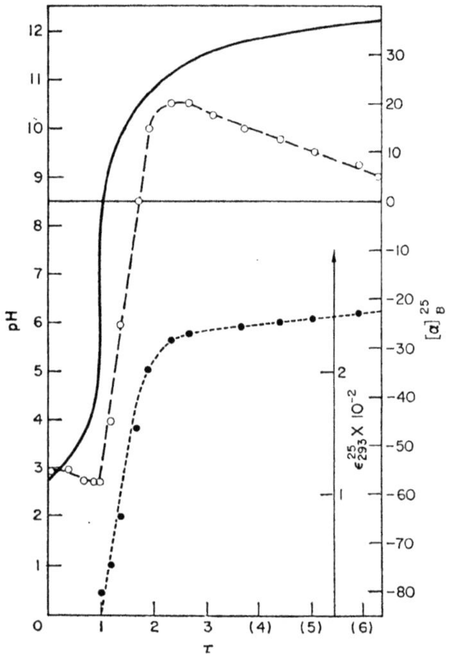

Fig. 4. pH, $[\alpha]_{436}^{25}$ and ε_{293} in function of the degree of neutralisation τ of N-tosyl L-tyrosine (III:pH:$f(\tau)$ —— $[\alpha]_{436}^{25} = f(\tau)$ -·--- $\varepsilon_{293} = f(\tau)$ ----- (τ as defined in footnote*).

Even in the more complex cases [16] if the optical rotary powers of the limiting forms are known then the pK's of the functions can be calculated from these straight lines by using the additivity law of contributions of different optically active species i:

$$(\alpha)_{lu} = l \cdot \sum_i [\alpha]_i \cdot c_i$$

in which: l = path length; $[\alpha]_i$ = specific rotation of i; c_i = concentration of i.

* τ is the degree of neutralization defined as

$$\frac{\text{number of neutralized equivalents}}{\text{number of molecules}}$$

in order to have whole values for each ionizable function.

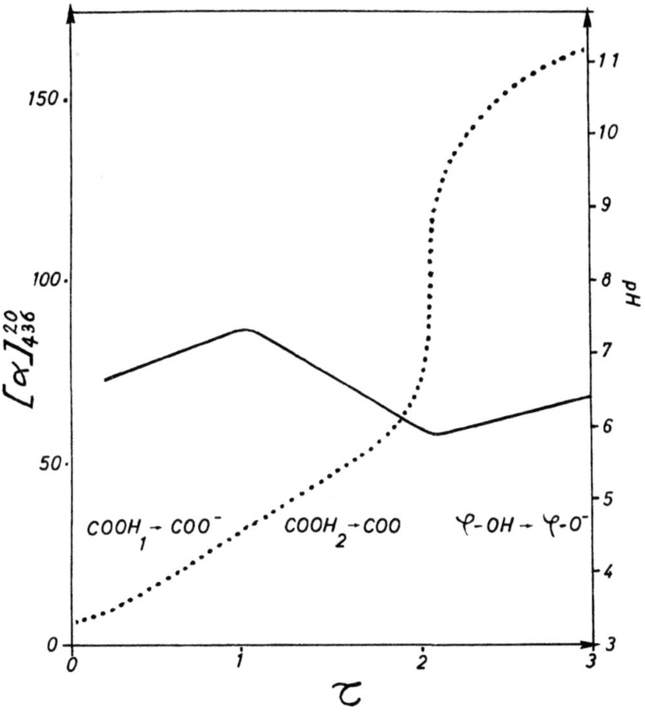

Fig. 5. pH and [α] in function of the degree of neutralisation τ of R(+) p-hydroxybenzyl succinic (IV). (pH $= f(\tau)$ .....; $[\alpha]_{436}{}^{20} = f(\tau)$ ——) (τ as defined in notes p. 354).

Sometimes the optical activity enables one to distinguish between the successive ionisations of functions of similar natures having pK's very close to one another and which are not separated on the potentiometric curve. In Figure 5, the two carboxilic acidities of R(+) p-hydroxybenzyl succinic acid indistinguishable on the potentiometric curve, clearly appear on the polarimetric curve in function of the degree of neutralization [2].

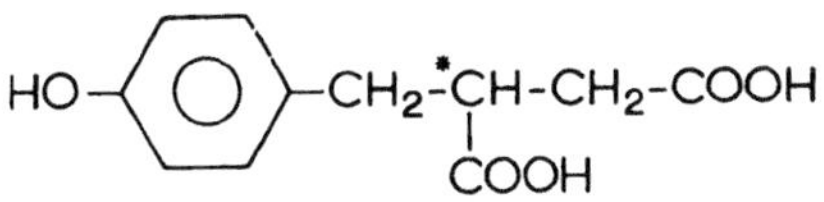

IV

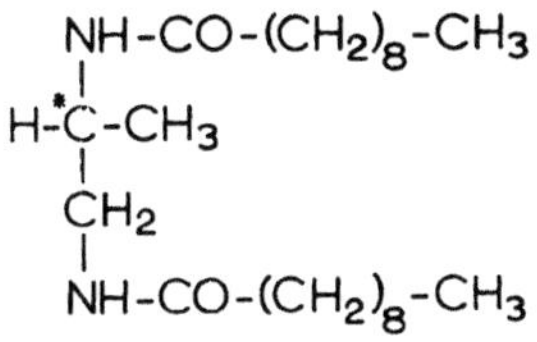

V

Chemical interactions (solvatation, complexation) give rise to similar modifications. Figure 6 shows the evolution of the ORD curve of a diamide: N—N'—dicaproyl(—) 1,2-diamino propane (V) in the presence of increasing quantities of calcium chloride in solution in methanol; it is the result of the formation of an adduct between (V) and

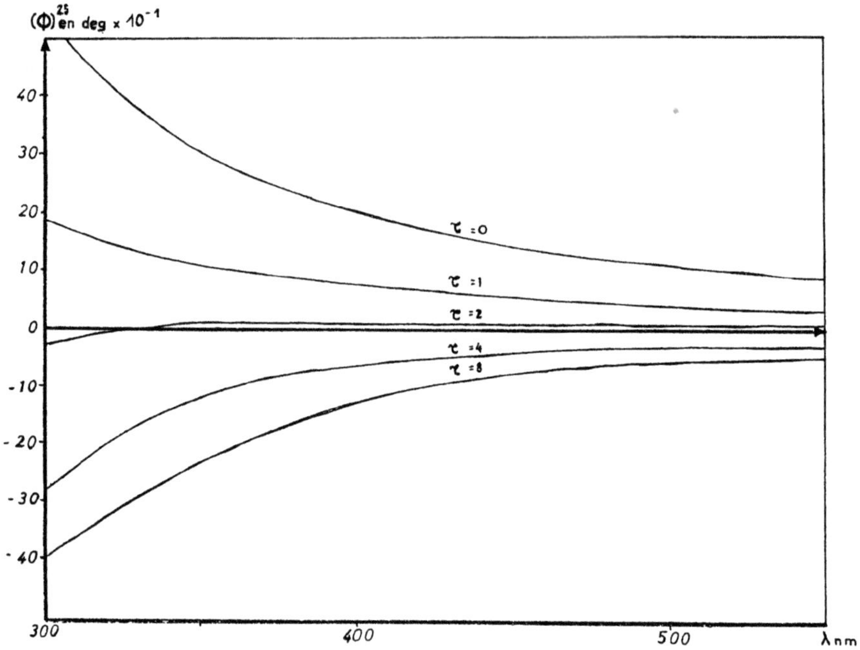

Fig. 6. Evolution of the ORD of N-N-dicaproyl(—) 1, 2-diamino propane (V) in methanol ($c = 6 \times 10^{-3}$ M l^{-1}) solution with increasing amounts of $CaCl_2$. ($\tau = (CaCl_2)/(V)$) M/l^{-1})

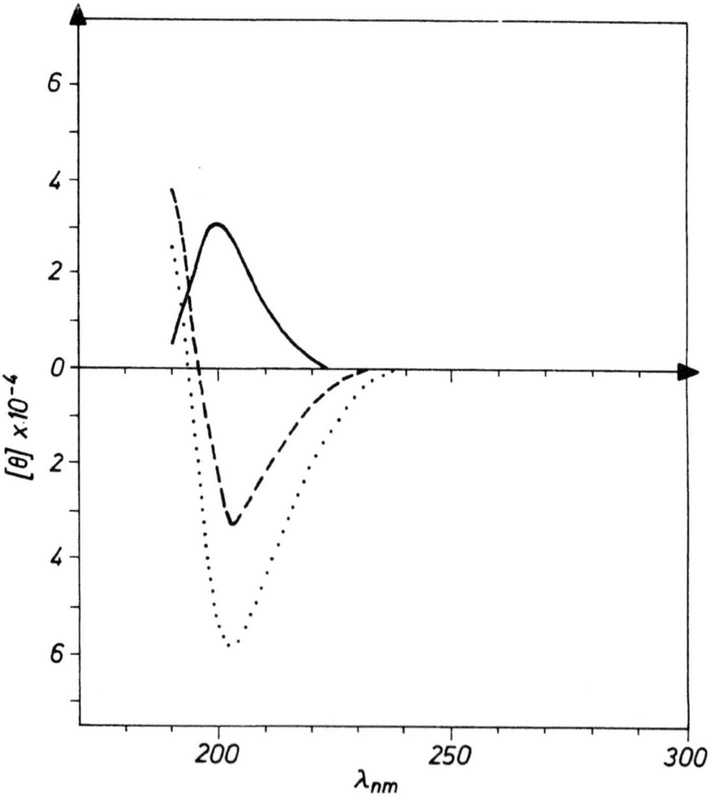

Fig. 7. CD of (*V*) in methanolic solution ——— and in methanolic $CaCl_2$ solutions ($c_{CaCl_2} = 0.15$ M/l^{-1} ------, $= 0.5$ M/l^{-1}......).

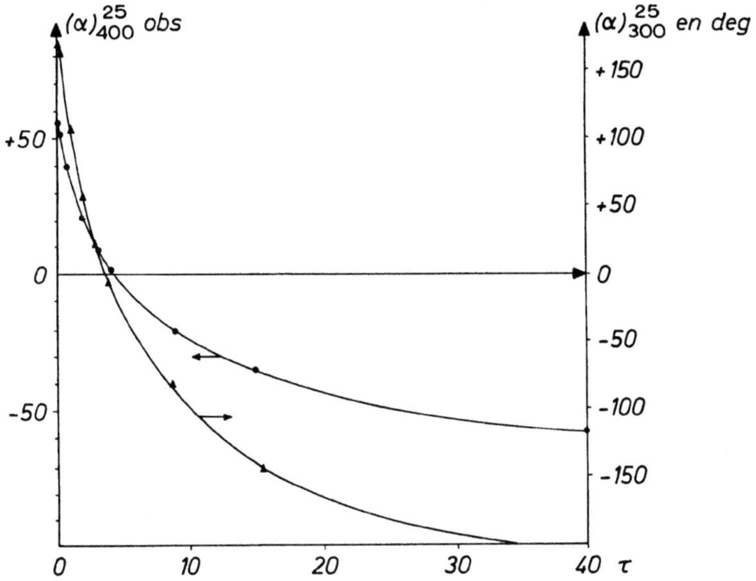

Fig. 8. Variation in the optical rotary power of a 6.10^{-3} M/1 solution of (*V*) in methanol + $CaCl_2$ in function of the proportion of $CaCl_2$ ($\tau = CaCl_2/(V)$) at 300 nm and at 400 nm.

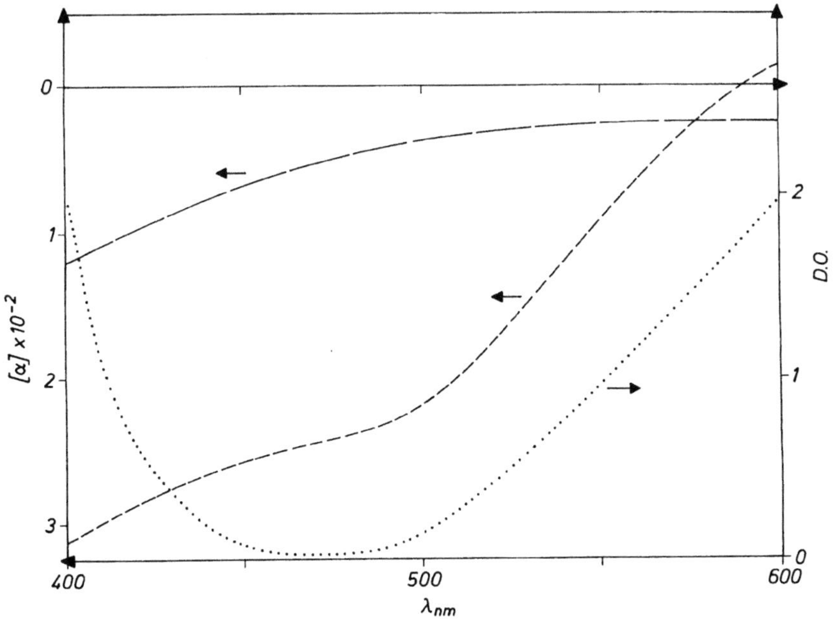

Fig. 9. ORD of an aqueous solution of N-tosyl L-alanine (VI) in the presence of $Cu(NO_3)_2$ at pH 2.6——— (Cu^{++} not complexed) and at pH 10 ---- (Cu^{++} complexed) (dotted line: UV curve of complexed Cu; free Cu^{++} do not absorb in this region of the spectrum).

the salt, and this causes a change in the sign of the COTTON effect at 200 nm (Figure 7).

By applying the law:

$$(\alpha)_{lu} = l \cdot \sum_i [\alpha]_i \cdot c_i$$

to an equilibrium of the form:

$${}_x(V) + {}_y\mathrm{CaCl_2} \overset{K}{\rightarrow} [(V)_x \cdot (\mathrm{CaCl_2})_y]$$

we have obtained, from optical rotary powers measured in the visible spectrum (Figure 8) an apparent K value of 42 ± 2 l-mole^{-1} which is constant along the greater part of the curves for $x = y = 1$ [17].

In the case when the compound reacting with the optically active molecule also possesses one or more electronic transitions the evidence of the chemical interaction can become indiscutable. The very existence of an association submits the electronic transitions of the compound to the dissymmetry of the chiral molecule. They become optically active and, consequently, detectable by the intermediary of the corresponding induced COTTON effects. Figure 9 shows the induction of optical activity in the d–d transition of the Cu II ion complexed with the optically active ligand N-tosyl L-alanine (VI) in aqueous solution at pH 10 [18].

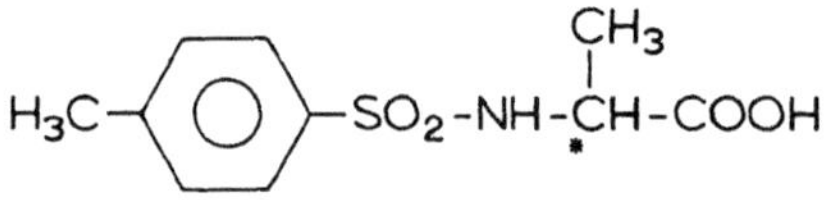

VI

3.2. Case of macromolecules

Macromolecules are constituted by a number of small monomeric units that are often considered as independent molecules. Their optical activity should thus depend on the configuration, on the conformation and on the chemical reactivity of these monomeric units. However the situation is complicated by a further factor shown at first for biopolymers and then considered of importance for all polymers, that is to say by *the molecular conformation.* For highly stereoregular biological molecules the existence of sequences in which the chromophores are distributed in an ordered array in space is revealed by a coupling of the induced dipole moments. This excitonic-type coupling is able to provoke modifications in the COTTON effects that depend on the spatial disposition of the chromophores [19]. In the case of poly-α-amino-acids in the α-helix form one can observe a splitting of the π-π^* COTTON effect of the $C{=}O$ of the peptide in random coil into two COTTON effects $\pi_{\parallel} - \pi^*$ and $\pi_{\perp} - \pi^*$ of opposite signs and polarized respectively parallel and perpendicular to the axis of the helix (20). Moreover, a negative COTTON effect usually attributed to the transition $n - \pi^*$

appears. This splitting is shown in ORD by the passage from a simple curve to a complex one [7].

Similar results have been recorded for stereoregular synthetic polymers of other types: polyolefines [21], polyisocyanates [22] and heavily encumbered polyamides [22].

The question is to determine how far these different factors (configuration, conformation and chemical reactivity of the monomeric unit and macromolecular conformation) intervene in the explanation of the modifications of the COTTON effects resulting from changes in external conditions.

As with the small molecules, experience has lead us to distinguish between two types of synthetic polymers.

4. Stereoregular Polymers Able to Take on Preferential Ordered Conformations in Solution

The behaviour of biopolymers being known, it is necessary to take into account the eventuality of ordered secondary structure in the interpretation of the variations in the COTTON effects following a change in the external conditions. There is often no ambiguity since the existence of a preferential conformation is shown by other ways. This is not always the case particularly, for new polymers the behaviour of which is discussed sometimes only from optical activity data. Under these conditions, the chemical reactivity is a factor all the more delicate to discuss since it can modify COTTON effects directly by chromophors or by the intermediate of the conformational equilibrium.

A wellknown example is given by the ionisation of poly L-glutamic acid. Inversion of the size of the rotatory power in the visible region [24], the change from complex ORD to quasi simple ORD [25] and the change from three main dichroic bands to one only [26] result from the conformational transition from the helix-α unionized form to the random coil ionized form. A similar behaviour has been observed for a stereoregular polybase: poly L-lysine [26].

As for interactions with ions or small molecules in solution, ORD and CD are effective only in the cases where these ions or small molecules show absorption bands in the visible region since one observes induction of optical activity in these bands. For example, the CD curve of poly L-lysine in helix at pH = 10.5 with Cu^{++} shows a induced COTTON effects at the d-d transition of the Cu^{++} ion near 500 nm. The decrease of pH to 8.5 suppresses interactions with the main chain and consequently the induced optical activity [27]. Similarly, for degrees of neutralization below 0.7, the interaction of acrydine orange with poly-S-carboxymethyl L-cysteine VII is demonstrated by the induction of a COTTON effect at the level of the absorption band of the dye. Above 0.8, induced COTTON effects do not appear any more, the variation of the ORD of the polymer alone during the neutralization is then considered as resulting from a β-structure-random coil transition [28]. It may be noted that induced optical activity in the side chain chromophores could also explain such behaviour.

For the interactions with ions or molecules without absorption bands in the visible region, the discussion is based only on the optical activity of the macromolecule. The interpretation is often ambiguous since the ORD and CD changes may be interpreted as conformational change or chemical reaction or both. An example is given by the different salts of a polypeptidic polycation: the poly S-methylsulfonium L-methionine VIII.

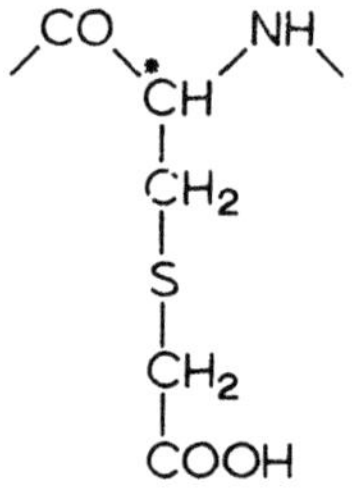

VII

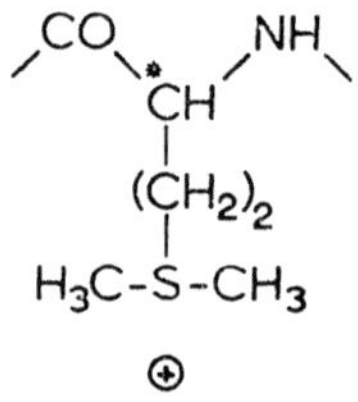

VIII

ORD curves of the polyiodide and the polythiocanate are concentration-dependent in opposition to these of polychoride and polybromide [29]. The given interpretation uses a β-structure-random coil conformational transition due to the site-binding of iodide and thiocyanate ions. We must note that this site-binding could be sufficient to explain the differences of ORD curves of the salts if the side chain chromophores appear optically active which is not shown in the paper.

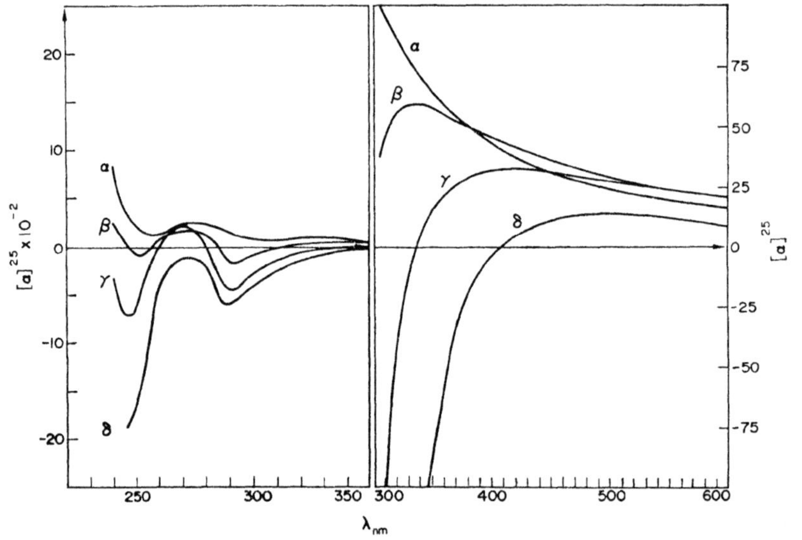

Fig. 10. ORD curves of polycondensates of N-tosyl L-tyrosine (III) with HCHO prepared in acidic medium (IX) with different initial proportion of HCHO (solvent: dioxanne)

$$\alpha = \frac{0.5}{1}; \quad \beta = \frac{1}{1}; \quad \gamma = \frac{1.5}{1}; \quad \delta = \frac{2.5}{1}.$$

5. Stereoirregular Optically Active Polymers

The special case of the formophenolic polycondensate of N-tosyl L-tyrosine prepared in acidic medium suggested to us a possible simplification in the interpretation of the optical activity of polymers in random coils.

Actually, the ORD of this polymer in dioxane has been found complex and drastically different from that of its unit model molecule. Furthermore, it obeys Moffitt's equation with a_0, b_0 and λ_0 constants very similar to those characteristic of the helical conformations of polypeptides [3]. At a first glance this could be attributed to the influence of a preferential macromolecular conformation. Yet we have shown that formaldehyde does not only react in α with the phenol giving a formophenolic polycondensate, but also with some of the side-chain sulfonamide functions giving a copolymer (IX) the composition and the ORD of which depend on the initial proportion of HCHO (Figure 10).

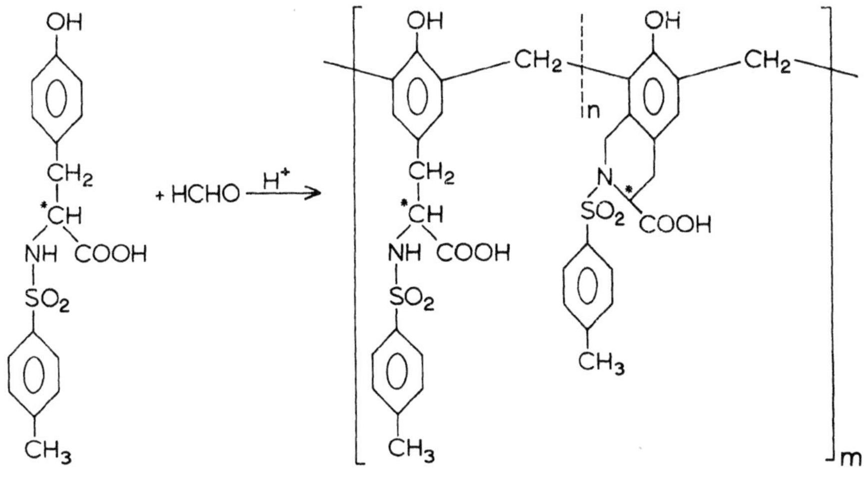

III IX

The similarity with the ORD's obtained by mixing the appropriate models of each unit model molecule of this copolymer: N-tosyl L-phenylalanine (X) and N-tosyl tetrahydro-1, 2, 3, 4 isoquinolinoic-3 acid (XI) (Figure 11) has shown that for this irregular polymer the influence of the conformational equilibrium is only of secondary importance when compared with that of the chemical modification [30].

From now on we will consider first of all that any appreciable perturbation of the COTTON effects consecutive to a modification in the external conditions is the result of a chemical reaction before envisaging the influence of conformations. Up til now, we have not needed to take into consideration order-disorder transitions to interpret the behaviour of the optically active polymers that we have synthesized.

The evolution of the optical activity of a polyacid of the formophenolic type (XII)

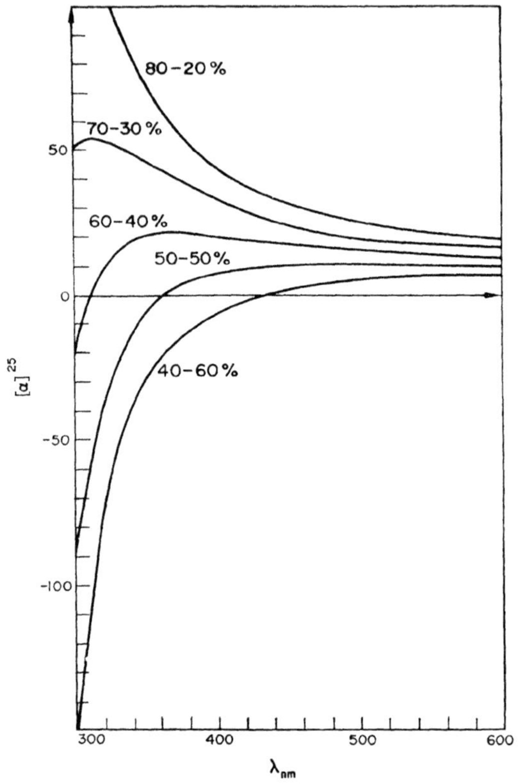

Fig. 11. ORD curves of mixtures of N-tosyl L-phenylalanine (*X*) and of tetrahydro-1, 2, 3, 4 isoquinolinoic-3 acid (XI) used as models of the monomeric units of the copolymer (IX) (solvent: dioxanne).

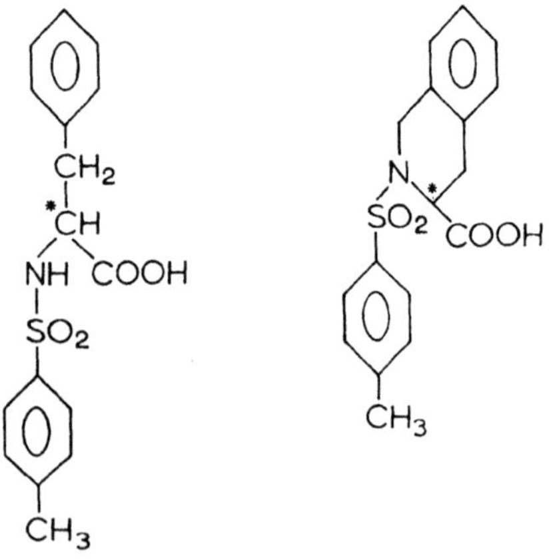

X XI

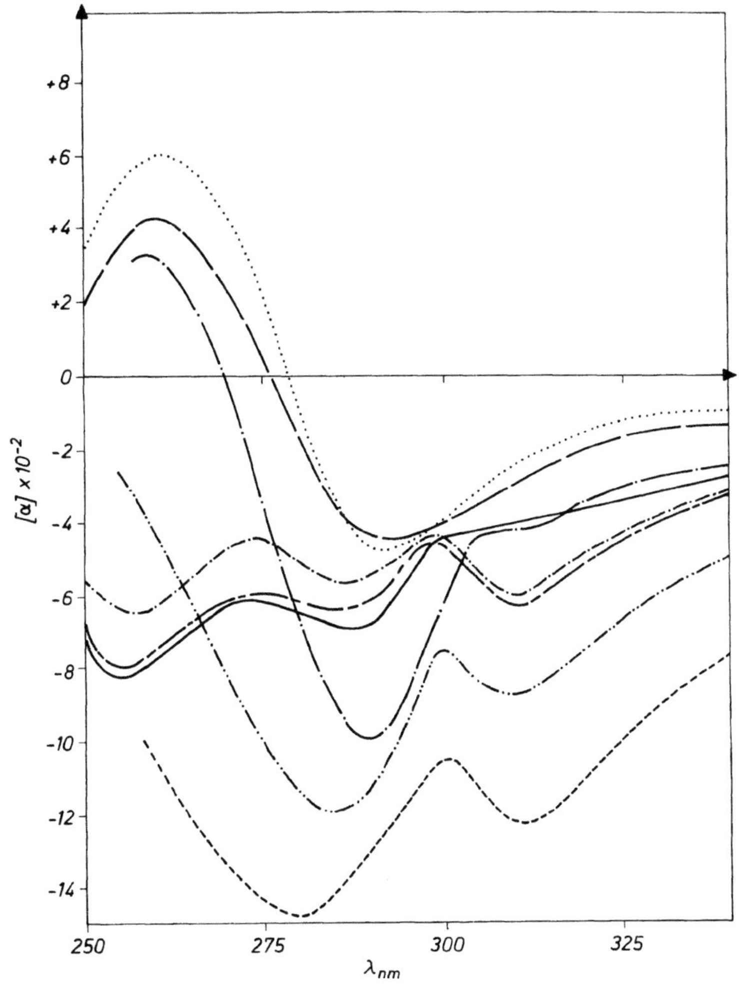

Fig. 12. ORD curves of an N-tosyl L-tyrosine (III)-HCHO polycondensate prepared by acidic catalysis (XIII) in function of the pH of the excess potassium hydroxide concentration:

pH = 4.8 ————; pH = 6.3 — — — — —; pH = 8,8 —·—·—·;
pH = 11.6 —— —— ; KOH = $M/50$ .... ; KOH = $M/5$ ——·—— ;
KOH = $M/2$ —··—·· ; KOH = M --------

during neutralisation by potassium hydroxide [15] can be cited as a first example. The ORD curves of this multifunctional polyelectrolyte depend on a large number of COTTON effects (Figure 12). Qualitatively the same variations are found as for N-tosyl L-tyrosine used as model molecule monomeric unit (Figure 3).

Between pH.8.8 and KOH = $M/50$ the appearance of a positive contribution around 260 nm can be clearly seen, and for higher potassium hydroxide concentrations, a

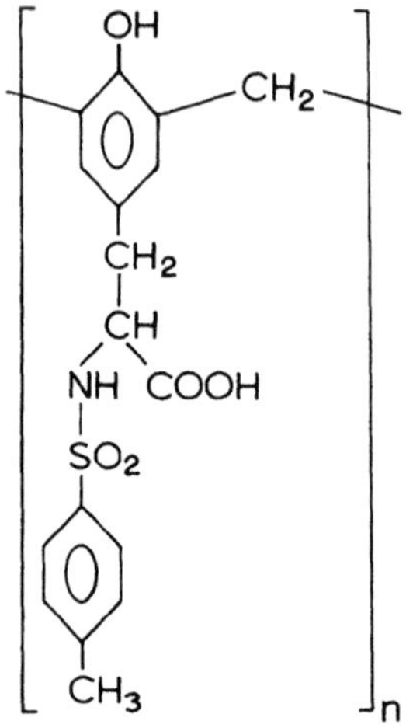

XII

negative one appears around 270 nm; they are due to the respective ionisations of the phenol and sulfonamide functions.

The modifications in the COTTON effects also affect the optical rotary power that is measured in the visible region in function of the degree of neutralisation (Figure 13).

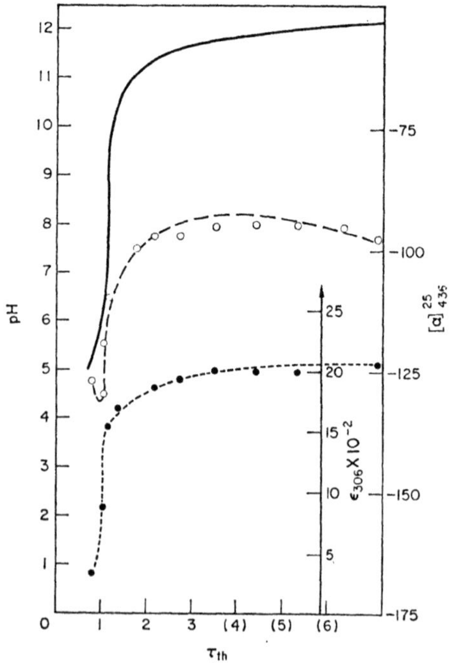

Fig. 13. pH, $[\alpha]_{436}{}^{25}$ and ε_{306} in function of the degree of neutralisation τ of XII. (pH $= f(\tau)$ ——; $[\alpha]_{436}{}^{25} = f(\tau)$ – – – –; $\varepsilon_{306} = f(\tau)$ - - - -).

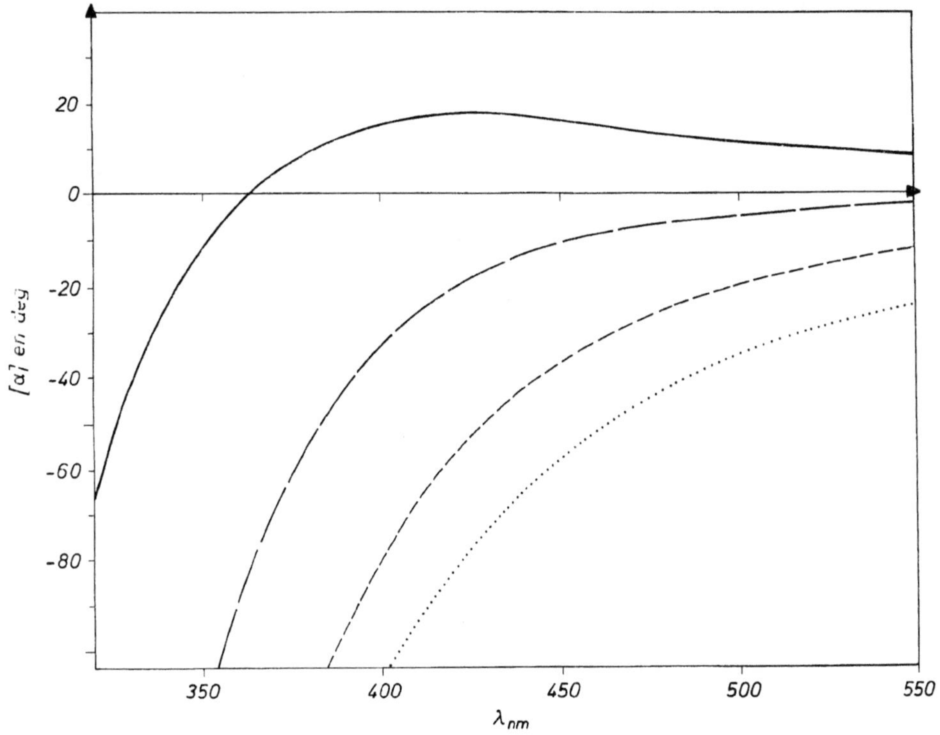

Fig. 14. ORD of poly(−) 1, 2-diamino propane sebacamide (XIII) in methanolic solution ($c = 0.5$ g 100 cm^3) in the presence of $CaCl_2$.
($c_{CaCl_2} = 2$ M/l^{-1}——; $= 1.5$ M l^{-1}— —; $= 1$ M l^{-1}-----; $= 0.5$ M l^{-1}......).

The shapes of the curves are the same as for the model (Figure 4), but it must be noted that variations in the optical rotary power and in the absorption of the phenate ion do not vary in a linear fashion with the degree of neutralisation of the totality of ionizable groups and that the extreme values do not correspond to the theoretical equivalences of each of the functions.

Similar differences have already been noted for other polyelectrolytes [2, 31]. Apart from any direct conformational effect, there are at least two reasons that can explain these differences:

- the ionisation constant depends on the degree of ionisation, and there is consequently no simple equilibrium.
- as with all other specific parameters of a monomeric unit, the optical rotary power per repeating unit depends on the state of the neighbouring units and, consequently, on the stage of the reaction.

Under these conditions the additivity law of the contributions of different species in solution cannot be applied strictly to the units of a macromolecule. Nevertheless, qualitatively one can take account of the modifications in the COTTON effects, especially for detecting reversible or irreversible chemical reactions [32].

This argument has led us to suspect the reaction of formaldehyde on the sulfonamide functions during the formophenolic polycondensation of N-tosyl L-tyrosine in the acidic medium mentioned previously [15].

It is also by this means that we have studied the interaction of mineral salts with nylon-type polyamides in alcoholic medium using an irregular, optically active polymer: (—) 1, 2-diamino propane polysebacamide (XIII).

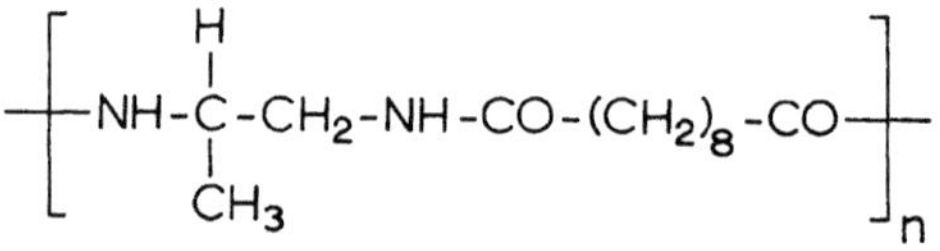

XIII

Nylons are insoluble in alcohols, but a large number of them dissolve when in the presence of mineral salts, in particular of calcium chloride.

We have seen that the ORD of (XIII) in methanol + $CaCl_2$ solution is complex and depends on the concentration of the salt (Figure 14).

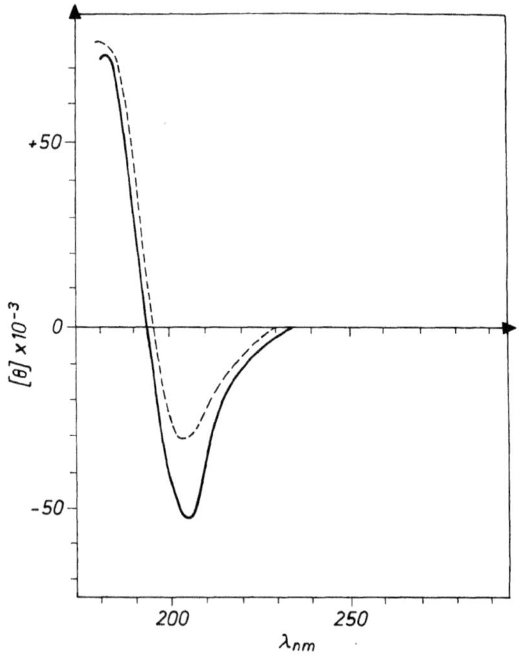

Fig. 15. DC of XIII in 2 M methanol + $CaCl_2$ solution ---- and IM methanol + $CaCl_2$ solution ——.

One of the COTTON effects situated in the region of the spectrum of $\pi-\pi^*$ transitions of C=O of amides decreases when the concentration of the salt increases (Figure 15). By comparison with the behaviour of a model molecule placed under the same conditions (*V* described above), this decrease has been attributed to the formation of adducts between the salt and the amide functions of the macromolecule explaining also the solubilisation. The existence of adducts along the polymer chains has been confirmed by precipitation of polyamide with added and not included salt from polyamide$-CaCl_2$ methanolic solutions [33].

Another example is that of the formophenolic polycondensate of N-benzoyl L-tyrosine (XIV) which has a carboxilic, a phenol and an amide function per unit.

In 0.1 M potassium hydroxide solution the addition of lithium bromide causes an inversion in the sign of the ORD. Potassium chloride has the same effect, though to a less extent for the same salt concentration (Figure 16). This inversion comes about progressively without any manifestation of conformational transition (Figure 17). Since it does not happen for similar polymers without amide functions, we have at-

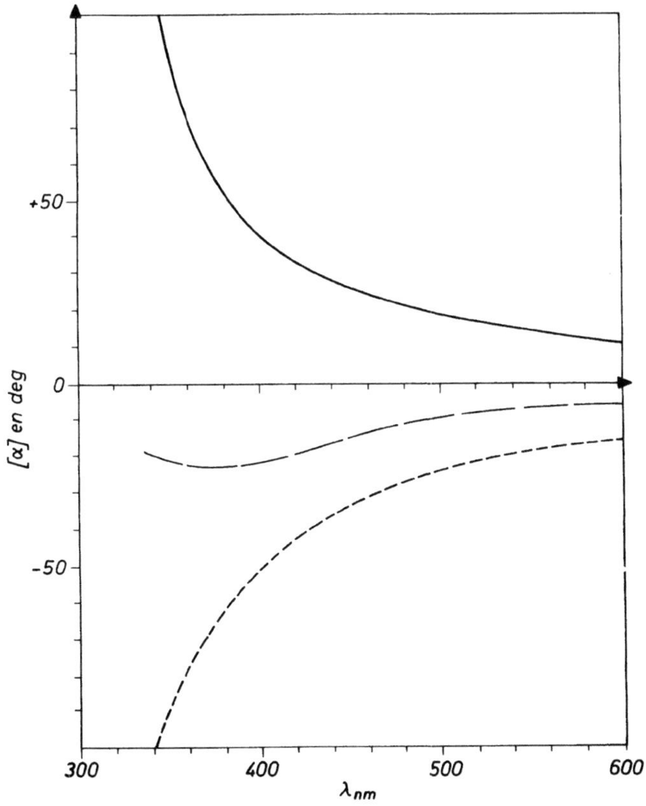

Fig. 16. ORD of N-benzoyl L-tyrosine and HCHO polycondensate (XIV) in 0.1 M KOH solution ——, KOH 0.1 M + KCl 3.6 M —— —— and KOH 0.1 M + LiBr 3.6 M ------.

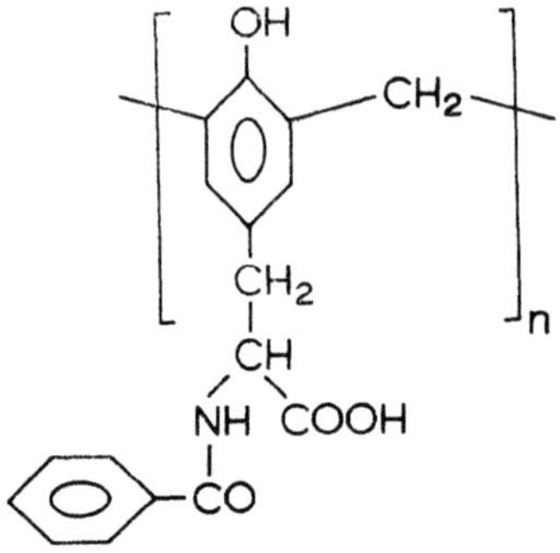

XIV

tributed it to an interaction between these functions and the salt without stating the exact nature of this interaction just from the ORD curves [34].

Although not directly contained within the framework of this paper, before finishing, I should like to mention the case when it is the ion or the small molecule that is optically active instead of the macromolecule because this case offers an interesting approach to ion-polyion interactions through the optical activity. One can mention for example the variations of the ORD of complex Co III ions in the presence of sodium

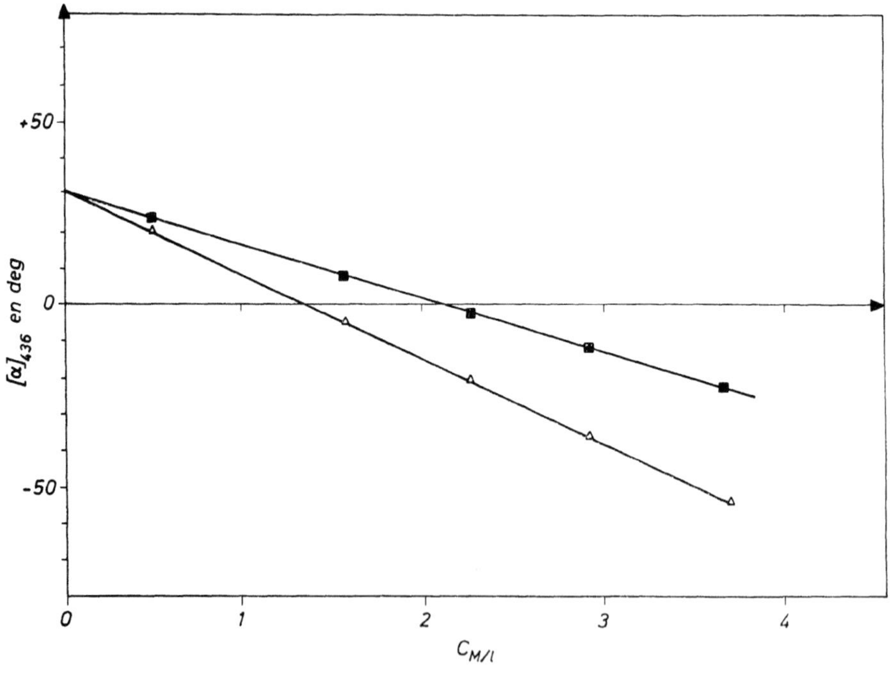

Fig. 17. Variation of the optical rotary power of XIV at 436 nm in 0.1 M KOH solution in function of the concentration of KCl ■ and of Li Br △.

salts of various polyelectrolytes (polyphosphates, polymethacrylates etc...) that have been attributed to selective fixation of the complex ion on sites [35].

6. Conclusions

We hope that these few examples have shown that in certain cases the optical activity which has been widely used for conformational analysis can also provide excellent spectral methods of detecting and studying reactions and interactions of polymers, and especially of polyelectrolytes, in solution.

The limits are first of all of a technical nature because of the imprecision in measuring the COTTON effects on the absorption bands and because of the difficulty in exploiting the curves quantitatively. They are secondly of a theoretical nature because in the majority of cases simply additive spectrophotometric laws cannot be applied to monomeric units, nor can the optical activity of molecules be calculated from molecular parameters except in a few cases when the structures are rigid and well defined.

One of our immediate objectives is to evaluate how far the neighbouring units intervene on the specific parameters of another given unit in order to circumvene the impossibility of applying spectrophotometric laws by statistical calculations. We think too that the ORD and, most especially, the CD of irregular optically active polymers should provide interesting means of studying the problems of identification of ion-polyion interactions and of the interpretation of the denaturing action of different agents of biopolymers such as urea, salts, etc....

Acknowledgements

I should like to express my gratitude to Professor E. Sélégny for having initiated this line of research and to my fellow-workers Dr J. Huguet, Mr G. Andrieux, Dr J. Beaumais, Dr C. Braud, Dr J. C. Fenyo and Dr M. R. Hamoud for having helped me to realize it.

References

1. Pino, P., Ciardelli, F., and Zandomenghi, M.: *Ann. Rev. Phys. Chem.* **21**, 561, (1970).
2. Sélégny, E., Vert, M., and Thoaï, N.: *Compt. Rend. Acad. Sci.* **262**, 189 (1966).
3. Vert, M. and Sélégny, E.: *Bull. Soc. Chim. Fr.* **663** (1971).
4. Caldwell, D. J. and Eyring, H.: *Theory of Optical Activity*, J. Wiley and Sons, New York, 1971.
5. Crabbe, P.: *ORD and CD in Organic Chemistry*, Holden Day, San Francisco, Calif., 1965.
6. Djerassi, C.: *ORD Applications to Organic Chemistry*, MacGraw-Hill, New York, 1960.
7. Fasman, G. D.: *Poly-α-amino-acids*, M. Dekker Inc., New York, 1967.
8. Jirgensons, B.: *Optical Rotatory Dispersion of Proteins and Other Macromolecules*, Springer-Verlag, Berlin, 1969.
9. Snatzke, G.: *ORD and CD in Organic Chemistry*, Heyden and Sons, Ltd., 1967.
10. Velluz, L., Legrand, M., and Grosjean, M.: *Optical Circular Dichroïsm*, Verlag Chemie Acad. Press, 1965.
11. Moscowitz, A.: in G. Snatzke (ed.), *ORD and CD in Organic Chemistry*, Heyden and Sons Ltd., 1967, p. 329.
12. Vert, M.: Dr. Es Sci. Thesis; Rouen, june 1969, No. CNRS AO 3224.
13. Parriaud, J. C.: *Bull. Soc. Chim. Fr.*, 103 (1950).

14. Gervais, H. P. and Rassat, A.: *Bull. Soc. Chim. Fr.*, 743 (1961).
15. Vert, M. and Sélégny, E.: *Europ. Polym. J.* **7**, 1321 (1971).
16. Fenyo, J. C., Beaumais, J., Sélégny, E., Petit-Ramel, M., and Martin, R. P.: *J. Chim. Phys.*, 299 (1973).
17. Hamoud, M. R.: Dr. es Sci. Thesis, Paris, July 1971, No. CNRS AO 5941.
18. Résultats non publiés.
19. Kasha, M.: *Radiat. Res.* **20**, 55 (1963).
20. Moffitt, W.: *J. Chem. Phys.* **25**, 467 (1956).
21. Pino, P., Ciardelli, F., Lorenzi, G. P., and Motagnoli, G.: *Makromol. Chem.* **61**, 207 (1963).
22. Goodman, M. and Chen, S. C.: *Macromol.* **4**, 625 (1971).
23. Overberger, C. G., Ohnishi, A., and Gomes, A. S.: *J. Polymer Sci.* A1, **9**, 1139 (1971).
24. Doty, P., Wada, A., Yang, J. T., and Blout, E. R.: *J. Polymer Sci.* **23**, 851 (1957).
25. Blout, E. R.: in C. Djerassi (ed.), *ORD Application to Organic Chemistry*, McGraw-Hill, New York, 1960, p. 248.
26. Holzwarth, G. and Doty, P.: *J. Am. Chem. Soc.* **87**, 218 (1965).
27. Nozawa, T. and Hatano, M.: *Makromol. Chemie* **141**, 21 (1971).
28. Makino, S., Murai, N., and Sugai, S.: *J. Polymer Sci. B*, **6**, 477 (1968).
29. Makino, S. and Sagai, S.: *J. Polymer Sci. A2*, **5**, 1013 (1967).
30. Vert, M.: *Europ. Polymer J.* **8**, 513 (1972).
31. Morawetz, H.: *Macromolecules in Solutions*, Inters. Publ., New York; *High Polym.*, Vol. 21 (1965), p. 243.
32. Sélégny, E. and Vert, M.: IUPAC Inter. Symp. Macromol., Helsinki 1972, prep. II-54.
33. Hamoud, M. R., Vert, M., and Sélégny, E.: *J. Polymer Sci. B*, **10**, 361 (1972).
34. Beaumais, J., Vert, M., and Sélégny, E.: *Makromol. Chem.* **170**, 23 (1973).
35. Crescenzi, V. and Pispisa, B.: *J. Polymer Sci. A2*, **6**, 1993 (1968).

SOME POSSIBILITIES OF NUCLEAR MAGNETIC RESONANCE IN THE STUDY OF POLYELECTROLYTE SOLUTIONS

G. WEILL and P. SPEGT

Centre de Recherches sur les Macromolécules, C.N.R.S., 67083 Strasbourg, France

The use of nuclear magnetic resonance (NMR) in problems of molecular interaction has taken an increasing importance. In the case of simple electrolytes it has provided particularly detailed informations on the hydration shell and the exchange time of water [1, 2]. There is a number of problems in the study of polyelectrolyte solutions for which NMR techniques offer new possibilities. Let us mention in particular:

- The stoechiometry and constants of binding.
- The characterisation of the binding sites and the state of hydration of bound ions.
- The mobility of counter-ions.
- The state of water in concentrated solutions.

The development of these techniques has been mostly stimulated by problems involving biopolymers with specific sites but they are directly applicable to synthetic polyelectrolytes. They use the changes, induced by the addition of a polyelectrolyte to an ionic solution, in the three main quantities which characterize the magnetic resonance [3]:

- The resonance frequency.
- The line shape of the resonance signal.
- The relaxation times T_1 and T_2.

The observed signal can either be that of the protons of water, or that of less classical nuclei as ^{205}Tl (an ion which has the advantage of a spin 1/2, a large natural isotopic abundance of 70.5%, and a large gyromagnetic ratio, of the order of only 1/2 of the proton) or nuclei of higher spin I like D ($I=1$) and ^{17}O ($I=5/2$) for the observation of water, or ^{7}Li, ^{23}Na, ^{39}K, ^{43}Ca ($I=1$) for the observation of the counter-ion itself. In the case of nuclei with spins of higher multiplicity the nuclear resonance is complicated by the existence of a nuclear electric quadrupole moment Q which interacts with field gradients.

The preceding enumeration reveals the necessity, for those who are only familiar with the 'high resolution proton magnetic resonance' and its application to the analysis of the structure and conformation of molecules, to go into some of the fundamentals of NMR which are necessary for the understanding of more general NMR methods.

1. Energy Levels of a Nucleus with a Quadrupole Moment

A spin I (which is supposed to be uncoupled to other spins) when put in a magnetic

Eric Sélégny (ed.), Polyelectrolytes, 371–382.

field H can take $2I+1$ energy states characteristic of the Zeeman levels corresponding to the different orientations (value m of the z component) of the spin with respect to the external field. If the nucleus carries an electric quadrupole moment Q, the spin degeneracy is already raised, in a zero external magnetic field, by its interaction with the local electric field gradient eq into $2I$ quadrupolar sublevels. In the presence of an external magnetic field the Zeeman levels are therefore no more equidistants and selection rules indicate that there are $2I$ allowed transitions of energy:

$$h\nu_{m-1\to m} = \gamma h H_{\text{local}} - \tfrac{3}{8} e^2 qQ \frac{2m-1}{3I^2 - I(I+1)} (3\cos^2\theta - 1), \tag{1}$$

where θ is the angle of the electric field gradient with the direction of H. One sees that the distance between the lines corresponding to different m values depend strongly on the orientation of the molecule. If all orientations are equally probable (powder diagram) the superposition of individual lines leads to a characteristic line shape (Figure 1) from which the 'quadrupolar splitting' e^2qQ can be measured. In a rigid lattice quadrupolar interaction generally dominates the line shape for nuclei with $I>1$. For spin

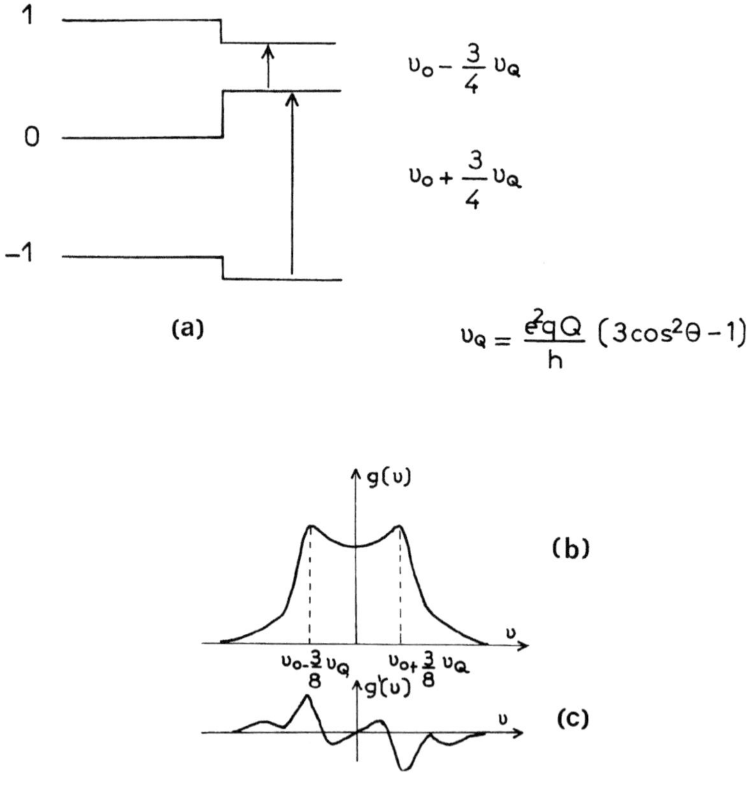

Fig. 1. Quadrupolar splitting for $I=1$. – (a) energy levels and allowed transitions; (b) powder spectrum; and (c) derivative of the powder spectrum.

1/2 nuclei, the line shape is dominated by variations of the local field, due to the variation of the dipolar interaction of the neighbouring spins, with the angle of the internucleus vector with the direction of H.

2. Relaxation Times and Relaxation Processes [4]

The energy $h\nu$ of allowed transitions is always very small compared to kT (typically $\nu = 40$ Mc for protons in a 10^4 Oe magnetic field, i.e. $h\nu/kT \sim 10^{-6}$). According to Boltzmann statistics, the difference in population of the Zeeman levels is therefore extremely small. The absorption of photons of energy $h\nu$ would rapidly equalize these populations (bringing the 'spin temperature' to infinity) if no other process than Einstein's spontaneous emission would contribute to restore the equilibrium Boltzmann distribution. Indeed, the 'natural life time' at these frequencies is of the order of 10^{12} s, while in practice the equalization of the spin temperature with the lattice temperature requires time of the order of the second in liquids. The relaxation mechanisms which act in these circumstances are linked to the fluctuation of the magnetic field (or quadrupolar coupling) inside of the sample, more precisely to the Fourier component of this fluctuation at the NMR frequency. The relaxation so far described correspond to the recovery of the z component of the magnetisation which measures the actual difference of population of the Zeeman levels. Therefore it is called longitudinal or spin lattice relaxation time T_1.

It must be realized that bringing the magnetisation out of equilibrium leads to the appearance of a transverse component of the magnetisation, i.e. in the plane containing the magnetic vector H_1 of the RF field at a rotating frequency ω_0. If as in Figure 2 the RF field is applied a time just sufficient to rotate the magnetisation vector M from the z direction into the H_1 plane ($\pi/2$ pulse) it is easy to understand that two relaxation times are necessary to describe the evolution of the magnetisation vector: T_1 which characterize the recovery of the z component and T_2 which characterize the decay

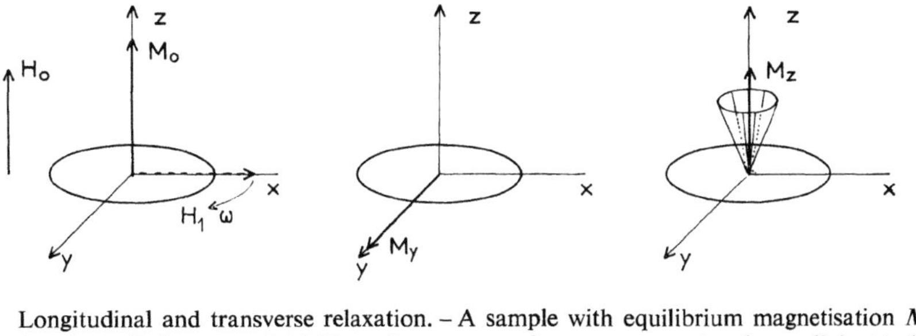

Fig. 2. Longitudinal and transverse relaxation. – A sample with equilibrium magnetisation M_0 is submitted to the radiofrequency field H_1, at the resonance frequency ω_0, a time just sufficient to bring the magnetisation in the xy plane. In the 'rotating frame' at frequency ω_0 it is then perpendicular to the direction H_1 (y axis Figure b). Due to the spin spin interactions individual spins spread so that the resultant of their z components increases towards M_0 with a time constant T_1 while the resultant of their transverse components decays to zero with a time constant T_2. Figure c shows how T_2 can be smaller than T_1.

of the transverse component (spin-spin or transverse relaxation time). This decay can be much faster than the recovery of the longitudinal component, due to the progressive dephasing of individual spin components as represented in Figure 2. This dephasing arises from the zero frequency component of the magnetic field fluctuation. Field inhomogeneities throughout the sample play a similar role but can be eliminated by the 'spin echo' technique [4] used for the measurement of T_2. If pulse techniques are generally needed for the determination of T_1 and T_2, they can be measured, in the 'high resolution' spectra, from line width determination and saturation studies (variation of the intensity of the signal with the power of the RF field).

In many cases, in liquids, the fluctuations of the local field are mainly due to the modulation of the spin-spin dipolar interaction through the rotational brownian motion of the molecule. If it is assumed to be described by a single correlation time τ, as for a macroscopic sphere, analytical expressions of T_1 and T_2 can be derived and take a rather simple form.

$$\begin{aligned} T_1^{-1} &= \frac{3}{10}\frac{\gamma^4 h^2}{r^6}\left|\frac{\tau}{1+\omega^2\tau_c^2}+\frac{4\tau}{1+4\omega^2\tau^2}\right| \\ T_2^{-1} &= \frac{3}{20}\frac{\gamma^4 h^2}{r^6}\left|3\tau+\frac{5\tau}{1+\omega^2\tau^2}+\frac{2\tau}{1+4\omega^2\tau^2}\right|, \end{aligned} \tag{2}$$

where r is the internuclear distance and ω the resonance circular frequency. If $\omega\tau \ll 1$, $T_1 = T_2$ as it should from the preceding discussion. T_1 has a minimum for $\omega\tau \sim 1$ while T_2 is a monotonously decreasing function of τ.

For spins of higher multiplicity, the dominant relaxation mechanism in fluids is also the rotational brownian motion which modulates the angle θ of expression (1). For fast isotropic rotations the relaxation times become also inversely proportional to the correlation time.

$$T_1^{-1} = T_2^{-1} = \frac{3}{8}\left[\frac{e^2 qQ}{\hbar}\right]^2 \tau. \tag{3}$$

3. Motional Narrowing

The onset of molecular motion which modulates the local interaction will not only govern the relaxation time. If fast enough, it will average out the local field and quadrupolar splitting, leading to a narrowing of the resonance signal. This averaging process will become effective for correlation times small compared to the inverse of the rigid lattice line width ΔH in a frequency scale

$$\tau \ll (\gamma \Delta H)^{-1}.$$

Typically it corresponds for protons to $\tau \ll 10^{-4}$ s. In the 'extreme narrowing' limit the actual line width becomes inversely proportional to T_2.

4. Chemical Exchange

A rather similar averaging process can take place when the nuclei can exist in two different chemical or physical environment in the sample (for example protons in a partly dissociated acid, water in the hydration shell and in the bulk solvent ...). To each environment will correspond:

– a definite resonance frequency (chemical shift)
– a given relaxation time T_1 and T_2.

If the *exchange* between the two environments *is slow*, which means a long residence time compared to the inverse of the difference in resonance frequencies and to the relaxation times, two resonance signals can be observed, corresponding to the nuclei in their two specific environments a and b.

If the *exchange is fast*, there will be only one resonance signal whose frequency and relaxation time are the weighted average of the values in the specific environments. If x is the stationary number of nuclei in environment a

$$\nu = x\nu_a + (1 - x)\,\nu_b$$
$$\frac{1}{T_{1,2}} = \frac{x}{T^a_{1,2}} + \frac{1 - x}{T^b_{1,2}}. \qquad (4)$$

This averaging effect can be very interesting since if $\nu_a - \nu_b$ and T^b/T_a are large it will be possible to detect the characteristics of nuclei in environment a even for very small values of x.

The principle of most of the experiments designed for the study of polyelectrolytes can now be understood on the basis of these qualitative explanations. We shall restrict ourselves to four types of experiments, with a special emphasis on our own work on the use of paramagnetic counter-ions.

4.1. Direct study of the counter-ion

A few studies of the modification of the counterion magnetic resonance signal upon binding to polyelectrolytes have been performed. James and Noggle [5] have measured the binding constant of ^{23}Na to s. RNA and the ratio of the relaxation times of free and bound Na. The large value of 12.5 reflects either a large increase of the field gradient or more probably a decrease of the rotational correlation time. Further example of the use of ^{23}Na relaxation time will be found in Dr Leyte's paper in this volume. The use of ^{205}Tl which has no electric quadrupole moment has allowed an approximate determination of its residence time on the enzyme pyruvate kinase, from line shape analysis. [6].

4.2. Paramagnetic counter-ions – water relaxation enhancement and chemical shifts

The use of paramagnetic counterions is of particular interest since they affect considerably the chemical shift and relaxation of the protons of water in their first hydration shell. In the fast exchange limit the intense signal of water carries all the interesting information. Many terms govern the relaxation and chemical shift of water, among which the most important are: [7, 8, 9, 10]

– the number p and distance of the exchangeable water molecules in the first hydration shell.

– the correlation time τ_c for the rotational motion of the ion.

– the electronic correlation time τ_s of the electronic spin.

According to the large or small value of τ_s, a given counter-ion will be more or less effective in the relaxation process: Mn^{++} has a large τ_s and will be mainly used for 'relaxation enhancement' studies, Co^{++} has a very small τ_s and will be used for chemical shift determinations since the decrease of T_2 upon binding remain sufficiently small to limit the broadening of the water signal.

Site binding of a counterion to a polyelectrolyte involves the replacement of a given number of the p water molecules of the first hydration shell by ligands from the macro-

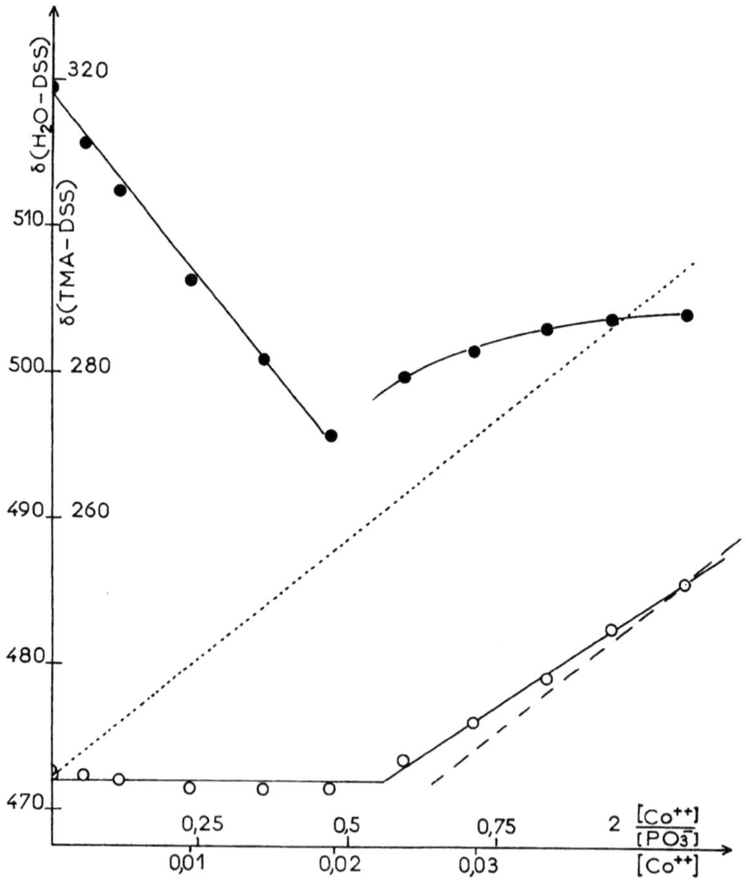

Fig. 3. Chemical shift of the water (○) and TMA signals (●) upon addition of Co^{++} to a polyphosphate of TMA solution. ------ Chemical shift of a Co^{++} solution without polyelectrolyte.

molecule. Both p and ζ_c are modified. Assuming that the distance between the ion and the remaining water of the first hydration shell is unchanged in average, a comparison of the chemical shift of water in the presence and absence of polyelectrolyte gives directly the change in p. Changes in both p and ζ_c are involved in the variation of the relaxation times. Defining the 'relaxation enhancement' ε as [7]:

$$\varepsilon = \frac{T_i^{-1}(M, P) - T_i^{-1}(O, P)}{T_i^{-1}(M, O) - T_i^{-1}(O, O)}, \tag{5}$$

where $T_i\,(M, P)$ is the relaxation time in the presence of a concentration M of counter-ion and a concentration P of polyelectrolyte, it is easy to derive from the variation of ε with M and P:

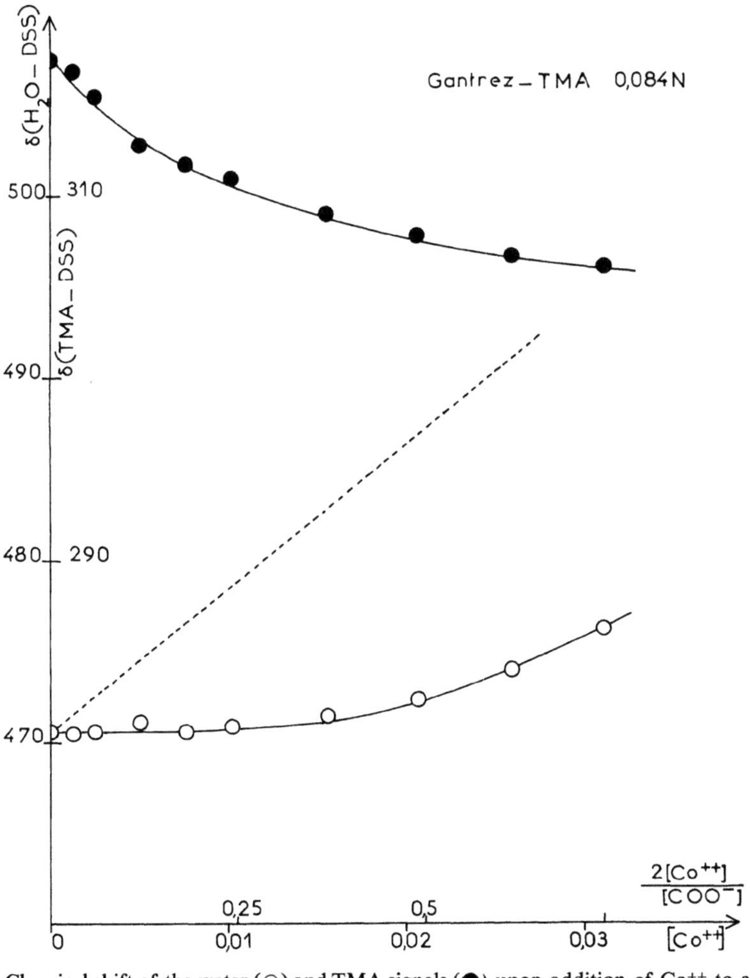

Fig. 4. Chemical shift of the water (○) and TMA signals (●) upon addition of Co^{++} to a solution of the TMA salt of Gantrez.

– the number of binding sites
– the binding constant
– the characteristics of the binding site,

namely the number of exchangeable water molecules left in the first hydration shell of the counter-ion. It is generally necessary in order to obtain a more complete information to measure both T_1 and T_2 in a range of NMR frequencies and to perform EPR determination of the binding. An excellent example of this type of work is given by Danchin et Gueron in the case of Mn^{++} binding to tRNA [11].

We have performed a study of the chemical shift of water in Co^{++} solutions in the presence and absence of the tetramethylammonium (TMA) salt of different polyelectrolytes [12]. A very typical example is given in Figure 3 relative to polyphosphate. The binding is characterized by a complete loss of exchangeable water (no shift in the presence of polyelectrolyte) and a sufficiently high constant to be nearly stoechiometric up to the total number of binding sites which correspond to 1 Co^{++} for 4 PO^{4-} groups. Other polyelectrolytes, like the alternated copolymer of vinylether and maleic anhydride (Gantrez, Figure 4) reveal a similar dehydration but the binding constant is smaller. In this case the relaxation enhancement has been measured and is of the order

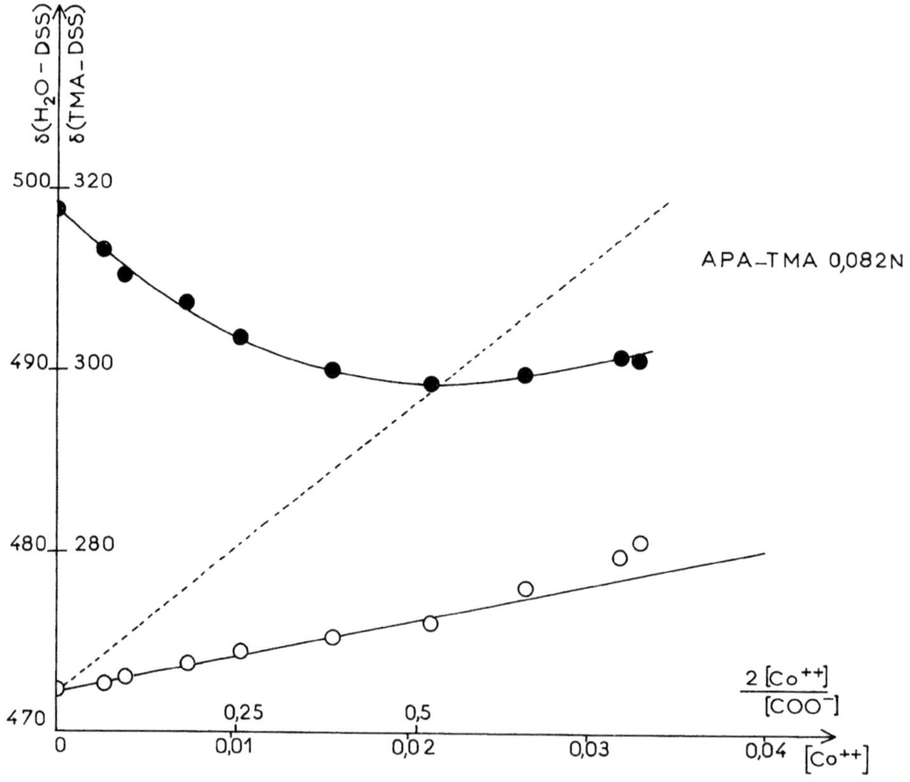

Fig. 5. Chemical shift of the water (○) and TMA signals (●) upon addition of Co^{++} to a solution of the TMA salt of polyacrylic acid.

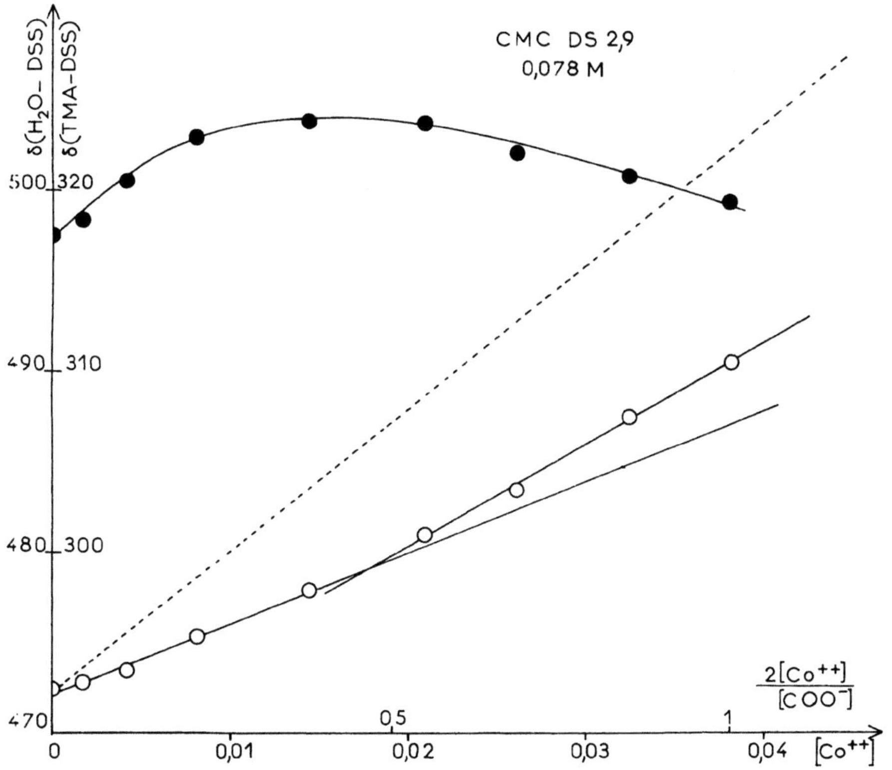

Fig. 6. Chemical shift of the water (○) and TMA signals (●) upon addition of Co^{++} to a solution of the TMA salt of a CMC of degree of substitution 2,9.

of 5. Polyacrylic acid (Figure 5) shows only a partial dehydration while carboxymethylcellulose (CMC) of high degree of substitution shows a complicated behaviour related to the presence of several binding sites (Figure 6).

The use of TMA salts, with its easy observable proton signal, allowed the observation of the counterion chemical shift. Its unexpected behaviour which is represented on top of Figure 3 has led to a promising technique for the study of the counter-ion distribution around the polyion. It is due to the 'pseudo contact shift' resulting from anisotropic g factor of the paramagnetic ion. This effect is now in fact widely used in high resolution NMR where lanthanides chelates are used as 'shift reagents' [13]. Due to the anisotropy of the electron spin gyromagnetic tensor, the electron spin-nuclear spin dipolar interaction does not average to zero in the fast rotation limit. The remaining shift is function of the relative position of the ion and nucleus. Its analytical expression is given by:

$$\frac{\Delta H}{H} = \frac{\beta^2 S(S+1)}{27kT}(g_{\parallel} + 2g_{\perp})(g_{\parallel} - g_{\perp})\frac{3\cos^2 X - 1}{\mathbf{r}^3}, \tag{6}$$

where β is Bohr's magneton; S the spin of the paramagnetic ion; $g_{\parallel}$ and $g_{\perp}$ the principal values of the g tensor; X the angle of the axis of symmetry of this tensor with the ion–nucleus vector **r**.

For reasonable values of the anisotropy the shift can reach 4 ppm at 5Å. In the case of a polyelectrolyte the effective chemical shift will be averaged over all possible positions of the diamagnetic counterion bearing the observable nucleus, according to the distribution function around the polyion. The characteristic behaviour of the TMA chemical shift, in Figure 3, is now easily understood. In the first part of the variation the shift is proportional to the amount of bound Co^{++} which binds to the polyelectrolyte, increasing the number of chelated anisotropic centers along the polymer. At the same time, a part of the TMA is released from the neighbourhood of the polyelectrolyte due to the reduced charge of the Co^{++} polyelectrolyte complex. At the point of equivalence, where all possible chelated Co^{++} is site bound, a further addition of Co^{++} displaces some TMA from the polyelectrolyte so that its average chemical shift tends to its value in the absence of Co^{++} and polyelectrolyte. However there is a competition between monovalent and divalent ions trapped in the electrostatic potential of the polyion. The rather low plateau value of the TMA chemical shift observed upon addition of quantities of Co^{++} close to the total concentration of TMA, reflects the fact that the layer of trapped counter-ions remains rich in monovalent ions. This is not unexpected due to the balance between polyelectrolyte–counter-ion attraction and counter-ion – counter-ion repulsion [14]. Figures 4 to 6 show that this behaviour is general. In the case of CMC the pseudo contact shift is of opposite sign. This technique seems therefore rich in information and promising for the study of short range interaction.

If the need of paramagnetic counterion may seem a stringent limitation, it must be noted that measurements involving the competition between diamagnetic and paramagnetic ions can help to characterize and explain the specificity of ion binding.

4.3. Measurements of diffusion coefficients

NMR offers an interesting method for the measurement of self diffusion coefficients D when performing a classical T_2 determination by spin echo in the presence of a magnetic field gradient G. The translatory diffusion of the nucleus in this gradient modulates the local field in such a way that the amplitude of the echo decays like $\exp(-t/T_2 - \frac{1}{12}\gamma^2 G^2 D t^3)$. The use of pulsed gradients allow measurements down to $D = 10^{-9}\ \mathrm{cm}^2\ \mathrm{s}^{-1}$. An important feature of this technique is that the diffusion is measured over times of the order of a few milliseconds. In cases of restricted diffusion, the apparent diffusion coefficient will be a more complicated function of the time between pulses and of the duration of the field gradient, from which models of restricted diffusion can be tested and barriers to diffusion evaluated [16]. Applications close to the field of polyelectrolyte have been published in the case of adsorbed water [17]. An interesting possible measurement could be performed on the self diffusion of the counterion (^{205}Tl would be well suited) in dilute and concentrated polyelectrolyte solutions. The mobility of the counterion trapped in the electrostatic potential of the polyion could probably be obtained directly.

4.4. Water mobility in concentrated solutions

The preceding method can be applied to the measurement of the diffusion of water. However, in concentrated phases, the mobility can be high without involving an isotropic rotational movement of the water molecule. The use of D_2O will give a direct information on the anisotropy of the water rotational motion through the study of the deuteron quadrupolar splitting [18]. This method has been used by Charvolin and Rigny [18] in the study of the mesophases of the water-sodium laurate system. Each anisotropic mesophase is characterized by the presence of a narrow signal due to 'free water' and a broad signal from which the apparent quadrupolar splitting of bound water can be calculated. The presence of both components can also be deduced from the saturation behaviour of the signal of protons. For the isotropic cubic phase, the averaging is naturally complete and there is only one narrow signal.

At the present stage of the study of polyelectrolytes solutions, where the gross features of the thermodynamic properties are well understood, progress concerning the details and specificity of counterion binding can only arise from experiments dealing with short range interactions. We hope that this rapid survey has made clear the physical basis of some NMR techniques which have proved useful for this purpose.

References

Reviews on the use of NMR in ionic solutions

1 Hinton, J. F. and Amis, E. S.: *Chem. Rev.* **67**, 367 (1967).
2 Hertz, H. G.: *Progr. NMR Spectrosc.* **3**, 159 (1967).

For a simple introduction to NMR in view of the following applications

3 Carrington, A. and Mc Lachlan, A. D.: *Introduction to Magnetic Resonance*, Harper International Edition, 1967.

For a simple introduction to relaxation and pulse techniques in NMR

4. Farrar, T. C. and Becker, E. D.: *Pulse and Fourier Transform NMR*, Academic Press, 1971.

Studies of the counter-ion resonance

5. James, T. L. and Noggle, J. H.: *PNAS* **62**, 644 (1969).
6. Kayne, F. J. and Revben, J.: *JACS* **92**, 221 (1970).

Use of paramagnetic counter-ions

7. Eisinger, J., Fawaz Estrup, F. and Shulman, R. G.: *J. Chem. Phys.* **42**, 43 (1965).
8. Mildvan, A. and Cohn, M.: *Biochem.* **2**, 911 (1963).
9. Dwek, R. A.: *Adv. Molecular Relax. Processes* **4**, 1 (1972).
10. Luz, Z. and Shulman, R. G.: *J. Chem. Phys.* **43**, 3750 (1965).
11. Danchin, A. and Gueron, M.: *J. Chem. Phys.* **53**, 3599 (1970).
12. Spegt, P. and Weill, G.: *Compt. Rend. Acad. Sci.* **274C**, 587 (1972).
13. Fischer, R. D. and Von Ammon, R.: *Angew. Chemie* **11**, 675 (1972).
14. Minakata, A., Imai, N., and Oosawa, F.: *Biopolymers* **11**, 347 (1972).

Self diffusion measurements

15. Stejskal, E. O.: *J. Chem. Phys.* **43**, 3597 (1965).
16. Stejskal, E. O.: *Adv. Molecular Relax. Processes* **3**, 27 (1972).
17. Kärger, J.: *Z. Physik. Chem.* **248**, 27 (1971).

Mobility of water in concentrated solution

18. Lawson, K. D. and Flautt, T. J.: *J. Phys. Chem.* **72**, 2066 (1968).
19. Charvolin, J. and Rigny, P.: *J. Phys.* **30**, C4–83 (1972).

NUCLEAR MAGNETIC RESONANCE RELAXATION OF COUNTER-IONS IN POLYELECTROLYTE SOLUTIONS

J. C. LEYTE, L. H. ZUIDERWEG, and J. J. VAN DER KLINK
Gorlaeus Laboratoria, Afdeling Fysische Chemie III, Rijksuniversiteit, Leiden, The Netherlands

1. Introduction

Experimental results on ^{23}Na relaxation in poly(acrylic acid) (PAA) solutions will be presented and discussed. As the ^{23}Na ion relaxation occurs by way of a quadrupolar mechanism the discussion of NMR relaxation in polyelectrolyte solutions will be restricted to quadrupolar relaxation.

The electric quadrupole of the ionic nucleus interacts with the electric field gradient at the site of the nucleus. In the liquid state this field gradient will be time dependent due to the molecular motion in the solution. If the fourier spectrum of the time correlation function of the field gradient contains sufficient amplitude at the right frequencies nuclear spin transitions may be induced. These bath induced transitions will eventually drive a non equilibrium spin system to thermal equilibrium. According to Herz and coworkers [1] the field gradient at a given ionic nucleus in an electrolyte solution is mainly due to the solvent dipoles and the charges on the surrounding ions.

In the following presentation the work of several authors has been used and to avoid continuous citation, reference to these important works is made here [2, 3, 4].

2. Relaxation in Polyelectrolyte Solutions

The Hamilton operator $H_1(t)$ describing the interaction of the spin system with the bath can, for the quadrupolar relaxation mechanism, be written as a function of space quantized operators $A^{(p)}$ and time dependent bath functions $V^{(p)}(t)$.

$$H_1(t) = \sum_p (-1)^p V^{(p)}(t) A^{(-p)}. \tag{1}$$

In Table I these functions and operators are defined explicitly in terms of the spatial derivatives $V_{\alpha\beta}$ of the electrostatic potential, the angular momentum operators I_α and the nuclear quadrupole moment Q.

The orientation of the coordinate system in (1) is laboratory fixed with the z-axis along the static magnetic field H_0. The origin is chosen to coincide with the relaxing nucleus. In this coordinate system the components $V^{(p)}(t)$ of the electric field gradient fluctuate due to the motion of the environment relative to the nucleus considered.

The time dependence of the magnetization is then calculated by solving the equation of motion for the density matrix σ to the second order in $H_1(t)$. The results are

Eric Sélégny (ed.), Polyelectrolytes, 383–389.

TABLE I

p	$V^{(p)}$	$A^{(p)}$
0	$\frac{1}{2}V_{zz}$	$\frac{1}{2}\frac{eQ}{I(2I-1)}(3I_z^2-I^2)$
± 1	$\mp\frac{1}{\sqrt{6}}(V_{zx}\pm iV_{zy})$	$\mp\frac{1}{4}\sqrt{6}\frac{eQ}{I(2I-1)}(I_zI^{\pm}+I^{\pm}I_z)$
± 2	$\frac{1}{2\sqrt{6}}(V_{xx}-V_{yy}\pm 2iV_{xy})$	$\frac{1}{4}\sqrt{6}\frac{eQ}{I(2I-1)}I^{\pm 2}$

conveniently expressed using the spectral density $J_{qp}(\omega)$. It is defined as the Fourier transform of the time correlation function of the field gradient.

$$J_{qp}(\omega)=\int_0^{\infty}\langle V^{(q)}(t)\,V^{(-p)}(t+t')\rangle\,e^{iq\omega_0 t'}\,\mathrm{d}t'. \tag{2}$$

The brackets indicate an equilibrium average. In terms of the quantities defined so far the equation for the time dependence of $\langle I_z\rangle$, which is proportional to the longitudinal magnetization is shown in (3)

$$\frac{\mathrm{d}\langle I_z^*\rangle}{\mathrm{d}t}=-\sum_{p,q}(-1)^q\,e^{-i(p+q)\omega_0 t}\times$$
$$\times\,\mathrm{Tr}\left[A^{(-p)},\left[A^{(-q)},\sigma^*(t)-\sigma_0\right]\right]I_zJ_{q,p}(\omega). \tag{3}$$

Here, σ_0 refers to thermal equilibrium and the starred quantities are to be taken in the interaction representation. Neglecting nonsecular terms introducing the narrowing limit (a small correlation time τ compared to the inverse Larmor frequency, $\omega_0\tau\ll 1$) and using the rotational invariance of the Hamiltonian describing the molecular motions, Equation (3) simplifies as follows.

$$\frac{\mathrm{d}\langle I_z\rangle}{\mathrm{d}t}=-J_{00}(0)\sum_q\mathrm{Tr}\left[A^{(-q)},\left[A^{(q)},I_z\right]\right](\sigma^*(t)-\sigma_0)=$$
$$=-T_1^{-1}(\langle I_z\rangle-\langle I_z\rangle_0) \tag{4}$$

$$T_1^{-1}=\frac{3(eQ)^2(2I+3)}{2I^2(2I-1)}J_{00}(0). \tag{5}$$

It may be noted that the bath properties are collected in the spectral density. To allow a more detailed discussion the structure of $J(\omega)$ has to be considered now.

The field gradient components $V^{(p)}(t)$ at a ^{23}Na ion nucleus in an aqueous solution containing negatively charged polyions, sodium ions and water molecules contains contributions from the ionic charge distribution, $V_i^{(p)}(t)$, and the solvent molecules, $V_s^{(p)}(t)$.

$$V^{(p)}(t)=V_i^{(p)}(t)+V_s^{(p)}(t). \tag{6}$$

If V_i and V_s are taken to be uncorrelated the same separation can be made for $J(\omega)$ and consequently the relaxation rate T_1^{-1} will contain independent contributions from the ionic charges and the solvent molecules.

$$T_1^{-1} = T_{1,i}^{-1} + T_{1,s}^{-1}. \tag{7}$$

In the discussion of the ionic contribution the rod-like model for polyelectrolytes will be kept in mind. In addition to the laboratory fixed coordinate system S a second frame S' will be introduced. The origins of the systems coincide but the z' axis of S' is parallel to the nearest polyion and the x' axis passes through the rod radially. This system should be particularly useful if deviations of the charge distribution from cylindrical symmetry (around the rod) are small and/or short lived with respect to the correlation time of $V^{(p)}(t)$. At the origin of S$'$ the field gradient will have C_{2V} symmetry and we have $V'_{zz}=0$, only V'_{xx} and V'_{yy} being not necessarily zero there. The gradient components in S', $V'^{(p)}(t)$ are time dependent only through their dependence on ρ, the distance between the ion and the rod-like molecule. The time dependence of $V^{(p)}(t)$ therefore arises from the time dependence of the orientation of S' with respect to S and from the ρ-dependence of the components in S'. The components $V^{(p)}(t)$ in S can now be expressed as a function of the $V'^{(p)}(t)$ in S' by means of the Wigner rotation matrices $D^{(2)}_{q,m}(t)$.

$$V^{(q)}(t) = \sum_m D^{(2)}_{m,q}(t)\, V'^{(m)}(t). \tag{8}$$

The ionic contribution to $J_{00}(0)$ can now be formulated as shown in (9)

$$\begin{aligned}(J_{00}(0))_i &= \int_0^{\tilde{}} \mathrm{d}t' \sum_{m,m'} \langle V'^{(m)}(t)\, V'^{(m')}(t+t')\, D^{(2)}_{m,0}(t)\, D^{(2)*}_{m',0}(t+t')\rangle = \\ &= \tfrac{1}{24} \sum_{m,m'} \delta_{m,\pm 2}\, \delta_{m',\pm 2} \times \\ &\quad \times \int_0^{\tilde{}} \mathrm{d}t' \langle e^2 f(t)\, f(t+t')\, D^{(2)}_{m,0}(t)\, D^{(2)*}_{m',0}(t+t')\rangle .\end{aligned} \tag{9}$$

Here f is defined by $ef = V'_{xx} - V'_{yy}$. The ionic contribution to the relaxation rate is obtained by combining (9) and (5).

$$\begin{aligned}T_{1,i}^{-1} &= \tfrac{1}{16} \frac{2I+3}{I^2(2I-1)} (e^2Q)^2 \sum_{m,m'} \delta_{m,\pm 2}\, \delta_{m',\pm 2} \times \\ &\quad \times \int_0^{\tilde{}} \mathrm{d}t' \langle f(t)\, f(t+t')\, D^{(2)}_{m,0}(t)\, D^{(2)*}_{m',0}(t+t')\rangle \times \\ &\quad \times D^{(2)}_{\pm 2,0}(\theta,\phi) = (\tfrac{3}{8})^{1/2} \sin^2\theta\, e^{\pm 2i\phi}.\end{aligned} \tag{10}$$

To allow a qualitative discussion of ^{23}Na relaxation in aqueous PAA solutions some further simplifications have to be introduced. The time dependence of $f(t)$ is due to the radial diffusion of the counter-ion relative to the polyelectrolyte molecule. As translation of the counter-ions on the cylindrical equipotential surfaces is generally considered to be a much faster process than radial diffusion it will be assumed that the time correlation function in (10) will tend to zero due to the angular time dependence before $f(t)$ has changed significantly. In other words, the gradient magnitude f is taken to be effectively time independent.

The time dependence of θ and ϕ are due to the reorientation of the rod and angular diffusion of the counter-ion with respect to the rod. At this stage the reorientation of the polymer rod is neglected as a slow process. Then, $T_{1,i}^{-1}$ is determined by the modulation of the field gradient due to the time dependence of ϕ. The spectral density now reduces to

$$(J_{00}(0))_i = \tfrac{1}{64} \int_0^{\infty} \mathrm{d}t' \sum_{m,\, m'} \delta_{m,\, \pm 2}\, \delta_{m',\, \pm 2} \times \times \langle e^2 f^2 \sin^4 \theta\, e^{im\phi(t)}\, e^{-im'\phi(t+t')} \rangle . \tag{11}$$

After averaging the θ dependence and introducing random diffusion for ϕ at constant ρ there will be only ρ dependent quantities left within the brackets. The result is given in (z) for ^{23}Na relaxation ($I = {}^3/2$).

$$T_{1,i}^{-1} = \tfrac{1}{40} \frac{(e^2 Q)^2\,(2I+3)}{I^2\,(2I-1)} \langle f^2 \tau_2 \rangle = \tfrac{1}{120}\,(e^2 Q)^2 \langle f^2 \rho^2 D^{-1} \rangle . \tag{12}$$

Here $\tau_2 = \rho^2 (4D)^{-1}$ is the correlation time for the ϕ-diffusion and D should essentially be the selfdiffusion coefficient for the counter-ions. The contribution of the solvent molecules to the ionic relaxation rate has been extensively discussed in the literature cited in the introduction. The electric dipole moment of a water molecule induces a field gradient at the nucleus of a solvated ion. This gradient is time dependent due to the reorientational movement of the H_2O molecule. In the narrowing limit and assuming random reorientation of the solvent molecules the contribution to T_1^{-1} is

$$T_{1,\omega}^{-1} = \tfrac{3}{2} \frac{2I+3}{I^2\,(2I-1)} W \tau_c , \tag{13}$$

w is the mean squared interaction of the nuclear quadrupole moment with the solvent field gradient and τ_c is the correlation time for the reorientation of the H_2O molecules.

3. Some Results

The longitudinal relaxation time T_1 of ^{23}Na has been determined in aqueous PAA solutions as a function of the degree of neutralization α and as a function of polymer concentration. These results have been collected in Figure 1 and Figure 2. All relaxation times were obtained with a Bruker pulsed NMR spectrometer operated at 16 MHz and using (π, $\pi/2$, π) pulse sequences. For a few solutions it was verified that the narrowing limit ($T_1 = T_2$) was valid.

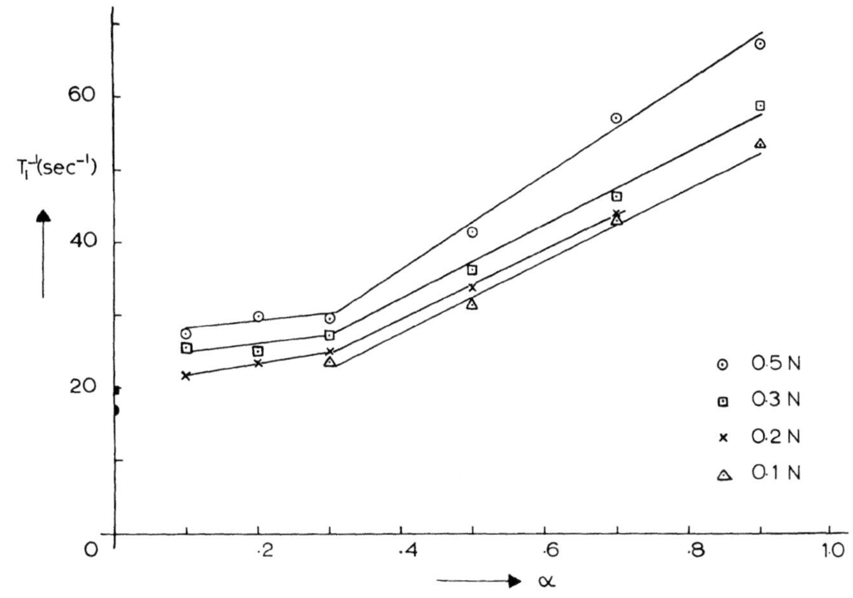

Fig. 1. Longitudinal relaxation rate of $^{23}Na^+$ ions in aqueous polyacrylic acid solutions as a function of the degree of neutralization for different polyacid concentrations. ●: Infinite dilution limit for $^{23}Na^+$ relaxation in H_2O. ■: 0.05 M NaCl in 0.5 N polyacrylic acid at $\alpha = 0$.

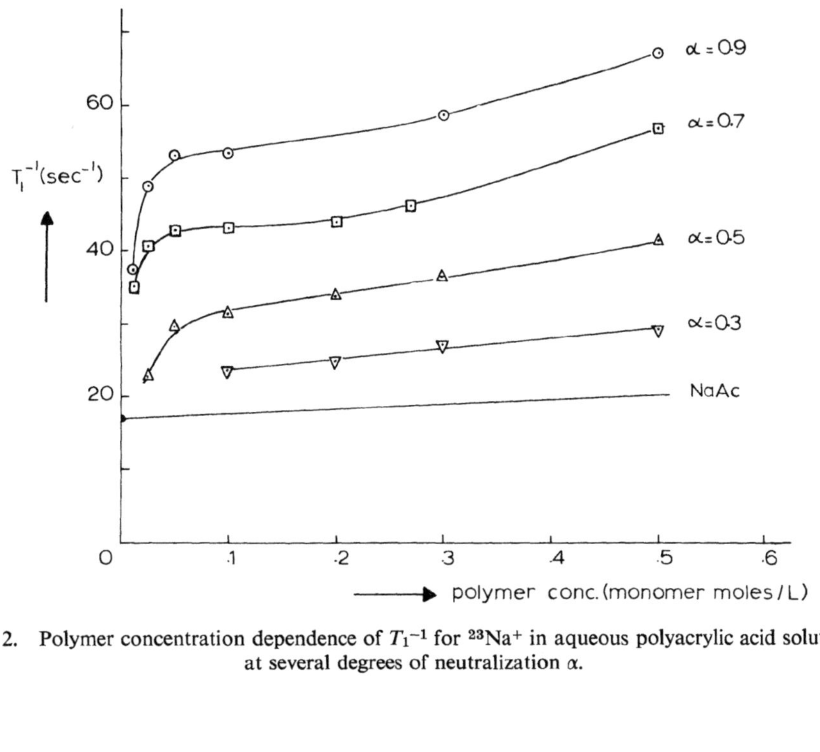

Fig. 2. Polymer concentration dependence of T_1^{-1} for $^{23}Na^+$ in aqueous polyacrylic acid solutions at several degrees of neutralization α.

STUDIES ON THE TRANSITIONAL FORMATION AND MOLECULAR INTERACTIONS OF A POLYELECTROLYTE

KANG-JEN LIU

Corporate Chemical Research Center, Allied Chemical Corporation, Morristown, N.J., U.S.A.

Abstract. Polyelectrolytes are generally formed through molecular associations or dissociations. They deviate one way of other from their corresponding simple electrolytes. The question that how and when these deviations appear as a function of molecular size is a most interesting subject to polymer scientists. Information of this nature is important for understanding the polymeric effect on molecular interactions in systems involving charged, reactive and catalytic polymers.

We have systematically studied a well defined model system, potassium iodide-poly(ethylene oxide) in methanol, by using mainly high resolution NMR technique. Oligomers from monomer to heptamer were studied in addition to high polymers of different average molecular weight. Experimental results indicate that the interaction of potassium iodide with low oligomers ($n \leqslant 6$) is very weak, and become strong as $n \geqslant 7$. The polymeric-type interaction sets in at $n = 7$, suggesting that a cooperative effect of this interaction appears only when $n \geqslant 7$. The results from electrophoresis experiments indicate that the positively charged potassium ion is actually the species which directly interacts with the polymer. Once the polymer associates with potassium ions in methanol, it behaves as a polyelectrolyte. The apparent association constant of the potassium iodide to PEO, obtained from the data of chemical shifts, decreases as one increases either the salt to polymer ratio at constant polymer concentration or the polymer concentration at constant salt to polymer ratio.

Further studies have also been made on the effects of this short-range, ionic association and the well-known charge-charge repulsive long-range interactions on the segmental motion of the polyelectrolyte. Addition of potassium iodide to poly(ethylene oxide) in methanol produces large changes in the nuclear spin-lattice relaxation (T_1) of the polymer protons. These changes can be described quantitatively in terms of free and charged units along the polymer chain. Long-range interactions, as evidenced in the macroscopic solution viscosity, are not reflected in T_1. An important distinction between macroscopic and microscopic properties is clearly shown.

Solvent effects on the ionic association have also been studied. Competition between the ionic association and the solvation of the polymer segments are discussed in terms of conformational changes in various systems.

1. Introduction

Polyelectrolytes are generally formed through molecular associations or dissociations. They deviate one way or other from their corresponding simple electrolytes. The question that how and when these deviations appear as a function of molecular size is a most interesting subject to polymer scientists. Information of this nature is important for understanding the polymeric effect on molecular interactions in systems involving charged, reactive and catalytic polymers.

Generally, the interactions of chain molecules with low molecular weight species may be of two types [1]. In the first, the polymer interact with various reagents in much the same manner as would be expected of low molecular weight analogs of the repeating unit of the macromolecule. Of much more interest are macromolecular interactions of the second type, in which cooperative phenomena are essential to the stability of the complex, so that the polymer behaves in a manner which is qualitatively different from that of its low molecular weight analogs. The nature of the polymeric cooperative effect, if any, may be different for various systems, depending on

Eric Sélégny (ed.), Polyelectrolytes, 391–418.

the properties of interacting species. Generally, one might always obtain this type of information by studying the polymeric complexes as a function of molecular weight in the range where the cooperative effect is most sensitive to the chain length. Even the well-known starch-iodine complex has not been examined completely in the interesting molecular weight range [2, 3]. It is one of the purposes of this presentation to discuss results of this type. The effects of the ionic association (shortrange interaction) and the well known charge-charge repulsion (long-range interaction) on the segmental motion of this polyelectrolyte are compared. Solvent effects on the ionic association have also been studied. Poly(ethylene oxide), PEO, and potassium iodide were chosen in the present study. Various physical techniques – NMR, IR, Electrophoresis, Viscometry, and Relaxation measurements – were used in this study.

2. Preparation of Highly Purified Oligomeric PEO

Penta- and heptaethylene glycols were prepared by condensation of the monosodium salts of mono- and diethylene glycols, respectively, with the dichlorides of the triethylene glycol [4].

$$2HO(-CH_2CH_2O-)_p\,Na + ClCH_2CH_2OCH_2CH_2OCH_2CH_2Cl \xrightarrow[16hr]{95°C} HO(-CH_2CH_2O-)_{2p+3}H + 2NaCl,$$

where $p=1$ and 2 for penta- and heptaethylene glycols, respectively. Hexa- and octaethylene glycols were similarly synthesized.

$$2HO(-CH_2CH_2O-)_q\,Na + ClCH_2CH_2OCH_2CH_2Cl \xrightarrow[16hr]{95°C} HO(-CH_2CH_2O-)_{2q+2}H + 2NaCl,$$

where $q=2$ and 3 for hexa- and octaethylene glycols, respectively. The reaction time of 16 h was found to be adequate. Prolonged standing at 95 °C for 3 days as previously reported in the literature [4] is not necessary, and in fact, makes purification more difficult.

These products were isolated by high vacuum distillation, giving sharp boiling points as shown in Table I. The high resolution NMR spectra of the hydroxy protons

TABLE I

Boiling points of poly(ethylene oxides)

	Bp, °C	Pressure mm Hg
$HO(-CH_2CH_2O-)_3H$	116 ± 1	1.5
$HO(-CH_2CH_2O-)_4H$	105	0.01
$HO(-CH_2CH_2O-)_5H$	134	0.02
$HO(-CH_2CH_2O-)_6H$	168	0.07
$HO(-CH_2CH_2O-)_7H$	186	0.07
$HO(-CH_2CH_2O-_8H$	213	0.07

and of the ethylene protons are quantitatively in accord with the concentrations expected. Pure penta-, hexa-, hepta-, and octaethylene glycols were obtained. Tri- andtetraethylene glycols were obtained from Aldrich and Dow Chemicals, respectively. They were purified by distillation.

Fractions of high molecular weight polyethylene oxide were obtained from Union Carbide and used without further purification. They are essentially pure compounds without any contaminating impurities which may affect the high resolution NMR measurements.

3. Cooperative Interactions Between Potassium Iodide and PEO

3.1. Experimental section

Fractions of high molecular weight PEO and various pure oligomers were described in the previous section. Deuterated methanol was obtained from Merck Sharpe and Dohme of Canada. Potassium iodide was obtained from Baker and Adamson. Dimethoxyethane was obtained from Matheson Coleman and Bell, purified by distillation, and used as a monomeric analog.

Stock solutions were made either in pure deuterated methanol or in methanol containing 1% of cyclohexane (internal NMR reference). Various sample solutions were then mixed in the NMR sample tube. The spectra were obtained using a Varian A-60 spectrometer (Varian Associates, Palo Alto, Calif.). Dioxane (8%) in benzene was used as an external standard. The chemical shift is expressed cycles per second (cps) downfield. Electrophoresis experiments were measured in a Perkin-Elmer Model 238 electrophoretic apparatus. Both ascending and descending boundaries were observed.

3.2. Results and discussion

A PEO polymer molecule may generally be represented by the structural formula

$$HO-CH_2CH_2O-(CH_2CH_2O)_{n-2}-CH_2CH_2OH.$$

The ethylene groups in the parenthesis are defined as internal ethylenes of the ether type (E_i), which are different from the end ethylene groups. All the E_i protons of PEO in methanol show a single NMR peak which overlaps the splitting NMR peaks of the end ethylene protons. For the monomeric analog, di-methoxyethane, the NMR peak of the two methyl end group is displaced from that of the ethylene protons, and will not be shown in the present spectra.

A set of representative NMR spectra of the ethylene protons of PEO's, 1% in deuterated methanol, with and without the presence of 0.05 M potassium iodide is presented in Figure 1. The introduction of potassium iodide into the PEO solutions' has little effect on the NMR spectra of the E_i protons for oligomers ($n \leqslant 6$), but the spectrum of the heptamer is significantly changed into two partially resolved peaks. The NMR spectra of the higher members ($n > 7$) of the series are all affected by potassium iodide, although the apparent influence becomes less pronounced for high

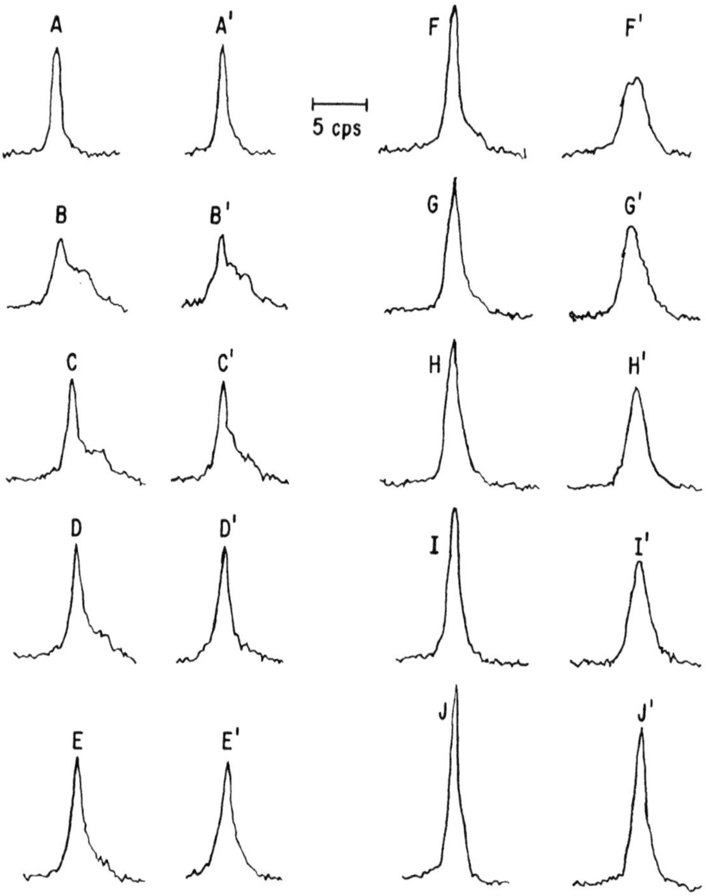

Fig. 1. Representative NMR spectra of the ethylene protons of 1% PEO in deuterated methanol: (A) Dimethoxy ethane, (B) Tri-ethylene glycol, (C) Tetra-ethylene glycol, (D) Penta-ethylene glycol, (E) Hexa-ethylene glycol, (F) Hepta-ethylene glycol, (G) PEO-400 ($\bar{n}=9$), (H) PEO-600 ($\bar{n}=13$), (I) PEO-1000 ($\bar{n}=22$), (J) PEO-6000 ($\bar{n}=140$); (A′–J′) are the corresponding samples of (A–J) with the presence of 0.05 M potassium iodide.

polymers, because most of the E_i groups of a high molecular weight PEO are similarly affected. Therefore, it seems clear that there exists a transition of interactions between PEO and potassium iodide as the polymer chain grows to a length of seven monomeric units.

The differences (Δ) of chemical shifts of PEO with and without the potassium iodide are plotted as a function of the chain length (n) in Figure 2. The concentration of PEO and potassium iodide were kept constant for all measurements, and the bulk magnetic susceptibility correction (see later) was also applied. Again, a sharp change is observed at $n=7$.

These results seem to suggest that the conformational structures of PEO's in these

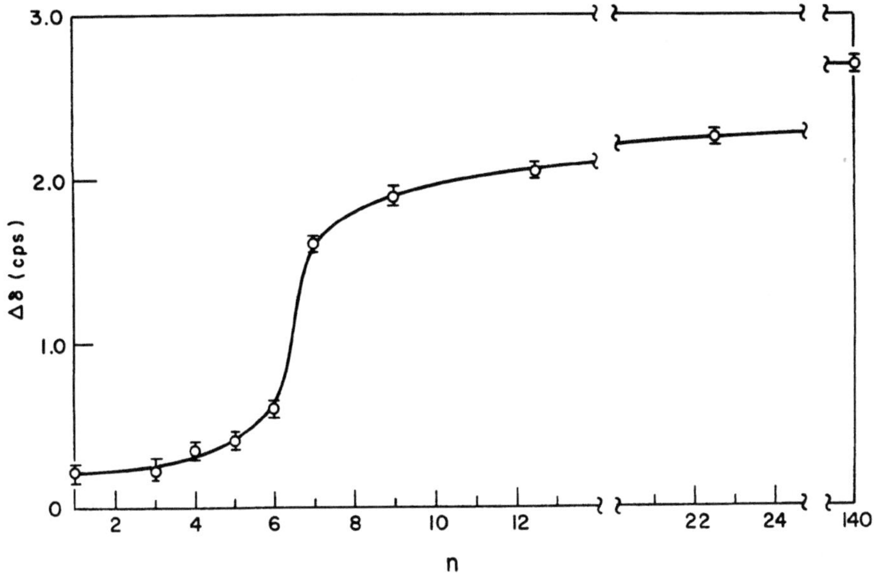

Fig. 2. Plots of the difference of chemical shifts (with bulk magnetic susceptibility correction) of PEO (1 %) in methanol with and without the potassium iodide (0.05 M) as a function of the chain length (n) at 35°C.

solution systems may change drastically at $n=7$. It has been proposed that the association between PEO and potassium iodide is due to an ion-dipole interaction [5], and only a drastic conformational change of PEO would cause a change of the dipole moment of the polymer segments and, consequently, of the interaction. From the present results, one cannot discriminate whether this conformation change at $n=7$ is provoked by the interaction of potassium iodide with PEO, or it is originally there even in the absence of the salt. We believe that the latter is more likely to be true, because our previous NMR results on the structures of oligomeric PEO's in some other solvents without the presence of salt also indicated a drastic structural change at $n=7$ [6].

The positive Δ values (Figure 2) means that the NMR shift of PEO moves toward a lower magnetic field as a result of interaction with potassium iodide. For the present proton system, this seems to indicate that the charge density in the neighborhood of the E_i protons is reduced [7]. This can happen only when the positively charged potassium ion interacts with the etherate oxygen of PEO. From the chemistry point of view, it also seems reasonable to expect that a dimethoxyethane (the monomeric analog) should interact with the positive potassium ion rather than the negative iodide ion, and yet the general nature of the ion-dipole interaction in the present system is similar for all PEO's. Therefore, we believe that the potassium ion is actually the species which directly interacts with the polymer. We do not think that the negatively charged iodide ion would directly interact with the etherate oxygen of

PEO as previously suggested [5]. Although it is not important for the purpose of present investigation to identify the exact species which interacts with the polymer, we would still like to clarify this problem. We have carried out further electrophoretic measurements on a sample solution containing 2% PEO (3×10^5) and 0.10M potassium iodide in methanol, using a Perkin-Elmer 238 electrophoretic apparatus. Our preliminary results indicate unambiguously that the polymer in this solution is a polycation in nature. Of course, once the PEO becomes a polycation through its interaction with the potassium ion, its solution properties should also be expected to depend on the counteranions. Although it is not the purpose of the present paper to discuss the subject of polycations, we would still like to point out that the effects of counteranions on the solution properties of polycations are sometimes very specific and they are not as well understood as polyanions [8, 9].

Plots of Δ values (after the bulk magnetic susceptibility corrections) of 1% PEO-6000 and its monomeric analog, dimethoxyethane, as a function of potassium iodide in methanol are shown in Figure 3. It is obvious that the polymer and its monomeric analog interact differently with the potassium ion when the salt concentration is low; further associations of potassium ion with the polymer are reduced by the existing bound ion at higher salt concentrations. In other words, the apparent association

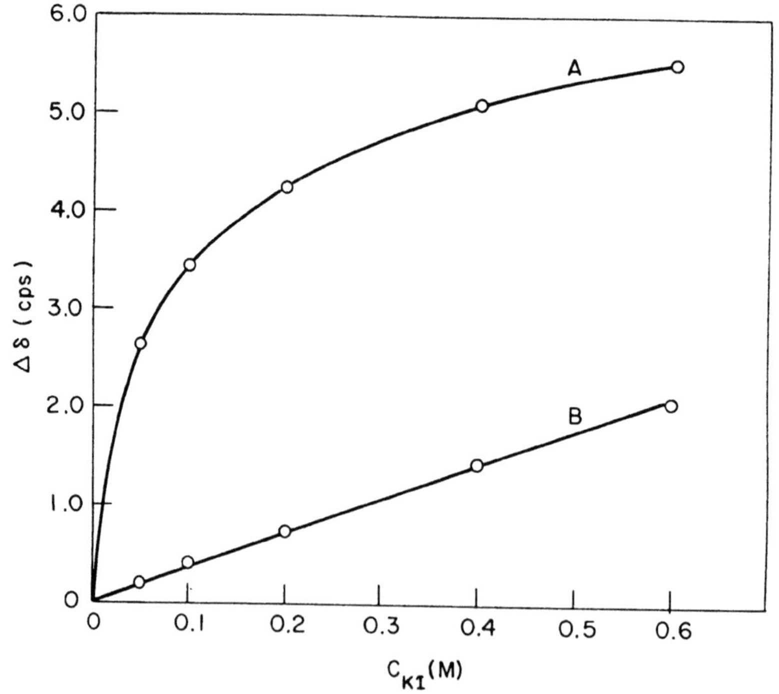

Fig. 3. Plots of the difference of chemical shifts (with bulk magnetic susceptibility correction) of 1% PEO-6000 (*A*) or ethylene protons of dimethoxy ethane (*B*) as a function of the concentration of potassium iodide in methanol at 35°C.

constant of the potassium ion and PEO decreases as one increases the salt concentration. For the monomeric analog, the association is rather weak at low salt concentrations, but remains constant when the salt concentration is increased. The ratio of the Δ value of the polymer to that of the monomeric analog is more than 12 when the potassium iodide is 0.05 M; it decreases to less than 3 when the salt concentration is 0.5 M. Measurements were also made on PEO-6000 at other concentrations and the results are summarized in Table II.

TABLE II

The difference (Δ) of chemical shifts of PEO-6000 in methanol in the presence and absence of potassium iodide at 35°C

Polymer concentration, %	Potassium iodide concentration, M	Δ value, cps
2	0.05	2.7
2	0.10	3.7
2	0.20	4.5
2	0.40	5.5
2	0.60	5.7
5	0.05	1.5
5	0.10	2.7
5	0.20	3.9
5	0.40	4.8
5	0.60	5.2
10	0.05	0.9
10	0.10	1.7
10	0.20	2.8
10	0.40	4.3
10	0.60	5.1

Measurements of chemical shift were made on 1% cyclohexane, which was mixed with methanol and used as a solvent for preparing the previous samples. This was done because no association between cyclohexane and potassium iodide takes place in the present systems. The results are shown in Figure 4. No detectable difference was observed for Δ values of the PEO obtained in either pure methanol or methanol containing 1% cyclohexane. Therefore, it is reasonable to use the Δ values of the cyclohexane in various samples to correct the bulk magnetic susceptibilities.

The apparent association constant (K) of the potassium ion to PEO in methanol may also be computed from their chemical shift data. The free and bound units of PEO exchange rapidly to yield a single NMR reasonance adsorption which is an average of the chemical shifts of PEO in the two states. This can be expressed by Equation (1)

$$\delta_{\text{obs}} = X_f \delta_f + X_b \delta_b , \tag{1}$$

where δ_{obs} is the observed chemical shift of PEO-6000; δ_f and δ_b are respectively, chemical shifts of free and bound PEO; X_f and X_b are mole fractions of free and bound

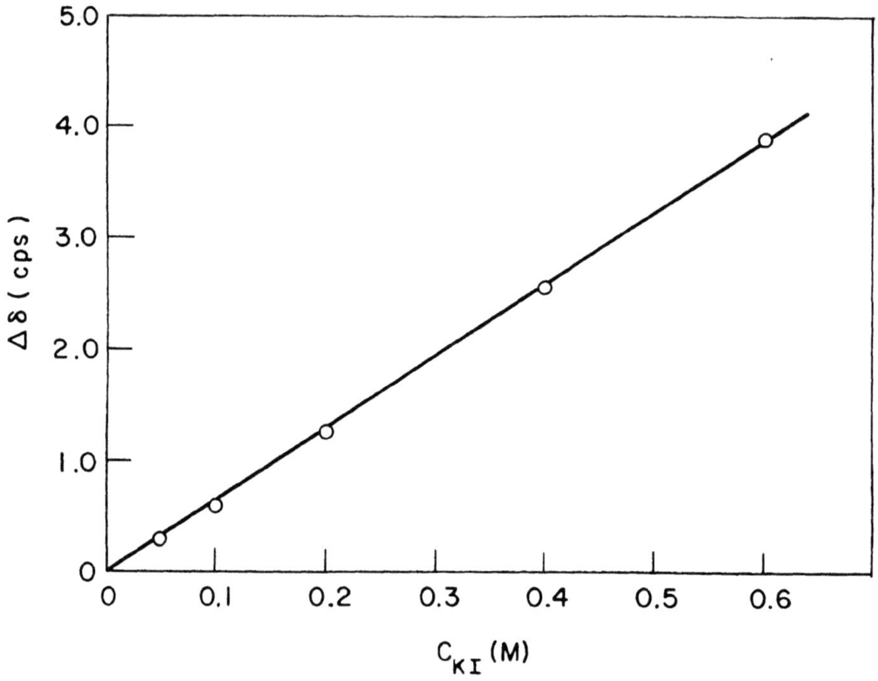

Fig. 4. Plots of the difference of chemical shift (without bulk magnetic susceptibility correction) of 1% cyclohexane in methanol containing 1% PEO as a function of the concentration of potassium iodide at 35°C.

PEO, respectively. The δ_{obs} and δ_f are obtained directly from experimental measurements; δ_b can be calculated by assuming that all the potassium ions are bound to the polymer when a large excess of the polymer is present [5], an example of this excess is 10% PEO (2.27) M) and 0.05 or 0.10 M potassium iodide. From these values, X_f and X_b may be calculated for all systems, and the K value may be obtained from Equation (2), where $[K^+]$ is the concentration

$$K = \frac{[\text{complex}]}{[\text{PEO}]_f \, [K^+]} \tag{2}$$

of the free potassium ion; the [complex] and $[\text{PEO}]_f$ are, respectively, the concentrations of the complexed and free PEO. These calculated formation constants are summarized in Table III. It is clearly shown that the apparent formation constant decreases as the ratio of the potassium iodide to PEO increases, while the total polymer concentration was kept constant. The K value also decreases as the polymer concentration increases at the same salt to polymer ratio. These phenomena are commonly found in polyelectrolyte solutions [1, 5, 10].

The Δ values of PEO-6000 were also measured in aqueous solutions. They are all zero after the bulk magnetic susceptibility corrections using cyclohexane as standard.

TABLE III

Apparent association constant of potassium iodide and PEO-6000 in methanol at 35°C

Polymer concentration, %	Ratio of potassium iodide to PEO	Apparent association constant (K)
1	0.22	2.28
1	0.44	1.29
1	0.88	0.74
1	1.76	0.43
1	2.64	0.31
2	0.11	4.20
2	0.22	1.76
2	0.44	0.88
2	0.88	0.47
2	1.32	0.37
5	0.09	3.78
5	0.18	1.39
5	0.35	0.54
5	0.53	0.34
10	0.04	Very large
10	0.09	1.84
10	0.17	0.84
10	0.27	0.50

On this basis, it may be concluded that PEO does not form a complex with potassium iodide in water. Previous viscometric measurements in aqueous systems led to the same conclusion [5].

4. Nuclear Magnetic Relaxation in PEO-Salt Solutions

4.1. Experimental section

PEO samples, having molecular weights of 4000, 6000 and 300000 were used in this study. These materials, designated PEO-4000, PEO-6000 and PEO-300000 have been described in a previous article [11]. Deuterated methanol – d_4, water and chloroform were obtained from Merck Sharpe and Dohme of Canada, and used without any further purification. All the other halogenated solvents were purified according to established procedures. Potassium iodide and potassium sulfate were obtained from Baker and Adamson and were used without further purification.

The sample solutions were prepared in standard NMR tubes. Each solution was subjected to four freeze-pump-thaw cycles to remove dissolved oxygen, and sealed under vacuum. T_1 measurements were performed at 34.5±0.5°C on a modified Varian DA-60 spectrometer, using the 'adiabatic passage with sampling' technique [12]. The relaxation data were time-averaged in a Varian C1024 Computer. At least 25 measurements of each relaxation time were used. The data appear reproducible within ±5%.

Viscosity was measured with a three-bulb Ubbelode dilution viscometer at 34.5± ±0.02 °C. The solvent flow time was such that kinetic energy corrections were negligible. Solvent viscosities were obtained by comparing flow times with that of a standard (water).

4.2. Results and discussions

Our experiments contrasted the proton relaxation of PEO in KI-CD_3OD solutions with that observed in KI-D_2O. Results for PEO-6000 are plotted in Figures 5 and 6. When KI is added to CD_3OD solutions of PEO, it has a pronounced effect on T_1. Experimental T_1 values change from 1.9 s with no added salt to 0.50 s in 0.6 M KI solution. In contrast, KI has no apparent effect on the relaxation of PEO in D_2O. We observed almost identical behavior for KI solutions of PEO-300000.

It has been shown that a definite association complex is formed between K^+ and PEO in methanol. This complex has not been detected in KI-PEO-water solutions. One can reasonably seek to explain the observed differences in relaxation behavior in terms of this complex. We pursue this hypothesis in subsequent paragraphs. Alternatively, the effect of KI on the two solvents must be considered. One could speculate that the added salt drastically changes the solution viscosity of methanol, through ion-solvent interactions, without interacting with PEO. Under these circumstances, a

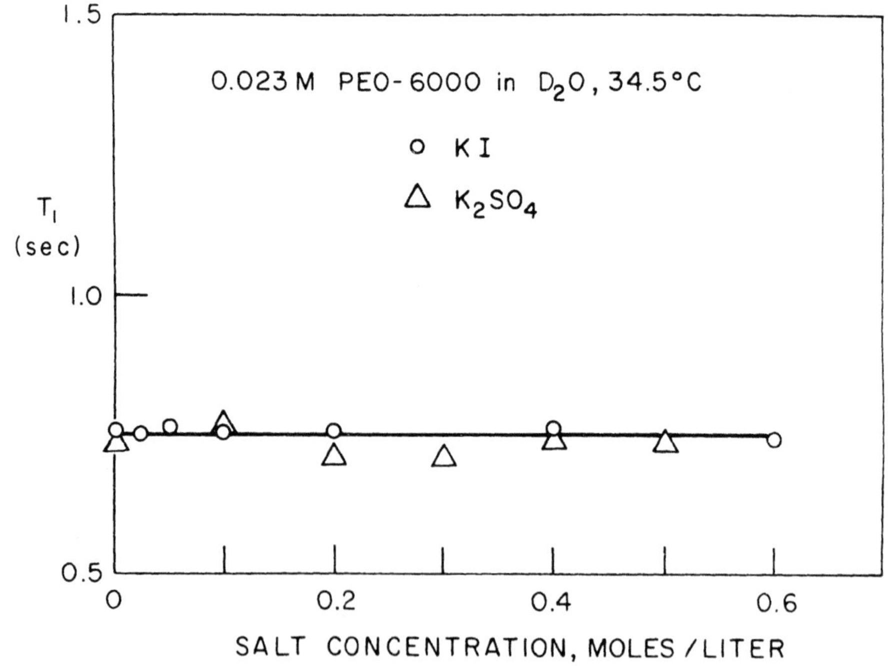

Fig. 5. T_1 measurements on 0.023 M PEO solutions in D_2O. The effect of added KI and K_2SO_4 is illustrated.

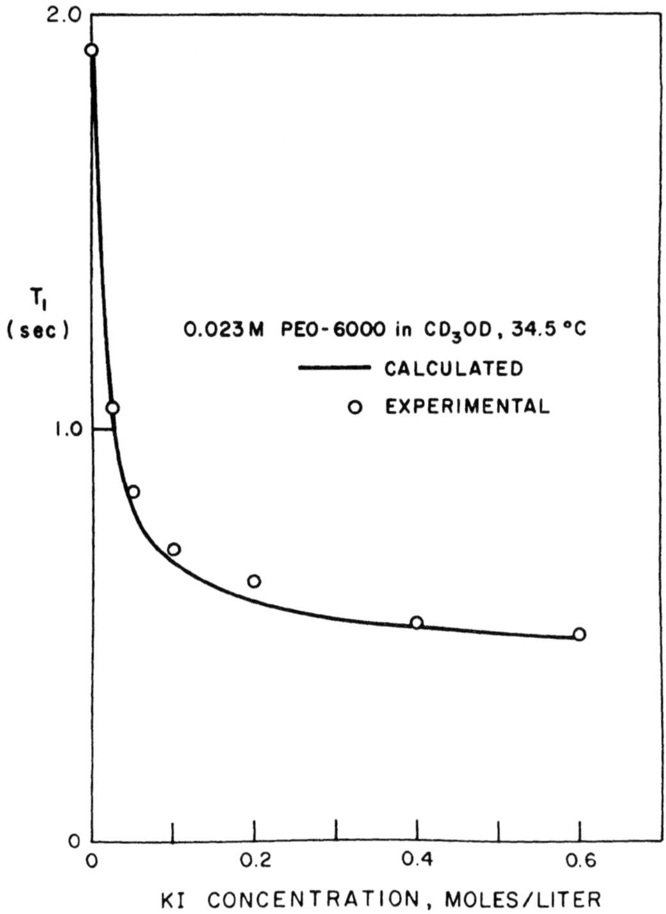

Fig. 6. T_1 measurements on 0.023 M PEO solutions in CD_3OD. The concentration of KI was varied. Curve is calculated by the method described in the text.

large effective solvent viscosity would presumably impede polymer mobility, and lower the T_1 values. Experimentally, we found the solution viscosity of 0.6 M KI in methanol to be 30% greater than the viscosity of methanol. The solution viscosity of KI methanol solutions varied linerarly with added KI over this concentration range. The small variation of η and its linear KI dependence argue strongly against any pronounced ion-solvent effects on the polymer T_1 values, when these factors are contrasted with the large, nonlinear concentration dependence of T_1. For completeness, we also examined the viscosity of KI-water solutions. The viscosity of a 0.6 M KI-water solution is 4% lower than the viscosity of water.

We now examine the effect of complex formation of PEO relaxation. It is reasonable that coordination of K^+ with a PEO chain segment would reduce the mobility of that segment relative to the mobility of the uncoordinated units. Accordingly, the effective relaxation time of the coordinated (or bound) species, $(T_1)_b$, should be shorter than the relaxation time of the free units, $(T_1)_f$ [14]. The T_1 data can be treated in terms of an exchange process between free and bound states. Chemical exchange enters because the mean lifetime of a particular 'bound' state is short compared to T_1 [15].

$$(T_1)_{\text{calcd}}^{-1} = (1 - \alpha)(T_1)_f^{-1} + \alpha (T_1)_b^{-1}. \quad (3)$$

The quantity α in Equation (3) is the fraction of bound PEO units. This quantity can be calculated from the known formation constants [13]. The experimental $(T_1)_f$ was 1.92 s $(T_1)_b$ was chosen to be 0.089 s. These quantities, together with values of α, generate the curve shown in Figure 6. Agreement between experiment and the calculated curve is excellent. This analysis implies that the mobility of individual PEO segments is strongly restricted through association with K^+. It is perhaps restricted as much as $(T_1)_f/(T_1)_b = 21$ times.

Figure 7 illustrates further T_1 measurements on PEO-6000 in CD_3OD solution. In

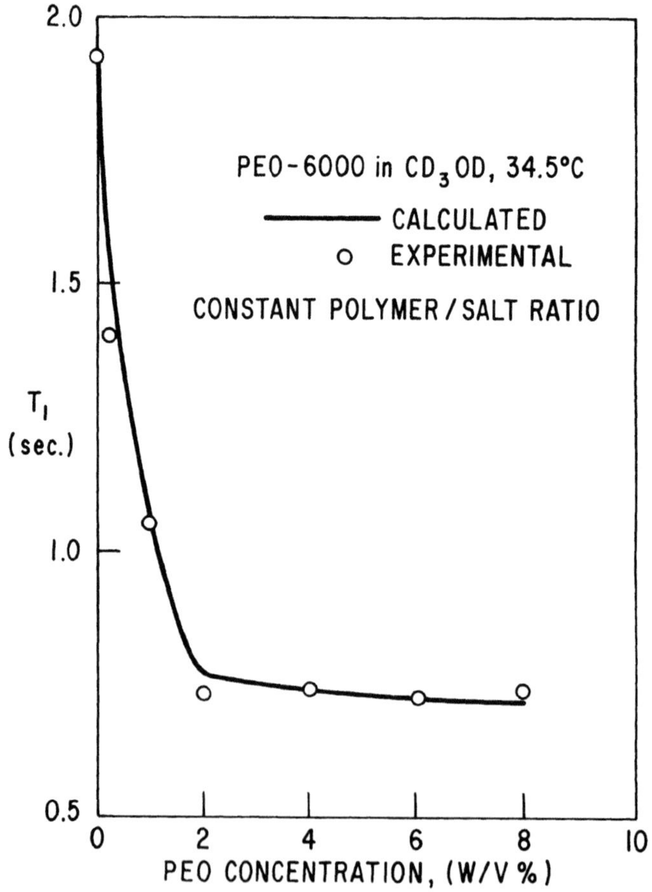

Fig. 7. T_1 measurements on PEO solutions in CD_3OD. The KI molarity was one-ninth that of PEO in all solutions. Curve is calculated by the method described in the text.

these experiments, the polymer concentration was varied while the polymer/salt ratio remained constant. The curve in Figure 7 was calculated from Equation (3), using the same values of α, $(T_1)_f$, and $(T_1)_b$. Once again the agreement is satisfactory. These curves show that the dependence of T_1 upon KI concentration in CD_3OD can be simply explained in terms of complex formation between the salt and the polymer. This interpretation is also consistent with the insensitivity of these experiments to the MW of PEO.

It is known that addition of neutral salts to polymer solutions reduces the over-all dimensions of the polymer chains (the 'salting-out' effect). The reduction in chain dimensions is reflected in the polymer viscosity. In contrast, nuclear relation is primarily concerned with the relative motions of neighboring nuclei. It deals with distances the order of angstroms, rather than thousands of angstroms. We used T_1 measurements to study salting-out effects at this local level. We confined our attention to aqueous PEO solution of K_2SO_4, since this salt is known to have large salting-out effects on PEO. Salting out undoubtedly occurs in KI-CD_3OD as well. We felt its effect on T_1 would be masked by the K^+-polymer coordination complex described earlier.

The results of this study are presented in Figures 5 and 8. The data indicate that added K_2SO_4 has no effect on PEO relaxation in D_2O. In contrast, this salt has a profound effect on the polymer viscosity. Particularly striking results were found in the system containing PEO-300000. As shown in Figure 8, the reduced viscosity of the polymer decreases more than 50 times passing from aqueous solution to 0.5 M K_2SO_4, where the polymer is on the verge of phase separation. No appreciable change in T_1 was observed over the same concentration range. These results indicate that the local segmental motion of the polymer chain is not affected by the presence of unbound ions.

We are now in possession of the following information: (1) the previously detected complex between K^+ and PEO in CD_3OD has a pronounced effect on T_1; (2) a looser polymer-salt interaction, such as that responsible for 'salting out', in K_2SO_4-PEO-D_2O solutions, has no apparent effect on T_1; (3) the salting-out interaction has a pronounced effect on solution viscosity. Complex formation may also affect polymer viscosity but it is difficult to separate its influence from the salting-out process.

Of the above, the relationship between salting out T_1, and polymer viscosity merits further comment or speculation – for it is, of course, extremely speculative to infer anything of a general nature from such meager knowledge. It may be somewhat difficult to understand how salting out can have such a pronounced effect on viscosity and yet have no observable influence on T_1. The answer, we feel, is that η and T_1 are basically sensitive to different things, although both measure transport properties of polymers. It is well established [11, 16] that nuclear relaxation of PEO in solution is mainly controlled by segmental rotameric transitions. For example, T_1 is influenced by the rates at which local trans-trans-trans, trans-gauche-trans, and other sequences interconvert. By altering these rates, through the introduction of a complexing agent like K^+, one produces changes in T_1. On the other hand, the viscosity of a polymer solution is governed primarily by polymer dimensions. Polymer dimensions can be

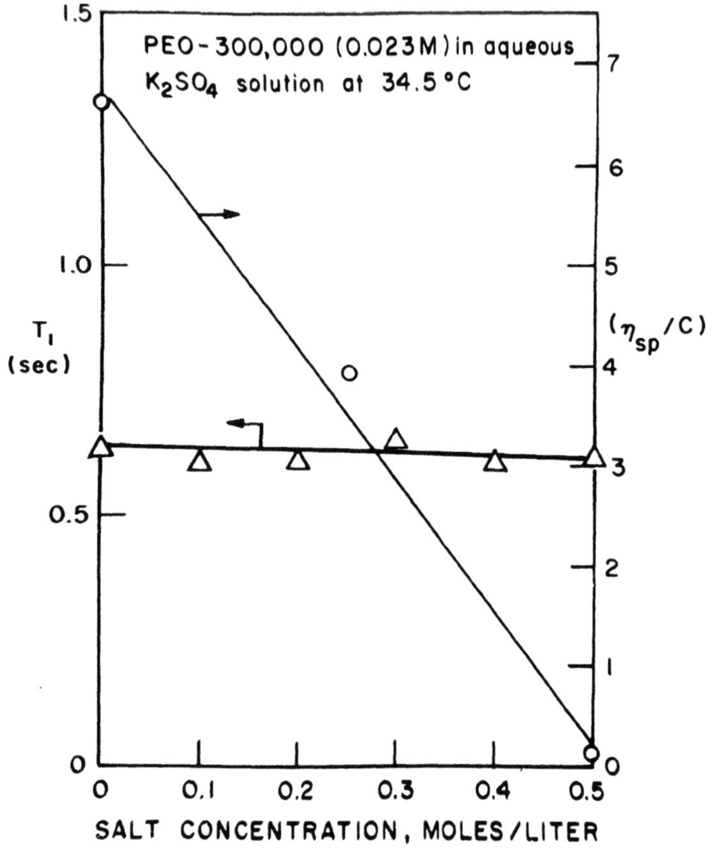

Fig. 8. T_1 and reduced viscosity data for D_2O solutions of PEO-300000.

theoretically expressed in terms of the number of trans-gauche-trans, trans-trans-trans sequences, etc.; and the way these sequences are arranged relative to one another along the macromolecular chain. Polymer dimensions do not reflect sequence interconversion rates directly although to be sure the equilibrium probability of each sequence is determined by the relative rates of transfer into, and out of, that local configuration. The present results suggest that addition of neutral salts to polymer solutions causes slight variations in the probabilities in these local sequences, or in their order along the chain. These changes need not be large to reduce the macroscopic chain dimensions, and lower the polymer viscosity. We can only speculate that any concurrent variations in the rates of rotameric transitions are too small to be detected in our T_1 measurements.

For further pursuing a better understanding of the role the solvent molecules play in these molecular interactions, we have carried out systematic studies on solvent effects on the nuclear magnetic relaxation of PEO. The proton spin-lattice relaxation of PEO in solution has been discussed preciously [11]. The relaxation appears to be mainly governed by segmental rotation of the polymer (rotameric transitions). These rotations may be influenced both by the local viscosity surrounding the individual segments and by the conformational populations of the chain units. Both these factors may be influenced by specific polymer-solvent interactions. In an attempt to minimize conformational differences in going from one solvent to another, and to emphasize viscosity effects, our first experiments studied in the T_1 of PEO in a series of halogenated alkanes. It was anticipated that specific polymer-solvent interactions would be similar in each of these solutions. The solvents chosen were CH_2Cl_2, $CHCl_3$, $CHCl_2CHCl_2$, $CHCl_2CCl_3$, and $CHBr_2CHBr_2$. These solvents have viscosities of 0.38, 0.49, 1.36, 1.85, and 7.20 cp, respectively (35 °C). The results of these experiments are shown in Figure 9, in a plot of T_1^{-1} vs. solvent viscosity, η. The slope of the line drawn through the experimental points is roughly two-thirds that predicted by the BPP-Debeye expression [17]. This is not too surprising, for in a polymer there are always nonsolvent units in the immediate neighborhood of each segment – the adjacent segments of the polymer chain! Only a fractional effect of the solvent viscosity should be expected to be transmitted to the polymer on this basis.

For purposes of comparison, the relaxation of *p*-dioxane, a small-molecule analog of PEO, was measured in chloroform, pentachloroethane, and tetrabromoethane. T_1^{-1} values of *p*-dioxane, obtained by extrapolation to 0% dioxane concentration, are plotted against solvent viscosity in Figure 10. The slope of the line through the experi-

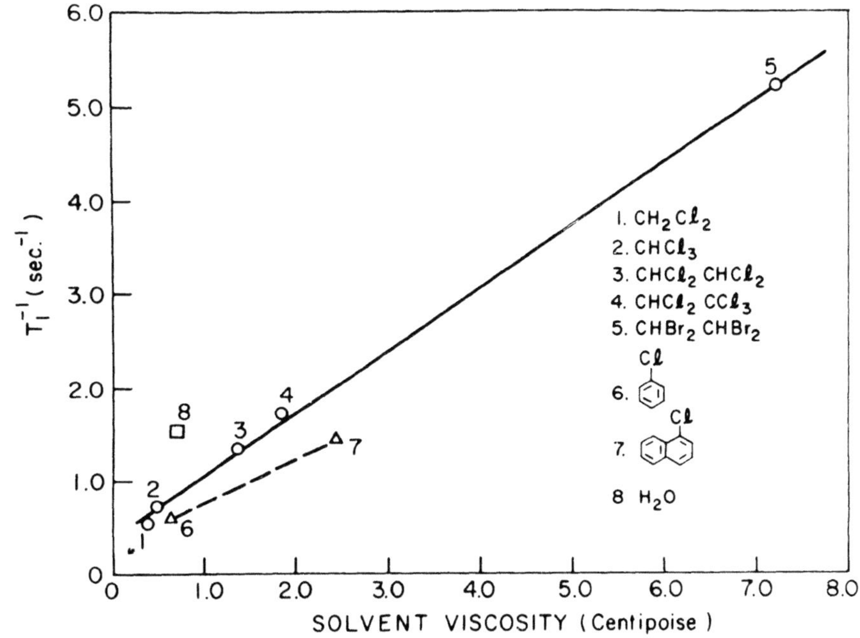

Fig. 9. Plot of T_1^{-1} versus η, solvent viscosity, for PEO-4000 in various solvents at 35°C.

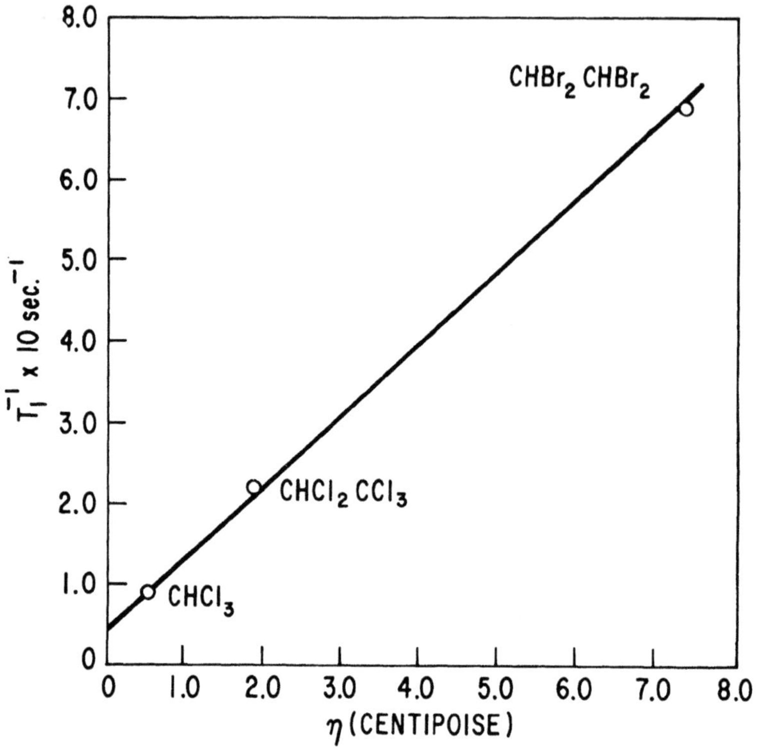

Fig. 10. Plots of T_1^{-1} versus η, solvent viscosity, for dioxane at zero concentration in several selected solvents at 35°C.

mental points is about 90% that predicted by the BPP-Debye expression. This suggests that solvent viscosity may be more important in determining the motions of small molecules than those of a polymer [18].

When a different class of solvents, such as α-chloronaphthalene and chlorobenzene, were used in preparing PEO solutions, the resulting PEO T_1 values exhibit a significantly different viscosity dependence from that observed in the halogenated alkanes. The viscosity of α-chloronaphthalene is 2.44 cp (35°C), yet the T_1 of PEO in this solvent is similar to that in 1,1,2,2-tetrachloroethane, whose viscosity is 1.36 cp. Results obtained in α-chloronaphthalene and chlorobenzene are shown in Figure 9. These T_1 values seem to be less viscosity sensitive than those in the halogenated alkanes This demonstrates that an experimental relationship between T_1^{-1} and η, established for one family of solvents, is not necessarily followed by solvents of a different type. These differences may result from variations in local viscosity owing to specific polymer-solvent interactions. Alternatively, the different solvents may induce difference preferred conformations of the chain units. Our present result cannot distinguish between these alternate explanations.

The influence of solvent-induced conformational changes in PEO may be studied with the aid of independent evidence. The T_1 of PEO in water is peculiarly short, 0.65 s. The viscosity of water is 0.723 cp (35°C). This T_1 result is shown in Figure 9, and it deviates significantly from the $T_1^{-1}-\eta$ relationship found for the halogenated alkanes. This difference may be attributed to a solvent-induced conformational change.

Further evidence for conformational changes is provided by the results shown in Figures 11 and 12. The solution viscosity of D_2O-CD_3OD mixtures passed through a maximum in the vicinity of the equimolar solution (Figure 11). In contrast, T_1^{-1} of PEO in these solvent mixtures first increases as small amounts of D_2O are added to CD_3OD. T_1^{-1} levels out as the D_2O content is increased and does not decrease, as does the solvent viscosity, when the D_2O concentration is increased beyond 60% (Figure 12). Similar T_1 measurements were made on *p*-dioxane in these solvent mixtures. In the case of this monomeric analog of PEO, the T_1^{-1} values parallel the viscosity results over the entire concentration range. A definite T_1^{-1} maximum is apparent near the equimolar CD_3OD-D_2O solution. These T_1 data strongly suggest a new solvent-induced conformational state of PEO in the water-rich solutions.

Further discussions on this subject will be presented in the next section, based on spectroscopic studies.

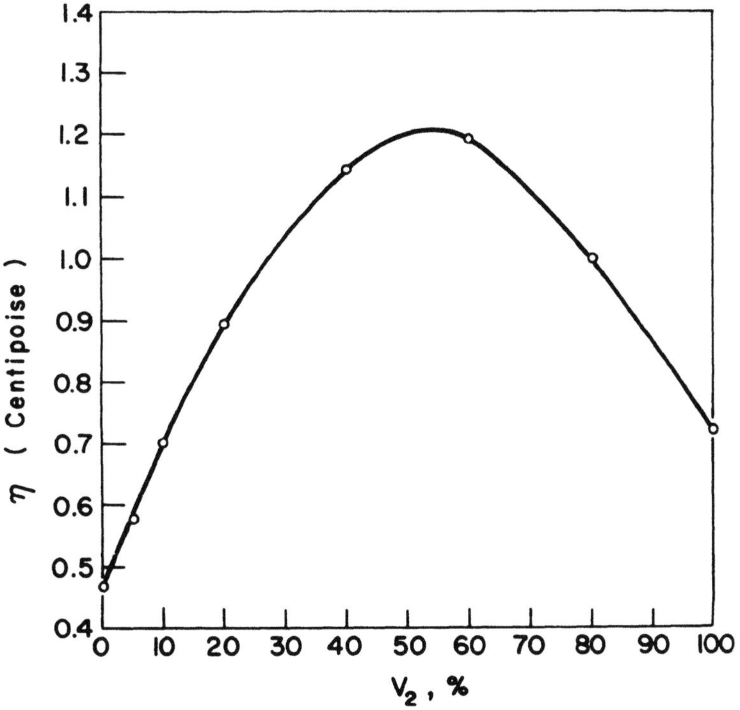

Fig. 11. Plots of solvent viscosity versus volume composition (V_2 for water) in water-methanol mixtures at 35°C.

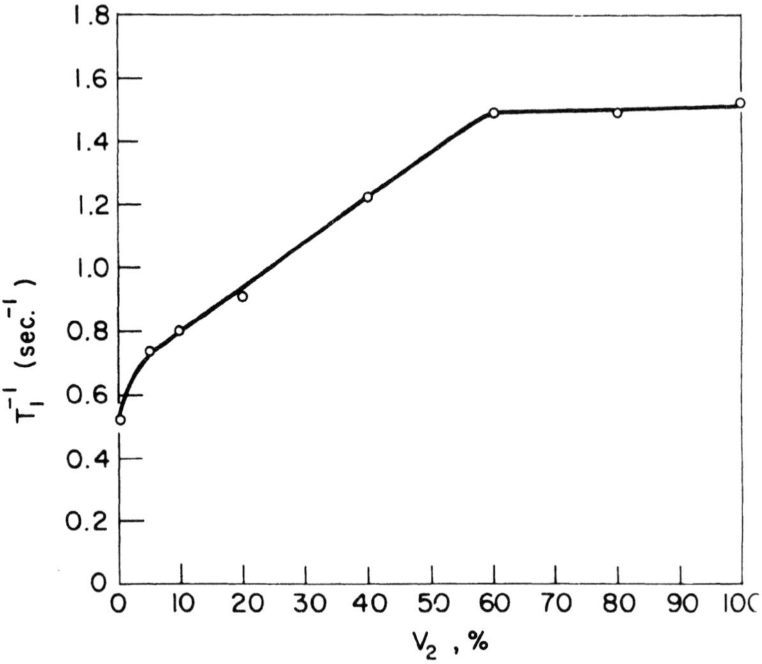

Fig. 12. Plots of T_1^{-1} versus V_2, volume composition of water, for PEO-4000 in water-methanol mixture at 35°C.

5. Solvent Effects on the Preferred Conformation of PEO

5.1. Experimental section

Fractions of high molecular weight PEO and various pure oligomers were described in Section 2. Deuterium oxide was obtained from Merck Sharpe and Dohme of Canada. All the other solvents used were purified in accordance with established procedures.

NMR spectra were measured by using a Varian A-60 spectrometer with a temperature control attachment (Varian Associates, Palo, Calif.). All chemical shifts reported are expressed in cycles per second (cps) relative to an external reference, cyclohexane (4%) in carbon tetrachloride.

The infrared spectra were recorded on a Perkin-Elmer model 237 grating spectrometer. Alkali halide and silver chloride cell windows were used with appropriate Teflon spacers to obtain comparable band intensities from several concentrations of PEO in benzene and deuterium oxide. A variable path cell was used in the reference beam to compensate for solvent absorption in the spectra of benzene solutions with low concentrations of PEO. Deuterium oxide was used instead of water as a solvent because, in the spectral regions of interest, it contributed less back-ground absorption. All solution compositions are given in volume per cent.

5.2. Results and discussion

The internal ethylene protons (i.e., all the ethylene groups, except for the two ethylene groups at the chain ends in the polymer) of all PEO's exhibit a single NMR peak in aqueous solutions. The chemical shift of this peak (δ_p) is not sensitive to the length of PEO chain [16]. Therefore, it was experimentally convenient to use low molecular weight liquid PEO for NMR studies of aqueous solutions. A representative plot of δ_p of PEO-400 at 35 °C as a function of water (D_2O) content is shown in Figure 13. The chemical shift of the OH proton is also included; δ_p increases with water concentration until a certain composition (about 50%), and then remains almost constant for any further addition of water. The chemical shift of the OH protons behaves similarly; it increases rapidly with water concentration up to about 50%, and then changes very slowly with further increasing of water. Similar measurements in other solvents, such as chloroform and benzene, give very different results. The chemical shift of the main polymer peak changes linearly with the solvent content for the complete composition range. The change of the chemical shift of the OH protons is very

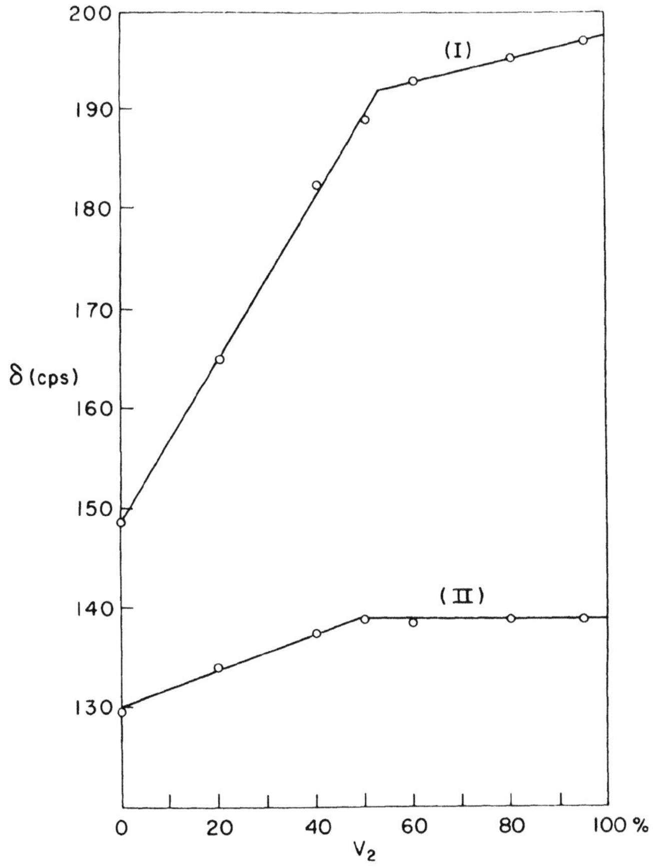

Fig. 13. A representative plot of δ_p (chemical shift of the ethylene protons) of PEO-400 at 35°C as a function of volume % water, V_2 (II). The chemical shift of the OH proton is also included. (I).

slow in concentrated polymer solutions, and becomes very rapid when the polymer concentration is further diluted. A representative plot is shown in Figure 14. All the chemical shift measurements (Figures 13 and 14) were also carried out at other temperatures (4°, 51°, and 80°C in aqueous system; 4° and 52°C in chloroform solutions); the general features were found to be very similar to those shown in Figures 13 and 14.

In order to have a better understanding of the detailed segmental environments of PEO under various circumstances, we have carried out further studies by using ir measuments. In particular, the ethylene rocking modes in the 700–1000-cm^{-1} region have been shown by experiment [19] and calculation [19–23] to be most sensitive to the conformation of the repeating unit in PEO. Therefore, the ir absorptions of PEO in this spectral region have also been carefully examined in the present investigation.

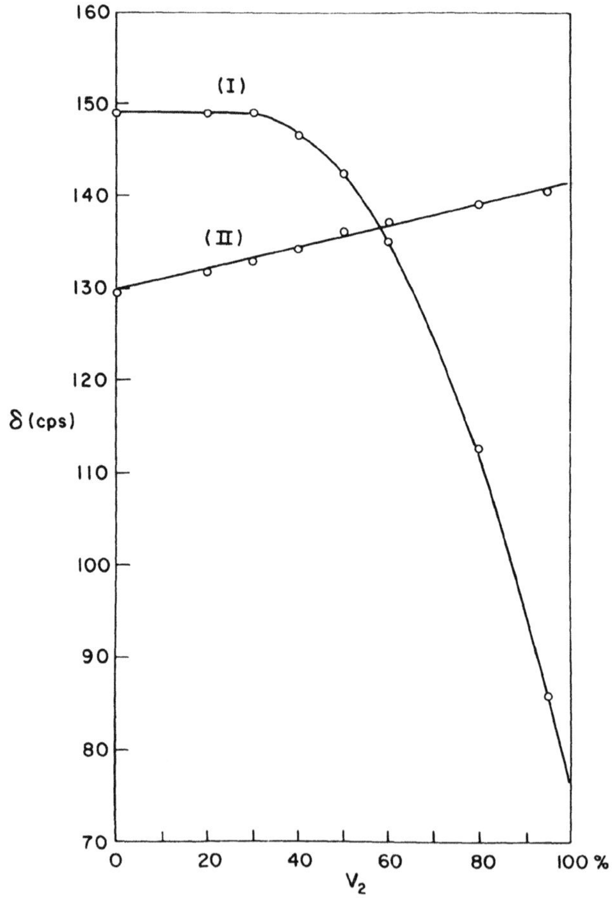

Fig. 14. Plots of chemical shifts of the ethylene and hydroxy protons of PEO-400 at 35°C as a function of chloroform content. The chemical shifts of OH and ethylene protons are represented by (I) and (II) respectively.

Three representative ir spectra of PEO-4000, which include spectra of a high molecular weight PEO in the crystalline state, aqueous and benzene solutions, are shown in Figure 15. The present ir spectra of PEO-4000 in the crystalline state and in the benzene solution are similar to those previously obtained by other investigators [20, 24, 27]. However, significant differences were observed in the IR spectrum of the aqueous PEO solution as compared with that of the corresponding benzene solution. The IR spectrum of crystalline PEO has strong absorptions at 844, 947, and 960 cm^{-1}. In the spectrum of PEO-4000 in benzene, which is also known to be very similar to that of the melt, the corresponding absorptions were found at 844 and 947 cm^{-1}. In addition, a week band of 886 cm^{-1} increases in intensity and new bands appear at 863, 993, and 1324 cm^{-1}. Most of the changes in the IR absorptions which appeared when the crystalline PEO was dissolved in benzene were not present in the spectrum of PEO in aqueous solution.

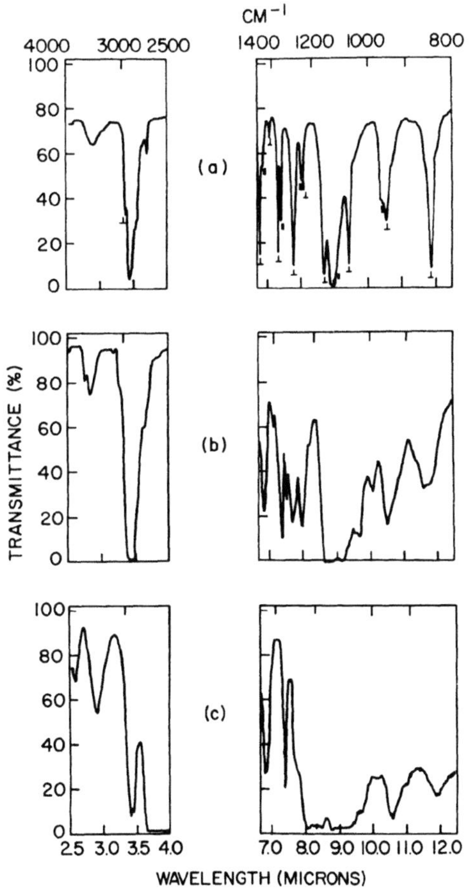

Fig. 15. Representative IR spectra of PEG-4000: (a) crystalline state, (b) 20% in benzene solution, (c) 20% in aqueous solution.

The effect of water molecules on the structure of PEO was systematically studied by measuring the ir spectra of PEO-600 in various aqueous solutions as shown in Figure 16. PEO-600 is a relatively low molecular weight liquid, and its ir spectrum is similar to that of high molecular weight PEO-4000 in benzene except for a stronger absorption at 886 cm^{-1}. The noncrystalline bands at 1324, 993, and 863 cm^{-1} may be observed to diminish in intensity or disappear as the D_2O content of the solution is increased. Residual absorption at 993 cm^{-1} is attributed to an OD bending vibration of $-CH_2CH_2OD$. At the same time, the decrease of the absorption at 886 cm^{-1} may clearly be observed. Also, the absorption at 2935 cm^{-1} was found to become stronger and narrower in the aqueous solutions.

Since it is known that the PEO molecules associate in concentrated solution

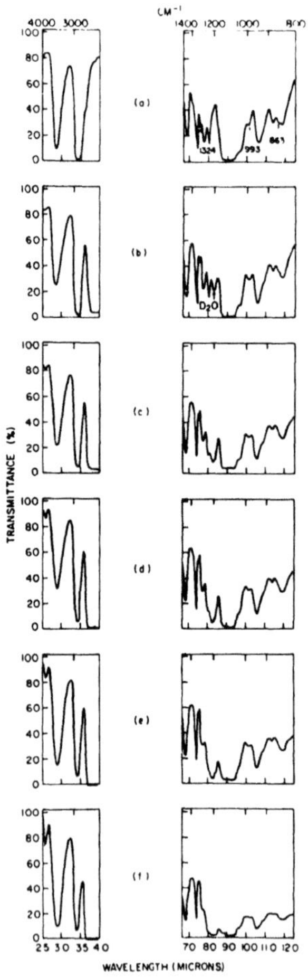

Fig. 16. IR spectra of PEO-600 in water with various concentrations: (a) 100% PEO, (b) 80%, (c) 60%, (d) 50%, (e) 35%, (f) 20%.

[6], we have also studied the concentration dependence of the ir spectrum of PEO in benzene. Three representative ir spectra of an oligomeric PEO, octamer, in benzene with various polymer concentrations (100, 50, and 5%) are shown in Figure 17. The spectrum of the concentrated solution (50%) is very similar to that of the pure liquid (100%). However, certain significant differences may easily be observed in the dilute solution (5%). The absorption peaks at 844 and 863 cm^{-1} merge into a broad unresolved peak at about 850 cm^{-1} in the dilute benzene solution. It is also clear that the intensity of the peak at 1324 cm^{-1} becomes stronger in the dilute benzene solution A weak peak at 866^{-1} and a sharp peak at 3600 cm^{-1} may also be observed in the dilute benzene solution. For high molecular weight PEO, the concentration dependence of the IR spectrum becomes less significant.

The chemical shift of PEO in aqueous solution shown in Figure 13 provides some

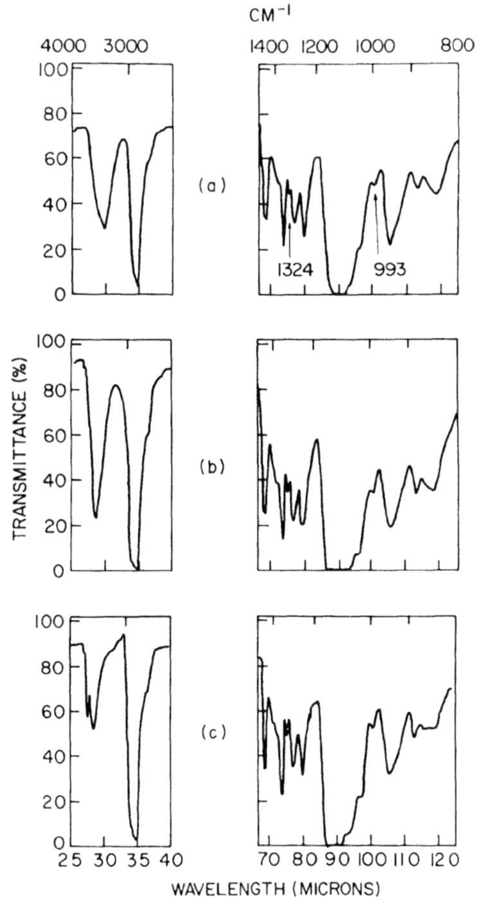

Fg. 17. IR spectra of oligomeric PEO, octomer, in benzene with various polymer concentrations (a) 100% PEO, (b) 50%, (c) 5%.

insight into the solvation of PEO in water. It seems to indicate that the average segmental environment of PEO changes linearly with water concentration until a certain amount of water is added. Any further addition of water causes a negligible effect on the segmental environment of PEO. These results suggest the formation of a stoichiometric hydrate containing about three molecules of water for each ethylene oxide unit in a PEO chain. This is consistent with previous speculation derived from viscosity results [28] and the change of chemical shifts of the OH protons solutions (Figure 13). The temperature dependence of the hydrate formation in these aqueous PEO solutions seems to be small, because no significant difference in plots such as that Figure 13 was observed when the temperature was changed from 4°to 35°, 51° and 80 °C.

Without specific interaction between the polymer and the solvent, the segmental environment changes linearly with the solvent content in chloroform or benzene over the entire composition range (Figure 14). The slow change of the chemical shift of the OH protons in the concentrated solutions is good evidence for the existence of strong molecular association through intermolecular hydrogen bonds. This dissociation would be expected to be favored by raising the temperature. The dissociation of PEO-400 was found to become rapid only after about 30, 40, and 55% of chloroform had been added to the solution at 51°, 35°, and 4 °C, respectively.

According to the present NMR results, the water molecules in aqueous PEO solution appear to exert a specific effect, in addition to that of dilution, on the segmental environment of PEO. The nature of this specific influence has been considered on the basis of valence theory [29]. We will base our interpretation on IR spectroscopy evidence.

The IR spectra of PEO in the liquid state or in solution may be interpreted by comparing them with the assigned bands in the vibrational spectra of crystalline PEO [20, 26, 27] and structurally related compounds such as 1,2-dichlorethane, [19, 21–23]. ethylene chlorohydrin [30], metal-ethylenediamine complexes [31, 32] ester and ether derivative of ethylene glycol [33], crystalline polyesters [34], and metal-ethylene glycol complexes [35].

For crystalline PEO-4000, the strong IR absorption at 844, 947, and 960 cm^{-1} are assigned to vibration involving principally rocking motion of the ethylene group in the PEO repeating unit in the TGT conformation of the COCCOC sequence [20, 26, 27]. In order to account for the multiplicity of absorption bands in the 800–1000 cm^{-1} region of the IR spectra of PEO, particularly for those found in the spectra of oligomers, in the liquid state of benzene solutions, the trans form of the OCCO group has been proposed [24, 33, 36, 37]. However, we do not believe that the trans form as a major structural unit has been well substantiated by the IR spectral evidence. Whenever observations have been made on OCCO groups definitely known to be in the trans form [19, 21–23, 32, 34, 38, 39] or calculations made for this conformation around the C–C bond [21–23], the infrared active rocking frequencies have always been found in the 700–800-cm^{-1} region. A band at 845 cm^{-1} in the IR spectrum of crystalline poly(ethylene terephthalate) has been attributed to a rocking vibration of

OCCO groups in the trans form [40], but Tadokoro has shown [41] that this absorption has the wrong dichroic behavior for the ethylene rocking mode, so that such an assignment is questionable. A very weak band can be observed near 810 cm^{-1} in the ir spectra of PEO obtained in this, as well as other investigations [24, 25]. This absorption could be assigned to ethylene groups in the trans form, but its persistence in the spectra of crystalline films [24] makes the assignment doubtful. In any case, since the absorption near 810 cm^{-1} is very weak when compared to the absorptions of the gauche form observed near 947 and 844 cm^{-1}, most of the ethylene groups in PEO, including both high and low molecular weight polymers, must be in the gauche conformation even in liquid or solution state.

The vibrational spectra of PEO in the liquid state or in solution should also be affected by the conformation around the C–O bond. Dioxane has the gauche conformation around the C–C bond as in PEO but the conformation around the C–O bond is gauche rather than trans as found for crystalline PEO. By using the same potential field to calculate the normal vibrations of dioxane as used for crystalline PEO, Miyazawa has shown that the conformation around the C–O bond can have a large effect on the ethylene rocking frequencies, and has emphasized the importance of this in the interpretation of PEO oligomer spectra [25]. These calculations and the infrared spectrum of dioxane show that the original bonds found at 1324, 993 and 863 cm^{-1} in the benzene spectra of PEO oligomers could well be caused by a gauche conformation of the COCC groups as in dioxane. This interpretation is in agreement with Davison's suggestion [26] that rotation on melting of crysstalling PEO occurs mainly at the C–O bonds where the energies of the trans and gauche conformation are closer to one another than for rotation around the C–C bonds.

The band at 886 cm^{-1} has been assigned to rocking vibration of the ethylene groups in both trans and the gauche conformations of PEO oligomers [24, 33, 36]. The band could, however, be associated with end groups since it disappears when the terminal hydroxyl is methylated [25] and becomes weaker as the chain length is extended (Figure 18). This band has been assigned to the trans conformation of the terminal $-CH_2CH_2OH$ [37], but we faver the assignment of the ethylene glycol spectrum [35].

The weak band near 866 cm^{-1} is assigned to CH_2–CH_2OH groups which are not hydrogen bonded since it appears only in the spectra of oligomeric PEO in dilute benzene solutions which contain non-hydrogen bonded OH groups as shown by the appearance of a sharp band at 3600 cm^{-1} in the OH stretching region. The shift from 886 cm^{-1} to a lower frequency is to be expected when the hydrogen bonds are destroyed [42].

The IR evidence of PEO in the crystalline form, the melt, and in benzene solutions indicated that (1) the conformation of the repeating unit in crystalline PEO is in the TGT form as previously suggested [19, 25, 26], (2) in the melt or in benzene solution most of the ethylene groups in PEO remain in the gauche conformation, and the 'disorder' of the polymer chain occurs mainly at the C–O bonds, (3) the conformations of the repeating unit of oligomeric PEO in benzene become more 'disordered' in the dissociated state than in the associated form, and (4) the random conformations of

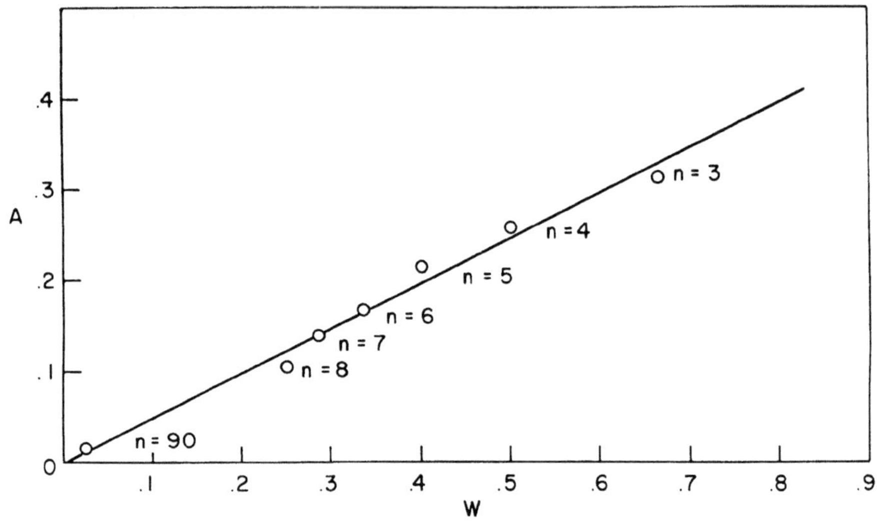

Fig. 18. A plot of the IR absorbance (A) at 886 cm^{-1} vs. weight fraction (W) of CH_2CH_2OH in PEO.

PEO, which have been observed in the liquid state or in benzene solution, are greatly diminished in aqueous solution, with the appearance of relatively sharp bands at 844 and 2935 cm^{-1} corresponding to the perpendicular bands at 844 and 2950 cm^{-1} in the spectra of crystalline PEO [20, 26, 27]. We believe that the conformation of PEO in aqueous solution retains to a large degree the TGT sequence characteristic of crystalline PEO. This is consistent with what we have found in our relaxation studies.

6. Conclusions

We have systematically studied a well-defined model polyelectrolyte system, potassium iodide-PEO in methanol. Our results clearly indicate that the interaction of potassium iodide with low oligomers ($n \leqslant 6$) is very weak, and becomes strong as $n \geqslant 7$. The polymeric-type interaction sets in at $n = 7$, suggesting that a cooperative effect of this interaction appears only when $n \geqslant 7$. The results from electrophoresis experiments indicate that the positively charged potassium ion is actually the species which directly interacts with the polymer. Once the polymer associates with potassium ions in methanol, it behaves as a polyelectrolyte. The apparent association constant of the potassium iodide to PEO, obtained from the data of chemical shifts, decreases as one increases either the salt to polymer ratio at constant polymer concentration or the polymer concentration at constant salt to polymer ratio.

The ionic association induced by the addition of potassium iodide to poly(ethylene oxide) in methanol produces large changes in the nuclear spin-lattice relaxation (T_1) of the polymer protons. These changes can be described quantitatively in terms of free and charged units along the polymer chain. Long-range interactions, as evidenced in the macroscopic solution viscosity, are not reflected in T_1. An important distinction between macroscopic and microscopic properties is clearly shown.

It is an obvious truth that a polymer molecule in solution does not exist as an isolated species (a 'polymer gas'). We have emphasized the important role of solvation on affecting molecular conformation and interactions [43–47]. Based on both our relaxation and spectrosopic results, a solvent induced conformational state of PEO in aqueous systems is suggested. Water molecules exhibit a specific solvation behavior for PEO and prohibit the ionic association of potassium iodide (or potassium sulfate) with PEO.

References and Notes

1. Morawetz, H.: *Macromolecules in Solution*, Interscience Publishers, Inc., New York, N.Y., 1966.
2. Rundle, R. E. and Baldwin, R. R.: *J. Amer. Chem. Soc.* **65** 554 (1943).
3. Swanson, M. A.: *J. Biol. Chem.* **223**, 825 (1948).
4. Fordyce, R., Lowell, E. L., and Hibbert, H.: *J. Amer. Chem. Soc.* **61**, 1905 (1939).
5. Lundberg, R. D., Bailey, R. E., and Callard, R. W.: *J. Polymer Sci.* **A1**, **4**, 1563 (1966).
6. Liu, K. J.: *Macromolecules* **1**, 213 (1968).
7. Abragam, A.: *The Principles of Nuclear Magnetism*, Oxford University Press, London, 1961.
8. Liu, K. J. and Gregor, H. P.: *J. Phys. Chem.* **69**, 1248 (1965).
9. Tanford, C.: *Physical Chemistry of Macromolecules*, John Wiley and Sons, Inc., New York, N.Y., 1962.
10. Liu, K. J. and Gregor, H. P.: *J. Phys. Chem.* **69**, 1252 (1965).
11. Liu, K. J. and Ullman, R.: *J. Chem. Phys.* **48**, 1158 (1968).
12. Anderson, J. E., Steele, J., and Warwick, A.: *Rev. Sci. Instr.* **38**, 1139 (1967).
13. Liu, K.: *J. Macromolecules* **1**, 308 (1968).
14. This point may not be entirely obvious. On the basis of earlier measurements, our present T_1 measurements are known to be in the region of 'extreme motional narrowing.' Under these conditions, T_1 is inversely proportional to molecular mobility.
15. Zimmerman, J. R. and Britten, W. R.: *J. Phys. Chem.* **61**, 1328 (1957).
16. Liu, K. J.: *International Symposium on Macromolecular Chemistry, Tokyo, Japan*, 1966, reprints **7**, 48 (1966).
17. Bloembergen, N., Purcell, E. M., and Pound, R. V.: *Phys. Rev.* **73**, 679 (1948).
18. One might worry about the influence of polymer-solvent nuclear dipole interactions on T_1. This intermolecular contribution can be safely neglected for the present purposes: its magnitude is roughly 0.03 s^{-1} for a proton-rich solvent like C_6H_6 and should decrease for solvents containing fewer protons. Experimental T_1 values of *p*-dioxane in dilute $CDCl_3$ and $CHCl_3$ solutions, and of PEO in H_2O and D_2O, are indistinguishable.
19. Brown, J. K. and Sheppard, N.: *Trans. Faraday Soc.* **48**, 128 (1952).
20. Miyazawa, T., Fukushima, K., and Ideguchi, Y.: *J. Chem. Phys.* **37**, 2864 (1963).
21. Nakagawa, I. and Mizushima, S.: *J. Chem. Phys.* **21**, 2195 (1953).
22. Nakagawa, I.: *J. Chem. Soc. Japan* **74**, 848 (1953).
23. Nakagawa, I.: *J. Chem. Soc. Japan* **76**, 813 (1955).
24. Kuroda, Y. and Kubo, M.: *J. Polymer Sci.* **36**, 453 (1959).
25. Machido, K. and Miyazawa, T.: *Spectrochem. Acta.* **20**, 1865 (1964).
26. Davison, W. H. T.: *J. Chem. Soc.* 3270 (1955).
27. Hoshihara, T., Tadokopo, H., and Murahashi, S.: *J. Chem. Phys.* **41**, 2902 (1964).
28. Rosch, M.: *Kolloid – Z.* **147**, 78 (1956).
29. Rosch, M.: *Kolloid – Z.* **150**, 153 (1957).
30. Mizushima, S., Shimanouchi, T., Miyazawa, T., Abe, K., and Yasumi, M.: *J. Chem. Phys.* **19**, 1477 (1951).
31. Powell, D. B. and Sheppard, N.: *Spectrochem. Acta* **17**, 68 (1961).
32. Powell, D. B. and Sheppard, N.: *J. Chem. Soc.* 3089 (1959).
33. Mayake, A.: *J. Amer. Chem. Soc.* **82**, 3040 (1960).
34. Davison, W. H. T. and Corish, P. J.: *J. Chem. Soc.* 2428 (1955).

studied (for example by Teorell [1], Sollner [2], or by the collaborators of Aharon Katchalsky's group, or by others) in order to test under simplified conditions the production of specific effects for example oscillations or simply generations of electric potentials by fixed charges due to polyelectrolytes, thus modelling biological phenomena and discussing the possible implications of their models in biology. It is due to the contribution of physical chemists that the charge effects have been predicted and many times verified by physiologists or molecular biologists and include, amongst others, the theories about cell or other separating membranes or nerve membranes [3].

But all these previous biophysical models took the necessary energy from external, synthetic sources or from the natural thermodynamic diffusive evolution of the system or more recently from photochemical reactions [4–6]. A necessary complementary step had to include chemical reactions in the models in order to better approach the major aspects of complex biochemical phenomena *in vivo*.

The compulsory energetic coupling of chemical (biochemical) reactions and of diffusive mass-transfers was phenomenologically identified and also introduced in the well-known case for example of the so-called active transport by biologists; it was explicited theoretically in several treatments using mostly calculations based on thermodynamics of non-equilibrium processes [7–11].

The accumulative pumping of ions or of some metabolites thus appears in frogskin [12], in excitable nerve or muscle membranes, in glucose or other sugar nutritions and in many other cases. A number of more mechanistical models which contain macromolecular or smaller unidentified transporters of the actively transported species have been drawn, but until recently none of these models could be tested or verified *in vitro*. Many of them are cited in Stein's book [13] or other collective works.

One of the main problems for the general acceptance of a model by biologists even for discussion is that it must contain biochemical molecules and reactions. Many of these reactions are catalysed by enzymes. The knowledge about enzymes has considerably increased during the last decade, and it is known [14] that these proteins add to the polyelectrolytic character specific reactivities, restricted sometimes to practically one species and then called electivity. Enzymes are active only in or near a given conformation and a charged or electronic state of the active site or of other groups; the activity is thus restricted to a limited pH zone and ionic strength can affect it just as well as the usual parameter: the temperature. Some external ionisable functions of the enzyme itself (amino, carboxy, phenol, thiol groups) and small or polyelectrolytic natural or synthetic buffers can regulate the enzyme activity in the solution; this is particularly useful in homogeneous enzymology when, for example, a neutral substrate (e.g. urea, glucose, esters, CO_2) is hydrolysed, oxidised, hydrated or split into charged species (ammonium carbonate, gluconic acid, acid and alcohol, bicarbonate.) This is also true when adenosine triphosphate (ATP^-) supporting one negative charge is split into diphosphate (ADP^-) and phosphate (P^-), also increasing the number of charged species. By this sort of feed-back action the corresponding change in pH generated by the product can retro-act and stop the enzyme activity if the solution is unbuffered. Thus we meet parameters that are already familiar to the physical chemist of polyelectrolytes.

It is fully demonstrated that important enzymes are present in membranes or other biological subunits in which, in opposition with situations studied in homogeneous enzymology in enzyme solutions, no significant convection (stirring) can take place. In this case diffusion processes can no longer be overlooked, just as in all heterogeneous systems including polyelectrolyte gels and membranes.

To realize sufficiently simple experimental systems in which the fundamental interrelations between the basic parameters could be studied, enzymes had to be immobilized in gels or membranes or on their surfaces and we come to the recent yet explosively expanding field of immobilized enzymes following the now generally accepted term. (More than 600 references on immobilized proteins, their applications and related fields are cited in reference [15]).

Here ways of immobilization, enzyme membranes and the mathematics of their diffusion-reactions will be introduced before treating regulations resulting from reaction-reaction or charge-reaction couplings in the stationary state or during evolutions.

It will be shown then how the convenient arrangement in space of the analysed subunits can produce transport selectivities or accelerations or can convert chemical energy into other forms of energy.

2. Immobilizing Enzymes

The immobilization of enzymes retaining at least part of their catalytical activity was successively attempted following four different lines:

(a) physical or physico-chemical adsorption;

(b) untrapping-inclusion in a gel;

(c) covalent bonding to reactive polymers or organic and inorganic supports;

(d) reticulation of the enzyme molecules alone or with other natural or synthetic macromolecules using at least bivalent crosslinking agents.

Along these same lines a large number of techniques using a great variety of enzymes or of materials has been reported in patents or in scientific literature especially since the early 1960's. Several reviews have resumed them [16], but new ones are continually being created. The immediately evident practical aim of these immobilizations is to obtain with good yields of conservation, particles, membranes and surfaces of high enzyme-activity utilizable in heterogeneous reactors or measuring instruments losing as little protein and activity as possible during the preparation as well as during the utilization, conservation or temperature variations.

For technical uses and even for some modelling, adsorption (e.g. on ion-exchangers [17], on aminated cellulosa [18]) and untrapping (e.g. in gels [19] or collodion membranes [20]) can be of interest. However, conservation of activity by *covalent bonding* can be considered by itself as a significant discovery.

One can easily understand that ten years ago namely because of the then recent success of the allosteric theory [21], well-known biochemists and biophysicists, jumping to some generalisations, sincerely doubted the possible conservation of activity of chemically modified enzymes. The opposite is widely demonstrated now with covalent bonding of many enzymes through their external carboxy, amino, phenol or thiol functions leaving the active site unchanged, and up to now its *intrinsic* selectivity and affinity unmodified. In some cases a nearly 100% conservation of activity was found [22], and more generally the covalent immobilization in some networks increases the thermal and chemical stability of enzymes. This leads to two other comments: firstly, up to about 1964 only the most stable hydrolytic proteases were linked chemically and it was not fully demonstrated until about 1968 that the other classes and much more fragile enzymes could be chemically bound *without losing a major part of their activity*. Secondly, the nature of the reactive functions can perhaps explain the historical evidence that many of the first chemists or membres of the groups actively engaged in preparing immobilized enzyme were initially interested in the synthesis of charged gels (ion-exchangers, ion-exchange papers). They used the same supports as, or reactions similar to those allowing the preparation of weak heterogeneous phase polyelectrolytes. [23–27].

Let us see a few examples.

Enzymes were first reacted in Germany with insoluble synthetic particles such as derivatives of polystyrene or of methacrylic acid (Manecke) [23]. They were combined in the famous group of Ephraïm Katchalski, in Israel, with synthetic polypeptides [24], then linked to ethylen-maleic anhydride copolymers [28] or included in collodion membranes [20] similar to those used earlier by Sollner [2] to prepare charged membranes.

In Sweden Axen and Porath *et al.* [25] activated sephadex gels namely by bromocyanamide, a reagent that is now used currently to link enzymes to other hydroxyl-containing polymers such as cellulose. In England Hornby *et al.* [29] transformed carboxymethylcellulose into its azide derivative, and the same group also used reagents of the cellulose dye industry [30].

Elsewhere several acrylamino-polymers have been diazotized and used to link proteins [31] or carboxyl groups of polyacids were activated by Woodwards's Reagent K [32] or carbodiimides [33].

The attachment of enzymes to porous glass bead surfaces activated by chlorosilanes was developed by Weetall [27] at Corning in U.S.A. Wichterle and Coupek *et al.* [34] have reported recently in Czechoslavakia on the preparation of poly (hydroxyethylmethacrylate) beads, a sort of hydrophilic organic glass, with calibrated pores and carrying a variety of chemical groups able to react with enzymes.

In France, the research was initiated in the mid-sixties in Rouen. The special technique developed [22, 35, 36, 37] uses the reticulating cross-links, with e.g. glutaraldehyde, between enzymes and other proteins such as human albumin (for medical uses) or bovin albumin or gelatin or collagen when mechanical properties are of more importance. Analogous methods have been employed by Quiocho and Richards [38] to link enzymes in the solid state for spectroscopic studies. It appears that at least one of the first *series* of a variety of non-proteolytic enzymes conserving a high fraction of their activity has been thus covalently bound. The versatility of this technique allows either variation of the enzymic activity by the amount of enzyme added or variation of the cross-linking by that of the reticulating agent. Thus particles, sponges or membranes are obtained directly, casted self-supporting or on different matrixes or supports [22, 37, 39]. Because of the easiness of this technique many models could and can be experimentally tested [40–45] and practical applications developed [46–48].

Illustrations have already been given [41, 44], but let us just remember that the desired quantities of enzyme and glutaraldehyde as cross-linking agents are mixed at ordinary temperature in a buffered solution of an inactive protein and abandoned; the solution will go into gel between 10 min and 24 h later, depending principally on the amount of reticulating agent and whether the temperature is room temperature or 4 °C. Before complete gelation the mixture can be casted as soon as sufficiently viscous into films; and this even on top of another proteinic film (multilayer membranes) [42].

There are also several other ways of linking inactive proteins and enzymes, for example insoluble chromic complexes have been prepared recently by Sélégny *et al.* [43] realizing a sort of tanning of proteins; Gautheron *et al.* [49] linked enzymes to collagen films. Brown *et al.* [50] used synthetic polyiodides and polyaldehydes, and surely many other ways will yet appear.

3. Variables and Pathways of Calculations

Several years of extensive research have, sometimes simultaneously, been devoted in a few groups to the explicitation of specific or more general mathematical models of systems including immobilized enzymes and to the corresponding experimental verifications.

Instead of an extensive and complete enumeration of calculations we shall try to regroup and establish a classification of problems, of the elements of logic according to which the representative models are or can be constructed, and to draw up a number of interesting conclusions before exposing some new findings.

Evidently the basic problem is to describe quantitatively, in a steady state or during time-dependent evolutions, the diffusion or flow of substrates or of products and the chemical enzymic reaction. The mass transfer and the transformation are coupled and interdependent; the question is *to explicate how and to what extent they influence each other* in function of the variables: the conditions and gradients existing in *space and time* in and at the boundaries of the system on the one hand and the characteristics of the reactions or of the medium and the geometry or symmetry of the latter on the other one.

At first sight mathematically these problems are not completely new; similar ones have been met in different fields and one can 'count one's weapons' before attacking the formulation and add the necessary complements: examples of diffusion of heat or matter (diffusion-absorptions) (diffusion-reactions) [51], heat transfer with reactions or propagation of 'vivid heat' ('chaleur vive'), diffusion of heat through media including heat sources or heat sinks [52], heterogeneous catalysis, hydrodynamics (flows through or along media exchanging heat or reacting), electrochemistry (reactions on electrodes) and even electrokinetics a.s.o. are familiar in physical chemistry or in engineering and are related in good text-books.

This formulation can be accomplished essentially by following two pathways: the kinetic one or the thermodynamic one by several techniques.

The physico-chemical kinetic calculation will proceed starting from the mechanism involved: if convection is absent in systems the different constituting parts of which are homogeneous in a sufficiently microscopic scale to admit at least the use of 'integrated or effective mean diffusion coefficients' and limit to limit integrations, the equations of *concentration profiles* will first be obtained by integration of the differential equation of diffusion-reactions which combines Fick's second law and the chemical rate equation. The local *diffusion fluxes* are calculated at the limits or, if so desired, at each point of the system by deriving the equation of concentration profiles in space and time and introducing the concentration gradient obtained in Fick's first law.

The local and general *reaction-rate* is given finally by differences in diffusion fluxes of reactants or products. The distribution of substrate (reactant) to product ratios is given by the local speed and stoichiometry of the reaction and the diffusion fluxes of these compounds.

Such formulations are known for reactions of first and even second order [46]. They have also been developed since 1968 ([36, 53]) for 'homogeneous' (gel-type) enzyme membranes in which the order of reaction varies from first to zero order when the substrate concentration increases (from $S \ll K_m$ to $S \gg K_m$) as will be shown in the following.

When the medium is heterogeneous one can also use this type of approach as a model [53]. In other cases a porous membrane model should be preferable.

But for very heterogeneous systems or for flow reactions (e.g. columns [54] or tubes [55]) the practically useful information such as the amount of substrate transformed in function of time can also be obtained by the *engineering approach*: one uses mass transfer equations (e.g. permeabilities or mass transfer coefficients instead of diffusion coefficients in membranes or unstirred layers and Reynolds, Nusselt and Schmidt numbers for the hydrodynamics [56, 57]) and the turn-over (or mean speed) of the reactions together with the dimensions of the parts constituting the system and obtains their inter-relations.

Such early calculations have been made for columns filled with enzymic particles [54]. However, it seems that the shape of the reaction pattern in the columns has not yet been calculated in the same detail as are the establishment and the steady state patterns of the ion-exchange fronts. Recent calculations for tubes bearing an enzyme layer on the wall [58] or even charged surfaces [56] have been made in more detail.

The *thermodynamic* approach must be a *non-equilibrium* one because of the continuous entropy production by the reaction in an open system; it will start by writing the coupled fluxes of transfer and transformation characterising the reaction by the affinity and explicate the *coupling* or *friction* coefficients. When the reacting medium is sufficiently homogeneous to be decomposed into elementary volumes, limit to limit integrations of gradients (dn/dx) are consistent; if this is not so, then concentration or other macroscopic differences at the external limits are employed ($\Delta n/\Delta x$). This is already similar to an 'engineering' approach.

In all cases the equations of the purely diffusive or convective mass transfer will be a part of equations describing transfer-reactions: they alone are obeyed in the limiting cases when the reaction stops. Inversely the laws of the reaction are obeyed locally as in a similar, stirred, homogeneous medium and are obeyed fully at 'saturation' when the mass transfer is faster than the chemical transformation. In the more general intermediate situation they are both of comparable rates.

A major advantage of the non-equilibrium thermodynamic formulation is that it can treat the reactive system as a 'black-box' and obtain external intercorrelations. Several systems differing largely in diffusion, reactive or geometrical properties or even in the number of reactions if some additive rules hold, can give identical external answers; this is the advantage, for example, as far as incompletely known biological systems are concerned. One can also look at the mechanisms in more detail inside a fairly homogeneous synthetic membrane, but then the procedure will not be very different from the kinetic pathway and mainly the formalism will differ.

Some time ago only the linear near-equilibrium systems of first order kinetics were sufficiently known to allow advanced calculations; now the non-linear analysis is well under way.

Some advantages of the kinetic treatment (e.g. of diffusion-reactions) are that its formulation is easy in differential equations, as is the understanding by biologists, and that it can be constructed progressively with previously tested procedures and analytical or numerical integrations.

Somehow it is perhaps not inexact to say that the kinetic theory goes from particular cases towards the general one and that the thermodynamic treatment chooses the opposite way. When the necessary information exists, the transcription of the kinetically calculated results in an irreversible thermodynamic formulation is evident; the opposite is true if the mechanisms are known.

The sketch of a logical analysis of reactive immobilized enzyme systems can be resumed by a few conclusions (Table I).

The results of all calculations are in quantitative agreement because of their necessary agreement with the second principle of thermodynamics, the conservation of momentum and unicity of time as well as with the generalized first principle (equivalence of all forms of energy) and the so-called Curie's law*.

Practically all calculations arrive directly or indirectly at the definition and the use of *dimensionless parameters* to characterize (and normalise) the reactive heterogenous system (Tables II and III). The parameters include *constants* of mass transfer, of reactions and of dimensions; they express a ratio of 'reaction time/mass transfer time'

* If the scalar chemical energy is transformed into *asymmetrical* (vectorial) works, then the system itself must be asymmetrical along this direction; in *symmetrical* systems only symmetrical work is obtained. Such an asymmetry can be due to membrane itself or to boundary conditions.

TABLE I

Principal factors ruling the properties of heterogeneous enzymic systems (Physical and chemical constitution, geometry, boundary conditions)

Material	– Permeable membranes; films, beads (grains, spheres, cylinders, discs, cubes) – Layers or surfaces on impermeable supports – Porous, non-porous material, sponges – Symmetrical, disymmetrical – Homogeneous, heterogeneous, structured	System	– Batch (compacted, suspensions, 'reactive walls') – 'Passing through' systems (diffusion or flow through membranes or columns) – 'Parallel flow' systems (planar contactors, tubes) – Symmetrical, assymmetrical – Finite, infinite
Nature of Convection	Diffusion, natural convection, forced convection (laminar, turbulent), unstirred layers	Mass transfer Characteristics	– Diffusion coefficients, permeabilities; Mass transfer coefficients, flow rates, Reynolds Nusselt Numbers – Gradients through medium and system; con- – centration or chemical potential differences between boundaries
Chemical Reaction	– Number of enzymes – Active enzyme site densities and distributions – Irreversible or reversible reactions – Parallel or consecutive reactions – Stationary state, quasi-stationary state – Evolution	Characteristics of Enzyme	Enzyme-substrate affinity (K_m, K_S) Michaelien, allosteric or other; necessity of co-substrate or co-factor Inhibited or not by excess of substrate or product. pH dependence of activity Stability (time, temperature, pH, chemical)
Enzyme-Regulatory Component of Material	– Buffering substances (fixed charges) } acting on enzymes – Substances or links stabilizing the active form } acting on enzymes – Components solubilizing or insolubilizing substrate, product, activators, inhibitors (fixed charges, lipids, complexes, clathrates, etc.) near enzyme site – External (selective) layers excluding some components.	Feed-back and Regulations by the Reaction	– Activation or inhibition by product (through pH or directly) of the producing enzyme or another enzyme – Induction, suppression or inversion of fixed charges by product – Inducing volume flows, contractions, expansions

TABLE II

Principal symbols

(*Note:* except for universally used symbols, capital letters denote qualities as well as 'quantities expressed in standard units'. Lower case or Greek letters (Table III) denote dimensionless normalized quantities or parameters) (units in c.g.s. are indicative of dimensions only).

A	activator or its absolute concentration (if not otherwise specified)
$C_1, C_2 \ldots$	integration constants
D, D_S	diffusion or 'effective diffusion coefficient' of substrate ($cm^2\ s^{-1}$)
D_p, D_{ST}	same for product or substrate-transporter complex
E, E_A, E_B	enzyme, enzyme A or B or concentration of corresponding 'free' active sites able to catalyse (e.g. not denaturated by reticulation) and not engaged in a complex (e.g. mole cm^{-3})
E_0	total concentration of not denaturated active sites
e	thickness of membrane or layer (cm)
ε	elementary charge of electron
F	Faraday
I	inhibitor or its concentration by unit volume
J	diffusion flux of a compound (mole $cm^{-2}\ s^{-1}$) in the steady state
J_D	pure Fickian diffusion *of substrate*
J_S, J_{DS}	*diffusion* fluxes of substrate with reaction at any point of coordinate X
J_1 or J_{s1};	diffusion flux of substrate with reaction at entering face (1)
J_2 or J_{s2}	or exit face (2) of membrane
J_p or J_{DP}	product diffusion fluxes at any point X
J_{P1} or J_{P2}	product diffusion fluxes with reaction at faces 1 or 2
k	kinetic constant of: $+1$ formation, -1 decomposition of complex ES; $+2$ (or cat.): of $ES \rightarrow P$, -2: of $P \rightarrow ES$
k'	apparent first order constant with $S \ll K_m$
K	equilibrium constant specified with subscript in text
K_m	Michaelis constant (mole 1^{-1} or m mole cm^{-3}); subscripts: app = apparent; f = forward reaction, b = backward reaction; i = in presence of inhibitor; pH: different from optimal pH
P	product or its concentration per unit volume (subscripts: same as for S).
Q	other product of successive reactions
R	'Ideal gas constant' or product of reaction if specified
R^-	volumic charge density (exchange capacity per unit volume)
S	substrate or its concentration per unit volume; $S(X)$ at point X, S_1 on face 1, S_2 on face 2...
T	absolute temperature; or 'transporter' and its concentration if specified
t	time (sec)
V or V_M	maximum rate of enzyme reaction with excess (non inhibiting) substrate; ($S \gg K_m$) (in mole $cm^{-3}\ s^{-1}$); subscripts: i = in presence of inhibitor; pH: at not optimal pH; A or B: of enzyme A or B.
v	rate of reaction in a given condition (subscripts as for V)
X	distance (cm) from origin along reference axis; (usually $X=0$ at face 1, $X=e$ at face 2; if different specified in text or table).
Z	total number of charges of a given type of species ($Z = \Sigma z\varepsilon$)
z	number of elementary charges ε of an ion or group (or specified)

TABLE III

Relative, dimensionless (normalized) variables or parameters

$i = I/K_I$	Normalized inhibitor concentration (see s)
$j = J_2/J_1$	Ratio of substrate fluxes through faces 2 and 1 (characterising the type of concentration profile)
J_1/eV	Dimensionless ratio of substrate flux through 1 cm^2 of face 1 to that transformed at maximum rate of reaction in volume element (e cm | cm^2)
$p = P/K_m$	normalized dimensionless product concentration (see s)
$s = S/K_m$	dimensionless substrate concentration per unit volume expressed in 'numbers of K_m of corresponding enzyme'; (same subscripts as for S in Table II)
σ or σ_s	diffusion-reaction parameter characterising the material of reactive membrane and substrate $[\sigma = [(e^2/D)$ sec. (V/K_m) sec$^{-1}]$
σ_p	same as preceding for membrane and product $\sigma_p = e^2 V/D_p K_m$
$t' = \dfrac{t}{e^2/D}$	dimensionless time variable (sec cm^{-2} cm^2 sec^{-1})
$V_A{}^B = V_A/V_B$	ratio of maximum speed per unit volume of membrane of enzyme A and B.
x	normalized distance along reference axis $x = X/e$; $x = 0$ at face 1, $x = 1$ at face 2 where $X = e$ (other notations specified).
	Parameters simplifying the writing of MacLaurin series expressions with irreversible reactions in other cases subscripts and formulations are specified: $\beta_1 = \dfrac{1}{12}\dfrac{\sigma}{(1+s_1)^2}$ $\quad \lambda_1 = \dfrac{s_1}{1+s_1}$ $\quad \varrho_1 = \dfrac{1}{1+2\beta_1}$
α	Thiele number for first order reaction or specified with subscripts.
γ	Thiele modulus: $\sqrt{\sigma}$ for first-order reaction.

in and through the same volume (surface). Such parameters already existed (Thiele or Damkeller numbers) and had to be adapted to enzyme kinetics. Homogeneous enzyme kinetics [14] or mathematics of diffusion [51] also use such 'relative' expressions.

The partial differential equations that are obtained in most formulations are solved *first in the particular simplified cases for which analytical solutions exist, and then more general approximate solutions are proposed through series; finally graphical or numerical integrations provide solutions of the general time-dependent cases* which are usually physically understandable after representation in graphs or tables.

Whole order irreversible or equilibrated reactions, stationary state assumptions, selected boundary conditions and geometries make simplifications possible.

I have been asked many times to *define the aims of calculations and modelling*; this is an occasion to list some of the major ones, which describe mono- or poly-enzymic systems in order to:

(a) be able to measure constants and parameters of materials and systems;

(b) detect intrinsic modifications or environmental effects on kinetics or specificities due to insolubilization of enzymes;

(c) predict and realize optimalizations, self or external regulatory effects, oscillations;

(d) analyse complex systems;

(e) create or invent transport phenomena;

(f) interconvert chemical and other forms of energy (osmotic, mechanical, hydrodynamic, electrical or radiative);

(g) invent, analyse, simulate and express the conclusions of biophysical models which can help the logical physical, biophysical, biological supramolecular understanding of nutritions, regulations, selectivities, allosteria, allotopies, energy conversions etc....;

(h) invent applications in measuring, electrochemistry, separations, industrial productions or curing;

(i) help to define the optimal characteristics of enzyme material for these applications or uses.

The systems can finally be treated on the basis or the combinations of:

- diffusion (convection) control of reaction;
- reaction control of heterogeneous reactions (parallel or consecutive reactions);
- reaction control of diffusion (convection);
- energy conversions through mediations or other consequences.

4. Heterogeneous Kinetics of Diffusion-Reactions with Homogeneous Enzyme Membranes

The basic equations of diffusion-reactions and their solutions are more easily presented in the geometrically one-dimensional case of enzyme membranes (Figure 1) separating two substrate (S) solutions of concentrations S_1 and S_2 or alternatively plunged into a substrate (and product P) solution; the total variation of substrate concentration with time is given at each point of the membrane situated on the space reference axis Ox perpendicular to each parallel face of the membrane by Equation (1) [26, 42, 53]:

$$\left(\frac{\mathrm{d}S}{\mathrm{d}t}\right) = \left(\frac{\partial S}{\partial t}\right)_{\text{diffusion}} + \left(\frac{\partial S}{\partial t}\right)_{\text{reaction}} + \left(\frac{\partial S}{\partial t}\right)_{\text{formation}} \tag{1}$$

As it will be seen, similar equations can be written for each reactive component of the system. The list of factors ruling the membrane and system are given in Table I.

Starting from simplifying assumptions, we shall now expose a few cases. Looking first into irreversible monoenzymatical homogeneous membranes, we shall present laws and calculations giving the diffusion control of reactions and inversely the modifications in diffusion due to reactions with symmetrical and unsymmetrical boundary conditions of the reactive layer, first in the steady state and then in function of time. Modifications in the law of kinetic reactions will follow. After the basically uncharged system has been understood, charge effects will appear as producing selectivities in partitions or mobilities of reactive or non-reactive species and also as regulators of enzyme activity or even of volumes.

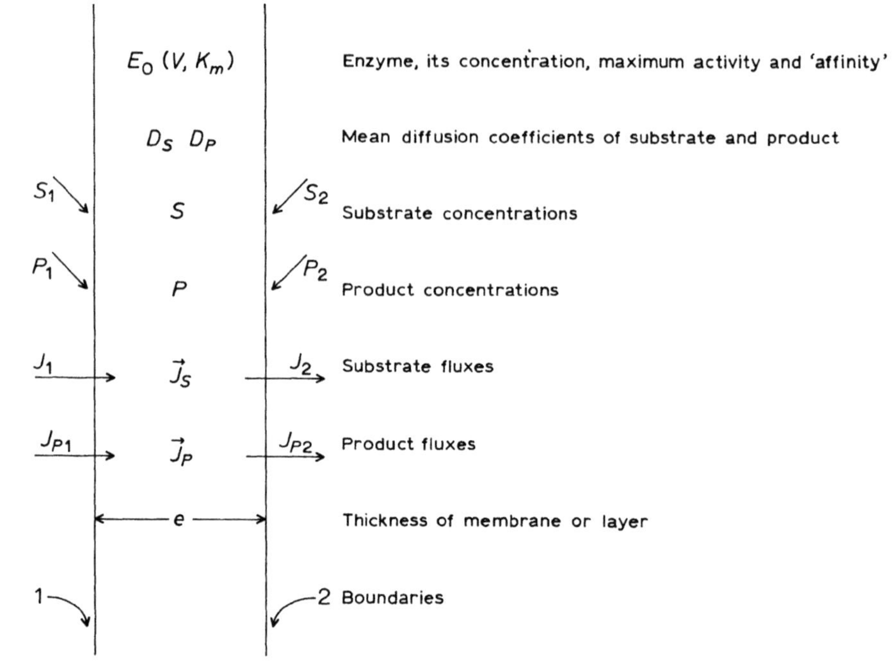

Fig. 1. Diagrammatic representation of membrane and system and of their characteristic constants.

4.1. Monoenzymatical membranes

4.1.1. *The Fundamental Case: Irreversible Reaction, Michaelian Kinetics*

4.1.1.1. *The Membrane and Assumptions.*

The simplest fundamental cases are obtained if the following conditions are assumed:

The membrane is homogeneous in regard to the diffusion of substrate and product and also of effectively active enzymatical sites; its faces are planar and parallel.

The unique monoenzymatic reaction is *kinetically* Michaelian of the irreversible one-substrate type

$$E + S \underset{k_{-1}}{\overset{k_{+1}}{\rightleftharpoons}} ES \xrightarrow{k_{+2}} E + P \tag{2}$$

in the following operating conditions:

All the distribution coefficients between membrane and solution are equal to one; the effective diffusion coefficients of substrate and product are identical and independent of concentration, there are no interactions between diffusing species or with the solvent. Volume flows or variations and unstirred layers are inexistent or neglected. Charge effects are inexistent or cancelled. Moreover pH and temperature are constant and there are no autocatalytical or inhibiting retro-actions (feed-back) of substrate or product either on the permeability or the activity of the membrane; the enzyme activity is not affected by the duration of the experiment.

With these assumptions the third term on the right-hand side of Equation (1) disappears; using Fick's second law for the diffusive term and denoting by v the reactive term, one obtains:

$$\frac{\partial S}{\partial t} = D_S \frac{\partial^2 S}{\partial X^2} - v\,. \tag{3}$$

4.1.1.2. *The Stationary State.*

In the stationary state Equation (3) equals zero and using the Michaelis-Menten formulation [14, 26, 53] (see Table IX):

$$D_S \frac{\mathrm{d}^2 S}{\mathrm{d}X^2} = V \frac{S}{K_m + S} \tag{4}$$

and

$$D_P \frac{\mathrm{d}^2 P}{\mathrm{d}X^2} = - V \frac{S}{K_m + S} \tag{5}$$

from (4) and (5) after integration:

$$D_S \frac{\mathrm{d}S}{\mathrm{d}X} + D_P \frac{\mathrm{d}P}{\mathrm{d}X} = \text{Const.} = -[J_{DS} + J_{DP}]\,. \tag{6}$$

Note that Equations (4) and (5) can be rewritten in dimensionless form [56, 40, 41] (see Tables III, IV, V, VI)

$$\frac{\mathrm{d}^2 s}{\mathrm{d}x^2} = \sigma \frac{s}{1 + s} \tag{4'}$$

$$\frac{\mathrm{d}^2 p}{\mathrm{d}x^2} = - \sigma \frac{s}{1 + s}\,. \tag{5'}$$

4.1.1.2.1. *Analytical and approximate solutions.* The stationary-state equation has been solved analytically with symmetrical or unsymmetrical boundary conditions with zero order ($S \ll K_m$) reaction [26, 40, 41, 44] or first order ($S \ll K_m$) kinetics [53, 41, 59, 60]; see Tables IV and V.

A particular analytical solution reported by Sélégny *et al.* [61] and given in Equation (7), has the advantage of linking the local K_m, characterising the active site, to measurable constants and thus allows the determination of the 'real' or 'intrinsic' K_m of the immobilized enzyme [36, 40, 41, 44, 62, 63].

$$J_1^2 - J_2^2 = 2VD\left[(S_1 - S_2) - K_m \ln \frac{S_1 + K_m}{S_2 + K_m}\right]. \tag{7}$$

This solution is valid if the substrate flux is never nil in the membrane or layer considered.

TABLE IV

Diffusion reactions in a membrane with a reaction of zero order

Unidirectional (membrane type) steady-state equations of diffusion-reactions with one Michaelis type enzyme. Fundamental case: no regulatory effects; *zero order reaction*, $v = V_M$, *assumption*: $S \geqslant 10\ K_m$ or $s \geqslant 10$ to 50 in all points of the membrane.

$[\sigma = Ve^2/DK_m;\ s = S/K_m;\ p = P/K_m;\ V = V_{\max};\ 0 \leqslant X \leqslant e;\ 0 \leqslant x = X/e \leqslant 1]$

Specifications	Absolute values	Relative, dimensionless equations
(a) *Unsymmetrical boundaries*	$S_1 > S_2$ $X = 0 \rightarrow S = S_1;\ X = e \rightarrow S = S_2$	$s_1 > s_2$ $x = 0 \rightarrow s = s_1;\ x = 1 \rightarrow s = s_2$
Concentration Profile	(1) $S(X) = \dfrac{V}{2D} X^2 - \left(\dfrac{S_1 - S_2}{e} + \dfrac{Ve}{2D}\right) X + S_1$	(1′) $s(x) = (\sigma/2)\, x^2 - (s_1 - s_2 + \sigma/2)\, x + s_1$
Fluxes Pure diffusion of S	(2) $J_D = -\dfrac{D}{e} \displaystyle\int_{S1}^{S2} \mathrm{d}S = D \dfrac{S_1 - S_2}{e}$	(2′) $\dfrac{J_D}{Ve} = \dfrac{s_1 - s_2}{\sigma}$
Diffusion of S with reaction	(3) $J_S = D \dfrac{\mathrm{d}S(X)}{\mathrm{d}X} = D \dfrac{S_1 - S_2}{e} + \dfrac{Ve}{e} - VX$	(3′) $\dfrac{J_s}{Ve} = \dfrac{s_1 - s_2}{\sigma} + 1/2 - x$
Ingoing substrate (face 1; $X=0$; $x=0$)	(4) $J_1 = D \dfrac{S_1 - S_2}{e} + \dfrac{Ve}{2}$	(4′) $\dfrac{J_1}{Ve} = \dfrac{s_1 - s_2}{\sigma} + 1/2$
Outgoing substrate (face 2; $X=e$; $x=1$)	(5) $J_2 = D \dfrac{S_1 - S_2}{e} - \dfrac{Ve}{2}$	(5′) $\dfrac{J_2}{Ve} = \dfrac{s_1 - s_2}{\sigma} - 1/2$

Reaction Total: Effect of reaction on diffusion of S	(6) $J_1 - J_2 = Ve$	(6′) $\dfrac{J_1 - J_2}{Ve} = 1$
Face1	(7) $\dfrac{J_1}{J_D} = 1 + \dfrac{Ve}{2D(S_1 - S_2)}$	(7′) $\dfrac{J_1}{J_D} = 1 + \dfrac{\sigma}{2(s_1 - s_2)}$
Face 2	(8) $\dfrac{J_2}{J_D} = 1 - \dfrac{Ve}{2D(S_1 - S_2)}$	(8′) $\dfrac{J_2}{J_D} = 1 - \dfrac{\sigma}{2(s_1 - s_2)}$
Product Flux Ingoing face 1	(9) $(J_P)_1 = D_P \dfrac{P_1 - P_2}{e} - \dfrac{Ve}{2}$	(9′) $\dfrac{(J_p)_1}{Ve} = \dfrac{p_1 - p_2}{(D_S/D_P)\sigma} - \dfrac{1}{2}$
Out of total flux at face 2	(10) $\dfrac{(J_P)_2}{(J_P + J_S)_2} = \dfrac{(J_D)_P + Ve/2}{(J_D)_S + (J_D)_P} = \dfrac{D_P(P_1 - P_2) + Ve^2/2}{D_S(S_1 - S_2) + D_P(P_1 - P_2)}$	(10′) $\dfrac{(J_p)_2}{(J_s + J_p)_2} = \dfrac{D_P/D_S(p_1 - p_2) + \sigma/2}{(s_1 - s_2) + D_P/D_S(p_1 - p_2)}$
if $D_S = D_P = D$ thus $\sigma_S = \sigma_P = \sigma$	(11) $\dfrac{(J_P)_2}{(J_P + J_S)_2} = \dfrac{P_1 - P_2 + Ve^2/2}{(S_1 + P_1) - (S_2 + P_2)}$	(11′) $\dfrac{(J_p)_2}{(J_s + J_p)_2} = \dfrac{(p_1 - p_2) + \sigma/2}{(s_1 + p_1) - (s_2 + p_2)}$
moreover if $P_1 = P_2$	(12) $\dfrac{(J_P)_2}{(J_P + J_S)_2} = \dfrac{Ve^2}{2(S_1 - S_2)}$	(12′) $\dfrac{(J_p)_2}{(J_s + J_p)_2} = \dfrac{\sigma}{2(s_1 - s_2)}$
	(13) $\boxed{S_1 = S_2}$	(13′) $\boxed{s_1 = s_2}$
(b) *Symmetrical boundaries*	$J_1 = -J_2 = \dfrac{Ve}{2}$	$J_1/Ve = -J_2/Ve = 1/2$
	$2J_1 = Ve$	$2J_1/Ve = 1$

TABLE V

Diffusion reactions in a membrane with a reaction of first order

Equations of (membrane type) unidirectional diffusion reactions in the steady state with one Michaelis type enzyme. Fundamental cases: no regulatory effects; *firsts order reaction*: $v \sim k'S$; $k' = V/(K_m + \varepsilon)$; assumption: $S < K_m/10$ or $s < 10$ in all points of the membrane.

$|\sigma' = \sqrt{\sigma} = \sqrt{Ve^2/DK_m};\ s = S/K_m;\ p = P/K_m;\ V = V_{\max};\ 0 < X < e;\ 0 < x = X/e < 1|$

Specifications	Absolute values	Relative dimensionless equations
(a) *Unsymmetrical boundaries*	$S_1 > S_2$; $X = 0 \rightarrow S = S_1;\ X = e \rightarrow S = S_2$	$s_1 > s_2$; $x = 0 \rightarrow s = s_1;\ x = 1 \rightarrow s = s_2$
Concentration profile	(1) $S(X) = \dfrac{S_1 \sinh \sigma'(1-x) + S_2 \sinh(\sigma' x)}{\sinh \sigma'}$	(1′) $s(x) = \dfrac{s_1 \sinh \sigma'(1-x) + s_2 \sinh(\sigma' x)}{\sinh \sigma'}$
Fluxes Pure diffusion of S	(2) $J_D = -\dfrac{D}{e} \int_{S1}^{S2} \mathrm{d}S = D \dfrac{S_1 - S_2}{e}$	(2′) $\dfrac{J_D}{Ve} = \dfrac{s_1 - s_2}{(\sigma')^2}$
Diffusion of S with reaction	(3) $J_S = \dfrac{D}{e} \sigma' \dfrac{[S_1 \cosh \sigma'(1-x) - S_2 \cosh \sigma' x]}{\sinh \sigma'}$	(3′) $\dfrac{J_1}{Ve} = \dfrac{s_1 \cosh \sigma'(1-x) - s_2 \cosh \sigma' x}{\sigma' \sinh \sigma'}$
Ingoing substrate Face 1 ($X = x = 0$)	(4) $J_1 = \dfrac{D}{e} \sigma' \dfrac{S_1 \cosh \sigma' - S_2}{\sinh \sigma'}$	(4′) $\dfrac{J_1}{Ve} = \dfrac{s_1 \cosh \sigma' - s_2}{\sigma' \sinh \sigma'}$
Outgoing substrate Face 2 ($X = e$; $x = 1$)	(5) $J_2 = \dfrac{D}{e} \sigma' \dfrac{S_1 - S_2 \cosh \sigma'}{\sinh \sigma'}$	(5′) $\dfrac{J_2}{Ve} = \dfrac{s_1 - s_2 \cosh \sigma'}{\sigma' \sinh \sigma'}$

Total Reaction in Membrane: Effect of reaction on diffusion of S	(6) $J_1 - J_2 = \frac{D}{e}\sigma'(S_1 + S_2)\frac{\cosh\sigma' - 1}{\sinh\sigma'}$	(6′) $\frac{J_1 - J_2}{Ve} = \frac{s_1 + s_2}{\sigma'}\frac{\cosh\sigma' - 1}{\sinh\sigma'}$
Face 1 ($X = x = 0$)	(7) $\frac{J_1}{J_D} = \frac{\sigma'}{S_1 - S_2}\frac{S_1\cosh\sigma' - S_2}{\sinh\sigma'}$	(7′) $\frac{J_1}{J_D} = \frac{\sigma'}{s_1 - s_2}\frac{s_1 - s_2\cosh\sigma'}{\sinh\sigma'}$
Face 2 ($X = e$; $x = 1$)	(8) $\frac{J_2}{J_D} = \frac{\sigma'}{S_1 - S_2}\frac{S_1 - S_2\cosh\sigma'}{\sinh\sigma'}$	(8′) $\frac{J_2}{J_D} = \frac{\sigma'}{s_1 - s_2}\frac{s_1 - s_2\cosh\sigma'}{\sinh\sigma'}$
Product P flux Face 1 ($X = x = 0$)	(9) $(J_P)_1 = (J_D)_S + (J_D)_P - (J_1)_S = D_S\frac{S_1 - S_2}{e} + D_P\frac{P_1 - P_2}{e} - \frac{D_S}{e}\sigma'\frac{S_1\cosh\sigma' - S_2}{\sinh\sigma'}$	(9′) $\frac{(J_p)_1}{Ve} = \frac{s_1 - s_2}{(\sigma')^2} + \frac{p_1 - p_2}{(D_P/D_S)(\sigma')^2} - \frac{s_1\cosh\sigma' - s_2}{\sigma'\sinh\sigma'}$
(b) *Symmetrical boundaries*	(10) $S(X) = \frac{S_1[\sinh\sigma' x + \sinh\sigma'(1 - x)]}{\sinh\sigma'}$	(10′) $s(x) = \frac{s_1[\sinh\sigma' + \sinh\sigma'(1 - x)]}{\sinh\sigma'}$
	(11) $V = \frac{2J_1}{e} = 2S_1\frac{D}{e^2}\sigma'\frac{(\cosh\sigma' - 1)}{\sinh\sigma'}$	(11′) $\frac{2J_1}{Ve} = 2s_1\frac{(\cosh\sigma' - 1)}{\sigma'\sinh\sigma'}$

TABLE VI

Equations of (membrane type) unidirectional diffusion-reactions in the steady state with one Michaelis type-enzyme. Fundamental case: no regulatory effects; $v = VS/(K_m + S)$, order of reaction: 0 to 1; no assumption on S/K_m; approximation of Equation (1) by *McLaurin Series* $|\sigma = Ve^2/DK_m; s = S/K_m;$ $p = P/K_m;\ V = V_{\max};\ 0 \leqslant x \leqslant e;\ 0 \leqslant x = x/e \leqslant 1|$

Specifications: $S_1 \geqslant S_2;\ s_1 \geqslant s_2;\ X = 0 \to S = S_1;\ x = 0 \to s = s_1;\ X = e \to S = S_2;\ x = 1 \to s = s_2$

(a) *Unsymmetrical Boundaries Concentration profile*

(1) *Absolute values*

developing near $X=0$ or $x=0$

$$S(X) = S_1 - \frac{X}{1!}\frac{J_1}{D} + \frac{X^2}{2!}\frac{V}{D}\frac{S_1}{K_m + S_1} - \frac{X^3}{3!}\frac{V}{D^2}\frac{K_m}{(K_m + S_1)^2}J_1 + \frac{X^4}{4!}\frac{V^2}{D^2}\frac{K_m}{(K_m + S_1)^3}S_1 - \frac{2X^4}{4!}\frac{VK_m}{D^3(K_m + S_1)^3}(J_1)^2 - {} - \frac{6X^5}{5!}\frac{VK_m}{D^4}\frac{1}{(K_m + S_1)^4}(J_1)^3 + \frac{X^5}{5!}\frac{V^2K_m}{D^3}\frac{6S_1 - K_m}{(K_m + S_1)^4}J_1 + \cdots$$

(2) *Relative dimensionless equations*

$$s(x) = s_1 - \frac{(x)}{1!}\sigma\frac{J_1}{Ve} + \frac{(x)^2}{2!}\sigma\frac{s_1}{1 + s_1} - \frac{(x)^3}{3!}\sigma^2\frac{1}{(1 + s_1)^2}\frac{J_1}{Ve} + {} + \frac{(x)^4}{4!}\sigma^2\frac{s_1}{(1 + s_1)^3} - 2\frac{(x)^4}{4!}\sigma^3\frac{1}{(1 + s_1)^3}\frac{J_1^2}{V^2e^2} - {} - 6\frac{(x)^5}{5!}\sigma^4\frac{1}{(1 + s_1)^4}\frac{J_1^3}{V^3e^3} + \frac{(x)^5}{5!}\sigma^3\frac{6s_1 - 1}{(1 + s_1)^4}\frac{J_1}{Ve} + \cdots$$

(3) *Developing near x=1 gives:*

$$s(x) = s_2 - \frac{x - 1}{1!}\sigma\frac{J_2}{Ve} + \frac{(x - 1)^2}{2!}\sigma\frac{s_2}{1 + s_2} - \frac{(x - 1)^3}{3!}\sigma^2\frac{1}{(1 + s_2)^2}\frac{J_2}{Ve} + \cdots$$

Fluxes

(4) *For x=1 expression (2) allows calculations of J_1/Ve:*

$$s_2 = s_1 - \frac{1}{1!}\sigma\frac{J_1}{Ve} + \frac{1}{2!}\sigma\frac{s_1}{1 + s_1} - \frac{1}{3!}\sigma^2\frac{1}{(1 + s_1)^2}\frac{J_1}{Ve} + \frac{1}{4!}\sigma^2\frac{s_1}{(1 + s_1)^3} - {} - \frac{2}{4!}\sigma^3\frac{1}{(1 + s_1)^3}\left(\frac{J_1}{Ve}\right)^2 - \frac{6}{5!}\sigma^4\frac{1}{(1 + s_1)^4}\left(\frac{J_1}{Ve}\right)^3 + \frac{1}{5!}\sigma^3\frac{6s_1 - 1}{(1 + s_1)^4}\frac{J_1}{Ve} + \cdots$$

(5) *For x=0 expression (4) allows calculations of J_2/Ve:*

$$s_1 = s_2 + \frac{\sigma}{1!}\frac{J_2}{Ve} + \frac{1}{2!}\sigma\frac{s_2}{1 + s_2} + \frac{1}{3!}\sigma^2\frac{1}{(1 + s_2)^2}\frac{J_2}{Ve} + \frac{1}{4!}\sigma^2\frac{s_2}{(1 + s_2)^3} - {} - \frac{2}{4!}\sigma^3\frac{1}{(1 + s_2)^3}\left(\frac{J_2}{Ve}\right)^2 + \frac{6}{5!}\sigma^4\frac{1}{(1 + s_2)^4}\left(\frac{J_2}{Ve}\right)^3 - \frac{1}{5!}\sigma^3\frac{6(s_2)^2 - 1}{(1 + s_2)^4}\frac{J_2}{Ve} + \cdots$$

Table VI (Continued)

(6) *More limited expansions:*

and using dimensionless parameters of Table II

(a) Truncating after the first term in x^4 (terms linear in J_1) expressions (3) and (4) give:

$$\frac{J_1}{Ve} = \frac{s_1 - s_2}{\sigma}\varrho_1 + \frac{1}{2}\lambda_1\varrho_1(1+\beta_1)$$

$$\frac{J_2}{Ve} = \frac{s_1 - s_2}{\sigma}\varrho_2 - \frac{1}{2}\lambda_2\varrho_2(1+\beta_2)$$

(b) Truncating after both terms in x^4, from expressions (3) and (4) conserving terms in $(J_1)^2$

$$\frac{J_1}{Ve} = (1+s_1)\frac{6(1+s_1)^2(\sqrt{\Delta_1}-1)-\sigma}{\sigma^2}$$

$$\frac{J_2}{Ve} = (1+s_2)\frac{6(1+s_2)^2(1-\sqrt{\Delta_2})+\sigma}{\sigma^2}$$

$$\Delta_1 = (1+2\beta_1)^2 + 4\frac{\sigma\beta_1}{1+s_1}\left[\frac{s_1-s_2}{\sigma} + \frac{1}{2}\lambda_1(1+\beta_1)\right]$$

$$\Delta_2 = (1+2\beta_2)^2 + 4\frac{\sigma\beta_2}{1+s_2}\left[-\frac{s_1-s_2}{\sigma} + \frac{1}{2}\lambda_2(1+\beta_2)\right]$$

(7) *From expression (2) by derivation the local flux is:*

$$\frac{J(x)}{D} = -\frac{\mathrm{d}s(x)}{\mathrm{d}x} = \sigma A_1 - \sigma\lambda x + \sigma A_1(6\beta_1)x^2 + \left[\frac{\lambda^2 A_1^2(4\beta_1)}{1+s_1} - \lambda\sigma(2\beta_1)\right]x^3 \ldots$$

(b) Symmetrical boundaries ($s_1 = s_2$)
(in the mid-plane of membrane the substrate concentration is minimum ($S_{\min}$), it is taken as reference if it is not nil ($S_{\min} \neq 0$); there, all odd derivatives of $S(X)$ are nil: thus the odd terms of the series cancel; expressions (1) and (2) become:

$$(8)\quad S(X) = S_{\min} + \frac{X^2}{2!}\frac{V}{D}\frac{S_{\min}}{K_m + S_{\min}} + \frac{X^4}{4!}\frac{V}{2D}\frac{K_m}{(K_m+S_{\min})^2}\frac{S_{\min}}{K_m+S_{\min}} + \cdots$$

(9) Dividing by K_m (dimensionless form):

$$s(x) = s_{\min}\left(1 + \frac{x^2}{2!}\frac{\sigma}{1+s_{\min}} + \frac{\sigma^2}{4!(1+s_{\min})^3} + \cdots\right)$$

(10) Fluxes and reaction rate are related to $s_{\min}$:

$$\frac{2J_1}{Ve} = 2s_{\min}\sigma^2 D\left(\frac{1}{1+s_{\min}} + \frac{\sigma}{3!(1+s_{\min})} + \cdots\right)$$

In the basic conditions this intrinsic K_m was found to be the same as in solution and is defined in the same manner; it must be distinguished from the overall apparent K_m subject to diffusion limitations and environmental or micro-environmental effects as discussed later.

When the rates of reaction and of diffusion are comparable, no analytical solutions are known for Equation (4), but series can be used to approximate them with symmetrical or unsymmetrical boundary conditions: MacLaurin series for membranes ([36, 40, 41, 44, 64], Table VI) or Taylor series for spherical particles [44].

We have expressed some of the interesting forms of these solutions with dimensionless parameters in the cited tables.

4.1.1.2.2. *Applications and validity of the steady-state equation.* By a more detailed analysis of $S(X)$ and $J(X)$ steady-state equations with various boundary conditions one can draw a number of conclusions; but for now we will give examples concerning the determination of characteristic constants and of σ, the calculations of concentration profiles and the variation of the reaction rate with σ and the boundary conditions followed by a short discussion on the validity or applicability of each type of solution of Equation (4) or (4′).

4.1.1.2.2.1. *Experimental determination of each constant composing* $\sigma = e^2 V/DK_m$. Four determinations are necessary:

The thickness e of the membrane is determined by usual mechanical or optical methods. The effective diffusion coefficients are measured in absence of reaction in a diffusion cell: the enzyme activity can be annulled by eliminating from the system a necessary coenzyme, activator or cosubstrate or by adding an efficient non-competitive inhibitor.

V_M per unit volume of membrane is attained at high substrate concentration if no complications arise from inhibition by an excess of substrate; however this concentration becomes very high if σ has a high value. K_m can be obtained following Sélégny *et al.* [61] in a unique experiment with an accuracy equal to or better than in solution by measuring diffusion reaction fluxes of the substrate in a diffusion cell and by using Equation (7).

Thus σ values can be calculated writing $\sigma = (e^2/D)(V_M/K_m)$ from independent measurements (with no fudge factors), compared with diffusion-reaction experiments and the equations tested [40–42, 44].

Note also that some authors have made experiments in some inverse orders [62, 63] or used numerical interpolations of a series of experiments [58].

4.1.1.2.2.2. *Concentration profiles and pure substrate diffusion.* We have identified four different possible types of 'passive' concentration profiles of substrate or of product which are visualized in Figure 2. They are dependent on the parameter σ characterising the membrane, on $\Delta s = S_1 - S_2$ giving the boundary conditions and on $\Delta S/\sigma$ determining the order of the reaction or, in other words, the solution-membrane system. From the $S(X)$ and $J(X)$ equations or the corresponding dimensionless ones, we have explicated in Table VII the conditions and domains of existence of each profile *A*, *B*, *C*, *D*.

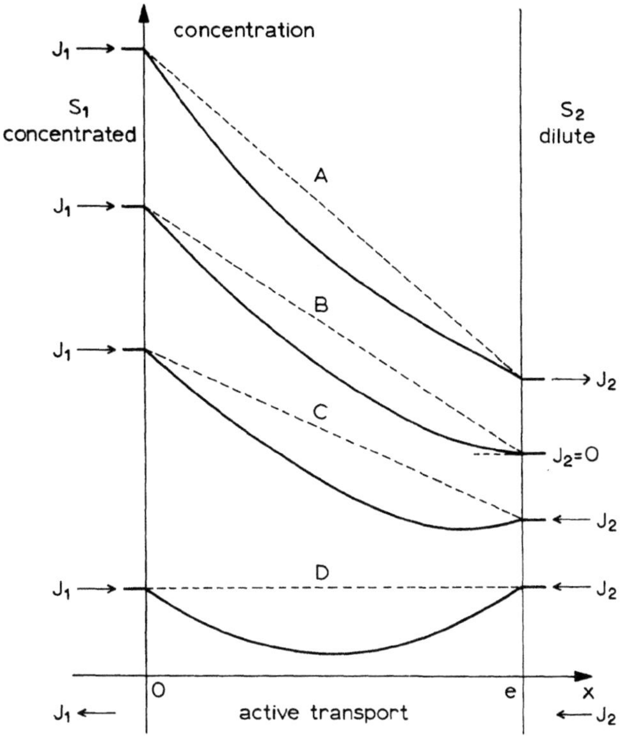

Fig. 2. Diagrammatic representation of steady-state concentration profiles in a membrane (from Sélégny *et al.* [40]). Full lines as well as fluxes J_1 and J_2 correspond to diffusion reactions; dashed lines illustrate the non-reactive Fickian diffusion profiles with the same boundary concentrations. Fluxes, conditions creating profiles of type A, B, C or D, position or *value* of the minimum can be calculated from Tables IV to VIII.

Characteristic profiles have also been calculated analytically [49] or numerically [54, 60] with various assumptions by several authors. Some illustrations are given in Figure 3.

It is good to remember that the slope of the tangent to these concentration profiles is proportional even in these diffusion reaction systems to the local pure diffusion through Fick's first law and that in the plane of a minimum or of a maximum in such a profile there is no net diffusion flux.

4.1.1.2.2.3. *The reaction rate.* The rate of the heterogeneous reaction is controlled again by the absolute or relative values of parameters σ (membranes), S_1, S_2 (boundaries) and $(S_1 - S_2)/\sigma$ (system) as in the previous case of concentration profiles. In a symmetrical system for example when pure diffusion fluxes are absent all fluxes are a consequence of the reaction and they give a direct measurement of this transformation. Two examples of the variation of v are given in Figure 4: one of them following Goldman *et al.* [53] corresponds to first order reactions in all the membrane; the other one has been calculated using the MacLaurin series by Sélégny and Le Guillon [64]

TABLE VII

Concentration profiles in the steady-state in the membrane

Characteristics and boundary conditions determining profiles *A*, *B*, *C*, or *D* of Figure 2, with zero, first or zero-to-first order reaction; calculations are made using the dimensionless $s(x)$ and $(J_S/Ve)\,(x)$ equations of Tables IV, V et VI.

($s_1 - s_2 = \Delta s > 0$; J_1 ingoing (face 1), J_2 outgoing (face 2) substrate fluxes; $j = J_2/J_1$)

Type of profile	Sign of		$j = J_2/J_1$	zero-order	First order	Zero-to-first order (limited to terms of McLaurin series first order in J_1)
A	J_1	J_2				
Both fluxes are positive	+	+	$0 < j < 1$	$\Delta s > \sigma/2$	$s_1 > s_2 \cosh\sigma'$	$\Delta s > \frac{1}{2}\sigma \frac{s_2}{1+s_2}\left(1 + \frac{1}{12}\sigma \frac{1}{(1+s_2)^2}\right)$ $(\Delta s < \frac{1}{2}\sigma\lambda_2(1+\beta_2)$
B Nil outgoing flux ($J_2 = 0$)	+	zero	$j = 0$	$\Delta s = \sigma/2$	$s_1 = s_2 \cosh\sigma'$	$\Delta s = \frac{1}{2}\sigma\lambda_2(1+\beta_2)$
C J_2 is negative	+	−	$-1 < j < 0$	$0 < \Delta s < \sigma/2$	$s_1 < s_2 \cosh\sigma'$	$0 < \Delta s < \frac{1}{2}\sigma\lambda_2(1+\beta_2)$
D Symmetrical profile	+	−	$j = -1$	$\Delta s = 0$	$\Delta s = 0$	$\Delta s = 0$

TABLE VIII

Minimum concentration and conditions of validity of analytical solutions of Tables IV and V

(a) *Position of minimum*

For profiles A and B (Figure 2) the minimum concentration is S_2 on face 2. For profile D (symmetrical) the minimum is situated in the mid-plane of the membrane ($x_{\min} = x/2$).

Profile C needs calculation; at this point J_s and $\mathrm{d}s/\mathrm{d}x$ equal zero:

Kinetic →	*Zero Order*	*First Order*	*Intermediate Order*
$0 < \varphi_{\min} < 1$:	$x_{\min} = \dfrac{s_1 - s_2}{\sigma} + \dfrac{1}{2}$	$\dfrac{\cosh \sigma' x_{\min}}{\cosh \sigma'(1 - x_{\min})} = \dfrac{s_1}{s_2}$	From Equation (7) in Table VI complete, numerically; or from its first three terms: $x_{\min} = \dfrac{\sigma\lambda_1 \pm \sqrt{\sigma^2\lambda_1^2 - 4\sigma A_1{}^2 12\beta_1)^2}}{2\sigma A_1(12\beta_1)^2}$

(b) *Value of* $s_{\min}$

From equation $s(x)$ using $x = x_{\min}$ in Tables IV, V and VI.

(c) *Conditions of validity:*

approximate error for 0 order equations at $x_{\min}$	
$s_{\min} \geqslant 50$	2%
$\geqslant 20$	5%
$\geqslant 10$	10%

approximate error for first order equation at $x_{\min}$	
$s_{\min} \leqslant 1/50$	2%
$\leqslant 1/20$	5%
$\leqslant 1/10$	10%

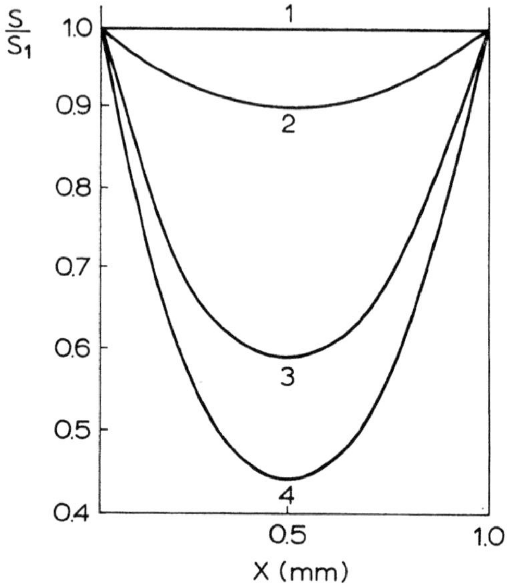

Figs. 3a–b. Concentration profiles calculated in a symmetrical membrane (a) of substrate with symmetrical boundary conditions and with $\sigma' = \sqrt{8.5}$; $S_1(\text{ml}^{-1}) = 10^{-1}(1)$, $10^{-2}(2)$, $10^{-3}(3)$, $10^{-4}(4)$ (from Sundaram *et al.* [60] and Equation (8)); (b) of substrate and of product with unsymmetrical boundary conditions ($S_1 \neq 0$, $S_2 = P_1 = P_2 = 0$ and $D_s = D_p$) (from Goldman *et al.* [53]). σ' values are indicated on the figure. Positions of $(P/S_1)_{\max}$:$X_{\max} = 0.406$ ($\sigma' = 1.5$); 0.375 ($\sigma = 3$); 0.334 ($\sigma' = 4.5$); 0.250 ($\sigma' = 10$).

correcting those reported before [36, 40, 41]. Previously we have reported a sigmoidal shape for the heterogeneous $v = f(s)$ calculated curves and this apparent regulatory effect has even been justified by experiments using a glucose-oxidase membrane. Moreover we have given a physical interpretation to explain this effect. Our more recent calculations have shown that the sigmoidal shape was due to a numerical error in the calculations and no sigmoid is predicted by using the first five terms of the series until quite high σ (≈ 800) values. In fact when the number of terms used increases the sigmoid shape appears for even higher values of σ. At these extreme limits the MacLaurin approximations must not be used directly, but only by sections of membrane which decreases σ. The experimentally observed sigmoid shape can be due to some secondary effects of the not so simply Michaelian glucose-oxidase catalysed reaction [65–67].

Note that the numerical solutions based on computerized linearisations have also been used recently by Horvath and Engasser [58], or previously to calculate such $v = f(s)$ graphs, (see Section 1.1.2.2.5.).

4.1.1.2.2.4. *Some remarks on the use of analytical solutions.* It must be understood that the equations of Table IV are only valid as long as the enzyme is saturated and thus the reaction is of zero order at all points of the membrane, even where the substrate concentration is the smallest. This $s_{\text{mini}} = S_{\text{mini}}/K_m$ can be located on face 2 with un-

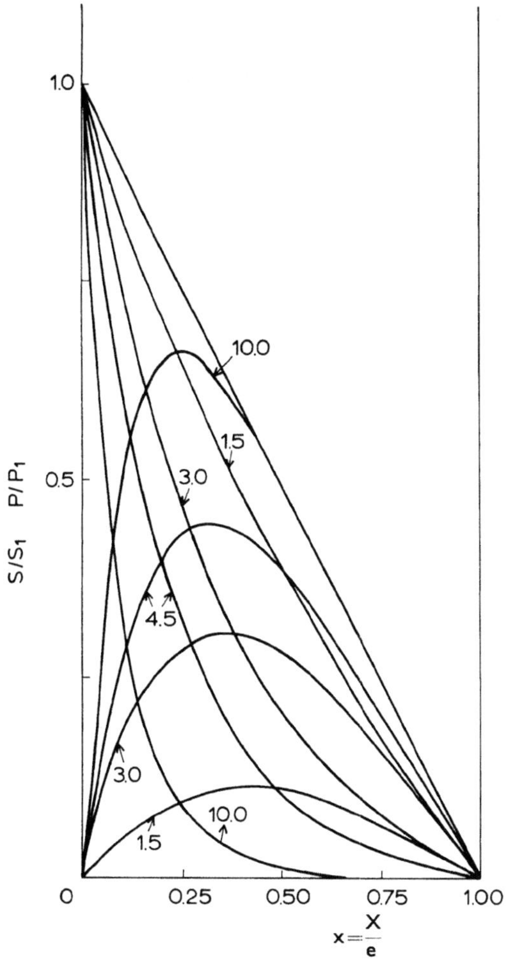

Fig. 3b.

symmetrical boundary conditions and no minimum in the concentration profile ($s_{\text{mini}} = s_2$). If there is a minimum, its location and value can be calculated (see Tables VII and VIII) from Tables IV, V, VI remembering that the substrate flux $J_S = 0$ at this point. Equations of Table V are valid as long as first-order kinetics can be admitted even at the point of highest substrate concentration in the membrane which will always be on face 1.

Following the tolerated error (2%, 5%, 10%), one can admit $s_{\text{mini}} \geqslant 50$, 20, 10 *for zero order* or, for the highest concentration: $s_1 \leqslant 1/50$, 1/20, 1/10 *for first-order* kinetics.

4.1.1.2.2.5. *Numerical solutions* of the steady-state equations (4) or (4′) can be obtained using the method proposed by Sundaram *et al.* [60] which reduces these equations to two first order ones. Writing $a = V_M/D$ and $b = K_m$ and also $z = \mathrm{d}S/\mathrm{d}X$ and integrating, one obtains:

$$z = \pm 2^{1/2}\left[aS - ab\ln\left(1 + \frac{S}{b}\right) + I\right]^{1/2}. \tag{8}$$

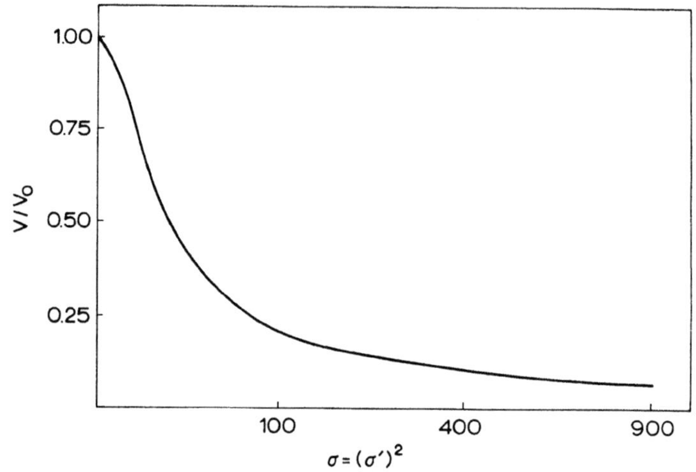

Fig. 4a.

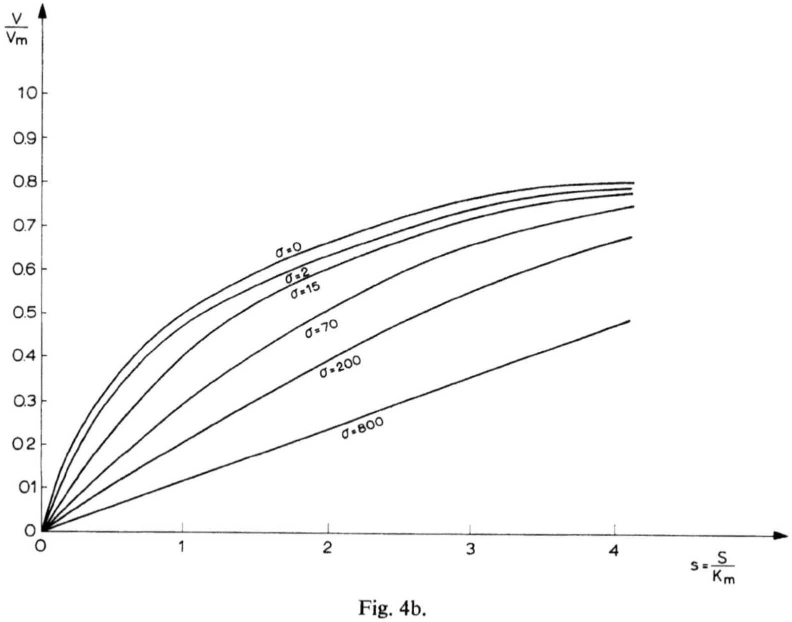

Fig. 4b.

Fig. 4a–b. Total relative activity of enzyme membranes calculated for different σ values in function of the normalized external substrate concentration (number of S_1/K_m) in symmetrical systems: (a) with first order reaction (from Goldman *et al.* [53]) (see Table IV), (b) with no assumption on the order of reaction (calculated by Sélégny and Leguillon using 5 term Mac Laurin series of Table VI).

Using symmetrical boundary conditions and successive approximations, by trial and error Equation (8) has been submitted to computer calculations with reasonable values for the constants in order to calculate steady-state substrate concentration profiles in the membrane. Consequently the variation of the rate of reaction of an enzyme-membrane in function of the symmetrical substrate boundary concentrations can also be obtained numerically because in the symmetrical case all diffusion fluxes are due to the reaction.

Similar calculations have been made by a few other authors [53, 58].

4.1.1.3. *Time-dependent Evolutions with Fundamental Assumptions*

Two types of evolutions can be considered: in the first case which is e.g. that of a membrane separating two reservoirs of limited volume the membrane boundary concentrations change with time. If this change is slow compared to the time needed for the establishment of the steady state in the membrane, we meet the case of successive quasi-stationary states [26, 40–41]. The succession of previously exposed steady-state solutions can be used to describe the progressive evolution of profiles from *A* to *D*.

A completely different problem is met when the evolution of concentration profiles and fluxes is considered in an enzyme membrane before the establishment of the steady state. In this case the non-linear partial differential Equation (3) must be solved numerically; this equation can be rewritten in dimensionless form:

$$\frac{\partial s}{\partial t'} = \frac{\partial^2 s}{\partial x^2} + \sigma \frac{s}{1+s} = 0. \tag{9}$$

Using the explicit method Sélégny *et al.* have reported [68] the calculated evolution of the substrate concentration profile in an originally void membrane and symmetrical boundary conditions. An illustration of the results is reproduced in Figure 5.

The evolution of the substrate flux entering the membrane is then easily obtainable from the slopes of the profiles at the internal membrane-solution interface, (Figure 6). The time necessary to attain the steady state of a diffusion-reaction has been calculated numerically both in the previous case and the one known as a 'reflection on the wall' in diffusion [14]. In this last case, for example an enzyme membrane covering a glass electrode impermeable to substrate as well as product, equations and solutions are simply those of a symmetrical half-membrane [45, 48]; effectively the impermeable boundary imposes $ds/dx=0$ and $dp/dx=0$, just as in the mid-point of a membrane with symmetrical boundary conditions (profile *D*).

4.1.2. *Extensions of the Basic Mono-Enzymatic Case*

Some changes in the fundamental assumptions given above will not modify drastically the mathematical treatment, but only the quantitative interrelations between the variables.

More explicitly, such changes can be due to modifications of the kinetic law or the distribution or diffusion of species. (see Table IX).

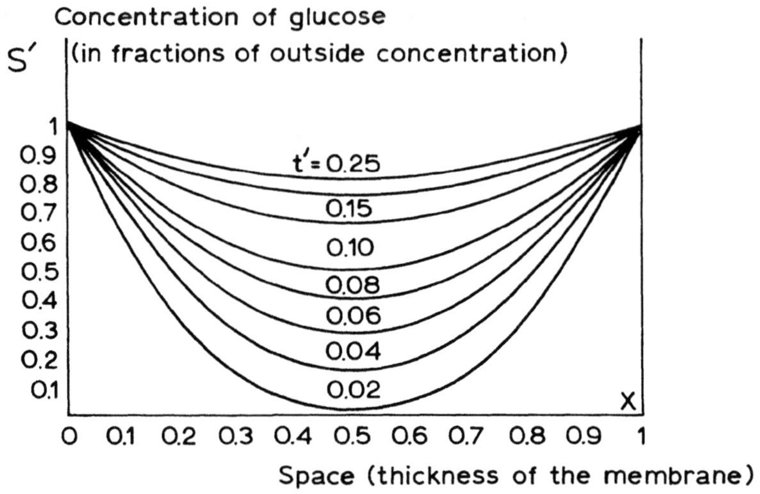

Fig. 5. Illustration of the calculated evolution of concentration profiles in an enzymatical membrane, with symmetrical and constant boundary substrate concentrations (S_1). The concentrations are expressed as fractions of external concentration ($s' = S/S_1$) and the distance as fractions of membrane thickness ($x = X/e$); the dimensionless time scale is $t' = t/(e^2/D)$ (t real time). The membrane characteristics correspond to a glucose-oxidase membrane ($\sigma = 1.4$; $V = 3.66 \times 10^{-3}$ mole cm^{-3} h^{-1}; $K_m =$ $= 1.3 \times 10^{-5}$ mole cm^{-3}; $e = 0.5 \times 10^{-2}$ cm; $D = 0.5 \times 10^{-2}$ cm^2 h^{-1}) (For computation 0X was divided in 20 intervals; 700 iterations were necessary to approach the stationary state, which needs 12.6 s in real time) (From Sélégny *et al.* [68]).

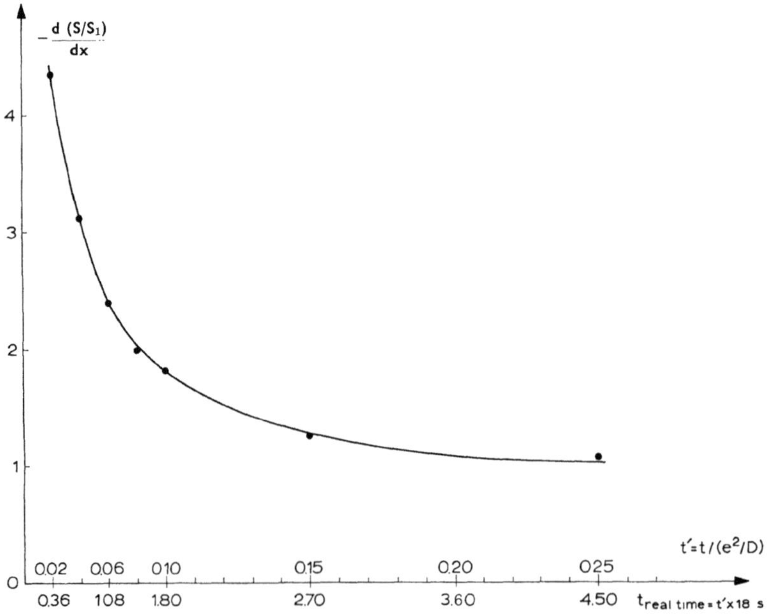

Fig. 6. Evolution of relative substrate diffusion-reaction fluxes in function of relative diffusion time; $[-D(\mathrm{d}s'/\mathrm{d}x)(t')$ the number of S/S_1 per cm^2 per hour in function of $t' = t/(e^2/D)]$. Curves and data of Figure 5 have been used for calculation.

TABLE IX

Some typical $v = f(S)$ kinetic expressions in enzyme solutions

[$V = V_{\max}$; v = actual rate; $s = S/K_m$; $i = I/K_i$; K_S = enzyme-substrate affinity constant; K_i = inhibitor-enzyme affinity constant]

	Absolute values	Relative dimensionless forms
(1) *Michaelis-Menten type*	$v = V \dfrac{S}{K_m + S}$	$\phi = \dfrac{v}{V} = \dfrac{s}{1+s}$
Integrated forms	$Vt = y + K_m \ln \dfrac{S_0}{S_0 - y}$	
	$pS = pK_m + \log \dfrac{V - v}{v}$	
(2) *Competitive inhibition*	$v_i = \dfrac{VS}{S + [K_m(1 + I/K_i)]}$	$\phi_i = \dfrac{s}{1+s+i}$; $\dfrac{1}{\phi} = \dfrac{1+s}{s} + \dfrac{i}{s}$
Effect on K_m		$(K_m)_{\text{inhibited}}/(K_m)_0 = 1 + i$
(3) *Non competitive Inhibition*	$v_i = \dfrac{VS}{(S + K_S)(1 + I/K_i)}$	$\phi_i = \dfrac{s}{(1+s)(1+i)}$
Effect on V		$\phi/\phi_{i(\text{inhibited})} = 1 + i$
(4) *Two substrates* (independent S_a and S_b)	$v = \dfrac{V(S_a S_b/K_a K_b)}{(1 + S_a/K_a)(1 + S_b/K_s)}$	$\phi = \dfrac{s_a}{1+s_a} \dfrac{s_b}{1+s_b}$
(5) Inhibition by (ES_2) excess of substrate	$v_{iS} = \dfrac{VS}{S + K_S + S/K'_S}$	$\phi_{iS} = \dfrac{s}{1 + s + s^2/K''_S}$; $\left(K''_S = \dfrac{K'_S}{K_S}\right)$
(6) Activation by excess substrate $(ES_{\text{inactive}} + ES_{a\ \text{active}})$	$v_{SS} = \dfrac{VS}{1 + K'_S(1 + K_{Sa}/K_S + K_{Sa}/S)}$	$\phi_{SS} = \dfrac{s'}{K_S^{Sa} + 1 + s' + \dfrac{1}{s_a}}$

In other terms one can change one of the constants D, e, V_M or K_m or their apparent values entering in σ or the geometry or the homogeneity of the enzyme layer or of solution.

Practically, we can divide the modifications into several classes:

(a) those due to regulating diffusible compounds different from the considered substrate (e.g. inhibitors, activators, cosubstrates, pH-active substances);

(b) those due to components of the membrane material itself (e.g. polyelectrolytes);

(c) those due to properties of the product or of the substrate giving rise to feedback effects and self-regulation (e.g. charged product, inhibition by product or substrate).

4.1.2.1. *Additional Diffusive Species*

4.1.2.1.1. *Inhibitors.* If the concentration of the inhibitor can be considered identical in the solution and in the membrane (just as in homogeneous enzyme kinetics) a competitive inhibitor will simply modify the K_m to introduce in all the previous equations, thus $K_{mi}=K_m(1+i)$; a non-competitive inhibitor will have an analogous effect on V_M and $V_{Mi}=V_M(1+i)$ will replace V_M in the calculations. $i=I/K_I$ is the normalised concentration of inhibitor compared to the corresponding affinity constant. (See Table IX, Equations (2) and (3).

4.1.2.1.2. *Effect of pH.* V_M varies also with the pH increasing first and decreasing afterwards; thus the $V_M=f(\text{pH})$ curve presents a maximum at the optimal pH. In a conveniently buffered system with uniform pH, any deviation from the optimal pH follows a law similar to the non-competitive inhibition and decreases σ. In many cases *the K_m also varies with the pH, which in turn can increase σ.*

As the interrelations between the enzyme activity and σ are not linear in all cases, deviations from the activity-pH curves can be expected in the heterogeneous system as compared to homogeneous enzymology. Especially the proportion of sites working at the same external substrate concentration depends on σ. Several observations and comments on these problems are cited in references [24, 40, 41, 44, 45, 52] etc.... and an illustration is given in Figure 7.

4.1.2.1.3. *Activators.* Many enzymes need the presence of small amounts of activators, for example of some metallic ions (Na^+, Mg^{++}, Ca^{++}, etc.) [14]. These activators usually associate with the enzyme following a Langmuir-type saturation curve and the effect on V_M is inverse to a non-competitive inhibition; one can use: $V_{Ma}=V_M(a/|1+a|)$ with $a=A/K_A$ being the normalised concentration of the activator.

4.1.2.1.4. *Co-substrate.* Two limiting cases are usually considered according to whether the two substrates are consecutive or in other words independent of each other or not. In the first of these cases, when the second substrate is in great excess (e.g. oxidation of glucose catalyzed by glucose-oxidase in excess oxygen) phenomenologically Michaelian-type kinetics are observed for the minority substrate. The inverse situa-

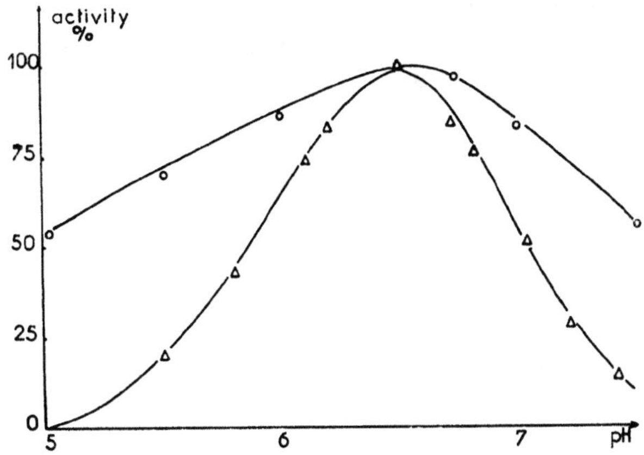

Fig. 7. Example of the variation of the activity in function of the pH. Enzyme: glucose oxidase in solution (0–0), and crosslinked on a cellophane matrix ($\Delta - \Delta$). From Sélégny *et al.* [40] and Broun *et al.* [37].

tion follows from the inversion of the substrate ratio. In the intermediate domain the V_M or the K_m of the reaction will be affected according to the independence or the interdependence of the substrates. (See Table X, Equation (4) and references [14, 40, 59] for example). (Note that due to these considerations a glucose-oxidase electrode can be used for detection of glucose or of oxygen depending on which of the two substrates is not in excess. The same comment is also valid for some other electrodes [69–71]).

4.1.2.1.5. *Other regulations by diffusible species.* Diffusible species can also affect the mass-transfer properties of the membrane by inducing charges or shrinkages (volume changes). The first of these effects will be better understood after the analysis of charge effects in a later section.

The shrinking affects both the accessibility or the diffusion properties of substrates or inhibitors or the thickness of the membrane and the enzyme concentration, and consequently V_M. σ will vary in the proportion of $e^2 V_M/D$. Such an interesting situation has been investigated experimentally by Horvath *et al.* [72], the shrinking being due to the effect of Ca^{++} ions on charged sites. This experiment is of great interest in the modelling exploration of *membrane allosteria* [73] and can be confronted with enzymatically inactive membranes such as ion-exchange membranes or phase-transitions of charged siccative oil membranes of Monnier recently studied by Goudeau [74] or of porous elastic membranes [75].

4.1.2.2. *Effects due to the Constituants of the Membrane*

We will consider successively diffusion-layers, partition coefficients, changes in enzyme activity and finally the combined effects.

4.1.2.2.1. *Diffusion-layers.* Such layers affect the boundary concentrations.

The enzyme membrane can be separated from the bulk solution by *a layer devoid of enzyme.* Such a layer of constant thickness can be added as a film on top of the enzyme-membrane or be due to unstirred layers in the solution. The substrate concentration on the boundary of the enzyme membrane to be introduced in the previous equations is related then to the bulk solution concentration by the purely Fickian (or passively permeating) substrate fluxes through the film. These fluxes must equal the diffusion-reaction flux entering the enzyme membrane in the stationary state. This problem has been considered for example by Goldman *et al.* [53] and Sundaram *et al.* [60] in the case of first order reactions and in our group [40, 42, 44, 76] with more general kinetics. An interesting property of such layers considered in the following is that they can introduce *passive selectivities* (based on solubility or charge differences or carriers) between different diffusing species such as substrates, inhibitors, activators, etc.

When the solution is not stirred or when the convection is insufficient, the *concentration-polarisation-layer* can be of variable thickness. Such problems are similar to those met with electrodes or ion-exchange membranes [46, 57, 60, 77]; I shall come back to this question with more detail in the section on 'facilitated transport'.

4.1.2.2.2. *Unequal distribution between solutions and membrane.* Due again to their solubility or sign of charge, diffusible species: substrate, product, inhibitors or activators can accumulate in the membrane or exclude themselves from it.

By generalizing the reasoning of Sundaram *et al.* [60], one can relate the internal concentration of species Q_{int} to the external one Q_{ext} by the partition coefficient K_D; thus $Q_{\text{int}} = K_D Q_{\text{ext}}$.

If Q is not reactive (inhibitor, activator or specifically H^+ ions) there is no need for other assumptions and increased or decreased enzyme activities are easy to predict quantitatively from K_D determinations and laws.

However, for the substrate itself the above relation can only be considered as a good approximation as long as *the reaction rate is slow compared to the diffusion.*

Until recently charge interactions have been considered with enzymes embedded in insoluble polyelectrolyte supports having ion-exchange properties. First the displacement of optimal pH of chymotrypsin adsorbed on kaolinite [78] was attributed to the effect of the charged support. Then polyanions such as poly(ethylen-maleic acid) copolymers [16c, 28, 79, 80] or carboxymethyl cellulose [29, 54, 81, 82, 83] or other derivatives or polyacid anhydrides [84], polysaccharides [85] some polybases (polyamines) [31, 79b] or amphoteric macromolecules (albumin [37, 76, 86], gelatin [43], collagen [49] or similar) have been used. Ions of opposite charge to the matrix will accumulate in the membrane; those of the same charge are excluded.

The K_D values can be predicted or obtained by the usual theoretical and experimental methods of polyelectrolyte and ion-exchange science. K_D is dependent on fixed charge density and thus also on degree of ionisation of the polyelectrolyte, the ionic strength of solution, the mole fraction of Q and its charge or eventual specific 'Q-fixed charge' interactions.

The school of Goldstein and of Professor Ephraim Katchalski *et al.* [79–80] used the expression of the equality of electrochemical potentials evaluated through the Maxwell-Boltzmann distribution function from which

$$K_D = \exp(Z\varepsilon\psi/kT), \tag{10}$$

$Z\varepsilon$ is the charge of the fixed groups, ψ the electrical potential and k Boltzmann's constant.

The ratio between the internal and external hydrogen-ion or substrate activity (concentration) will be modified in this ratio; consequently the optimal H^+ concentration will be modified in this same ratio and the apparent $(K_m)_{app}$, that is to say the external concentration for which the reaction rate is half of its maximum, in the reciprocal of K_D.

This means also that K_D and $(K_m)_{app}$ vary with the external salt concentration because of 'charge screening' with excess salt, $K_D \to 1$ at high ionic strength [29, 79–80, 81–82] and the effect of fixed charges on $(K_m)_{app}$ or on optimal pH diminishes or vanishes. The direct determination of the electrostatic term $\varepsilon\psi$ encounters the usual difficulties in gels; this term can be better *evaluated from enzyme activities* [79a] through variation of optimal pH and of $(K_m)_{app}$ than used to predict the reaction rate*.

Practically, just as for ion-exchangers the Donnan concept and distribution law give more immediate access to confrontation of behaviors; they were used by the English School. Expressing again that $(K_m)_{app}$ varies inversely with K_D and relating K_D to the effective volumic fixed charge density $R^- = Zm_c$ through the Donnan relationship, Wharton *et al.* [80] obtained:

$$(K_m)^2_{app} = \gamma^2 K_m [K_m - (K_m)_{app} (R^-)/I], \tag{11}$$

where I denotes the ionic strength of solution and γ the ratio of internal and external mean activity coefficients, selectivities between ions of the same charge (e.g. Na^+ and positively charged substrate) being neglected.

Binomial expansion, truncation after the first term and rearrangements give Equations (12) and (13):

$$\mathrm{I}/(K_m)_{app} = I/\gamma K_m + R^-/2K_m \tag{12}$$

or

$$(K_m)_{app} = \gamma K_m I/(\gamma R^-/2 + I).$$

It is seen that (if $O < [(K_m)_{app} R^-/2K_m I] < 1$) from Equation (12) the intercept of $I/(K_m)_{app} = f(I)$ is $R^-/2\, K_m$ and the slope equals $1/\gamma K_m$, assuming $\gamma = \text{Const.}$

Equation (13) predicts $(K'_m)_{app} = \gamma K_m/2$ when $I = \gamma R^-/2$; thus assuming $\gamma = 1$, $(K'_m)_{app}$ equals half of K_m when the external concentration equals half of the internal fixed charge concentration.

* Already Tsuruta and Inoue have estimated the local H^+ concentrations near a polyacrylic acid chain from sugar inversion.

For experimental verification the authors [80] used 0-(carboxymethyl)-cellulose onto which the enzyme bromelain had been grafted and hydrolysed α-N-benzoyl-L-arginine ethyl ester substrate. The dry-weight exchange capacity reported to the total volume occupied by the settled fibers in a measuring cylinder gave an evaluation of the charge density R^-; this approximation does not take into account the free-volume. Nevertheless good qualitative agreement was obtained in testing Equations (11) to (13).

$(K_m)_{app}$, however, did not regress to K_m even at very high ionic strength I. Not electrostatic interactions have been considered by the authors [80]; but it appears from their data that the variations in swelling with increased salt concentrations could also have been contributing.

At this stage, it was generally demonstrated that charged carriers do not affect the enzymic reactions of neutral substrates, enhance the overall affinity for substrates of opposite sign to the carriers and decrease the affinity when the substrate is of the same sign as the fixed charges.

Other relations characterising the distribution of the charged species such as the Sélégny, Korngold, Merle equation [88] or the Merle normalised equations [89] are also convenient, most especially if the external solution contains a polyelectrolytic co-ion. These are now being used by Vincent [85]. The increase of the enzyme activity due to the activating Mg^{++} counterions accumulated inside by negative charges in a membrane containing hexokinase has been shown by David [86].

The explicit interrelation between electrochemical activity, concentration and inhibiting or activating effects or reactivity of charged species still needs some complementary consideration and experimentation.

Optimal pH shifts due to charged matrixes taken from the works of Goldstein *et al.* [16c] are illustrated in Figure 8.

4.1.2.2.3. *Diffusion-reactions with charged substrate.* When the rate of the enzyme reaction is no longer negligible compared to that of the diffusion, the diffusion or mass-transfer equation must take account of the charge effects produced by a polyelectrolytic carrier.

Helfferich [90] had given a mathematical description of the diffusion of ions in ion-exchange beads based on the Nernst-Planck and the Nernst-Einstein equations, Hornby *et al.* [82] combined a case of this treatment with the Michaelis-Menten kinetic equation (Table IX) in order to calculate diffusion reactions with linear profiles in an idealized 'quasi-Nernst layer' of thickness x limited on one side by a bulk solution of concentration S_0 and on the other one by a charged enzyme-layer; the rate of reaction in the steady state equals the sum of J_e and J_D the fluxes of the charged substrate due respectively to the electrical and concentration gradients;

$$J_e = -U_s(zS_0)\,\mathrm{grad}\,\psi \tag{14}$$

$$J_D = -D\,\mathrm{grad}(S) \tag{15}$$

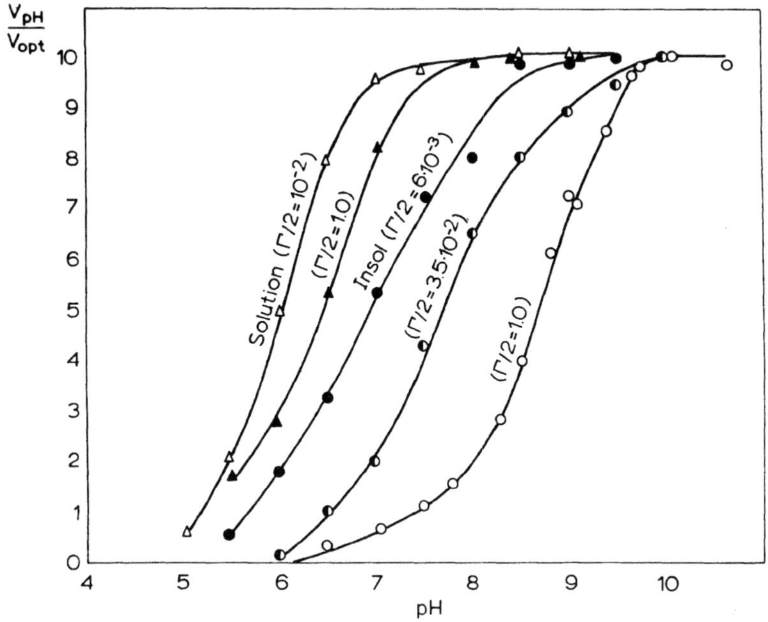

Fig. 8. Experimental shift of the 'activity-external pH' curve of an enzyme due to 'internal-pH-control' by a polyacid matrix, at various ionic strengths. ($\Gamma/2$) (From Goldstein *et al.* [16e]) Enzyme: trypsin, matrix: poly-(ethylene-maleic acid) copolymer, substrate: benzoyl-L-arginine ethylester.

the rate of the reaction is given by an approximate solution of the quadratic equation:

$$v = \frac{VS_0}{S_0 + \left(K_m + \dfrac{xV}{D}\right)\left(\dfrac{RT}{RT - zxF \operatorname{grad} \psi}\right)}. \tag{16}$$

From Equation (16) and the Michaelis equation the modification of the apparent $(K_m)_{\text{app}}$ due to the potential gradient is:

$$(K'_m)_{\text{app}} = \left(K_m + \frac{xV}{D}\right)\left(\frac{RT}{RT - zxF \operatorname{grad} \psi}\right). \tag{17}$$

This calculation, valid both for substrates of the same or opposite sign of charge as the carrier, neglects the concentration gradients inside the enzyme layer and is equivalent to assimilating this layer to an enzyme-surface. This last problem was approached in more detail by Shuler *et al.* [56] in a mathematical language of chemical engineering. They have expressed the 'effectiveness factor' giving the modification of the enzyme reaction-rate by diffusive and electrostatic effects as a function of K_m, an electrostatic parameter and the ratio of maximum reaction and transfer rates. Pointing out the difference of several orders of magnitude between the thicknesses of the electrostate and our group [44] [92] expressed the *general case* and approximated again the solutions through MacLaurin series. They pointed out especially that the effect of the product on pH on the one hand and the effect of this on the enzyme activity on the other followed individual laws; if only the variation of V_M resulted, one can write $V_{pH} = V_M(1 - f(p))$ when $f(p)$ gives the interrelation between product concentration and its non-competitive inhibiting effect on the enzyme (Table X). With these assumptions the calculations are straightforward. The *evolutions* in an originally void membrane have been calculated numerically by Kernevez [45] assuming in a first approximation that the inhibition followed a linear law from the optimal pH to the two limiting minimum and maximum pH values.

TABLE X

Feed-back regulations by the product in symmetrical membrane and systems. Approximations of the steady state solutions using MacLaurin series, limited to terms linear in J_1

(a) *Reversible reaction*

$$E + S \underset{k_{-1}}{\overset{k_{+1}}{\rightleftarrows}} ES \underset{k_{-2}}{\overset{k_{+2}}{\rightleftarrows}} E + P \qquad (\overset{f}{\longrightarrow},\ \underset{b}{\longleftarrow})$$

$$\frac{2J_1}{Ve} = -\frac{2J_2}{Ve} = \frac{s_{1f}K_{fb} - p_{1b}K_{eq}}{K_{fb}(s_{1f} + p_{1b} + 1)}\varrho_r(1 + \beta_r)$$

Notations: f refers to forward reaction of s witk K_{mf} Michaelis constant

b refers to backward reaction of P with K_{mb} Michaelis constant

$$s_{1f} = \frac{S_1}{K_{mf}};\ p_{1b} = \frac{P_1}{K_{mb}};\ K_r = \frac{K_{mf}}{K_{mb}};\ K_{eq} = \frac{S_{\text{equil}}}{P_{\text{equil}}};\ \varrho = \frac{1}{1 + 2\beta_r};\ \delta = \frac{D_P}{D_S};\ \sigma_p = \frac{k_2E_0e^2}{D_PK_{mb}}$$

$$\beta_r = -\frac{1}{12}\sigma_p\frac{(\delta + K_{eq})[K_r(1 + s_{1f}) + \delta p_{1b}] + (K_rs_{1f} + p_{1b})(K_r - \delta)}{K_r^2(s_{1f} + p_{1b} + 1)^2}$$

(b) *pH – active Product*

$$\frac{J_1}{Ve} = \frac{1}{2}(1 - f(p_1))\frac{1 + \beta_1}{1 + 2\beta_1} \quad \text{where: } s_1 = \frac{S_1}{K_m};\ p_1 = \frac{P_1}{K_m};\ \delta = \frac{D_S}{D_P};$$

$$\beta = \frac{1}{12}\left[\frac{\sigma_S(1 - f(p_1))}{(1 + s_1)^2} + \sigma_S\delta\frac{s_1}{1 + s_1}\right]$$

$f(p_1)$ is a global function of functions:

$\text{pH} = u(p)$; $V_{\text{pH}} = V_{\text{optim}}(1 - g(\text{pH})) = V(1 - g[u(p)])$

p is related to S_1, S and P by Equation (6) (page 437) combining: $V_{\text{pH}} = V(1 - f(p_1))$.

4.1.2.3.3. *Inhibition or activation by the substrate.* Detailed analysis of this case has received little attention until now according to my knowledge. The corresponding kinetic laws in a homogeneous solution are given in Table X. Interesting combinations include the increasing of the inhibiting or destructive substrate concentration by an increase in σ or alternatively under the protection of a diffusion layer; experimental observations have been made with urea – urease [48] and immobilized catalase – excess hydrogen peroxide [46, 47] respectively.

Protections by charge effects are discussed in other sections of this paper.

4.2. Reaction Control of Reactions

4.2.1. *Multi-enzymic Single-Layer Systems*

When two or more sequential enzymes are immobilized together in the same layer or membrane catalysing successive reactions:

$$S \xrightarrow{\text{enzyme } A} P \xrightarrow{\text{enzyme } B} Q \tag{19}$$

the activity of the second enzyme will depend on that of the first one and will deviate from that of homogeneous solutions containing the same amount of catalytic proteins.

Experimentally Mosbach *et al.* [93, 94] have linked sequential enzymes on insoluble particles; they have observed an increase in the activity of the second or third enzyme as compared to homogeneous solutions and this effect has been attributed to the shorter diffusion distance of the intermediate product from the first enzyme site to the second one; the external diffusion layer should and could also play a role in the observed phenomena by delaying the diffusion of the intermediate product into the solution.

We have also prepared, experimented and described mathematically some isotropic sequential bienzymatic irreversible gel membranes (β-galactosidase-glucose oxidase) in the stationary state [40, 41].

Equation (1) rewritten for the intermediate substrate-product P describes the coupling of the two enzyme reactions. If the first enzyme is assumed to be saturated and the second reaction is Michaelian, the dimensionless differential equation is as follows *in the stationary state:*

$$\frac{d^2 p}{dx^2} = \sigma_B \left[V_B^A - \frac{p}{1+p} \right], \tag{20}$$

where $p = P/K_{mB}$; $V_B^A = V_A/V_B$ (ratio of V maxes) and $\sigma_B = e^2 V_B / D\, K_{mB}$.

Concentration profiles calculated in first approximation from truncated Mac Laurin series, approximate solutions of Equation (4′), are illustrated in Figure 10. It is interesting to note now that the first of the coupled reactions regulates the second one (see caption of Figure 10) and also that the intermediate substrate-product has its concentration quite effectively stabilised within the membrane with regard to concentration variations of the first substrate outside. Densities of enzyme-sites and diffusion properties of the membrane control the stabilised level.

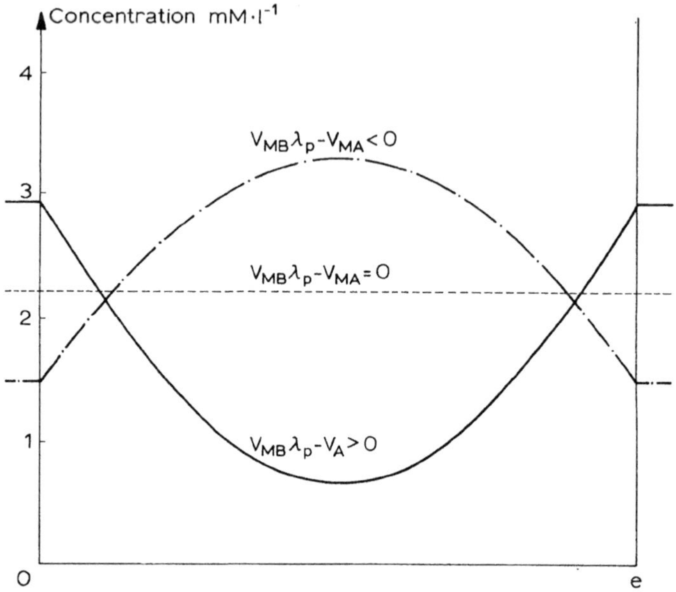

Fig. 10. Concentration profiles of intermediate product in a membrane containing two successive enzymes calculated using the simplified Equation (20). First enzyme at maximum rate; profiles and fluxes are dependent on the relative activity of second enzyme compared to first enzyme and diffusion fluxes; it is seen that if the value $V_{MB}\lambda - V_{MA}$ is negative, the system produces p; if this quantity is positive p is degraded by the system; if it is zero the concentration profile in the membrane is horizontal and the concentration of p remains unmodified. (From Broun *et al.* [42] and Thomas [44]).

There is no difficulty either in adapting the other steady-state solutions to such problems.

The *non-steady-state evolution* of reservoirs in contact with a more complicated hypothetical multienzymic membrane including successive, parallel and competitively inhibiting feed-back reactions corresponding to

$$S \to P \rightleftarrows Q \to R \quad \text{(inhibition of } S \to P \text{ by } R\text{)} \tag{21}$$

has been described mathematically and evaluated numerically by Sélégny *et al.* [68]. The possible solution of such a complicated system which contains many of the equations of basic situations in enzyme kinetics has shown the usefulness of this powerful tool. It has also demonstrated that systems of dimensionless non-linear partial differential equations constructed by the use of dimensionless parameters, dimensionless relative forms of Fick's law and of enzyme kinetics constitute an interesting field of the mathematics of numerical calculations. They give a great variety of problems of evolutions, optimalisations and regulations. Such solutions can be explicited and illustrated when numerical values of parameters and initial and boundary conditions are taken from experiment or reasonably stated and when the geometrical forms and dimensions are known. Logic, algorithms and validity of different methods of solutions of such problems and illustrative calculations have been reported by Kernevez in his thesis [45] and later.

4.2.2. *Time-Dependent Oscillations*

Another type of reaction-reaction interaction would be the possibility of attaining time-dependent oscillations of an intermediate product in a homogeneous enzyme membrane or solution such as are known for the Zhabotinsky oscillations [95, 96] in organic chemistry. Calculations have been done by Higgins [97] and Walter [98] and observed with natural extracts by Chance *et al.* [99]. However the necessary conditions seem to be quite strictly defined and until now bistabilities have been reported [100] but no completely synthetic reproduction of this phenomenon has been carried out with enzymatical systems; one can predict however that in the near future these difficulties will be overcome. Such oscillations should satisfy the ballistic equation which has been used with a changing of variables for exploring space oscillations by Sélégny *et al.* [76]. The high instability of the stationary state has shown us that some regulation is necessary to maintain the phenomenon in agreement with Walter [98] This should need for example coupled non linear kinetics in homogeneous systems and coupling between diffusion layers in heterogeneous ones [96]. An entertaining of damped oscillations observed in gel diffusion [101] is one of the possible solutions of this problem.

4.3. Facilitated transport (by carrier)

4.3.1. *Theory and Classifications*

Facilitated transport is a class of diffusion reactions in which the diffusing substrate reacts with the solvent or another *carrier* and consequently the diffusion flux of substrate following its chemical potential gradient is increased compared to the purely Fickian diffusion. In 1960 Scholander [102] found an increase of the oxygen transport through a water film when hemoglobin was added. Because of the importance of such phenomena in biology and also in the potential uses of facilitated transport in engineering, an increasing number of experimental and theoretical contributions has been devoted to the understanding of this phenomenon (see e.g. references [102–113]). In most cases only gas-liquid or liquid-liquid transport has been considered and the general equations of evolution had not yet been solved when our interest was attracted by the needs of a programme on artificial oxygenators. These studies and results relating enzyme-containing membranes using the concepts of intra- and extra-membrane transporters and of enzyme-catalysed facilitated transport have been given or commented in several papers [40, 46, 113] and in more detail in the thesis of Tran Minh Canh [47] and for the numerical calculations in that of Kernevez [45].

The interest of the facilitated transport is to decrease the resistances due to rate-limiting diffusion layers (or membrane) where forced convection is impossible. In other terms this means the diminishing of the concentration-polarization at the interface of two layers of different permeabilities. So facilitated transport is *the chemical way of fighting against such polarizations*. In some respects we meet a situation inverse to the one envisaged in the section on the limitations of diffusion-reaction fluxes by unstirred layers.

In practice a few supplementary limiting conditions are useful:
The reaction between the transported species S and the transporter T must be *reversible* if S must *enter and leave* a layer following:

$$S + T \underset{k_{-1}}{\overset{k_{+1}}{\rightleftharpoons}} ST\,. \tag{22}$$

One of these reactions could happen irreversibly if only the extraction (forward reaction) or the generation of S (decomposition of ST) is of importance. The evident role of the enzyme is to accelerate the reaction on which the facilitation depends; in the case of a reversible reaction this means the simultaneous increase of both constants k_{+1} and k_{-1} without changing their ratio which remains equal to the equilibrium constant.

The presence of the transporter T is limited to one of the layers; its flux as well as that of ST is nil at the interface(s); so are their concentration gradients. This gives the boundary conditions $dT/dx=0$ and $dST/dx=0$. T in excess gives: $k'_{+1}=Tk_{+1}$.

From the diagrams of Figure 11 and the equations of Table XI the understanding of the interrelations of reaction and transport will be straightforward even for those not familiar with this problem.

Four classes of problems are considered.

(1) An interface separates a medium of high permeability from an enzyme-layer of lower permeability to pure diffusion. For example, a gas-phase limiting an enzyme solution or membrane. The boundary-concentration on the gas-phase side will be constant; for both pure and facilitated diffusion the concentration gradient at the liquid side of the interface indicates the substrate-diffusion flux which is easy to calculate with the help of the already exposed equations.

(2) Insert between the very permeable phase (gas) and the less permeable enzyme solution or membrane, a membrane or unstirred layer devoid of enzyme of intermediate permeability with regard to the substrate (e.g. a passive silicone membrane). On the one side (the gaseous one) of the passive membrane the boundary concentration is imposed by the solubility of the permeant and is independent of the flux; on the other boundary of the same membrane the substrate concentration is controled by the diffusion-reaction. The Fickian diffusion flux through the passive membrane in the steady state equals the diffusion-reaction flux in the enzyme membrane. (This case has already been discussed, though in a different way, in the chapter entitled 'Diffusion-layers').

When the release of the substrate is considered instead of its entering, then situations, reactions and calculations are simply inversed.

(3) If a very permeable passive membrane is limited on both sides by passively less permeable enzyme layers, the steady state flux will be controled by the concentrations

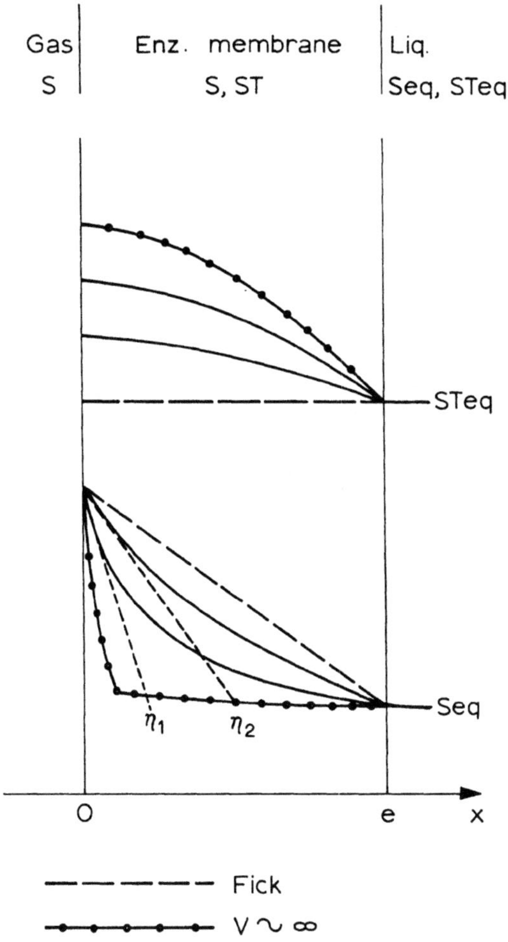

Fig. 11.I

Fig. 11.I–IV. Facilitated transports (concentration depolarisations by reactions): diagrams of the four types of systems analysed in text and Table XI. Figures correspond to assumption of perfect forced convection at extreme external boundaries and of reversible reactions.

I Gas (G) / enzyme membrane (EM) / Liquid (L)
(S) (S, St, T) (Seq., STeq., T)

II G / passive membrane (PM) / EM / L
(S) (S) (S, ST, T)(Seq. STeq. T)$_2$

III L / E M / PM / EM / L
(Seq. STeq. T)$_1$ (S, ST, T)$_1$ S (S, ST, T)$_2$ (Seq STeq T)$_2$

IV Reactive A Layer / Reactive B Layer
(T_A, ST_A, S) (S ST_B T_B)

Profiles: Pure Fickian diffusion (— — —); 'infinite' rate of reaction (–●–●–●–); intermediate ratios of diffusion / reaction rates (—). The tangents in Figure 11.I indicate the thickness η of 'equivalent Fickian (Nernst) layers' giving the same flux as the reactive membrane. With irreverrible and not equilibrated reactions (Seq)$_2$ should be replaced by $S_2=0$. Without forced convection S_2 becomes a function of σ and S_1 (Reference: Tran-Minh-Canh [46]).

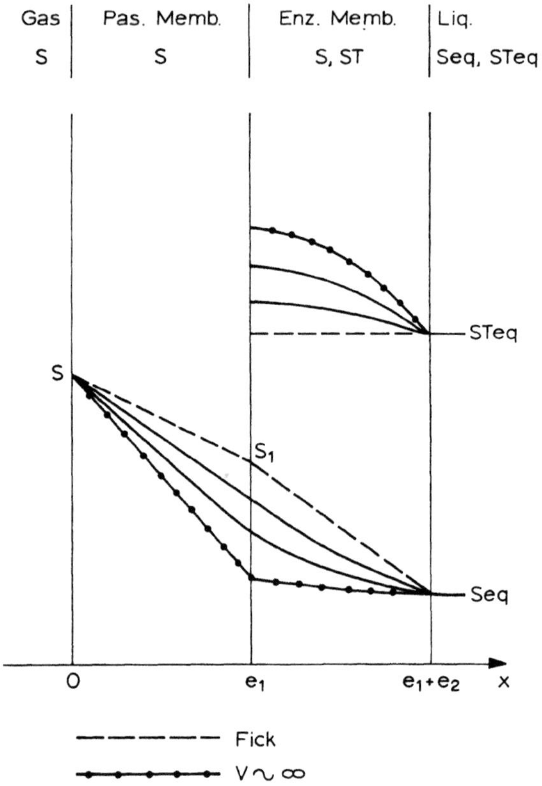

Fig. 11.II

at each boundary on the enzyme layer side and these boundary concentrations are in fact coupled.

(4) In another combination an interface separating two transporters can pose the problem of accelerating the slowest step between the decomposition of AT_1 and combination of $A+T_2$.

The limits of the phenomenon are:

(a) pure Fickian diffusion without reaction in which the interfacial concentration and the diffusion gradient depend only on the ratio of permeabilities on both sides of the interface;

(b) instantaneous reaction: the interfacial concentration of substrate becomes either nil ('total irreversible reaction') or that corresponding to the equilibrium value of the substrate-transporter reaction (Equation (22)); consequently the product-gradient replaces totally the substrate-gradient in the enzyme layer in agreement with Equation (6);

$$-[J_{DS}+J_{D(ST)}]=D_s\frac{\mathrm{d}S}{\mathrm{d}X}+D_{ST}\frac{\mathrm{d}(ST)}{\mathrm{d}X}=\text{Const.} \tag{23}$$

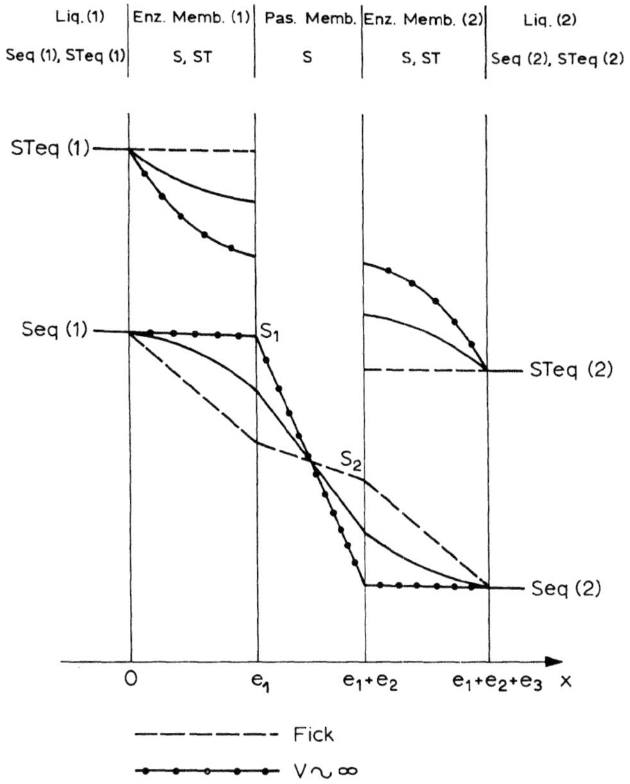

Fig. 11.III

In conclusion in case (1) in which the boundary concentration is held constant, the speed of the reaction can only influence the gradients in the enzyme layer: in case (2) with mobile boundaries, both gradients and boundary concentrations vary with the rate of reaction.

4.3.2. *Applications to Oxygenators*

Experimentally this facilitated transport has been tested [40, 46, 47, 113] in our group for the facilitation of CO_2 and oxygen transport through membranes.

CO_2 permeates through silicone membranes and such a membrane can be limited by a gas phase on the one side and a bicarbonate solution to which it is impermeable, on the other side, or by such a solution on both sides. (See cases (2) and (3) above). When carbonic anhydride is grafted as a very thin layer on the membrane, it accelerates the hydration of CO_2 or dehydration of bicarbonate in the unstirred layer (Figure 12). The transport of CO_2 is doubled at each liquid-membrane interface and thus quadrupled in liquid-membrane-liquid transport as is shown by experiments using ^{14}C tagged CO_2.

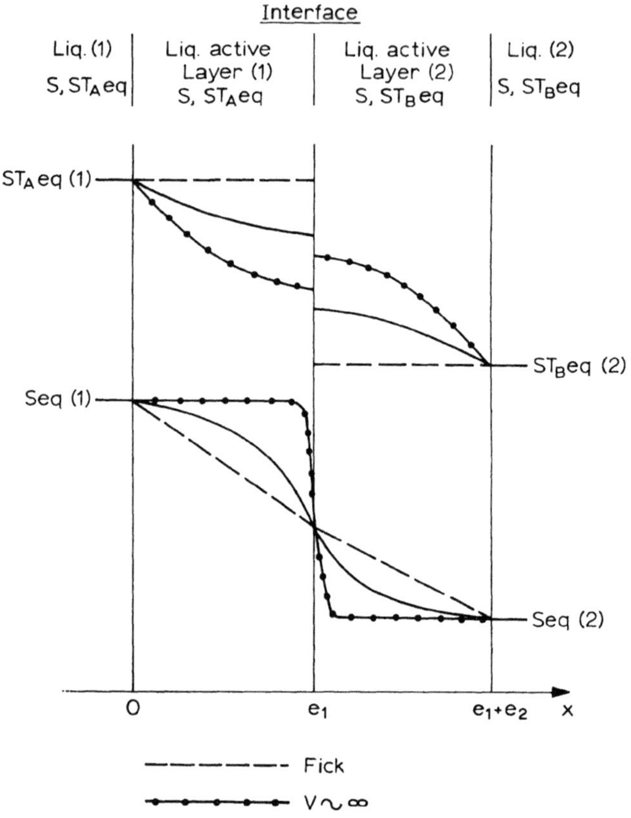

Fig. 11.IV

Hydrophylic films made from albumin for instance are permeable to H_2O_2 solutions which can be stabilised even by a weak acid such as uric acid. If catalase of high activity is grafted on the out-going face of the membrane, the totality of the oxygen will be liberated from its transporter, namely water.

Blood contains CO_2 stored as bicarbonate and hemoglobin will take up the liberated oxygen [46, 47]. Thus we have here the illustration of an *extra-membrane transporter* with CO_2, of an *intra-membrane transporter* with H_2O_2 and the convenient illustration of the theoretical cases cited above.

If this is an original approach to a biomedical use, it can also be remembered [40, 114] that permeating selectivities result from the facilitation of the transport of only one of two competing permeants; this can have some applications in engineering or separation science.

4.3.3. *Electrode Depolarisation*

There is a great analogy between electrode polarisation and membrane polarisation

TABLE XI

Facilitated transport

Assumptions: $VS + T \underset{k_{-1}}{\overset{k_{+1}}{\rightleftharpoons}} \dot{S}T$; T being in great excess $T \cdot k_{+1} = k'_{+1}$;

Steady-state equations and solutions in a few systems corresponding to Figure 11.
T and ST are present only in reactive or liquid layers or phases; all media are permeable to S. Equations and solutions can be taken from Tables IV, V, VI or Xa. The equations of this table are valid if the amount of enzyme added will only change k'_{+1} and k_{-1}, but not the rate law or $K_{eq} = k_{+1}/k_{-1}$.

Case Ia: Gas (containing S)/*reactive layer*/*efficiently convected liquid.*

Conditions at the boundaries of reactive phase:

Face 1: $S_1 = \text{Const.}$ (imposed by S_{gas}) $S_1 = K_D S_{gas}$

$$\frac{d(ST_1)_{x=0}}{dx} = 0 \text{ (gas impermeable to } ST)$$

Face 2: $S_2 = S_{equil}$; $(ST)_2 = ST_{equil}$ are constant, imposed by the liquid phase.
J_{ST} is due to the reaction only; $(ST_1)_{x=0}$ unknown is dependent on the rate of reaction and must be calculated.

(1) |From Equation (23)|: $-J_{(S+ST)} = D_S \dfrac{dS}{dX} + D_{ST} \dfrac{d(ST)}{dX}$

(2) By integration: $D_S S_X + D_{ST}(ST)_X = D_S(ST_1) - J_{(S+ST)} X$
The interrelation of S_X, $(ST)_X$ and $(ST_1)_{x=0}$ is obtained by substituting the value calculated for $X = e(x = X/e)$:

(3) $(ST)_x = \dfrac{D_S}{D_{ST}} [S_2 x + S_1(1-x) - S_x] + [(ST)_2 x + (ST_1)_{(x=0)}(1-x)]$

(4) (a) $\dfrac{\partial^2 S}{\partial x^2} - \alpha_1 S - \alpha_2 x + \alpha_3$

(b) $\dfrac{\partial^2 ST}{\partial x^2} = \alpha_1 (ST) - \alpha'_2 x + \alpha'_3$

(5) (a) $S_x = C_1 \exp(x\sqrt{\alpha_1}) + C_2 \exp(-x\sqrt{\alpha_1}) - \dfrac{\alpha_2}{\alpha_1} x - \dfrac{\alpha_3}{\alpha_1}$

(b) $ST_x = C'_1 \exp(x\sqrt{\alpha_1}) + C'_2 \exp(-x\sqrt{\alpha_1}) - \dfrac{\alpha'_2}{\alpha_1} x - \dfrac{\alpha'_3}{\alpha_1}$

(c) $\left(\dfrac{\partial(ST)}{\partial x}\right)_{x=0} = (C'_1 - C'_2)\sqrt{\alpha_1} - \dfrac{\alpha'_2}{\alpha_1} = 0$

Notations

$$a = \frac{k_{-1}e^2}{D_S}; \quad b = \frac{k'_{+1}e^2}{D_S}; \quad \delta = \frac{D_S}{D_{ST}}$$

$\alpha_1 = b + a\delta$; $\alpha_2 = a[\delta(S_1 - S_2) - ST_2] - a(ST)_{x=0}$;

$\alpha_3 = a\delta S_1 + a(ST)_{x=0}$;

$\alpha'_2 = b[\delta(S_1 - S_2) - (ST)_2] + b(ST)_{x=0}$; $\alpha'_3 = -bS_1 - b(ST)_{x=0}$

Solution:

Unknowns C_1, C_2, C'_1, C'_2 and $(ST)_{x=0}$ are calculated from Equation (5) with boundary conditions $x = 0$ and $x = 1$ and from α relations.

Table XI (Continued)

Efficiency of facilitation: J_{S_1}/J_D

$$(6)\qquad -\left(\frac{dS_1}{dx}\right)_{x=0}\Big/ J_D = e\frac{(C_2 - C_1)\sqrt{\alpha_1} + \alpha_{2/}\alpha_1}{S_1 - S_2} = \frac{1}{\eta}$$

ηe gives the reduction of the layer thickness necessary to produce the same flux in pure diffusion with the same boundaries.

Case IIa: Gas (containing S)/(I) *passive membrane* (permeable to S only)/(II) *reactive membrane/ convected liquid* (last two permeable to S, T and ST).

Conditions at the boundaries:

Gase-membr. I interface $(S_g)_{\mathrm{I}} = K_D \cdot S_g =$ Constant; imposed by gas; Membr. I/Membr. II: $(S')_{\mathrm{I}} = (S_1)_{\mathrm{II}} \cdot K_D$ and $\dfrac{d(St_1)_{x=0}}{dx} = 0$;

(K_D is the distribution coefficient between Membr. I and gas or II) Membr. II/liquid: same as in case Ia (S_{equil}; ST_{equil})

(S_1) is a function of diffusion-reaction characteristics and must be calculated; $J_{1\,(\mathrm{passive})} = J_{\mathrm{II}\,(\mathrm{facilitated})}$ steady-state fluxes give the necessary supplementary relation.

$$(7)\ (a)\qquad J_{\mathrm{I}} = D_{\mathrm{I}}\frac{(S_g)_{\mathrm{I}} - (S')_{\mathrm{I}}}{e_{\mathrm{I}}} = J_{(S_1)_{\mathrm{II}}} = -D_S\left(\frac{dS_1}{dx}\right)_{x_{\mathrm{II}}=0}$$

Again $(ST_1)_{x=0}$ is calculated from (5b), (5c), α_1, α'_2, α'_3 and reported in α_2 and α_3; fluxes and concentrations of S are then obtained from Equations (5a), (6) and (7).

Cases Ib and IIb. Ia and Ib describe the extraction of S from the gas. The passage of S from liquid and reactive layer are calculable with the same equations by changing the boundary conditions. (See reference [46].)

Case III: Passive membrane impermeable to ST and T separating two reactive layers (liquid/I reactive/II passive/III reactive/liquid).

Solutions are obtained as in cases II a and IIb, using equality of steady state fluxes $J = J_{\mathrm{I}} = J_{\mathrm{II}} = J_{\mathrm{III}}$ and adding relation (8) coupling the concentrations of S at interfaces I/II and II/III:

$$K_D[(S)_{x_{\mathrm{I}}=1} - (S)_{x_{\mathrm{III}}=0}]\frac{D_{\mathrm{II}}}{e_{\mathrm{II}}} = J$$

[57, 77]. The overpotential of the oxygen electrode of fuel cells in neutral media is largely due to the oxidation of water and formation of H_2O_2. In some test experiments undertaken on the proposal of Sélégny at the laboratory of Professor Bonnemay (Paris) in a cooperation of Kozel from Rouen and a very remarkable decrease of overpotential and a corresponding increase in closed circuit intensity have been observed when catalase is added to the electrolyte solution buffered at the optimal pH with an acetate buffer and thus oxygen is regenerated (unpublished).

5. Energy Conversions

Without a doubt one of the newest and most interesting fields of results concerns the conversion of the chemical energy of enzyme reactions into other forms: *for this the chemical reaction and another phenomenon are coupled kinetically and energetically. The coupling is mediated by the belonging of a compound (substrate or product) to both*

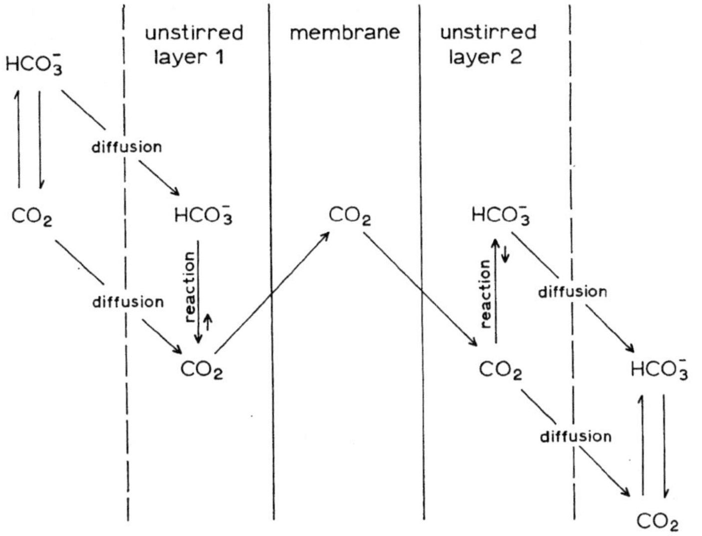

Fig. 12. Schematic representation of diffusion and diffusion-reaction paths of CO_2 transport in a system where a hydrophobic membrane separates two bicarbonate solutions of different concentrations. Carbonic anhydrase accelerates the two reactions in the unstirred layers and thus induces a greater HCO_3^- (extra-membrane transporter) diffusion on each side of the membrane. (From Sélégny *et al.* [40] and Broun *et al.* [113]).

phenomena; alternatively the reactivities of a given compound and one of its properties (e.g. charge) can belong to each phenomenon.

Accumulations and active transport can drive an osmotic flow and thus convert chemical into osmotic energy which will constitute the first two examples.

5.1. Accumulation of product; valves

We can mention only shortly that if a neutral permeant for example crosses the barrier of a charged membrane and transforms into a charged species through a chemical transformation such as splitting, oxidation or union with a charged species on the other side of the barrier, the back-diffusion becomes impossible and the charged membrane plays the role of a valve. Such situations [40, 42, 44, 76, 114, 115] have been observed and they are only the particular case of a class of systems in which the valve barrier is permeable to the substrate but not to the product.

An enzyme membrane itself can play the role of a valve if, because of its high activity, the substrate present on one side only cannot cross this membrane, and diffusion limitations favor the 'back-diffusion' of the product [40, 42, 44, 45, 53, 76, 114–115, 116].

A gradient of enzyme activity [114–115, 117] due to unsymmetrical distribution of the enzyme or asymmetrical limitations (activations) of its activity gives as well asymmetrical distributions of substrate and product in the two compartiments on each side of membrane even with symmetrical boundary concentrations.

Such accumulations are frequently called 'absorption' in cell-physiology [118]. This step seems to be also the one produced in several cases by plant-membranes for example by 'phosphorylation' leaving to an internal part of the cell or organel the decomposition of the substrate-carrier complex or compound; the total result is *then* active transport of the substrate which will be discussed in more detail in the next section.

5.2. Active transport substrate pumps

In the case of active transport a diffusible compound is transported from the lower concentration side of a membrane to the higher concentration one. Several theoretical or schematic models have been proposed for the explanation of this pumping observed in biological systems (mostly animal) [7–13, 119–121]. We have been interested in the generation *in vitro* of such transports by diffusion-reactions in which the energy necessary is taken from an energy-producing reaction. We have first identified [76] that in a general manner the wave-form, so-called *space-dependent oscillatory shape*, of the stationary state concentration profile of a substrate in a membrane is associated with diffusive active transport (Figure 13). The first mathematical treatment [76] made use of the ballistic equation. But one can remember the analogy between the heat equations (Fourrier) and the laws of diffusion (Fick) and the similar significance of concentration (or chemical or electrochemical potentials) and temperature [51]. It is evident that the temperature profile necessary to obtain a heat (or cold) pump needs the association of a heat sink and a heat source [52]. With enzyme reactions this means the association of two nearly inverse reactions, the first consuming and the second producing the same transported substrate. A number of different membrane systems can be conceived [40, 42, 76, 114–116] and tested, but before exposing them a few general remarks can be made.

Firstly the so-called Curie's law must be respected and the membrane system show dissymmetry in order to allow a vectorial phenomenon to appear through coupling of a chemical reaction which is scalar in its nature. Secondly, as for mechanical pumps, some sort of valve is useful in the system to increase the yield or add a complementary selectivity. Finally, calculations on some early systems [76] showed high instability which can be avoided by the creation of convenient geometrical or functional structures, for example.

Let us concentrate now on a variety of models using a small molecule or more specifically an ion as transporter and taking the energy from the decomposition of ATP.

5.2.1. *Permanent Multilayer Structures*

We have publsihed calculations and experimental demonstrations [40, 42, 44, 76] for the active transport of glucose produced by an artificial membrane composed of two active layers each containing a different enzyme: in the first one *hexokinase* catalyses the combination of ATP and glucose giving ADP^- + glucose-6-phosphate ($G\text{-}6\text{-}P^-$); in the second one *phosphatase* splits G-6-P into glucose and phosphate

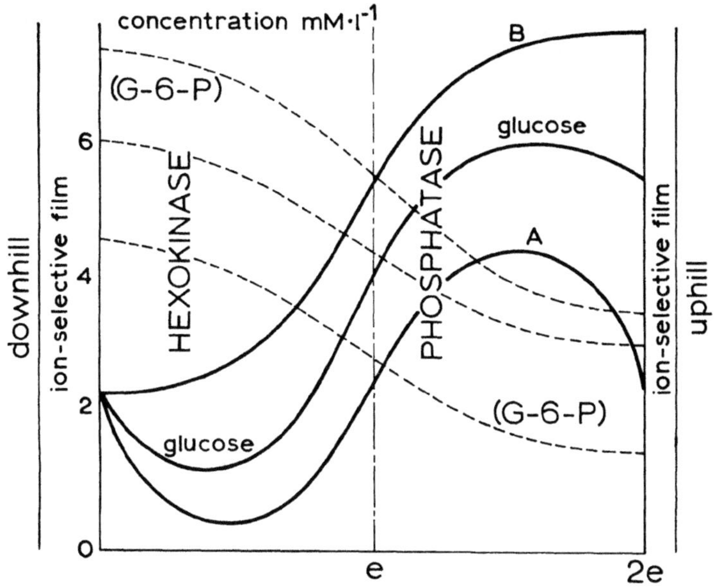

Fig. 13. Diagram of an active transport glucose-pump with *permanent structure*: ion-selective polyanionic external valve-membrane; hexokinase layer (heat sink equivalent for glucose: Glucose + $ATP^- \xrightarrow{Mg^{++}}$ Glucose-6-phosphate); phosphatase layer (heat source equivalent: glucose-6-phosphate → glucose + P^-). Typical 'space oscillatory profiles' (A) for maximum flux, nil external concentration difference ΔS; (B) for maximum ΔS and zero flux.

(P^-). Such a composite membrane behaves globally as a glucose-dependent ATPase. By transitory phosphorylation ATP^- furnishes both the *energy* necessary for the transport and the 'intramembrane phosphate *transporter*' of glucose. The negatively charged species are limited to the membrane volume by two external ion-selective negatively charged (albumin) valve-membranes.

The concentration profile of the actively transported species has two limiting shapes: a full wave for the maximum intensity of the net active transport (when concentrations are identical on both sides) or with an inflexion point for zero net flow but maximum difference between the substrate concentrations (chemical potentials) on both sides of the membrane (Figure 13).

The value of the concentration difference obtained at the plateau of the pump depends on the 'downhill' concentration (Figure 14).

For an ideal membrane of biological dimensions, a maximum increase of 130 times can be predicted by simulation on computer [42, 44, 45].

Fructose is a competitor of glucose (Figure 14); naturally the transport can be inhibited by any inhibitor of either of the two enzymes of the membrane, or by the absence of a cofactor (Mg^{++}). Transport will die if the available ATP is used up (Figure 14).

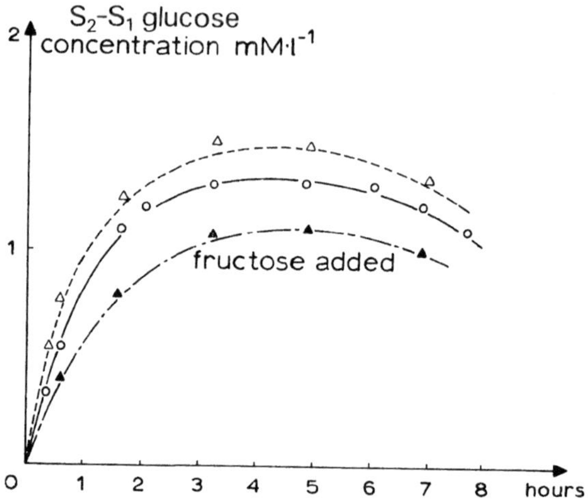

Fig. 14. Experimental results of a glucose pump working by active transport; energy source: ATP^-, substrate-transporter 'complex' G-6-P Glucose only (Δ/○) (from Sélégny *et al.* [40, 76b]); in presence of fructose-competitor (▲) (transposed from Thomas [44] and Broun *et al.* [42]). The final decrease of S_2 announces the 'death' of the pump by consumption of the ATP^- added in the membrane, and prevalence of the backdiffusion (characteristics: $e=3.10^{-3}$ cm; $V=10^{-3}$ mole cm^{-3} h^{-1}; $D=5\times10^{-3}$ cm^2 h^{-1}; $\sigma=20$, for each enzyme layer; compartment 1 virtually of infinite volume, compartment 2 of 6 ml; (same initial glucose concentration: $S_1=S_2$).

In conclusion *this pump is a permanent unidirectional one* working from downhill hexokinase→ phosphatase uphill. It works as soon as glucose arrives on the HK face from the outside or leaks from uphill.

Neglecting solvent flows, the overall conversion produces:

$$\Delta G_{\text{Reaction}} = |\Delta G|_1^2 + \Delta G_{\text{Back-diffusions}} + \Delta W_{\text{Transport frictions}} \tag{24}$$

$|\Delta G|_1^2$ is the work produced by the transfer of substrate from phase 1 to phase 2 on each side of membrane.

One can consider this type of pump as a logical combination of *facilitated transport* through membrane with *accumulation* including an *energetical reaction;* however different combinations of these elements and of enzyme reactions can give a number of different pumping mechanisms leading to other examples.

5.2.2. *Dissipative Functional Structures*

It can be predicted [114–115] that a pumping effect similar to the preceding one can be produced with the two sequential enzymes mixed together in one layer if their respective activities are different functions of substrate concentration or, more generally, if one of them is inhibited (or not activated) in the first part of the membrane and the other one in the other part. For the sake of simplicity, this can be illustrated with a *hydrogen ion gradient and two enzymes of different optimal pH.* When the hydrogen ion concentration is everywhere the same (or at least symmetrical) in the membrane and such that neither of the enzymes is working or both are working to the same extent, then there is no asymmetry (principle of Curie) and there is no pumping.

If however a pH gradient appears in the membrane for example by a modification of the acidity on both faces of the membrane (Figure 15) then one of the enzymes will be activated and the other one inhibited on the face of the higher pH; the inverse will happen on the side of the lower pH. We obtain a functionally asymmetric dissipative structure created by the pH gradient: the active transport will start and always pump from the first enzyme towards the second enzyme in the sequence of the reactions, as with the permanent structures, but here the absolute direction is controlled by the direction of the pH gradient. An inversion of the pH gradient can reverse the pump; the disappearance of this gradient will stop it.

One can observe that the hydrogen ion (or inhibitor or activator) gradient is not energetically coupled to the transport itself but behaves rather as an '*entropic switch-on signal*' while the chemical energy of pumping is still produced by ATP if, for example, the preceding glucose pump system is considered.

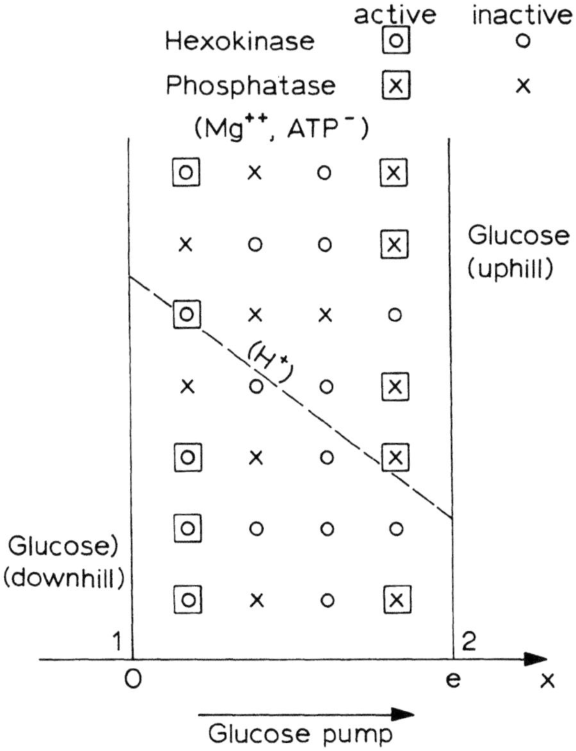

Fig. 15. Diagram of an active transport pump model based on a *dissipative functional structure*. The necessary dissipation to create this 'structure' is here the diffusion of an acid and its gradient; this creates inverse 'allotopic' activation or inhibition on face 1 and face 2 of the two enzymes of different optimal pH. (From Sélégny [115]).

The acidity gradient could be a consequence of the sequential reactions or be due to a temporary or a permanent dissymmetrical distribution of small electrolytes. Current research is under way in order to give 'experimental life quality' to this model [85].

Note that the idea of the formation of a concentration gradient of charged proteins under the influence of electrolyte diffusion potential has already been developed by Teorell [122].

5.2.3. *Permanent functional structures (Allotopia)**

The influence of polyelectrolytes on enzyme activity has been illustrated above.

Let us consider a membrane separating two solutions of the same pH and containing a homogeneous mixture of two enzymes. If a polyelectrolyte (or a conveniently selected charged lipid) is introduced into only half of this membrane or, alternatively, if two different polyelectrolytes are introduced into two different layers, then different local pH's will appear in both layers. This gives a new unsymmetrical structure in which, however, the active protein layer itself is of homogeneous enzymic composition, but if the two sequential enzymes (such as hexokinase and acid phosphatase) have sufficiently different optimal pH values, they will work individually in each layer: only one of them will be functionally present in each layer. Such a pump, represented diagrammatically in Figure 16 is a sort of combination of the two previous models and was proposed some time ago by Sélégny [114, 115]. It is now being tested experimentally by Vincent [85].

5.2.4. *Possible Combination of Ion-Pumps and of the Active Glucose Transport Model (Combination of Pumps, Charge Generation and 'Selective Valve-Membranes')*

In physiology coupled transport of stoichiometric amounts of positively charged ions (Na^+, K^+, Ca^{++}) is frequently observed together with sugar or amino-acid molecules. Such pumps can also be imagined *in vitro* by slight modifications of the previously described pump models.

(1) As already mentioned, the splitting of an ATP^- molecule into ADP^- and P^- generates a supplementary negative charge which must be compensated by a positive one; to avoid the consequent acidification of the reaction-medium which could inhibit the enzymes, the previously described glucose pump membranes have been buffered inside. But instead the necessary Na^+ or K^+ ions could be imported through the negatively charged ion-selective 'valve-membrane', from the down-hill solution for example.

(2) If a more elective result is sought for by associating the pump with another valve membrane, one can propose [114–115] to take advantage of the recent lipid membranes containing antibiotics through which only the cations complexed by the antibiotic-carrier can pass. If valinomycin is used only K^+ will pass and not Na^+: one K^+ for each split ATP^-.

* Allotopia covers the notion of the possibility or impossibility of activity of an enzyme in different locations, due to different local conditions.

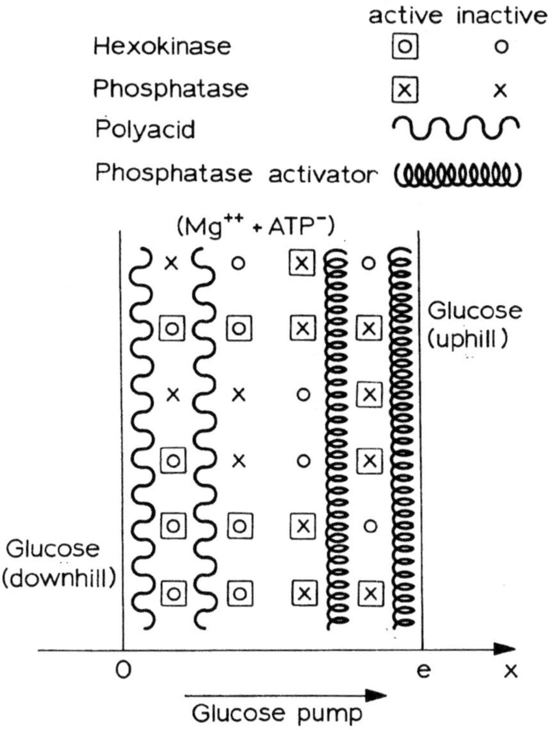

Fig. 16. Diagram of an active transport pump model with *permanent functional structure*. The local pH difference between faces 1 and 2 is due to different 'fixed chares' (Allotopia due to 'regulation by the matrix').

5.2.5. *Active Hydrogen-Ion-Gradient Amplificator (An Extension of the Law of 'Space-Oscillatory Stationary-State Profiles')* [114–115]

Another interesting system can show that the association of two enzymes and of the principles of oscillatory profiles and of Curie's law can create an equivalent of a hydrogen pump.

Consider that the two enzymes mixed in a layer are no more sequential but *independent*; take the case when one of them, of lower optimal pH, catalyses the transformation of a neutral substrate to an *acid product* and the other, of higher optimal pH, can *liberate a base* from another substrate. Consider again that the initial pH is uniform and intermediate between the two optimal pH values, thus rendering both enzymes inactive. I have submitted this model to the calculations of a mathematician [45].

Numerical solution on computer of the partial differential equations of these diffusion-reactions has shown that in such an enzyme membrane a small inverse shift of the pH on the limits of the enzyme layer (about 0.1 pH unit up on one side and about the same value simultaneously down on the other side) will rapidly generate a progressively self increasing activity for a different enzyme on each side of the membrane. A space oscillatory type profile appears rapidly through the membrane for the hydro-

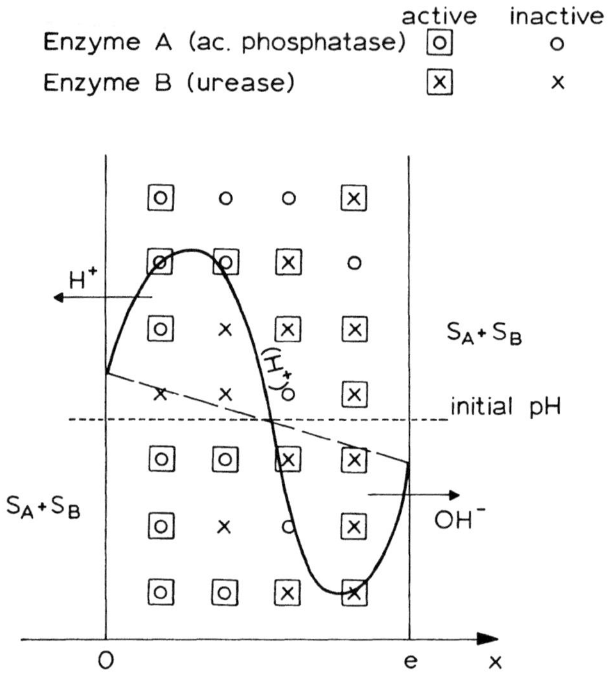

Fig. 17. Diagrammatic representation of a pH-amplificator model; membrane with two product-feed-back activated enzymes: (×) acid (H^+) producing enzyme A, (○) base producing (H^+ consuming, OH^- producing) enzyme B, with: optimal pH A<optimal pH B. Initially optimal-pH-A<initial pH<optimal-pH-B and no enzyme activity; induction of activity by a small decrease side 1 and increase side 2. Represented fluxes associated with the 'space oscillatory concentration profile' are progressively generated and correspond to the steady state with convenient boundary substrate concentrations and σ_{SA}, σ_{SB}, σ_{PA}, σ_{PB} values.

gen-ion concentration: consequently hydrogen ions will flow out of the membrane on the more acid side and flow in (or OH^- ions flow out) on the more basic side (Figure 17). Thus this pump is able to amplify a pH gradient applied to it.

5.3. Coupling of an Enzyme Reaction and the Mechanical Work of a Polyelectrolyte Gel or Fiber

The volume change of polyelectrolyte gels due to variations of their net charge density can be produced for example by interchanging bivalent and monovalent counter-ions or by modifications of the ionic strength. The volume change is able to produce mechanical work; this chemico-mechanical energy conversion has been used in a sort of a first model of muscle described namely by Katchalsky and Oplatka [6] employing a collagen fiber. A more direct coupling of an enzymatical reaction was however needed. In order to achieve this we have [43] used the properties of slightly crossed-linked amphoteric gelatin gels to undergo considerable deswelling when the pH is increased from neutral to alkalin by adding ammonium carbonate. In a series of experiments this ammonium carbonate was produced by the hydrolysis of urea catalysed by the enzyme

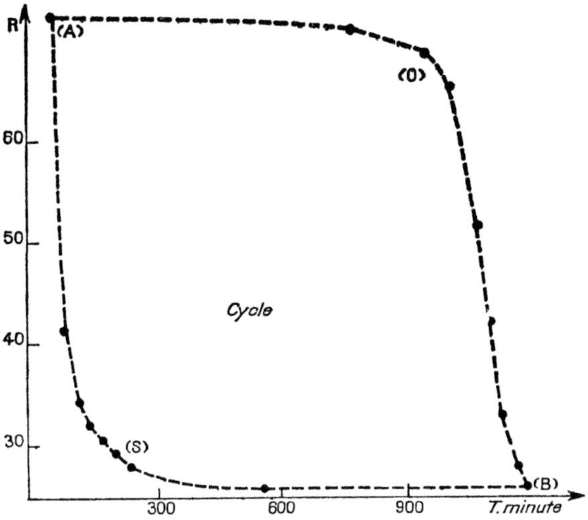

Fig. 18. Illustration of the coupling of an enzyme reaction to mechanical work: relative volume change of slightly reticulated gelatin in function of time (followed picnometrically): R = swelling water weight / dry weight; ASB = contraction due to production of ammonium corbonate (7‰) namely by urea + urease; BOA = reswelling in pure water (From Sélégny *et al.* [43]).

urease, added to the surrounding solution or linked into the gel. Results of Figure 18 show that cyclic swelling and deswelling can be generated by alternative enzyme reactions followed by washing with water. Here ammonium carbonate is a part of diffusion and of reaction and its property to act on pH *mediates*, by inducing the fixed charges, the transformation into mechanical work produced by the charge-repulsion or attraction effect. One could *change the mediating phenomenon* and decomplex calcium complexes or induce charge inversions or play with modifications of hydrophobic forces.

A different chemical reaction can naturally be used. This was done independently by Oplatka and his co-workers [123] who used the contraction of collagen induced by the acid-producing enzymatic hydrolysis of esters. They are now also developing systems using ATP. This mechano-chemical field will surely know an interesting development in connection with polyelectrolytes.

5.4. Production of hydrodynamic flows coupled to mechano-chemical enzyme reactions

The conversion of the chemical energy of the enzyme reaction into mechanical energy in the above example [43] is accompanied by a water uptake or release of the collagen or gelatin gel-membrane or fiber. If the addition of a convenient hydrodynamic valve allows the uptake of water on one side and its unsymmetrical release on the other side of the enzymatical gel or sponge, the energy can be spent in convective pumping of the liquid. One can imagine several ways of realizing such valves by taking examples from nature for instance.

5.5. Effect of higher cross-linking of the urease gel: the charge-generating membrane

Let us take a urease-protein gel similar to the one that we used in the mechano-chemical transformation but more cross-linked. An immediate consequence of higher cross-linking is that if activity starts on adding urea and ammonium carbonate is formed elevating the pH in the gel, then the protein gel becomes charged without significant change in swelling; the major effect is no more the mechano-chemical energy conversion, but the membrane will be able to develop membrane potentials and ion selectivities as other 'variable fixed charge membranes'. Equilibrium and diffusion properties of weak-polyelectrolyte gel membranes the charge of which is a function of pH and ionic strength have been studied intensively in our laboratory [124, 125].

The self-regulation of urease activity has already been exposed in the corresponding chapter of this paper. The studies on charge generation in urease membranes were undertaken by a juxtaposition of these two types of studies [86a, 86b]. The particularity of the system is that the nature of the ions generated contributes to the selectivities, remembering the properties of polyanion-polycation (for example snake-caged) gels and membranes [90, 126].

5.6. Conversion of chemical energy into electrical energy

We have already mentioned briefly the conversion of the formation of product into electrical signals by enzyme electrodes in a previous section. In correctly constructed fuel cells the chemical energy liberated gives rise to electric power-generation without production of heat. Attempts have been made to use enzymes in such fuel cells; some of them should be mentioned briefly.

Blumenthal *et al.* [10c] have generated an electric current through a solution where a salt resulted from hydrolysis of a neutral substrate; the necessary unsymmetry was introduced by an anion and a cation exchange membrane separating respectively the enzyme solution from reversible electrodes, constituting the electron collecting device and allowing the conversion from ionic to electron-flow.

More recently Wingard [127] proposed the use of glucose-electrodes in fuel cells.

The use of catalase in decreasing the oxygen electrode overpotential due to oxidation of water and formation of H_2O_2 is another illustration of the possible use of enzymes in electric generation (see section on facilitated transport).

Young *et al.* [128] have evaluated the different possible applications of enzymes in fuel cells. Namely: depolarization concentration cells, product cells and redox cells.

Bockris and Srinivasan [129] have analysed in a chapter of their book the electrochemical aspects of biological fuel cells.

The interest of such systems is twofold:

(1) Theoretically they can help in the analysis of electron transfer chains so abundant in biology (in chain reactions the electrodes are replaced by another reacting system).

(2) The extraordinary efficiency of enzyme catalysis can surely lead to practically

feasible and justifiable power sources; these could use fuels of biological origin or their synthetic equivalents.

I would like to predict that under the influence of general progress of immobilized enzymes associated with electrochemical studies, there will be a come-back of fuel cells and an extension in electrochemical applications. Elsewhere more detailed analysis and illustration justifying this prediction will be given.

6. Conclusion

My first hope is that the reader is persuaded that consecutive to the work of a few groups cited and also of those omitted through lack of space heterogeneous enzymology exists experimentally and theoretically.

Not only in the simplified basic case but also when there are kinetic complications, carrier or diffusion effects, several enzymes produce interdependent regulations, the phenomenon can be treated kinetically or by non-equilibrium thermodynamics in a quantitative manner and experimentally verified and illustrated.

Secondly, with the correct combinations of different enzyme-layers, charged gels, layers, membranes or fibers and so on, energy-converting systems can be constructed and described mathematically. In other terms, *structure-function-energy-conversion interrelations* are predictable, computable and produceable.

The most important general conclusion is perhaps that the logic is unique not only thermodynamically or kinetically but also in structures and compositions between phenomena catalysed in biology or created following this *synthetic pathway* for models or even engineering applications (e.g. artificial organs).

The analogy is not purely artificial or qualitative: optimalisation of yields, regulations, stability, selectivity or energy conversion couplings impose it spontaneously.

The terminology (allosteria, allotopia, facilitated or active transport etc....) issued from biology can also be illustrated by physico-chemical models.

Kinetic and experimental modeling is the field for translations from one discipline to another: from molecular to thermodynamic. In this respect the modeling as illustrated here plays the role of *supramolecular-chemio-mechanics* (or biochemio-mechanics).*

We have only touched on the practical aspects: uses of immobilized enzymes in analytical chemistry, bio and medical engineering, bio or chemical industries; they are strongly developing all over the world to-day.

I wish to finish this review by emphasizing that it is no longer a prediction but a simple fact that the investigation of interactions and interregulations of chemical enzymic reactions and other phenomena, in homogeneous or structured media, constitute one of the modern problems in all basic disciplines including those of electrochemistry and polyelectrolyte physical chemistry.

* Neither the energetic coupling between the fuel combustion and the kinetic energy of a car, nor the enumeration of its components would allow the invention of an automobile without mechanics and engineering.

Acknowledgements

I should like to express my thanks to Mrs G. Mallet for her constant help throughout the preparation of the English manuscript, to Mrs C. Sélégny and Mrs F. Cannic for typing and to Professor G. Stoner for reading it.

References

1. Teorell, T.: 'Introductory Lecture', in E. Sélégny, G. Boyd, and K. S. Spiegler (eds.), *Charged and Reactive Polymers*, vol. II, Forges les Eaux, Sept. 1973, D. Reidel Publ. Co., Dordrecht-Holland (to be published).
2. Sollner, K.: Introductory Lecture, in E. Sélégny, G. Boyd, and K. S. Spiegler (eds.), *Charged and Reactive Polymers*, vol. II, Forges les Eaux, Sept. 1973, D. Reidel Publ. Co., Dordrecht-Holland (to be published).
3. Tasaki, I.: *Nerve Excitation*, C. Thomas, Springfield, Illinois, 1968.
4. De Sorgo, M. and Prins, W.: *Mechanochemistry of a Photochromic Model Membrane*, Second Discussion Conference on Macromolecules, Prague (Aug. 1972).
5. (a) Van der Veen, G. and Prins, W.: *Nature* (London), *Phys. Sci.* **230**, 70 (1971); (b) Van der Veen, G.: this volume, p. 483.
6. Katchalsky, A. and Oplatka, A.: *Handb. Sens, Physiol.* **1**, (1971).
7. Prigogine, I.: *Thermodynamics of Irreversible Processes*, Wiley, New York, 1961.
8. de Groot, S. R. and Mazur, P.: *Non Equilibrium Thermodynamics*, North-Holland Publ. Co., Amsterdam, 1962.
9. (a) Katchalsky, A. and Curran, P. F.: *Non Equilibrium Thermodynamics in Biophysics*, Harvard University Press, Cambridge, Mass., 1965. (b) Katchalsky, A. and Oster, G.: 'Chemico-Diffusional Coupling in Membranes', in D. C. Tosteston (ed.), *The Molecular Basis of Membrane Function*, Prentice-Hall, Englewood Cliffs, New Jersey, 1969, p. 1.
10. (a) Kedem, O.: 'Criteria of Active Transport', in A. Kleinzeller and A. Kotyk (eds.), *Membrane Transport and Metabolism*, Academic Press, London, 1961, p. 87; (b) Kedem, O. and Caplan, S. R.: *Trans. Faraday Soc.* **61**, 1897 (1965); (c) Blumenthal R., Caplan, S. R., and Kedem, O.: *Biophys. J.* **7**, 735 (1967).
11. Sauer, F. *Habilitationsschrift*, Frankfurt, 1969; Meyer, J., Sauer, F., and Woermann: *Ber. Bunsenges. Physik. Chem.* **74**, 245 (1970).
12. Ussing, H. H.: *Acta Physiol. Scand.* **17**, 1 (1949); *Physiol. Veg.* **9**, (1), 1 (1971).
13. Stein, W. D.: *The Movement of Molecules across Cell Membranes*, Academic Press, New York, 1967; see also A. Kleinzeller and A. Kotyk (eds.), *Membrane Transport and Metabolism*, Academic Press, London, 1961; *Meeting on Membranes*, Rouen, 1970 in *Physiol. Veg.* 1971 (January Issue), ed. by E. Sélégny and M. Thellier.
14. Dixon, M. and Webb, E. C.: *Enzymes*, 2nd ed., Longmans, London, 1964; Laidler, K. J.: *The Chemical Kinetics of Enzyme Action*, Oxford at the Clarendon Press, London, 1958.
15. New England Research Application Center, University of Connecticut: *Immobilized Enzymes, a Compendium of References from the Recent Literature* (ed. by H. H. Weetall), Corning Glass Works, New York, 1973.
16. (a) Silman, I. H. and Katchalski, E.: *Ann. Rev. Biochem.* **35**, 873 (1966); (b) Wingard, L. B. and Finn, R. K.: *Chem. Eng. Prog. Sym.*, series **62**, (No. 69), 30 (1966); (c) Goldstein, L. and Katchalski, E.: *Z. Anal. Chem.* **243**, 375 (1968); (d) Goldstein, L.: *Use of Water Insoluble Enzyme Derivatives in Synthesis and Separation*, Fermentation Advances (ed. by D. Perlman), Academic Press, New York, 1969; (e) Melrose, G. J. H.: *Rev. Pure Appl. Chem.* **21**, 83 (1971); (f) Weetall H. H. and Messing, R. R.: *Insolubilised Enzymes on Inorganic Materials* in M. L. Hair: *The Chemistry of Biosurfaces*, Vol. 2, Marcel Dekker, New York, 1972; (g) Royer, G. P., Andrews, G. P., and Uy, R.: *Eng. Technol. Dig.* **1**, 99 (1973); see also ref. (30).
17. Grubhofer, N. and Schleith, L.: *Naturwissenschaften* **40**, 508 (1953).
18. Tosa, T., Mori, T., Fuse, N., and Chibata, I.: (a) *Enzymologia* **31**, 225 (1966); (b) *Enzymologia* **32**, 153 (1967); (c) *Biotech. Bioeng.* **9**, 603 (1967); (d) *Biol. Chem.* **33**, 1047 (1969).
19. (a) Bauman, E. K., Goodson, L. H., Guilbault, G. G., and Kramer, D. N.: *Anal. Chem.* **37**,

1378 (1965); (b) Guilbault, G. G. and Kramer, D.: *Anal. Chem.* **37**, 1675 (1965); (c) Hicks, G. P. and Updike, S. J.: *Nature* **214**, 986 (1967); (d) Mosbach, K. and Mosbach, R.: *Acta Chem. Scand.* **24**, 2804 (1970); (e) *Acta Chem. Scand.* **24**, 2084 (1970); (f) Nilsson, H., Mosbach, R., and Mosbach, K.: *Biochem. Biophys. Acta* **268**, 253 (1972). See also (f) Bernfield, P. and Wan, J.: *Science* **142**, 678 (1963); (g) Mitz, M. A. and Summaria, L. J.: *Nature, London* **189**, 576 (1961).

20. Goldman, R., Kedem, O., Silman, H., Caplan, S. R., and Katchalski, E.: *Biochemistry* **7**, 486 (1968).
21. (a) Monod, J., Changeux, J. P., and Jacob, F.: *J. Mol. Biol.* **6**, 306 (1963); (b) Monod, J., Wyman, J., and Changeux, J. P.: *J. Mol. Biol.* **12**, 88 (1965).
22. Avrameas, S., Broun, G., Sélégny, E., and Thomas, D. (Anvar): *French Patent* 2, 028, 722 (1968),
23. (a) Manecke, G. and Singer, S.: *Makromol. Chemie* **39**, 13 (1960); (b) Manecke, G. and Guenzel, G.: *Makromol. Chemie* **51**, 199 (1962); (c) Manecke, G. and Foerster, H. J.: *Makromol. Chemie* **91**, 136 (1966); (d) *Naturwissenschaften* **54**, 531 (1967); (e) *Naturwissenschaften* **54**, 647 (1967); (f) Manecke, G., Guenzel, G., and Foerster, H. J.: *J. Pol. Sci.* **30**, 607 (1970).
24. (a) Cebra, J. J., Divol, G., Silman, H. I., and Katchalski, E.: *J. Biol. Chem.* **236**, 1720 (1961); (b) Bar-Eli, A. and Katchalski, E.: *J. Biol. Chem.* **238**, 1690 (1963).
25. (a) Axen, R. and Porath, J.: *Nature* **210**, 367 (1966); (b) Axen, R., Porath, J. and Ernback, S.: *Nature* **214**, 1302 (1967); (c) Porath, J., Axen, R., and Ernback, S.: *Nature* **215**, 1491 (1967); (d) Axen, P. and Ernback, S.: *Europ. J. Biochem.* **18**, 351 (1971).
26. Sélégny, E., Avrameas, S., Broun, G., and Thomas, D.: *Compt. Rend. Acad. Sci. Paris* **266**, Série C, 1431 (1968).
27. (a) Weetall, H. H. and Hersh, L. S.: *Biochem. Biophys. Acta* **185**, 464 (1969); (b) Weetall, H. H.: *Prep. Charact. Sci.* **166**, 615 (1969); (c) *Nature* **223**, 959 (1969).
28. (a) Levin, Y., Pecht, M., Goldstein, L., and Katchalski, E.: *Biochem.* **3**, 1905 (1964); (b) Alexander, B., Rimon, A., and Katchalski, E.: *Biochem.* **5**, 792 (1966).
29. Hornby, W. E., Lilly, M. D., and Crook, E. M.: *Biochem. J.* **98**, 420 (1966).
30. (a) Kay, G. and Crook, E. M.: *Nature* **216**, 514 (1967); (b) Kay, G., Lilly, M. D., Sharp, A. K., and Wilson, R. J. H.: *Nature* **217**, 641 (1968); (c) Wilson, R. J. H., Kay, G., and Lilly, M. D.: *Biochem. J.* **108**, 845 (1968).
31. (a) Campbell, D. M., Leucher, E., and Lerman, L. S.: *Proc. Natl. Acad. Sci. U.S.* **37**, 575 (1951); (b) Grubhofer, N. and Schleith, L. Z.: *Z. Physiol. Chem.* **297**, 108 (1954); (c) Weliky, N. and Weetall, H. H.: *Immunochemistry* **2**, 293, (1965); (d) Surinov, B. P. and Manoilov, S. E.: *Biokhimiya* **31**, 387 (1966); (e) Silman, I. H., Albu-Weissenberg, M., and Katchalski, E.: *Biopolymers* **4**, 441 (1966); (f) see also ref. (24b); (g) Goldstein, L., Pecht, M., Blumberg, S., Atlas, D., and Levin, Y.: *Biochemistry* **9**, 2322 (1970).
32. Patel, R. P., Lopiekes, D. U., Brown, S. R., and Price, S.: *Biopolymers* **5**, 577 (1967).
33. (a) See ref. (31c); (b) Hoare, D. G. and Koshland, D. E.: *J. Biol. Chem.* **242**, 2447 (1967); (c) Weliky, N., Brown, F. S., and Dale, E. C.: *Arch. Biochem. Biophys.* **131**, 1 (1969).
34. (a) Wichterle, O.: *Synthetic Carriers of Bioactive Agents*, Second Discussion Conference on Macromolecules, Prague (Aug. 1972), Lecture L9; (b) Coupek, J., Torkova, J., and Krivakova, M.: *New Carriers of Bioactive Compounds, Synthetic Carriers of Bioactive Agents*, Second Discussion Conference on Macromolecules, Prague (Aug. 1972), Preprint, D 9/3.
35. Avrameas, S.: *Immunochemistry* **6**, 43 (1969).
36. Sélégny, E., Broun, G., and Thomas, D.: *Compt. Rend. Acad. Sci. Paris* **269**, Série D, 1330 (1969).
37. Broun, G., Sélégny, E., Avrameas, S., and Thomas, D.: *Biochem. Biophys. Acta* **185**, 260 (1969).
38. (a) Quiocho, F. A. and Richards, F. M.: *Proc. Natl. Acad. Sci. U.S.* **52**, 833 (1964); (b) *Biochem.* **5**, 4062 (1966).
39. Avrameas, S., Broun, G., Sélégny, E., and Thomas, D. (Anvar): *U.S. Patent*, 286.233 (5.9.1972).
40. Sélégny, E., Broun, G., and Thomas, D.: *Physiol. Veg.* **9**, (1), 25 (1971).
41. Thomas, D., Broun, G., and Sélégny, E.: *Biochimie* **54**, 229 (1972).
42. Broun, G., Thomas, D., and Sélégny, E.: *J. Membrane Biol.* **8**, 313 (1972).
43. Sélégny, E., Meffroy-Biget, A. M., and Labbé, M.: *Compt. Rend. Acad. Sci. Paris* **275**, Série C, 359 (1972).
44. Thomas, D.: *Elaboration de modèles biologiques structurés à l'aide de membranes porteuses d'enzymes réticulées*, Thèse d'Etat, No. CNRS: A. O. 5407 (Rouen 1971).
45. Kernevez, J. P.: *Evolution et contrôle de systèmes biomathématiques*, Thèse d'Etat de Mathématiques, Université de Paris VI, No. CNRS: A.O. 7246, 1972.

46. Tran-Minh-Canh: *Transfert et génération de gaz par les membranes; application aux oxygénateurs et aux capteurs*, Thèse d'Etat, No. CNRS A.O. 6262 (Rouen, 1971).
47. Broun, G., Canh Tran Minh, Thomas, D., Domurado, D., and Sélégny, E.: *Trans. Amer. Soc. Artif. Int. Organs.* **17**, 341 (1971).
48. Tran-Minh-Canh, Sélégny, E., and Broun, G.: *Compt. Rend. Acad. Sci. Paris* **275**, Série C, 309 (1972).
49. Julliard, J. H., Godinot, C., and Gautheron, D. C.: *FEBS Letters* **14**, 185 (1971).
50. (a) Brown, E. and Racois, A.: *Bull. Soc. Chim. Fr.* 4331 (1971); (b) *Bull. Soc. Chim. Fr.* 4337 (1971); (c) Brown, E. and Joyeau, R.: *Tetrahedron Lett.* **8**, 601 (1973).
51. (a) Jost, W.: *Diffusion in Solids, Liquids, Gases*, Academic Press Inc., New York, 1960; (b) Crank, J.: *The Mathematics of Diffusion*, Oxford at the Clarendon Press, London, 1967.
52. (a) Ribaud, M.: *Conduction de la chaleur*, Cours des Arts et Métiers, Paris, 1957; (b) Carslaw, H. S. and Jaeger, J. C.: *Conduction of Heat in Solids*, Oxford at the Clarendon Press, London, 2nd ed., 1971.
53. Goldman, R., Kedem, O., and Katchalski, E.: *Biochemistry* **7**, 4518 (1968).
54. Lilly, M. D., Hornby, W. E., and Crook, E. M.: *Biochem. J.* **100**, 718 (1966).
55. (a) Horvath, C., Sardi, A., and Woods, J. S.: *J. Appl. Physiol.* **34**, 181 (1973). (b) Horvath, C., Shendalman, L. H., and Light, R. T.: *Chem. Eng. Sci.* **28**, 375, (1973); (c) Horvath, C., Salimon, B. A., and Engasser, J. M.: *Ind. Eng. Chem. Fundamentals*, in press.
56. Shuler, M. L., Aris, R., and Tsuchiya, H. M.: *J. Theor. Biol.* **35**, 67 (1972).
57. Metayer, M., Bourdillon, C., and Sélégny, E.: *Compt. Rend. Acad. Sci., Paris* **272**, Série C, 1635 (1971).
58. Horvath, C. and Engasser, J. M.: *J. Theor. Biol.* in press; Horvath, C. and Engasser, J. M.: *Ind. Eng. Chem. Fundamentals* **12**, 229, (1973).
59. Thomas, D.: *Réalisation et étude de membranes enzymatiques en vue d'élaboration de modèles biologiques*, Thèse (Rouen, 1969).
60. Sundaram, P. V., Tweedale, A., and Laidler, K. J.: *Canadian J. Chem.* **48**, 1498 (1970).
61. Sélégny, E., Broun, G., Geffroy, J., and Thomas, D.: *J. Chim. Phys. et Physico-Chimie Biol.* **66**, 391 (1969).
62. De Simon, J. A. and Caplan, S. R.: *The Determination of Local Reaction and Diffusion Parameters of Enzyme Membranes from Global Measurements,* to appear in *Biochemistry*; De Simon, J. A.: *Transport and Reaction in Enzymatically Active Artificial Membranes: a Theoretical and Experimental Analysis*, Ph.D. Dissertation, Harvard University, Cambridge, Mass., 1970.
63. De Simon, J. A. and Caplan, S. R.: *Symmetry and the Stationary State Behaviour of Enzyme Membranes, J. Theoret. Biol.* **39** (1973) in press, and *Chem. Eng. Symp. Series* **68** (No. 99), 43 (1970).
64. Sélégny, E. and Leguillon, D.: (to be published).
65. (a) Swoboda, B. E. P. and Massey, V.: in: E. C. Slater (ed.), *Flavins and Flavoproteins*, Elsevier Publ. Co., 1966, p. 263; (b) Bright, H. J. and Gibson, Q. M.: *J. Biol. Chem.* **242**, 994 (1967); (c) Nakamura Satoshi and Ogura Yasuyuki: in K. Yagi (ed.), *Flavins and Flavoproteins*, University of Tokyo Press, 1968, p. 164,; (d) Duke, F. R., Weibel, M., Page, V. G., Bulgrin, V. G., and Lutny, J.: *J. Am. Chem. Soc.* **91**, 3904 (1969).
66. Weibel, M. K. and Bright, H. J.: *Biochem. J.* **124**, 801 (1971).
67. Sélégny, E., Lancon, M., and Vincent, J. C.: (to be published).
68. Sélégny, E., Kernevez, J. P., Broun, G., and Thomas, D.: *Physiol. Veg.* **9** (1), 51 (1971).
69. Guilbault, G. G.: *Anal. Chem.* (a) **38**, 527R (1966); (b) **40**, 459R (1968); (c) **42**, 334R (1970).
70. Guilbault, G. G.: *Biotech. Bioeng., Suppl.* **3**, 361 (1972).
71. (a) Guilbault, G. G. and Lubrano, G. J.: *Anal. Chim. Acta* **64**, 439 (1973); (b) Guilbault, G. G. and Nagy, G.: *Anal. Lett.* **6**, 301 (1973).
72. Horvath, C. and Sovak, M.: *Biochem. Biophys. Acta* **298**, 850 (1973).
73. Changeux, J. P., Thiery, J., Tung, Y., and Kittel, C.: *Proc. Natl. Acad. Sci. U.S.* **57**, 335 (1967).
74. Goudeau, H.: *Sélectivité ionique des membranes artificielles minces formées à partir de lipides non saturés*, Thèse d'Etat, Paris VI, 1973.
75. (a) Sélégny, E., Langevin, D., Tran-Minh-Cahs: 'Comportement d'alvéoles élastiques (caoutchouc naturel) en perméation gazeuse. Examen a la lumière de la théorie allostérique', *Compt. Rend. Acad. Sci. Paris*, in press. (b) Tran-Minh-Canh and Sélégny, E.: 'Detection of the Flowing Point of an Elastic Membrane by the Modifications of its Permeability to Gases under Bidemen-tional Stress', Polymer Letters (in press).

76. Sélégny, E., Broun, G., and Thomas, D.: (a) *Compt. Rend. Acad. Sci. Paris* **269**, Série C, 1377 (1969); (b) *Compt. Rend. Acad. Sci. Paris* **271**, Série D, 1423 (1970).
77. Metayer, M., Bourdillon, C., and Sélégny, E.: (a) *Desalination* **13**, 129 (1973); (b) *J. Chim. Phys.* **70** (5), 722 (1973).
78. McLaren, A. D. and Estermann, E. F.: *Arch. Biochem. Biophys.* **61**, 157 (1957).
79. (a) Goldstein, L., Levin, Y., and Katchalski, E.: *Biochem.* **3**, 1913 (1964). (b) Goldstein, L., in: *Methods in Enzymology*, vol. XIX (ed. by G. E. Perlmann and Lorand), 1970, p. 935.
80. (a) Katchalski, E., Silman, E., and Goldman, R.: *Adv. Enzymol.* **34**, 445 (1971); (b) Katchalski, E.: in P. Desnuelle, H. Neurath, and M. Ottesen (eds.), *Structure-Function Relationships of Proteolytic Enzymes*, Munksgaard, Copenhagen, 1970, p. 198.
81. Wharton, E. W., Crook, E. M., and Brocklehurst, K.: *European J. Biochem.* **6**, 565 (1968).
82. Hornby, W. E., Lilly, M. D., and Crook, E. M.: *Biochem J.* **107**, 669 (1968).
83. Patel, A. B., Pennington, S. N., and Brown, H. D.: *Biochem. Biophys. Acta* **178**, 626 (1969).
84. Kramer, D. M., Lehmann, K., Plainer, H., Reisner, W., and Sprossler, B. G.: in *Interaction of Ionic Gel Beads with Biomacromolecules*, 13th Prague Microsymposium on Macromolecules (IUPAC), preprint B6, Prague, 1973.
85. Vincent, J. C.: private communication.
86. (a) David, A.: *Effets des charges fixes ou mobiles sur le comportement d'une membrane artificielle porteuse d'enzymes greffées*, Thèse de Spécialité (Rouen, 1973); (b) David, A., Metayer, M., Broun, G., and Thomas, D.: *J. Membrane Biol.* (in press).
87. See ref. (82).
88. Sélégny, E., Korngold, E. and Merle, Y.: *Bull. Soc. Chim. Fr.* (5), 2252 (1968).
89. Merle, Y.: *European Polymer J.* **8**, 1265 (1972).
90. Helfferich, F.: *Ion-Exchange*, McGraw-Hill, New York, 1962.
91. Kozel, M.: *Enzymes insolubles à réactions réversible ou compétitivement inhibée*, Diplôme d'Etudes Approfondies, Chim. Macromol., Paris, 1969.
92. Broun, G., Sélégny, E., and Thomas, D.: *Feed-Back Regulation of Enzymes Covalently Bound in a Membrane*, 3rd Internat. Biophys. Congress of I.U.P.A.B., Cambridge, Mass, **IM6**, 82 (1969).
93. (a) Mattiason, B. and Mosbach, K.: *Biochem. Biophys. Acta* **235**, 253 (1971); (b) Gestrelius, S. Mattiason, B., and Mosbach, K.: *Biochem. Biophys. Acta* **276**, 393 (1972).
94. Mosbach, K.: *Biotech. Bioeng. Suppl.* **3**, 185 (1972).
95. Zhabotinsky, A. M.: (a) *Biofizika* **2**, 306 (1964); (b) *Acad. Sci. U.S.S.R.*, Moscow (Nanka), 1967; (c) *Russ. J. Phys. Chem.* **42**, 1649 (1968).
96. Glansdorff, P. and Prigogine, I.: *Structure, stabilité et fluctuations*, Masson et Cie, Paris, 1971.
97. Higgins, G.: (a) *Proc. Natl. Acad. Sci. U.S.* **51**, 989 (1964); (b) *Ind. Eng. Chem.* **59**, 19 (1967).
98. Walter, C.: (a) *Biophys. J.* **9**, (No. 7), 863 (1969); (b) *J. Theoret. Biol.* **23**, 23 (1969).
99. (a) Chance, B., Estabrook, R. W., and Ghosch, A.: *Proc. Natl. Acad. Sci. U.S.*, **51**, 1244 (1964); (b) Chance, B., Schoener, B., and Alsauer, S.: *J. Biol. Chem.* **240**, 3170 (1965).
100. Degn, H.: *Nature* (London) **217**, No. 5133, 1047 (1968).
101. Furmann, J.: *Oscillating Transport Phenomena at a Gel Liquid Interface* preprint II.21. Internat. Symp. on Macromolecules, IUPAC, Helsinki, 1972.
102. (a) Scholander, P. F.: *Science* **131**, 585 (1960); (b) Hemmingsen, E. and Scholander, P. F.: *Science* **132**, 1379 (1960); (c) Hemmingsen, E.: *Science* **135**, 773 (1962); (d) *Comp. Biochem. Physiol.* **10**, 239 (1963); (3) *Acta Physiol. Scand.* **62**, 246 (1965).
103. Wyman, J.: *J. Biol. Chem.* **241**, 115 (1966).
104. (a) La Force, R. C.: *Trans. Faraday Soc.* **62**, 1458 (1966); (b) Fatt, I. and La Force R. L.: *Science* **133**, 1919 (1961).
105. Hammel, H. T.: *U.S. Govt. Res. Dev. Rept.* **41**, (17), 26 (1966).
106. Ward, W. J.: *Science* **156**, 1481 (1967).
107. Bassett, R. J. and Schultz, J. S.: *Biochem. Biophys. Acta* **211** 194, (1970).
108. Otto, N. C. and Quinn, J. A.: (a) *Chem. Eng. Sci.* **26**, 949 (1971); (b) Quinn, J. A. and Otto, N. C.: *J. Geophys. Res.* **76** (6), 1539 (1971); (c) in: L. B. Wingard Jr. (ed.), *Enz. Eng.*, John Wiley, New York, 1972, p. 190.
109. Smith, K. A., Meldon, J., and Colton, C. K.: *A. I.Ch.E.G.* **19**, (No. 1), 102, (1973); Meldon, J.: *Mass Transfer with Reversible Chemical Reaction through Thin Liquid Films*, Thesis M.I.T., 1973.

110. Friedlander, S. K. and Keller, K. M.: *Chem. Eng. Sci.* **20**, 121 (1965).
111. Lanovicz, R. E. and Frazier, G. C. Jr.: *Chem. Eng. Sci.* **23**, 1335 (1968).
112. Blumenthal, R. and Katchalsky, A.: *Biochem. Biophys. Acta* **173**, 357 (1969).
113. Broun, G., Sélégny, E., Tran-Minh-Canh, and Thomas, D.: *F.E.B.S. Letters* **7**, 223 (1970).
114. Sélégny, E.: *Macromolecular Model Systems for Selectivity Transport and Work*, lecture L 11 presented at the Second Discussion Conference on Macromolecules (IUPAC, IUPAB), Prague, Aug. 1972.
115. Sélégny, E.: *Generation of Active Transports and Signals by Coupling Diffusion and Enzymatic Reactions in Permanent or Dissipative Structures*, Proceedings of the Battelle Memorial Meeting on Solid Phase Proteins, Seattle, Wash. Sept. 1971.
116. Thomas, D., Broun, G., Sélégny, E., and Kernevez, J. P.: Paper presented at the *First Meeting on Enzyme Engineering*, Hennicker, New England, Aug. 1971 and published by Thomas, D., Tran M. C., Gellf, G., Domurado, D., Paillot, B., Jacobsen, R., and Broun, G.: *Biotech. Bioeng. Suppl.* **3**, 299 (1972).
117. Kubin, M. and Spacek, P.: Second Discussion Conference on Macromolecules (IUPAC, IUPAB), Prague, 1972, Preprint No D.11/3.
118. *Handbook of Molecular Cytology* (ed. by A. Lima-De-Farina), North-Holland Research Monographs, Frontiers of Biology, vol. 15, Amsterdam, 1969.
119. Nims, L. F.: *Currents in Modern Biology* **1**, 353 (1968).
120. Pasynski, A. G., Moisseyeva, L. N., and Zvyagilskaya, R. A.: *Biokhimiya* **29**, 2 (1964).
121. Kepes, A.: *Physiol. Veg.* **9**, (1), 11 (1971).
122. Teorell, T.: (a) *Trans. Faraday Soc.* 940 (1937); (b) *Progress in Biophysics* **3**, 357 (1957).
123. Oplatka, A.: *Macromolecular Proteins in Three Dimensional Matrices*, lecture L 12, presented at the Second Discussion Conference on Macromolecules (IUPAC, IUPAB), Prague, Aug. 1972.
124. Metayer, M.: (a) *Etude Générale d'une membrane hydrophile, échangeuse d'anions, à ionisation variable*, Thèse d'Etat, No. CNRS A. O. 3339, Rouen, 1969; (b) Metayer, M. and Sélégny, E.: *J. Macromol. Sci. Chem.* **A(5)**, 633 (1971).
125. Sélégny, E. and Metayer, M.: *J. Macromol. Sci. Chem.* **A(5)**, 611 (1971).
126. Meffroy-Biget, A. M.: *Compt. Rend. Acad. Sci. Paris*, Série C, **266**, 1329 (1968).
127. (a) Wingard, Jr., L. B. and Liu, C. C.: *Proc. 8th Intl. Conf. Med. Biol. Eng.*, July 1969, Chicago, 26–10 (1969); (b) Nagda, N. L., Liu, C. C., and Wingard, Jr. L. B.: *Adv. Chem. Sci.* **109**, 655 (1972).
128. Young, T., Gray, Hadjipetrou, L. and Lilly M. D.: *Biotechnol. Bioeng.* **8** 581 (1966).
129. Bockris, J. O'm. and Srinivasan, S.: *Fuel Cells: Their Electrochemistry*, McGraw-Hill, New York, 1970.

PHOTOREGULATION OF POLYMER CONFORMATION

A Model Study for Biology

G. VAN DER VEEN* and W. PRINS
Dept. of Chemistry, Syracuse University, Syracuse N.Y. 13210, U.S.A.

1. General Introduction

The flow of energy from the Sun out into the universe, absorbed by day and radiated out at night, has brought about on Earth the molecular ordering from which life originated [1]. The same flow has since supplied the necessary energy for the continuation of life. The way in which living organisms utilize solar energy can be roughly divided in two classes. It is either trapped and subsequently stored in chemical bonds, to be used as source of energy, or it triggers a physiological response for information purposes. It is this second class of phenomena that provided the inspiration for the investigations reported in this article. To place this research in its proper context we will briefly discuss three systems of the type we are interested in that have been reported in the literature. These are the process of vision, the photoregulation of enzyme inhibition and the photoregulation of the viscosity of a polymer solution. Of these three the first one is purely biological, the last one completely artificial, while the second system is more or less in between.

1.1. The molecular basis of visual excitation

Research into the molecular events taking place in the process of vision, is closely connected with the name of George Wald. The lecture he gave when he received the Nobel price is a beautiful review of the subject [2]. (See Heller [3] for a recent review.)

The retinal rod- and cone-shaped cells in the eye account for our ability to see. Although the rods just record the incidence of light into our eye, while the cones function in the vision of colors, both types of cells contain essentially the same type of chromophore [2]. In the outer segments of the rods, numerous self-contained membrane sacks with the shape of a disc are stacked on top of each other. The planes of the discs are perpendicular to the long axis of the rod. Light enters the rod perpendicular to the discs also. The disc membranes are lipid bilayers of the Danielli-Davson type [4]. Rhodopsin molecules, the protein-chromophore combination that triggers visual excitation, are membrane proteins. They are amphiphilic in nature, since about half of the molecule is submerged in the lipid phase, while the other half sticks out into the aqueous surrounding. This arrangement may serve to keep the molecules oriented with respect to the incident light. The interior of the membrane sacks differs in ionic composition from the exterior. A sodium-potassium ionic gradient is maintained, which is upset after a permeability change of the membrane as a result of the action of

* The manuscript was prepared at the State University, Groningen, The Netherlands. We thank Dr H. J. C. Berendsen, Professor of Chemistry at this University for his many helpful suggestions.

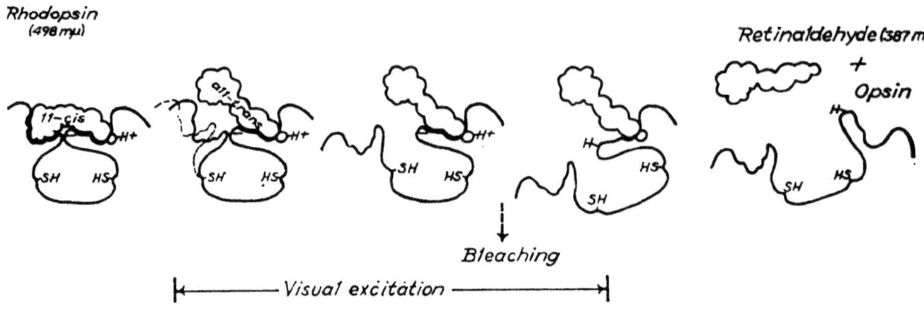

Fig. 1. Initial steps in the process of vision on a molecular level. The only action of light is to isomerize retinal from the 11-cis to the all-trans configuration (after Wald [2]).

light, leading in a manner not yet known to the initiation of a nervous impulse in the retina which is then carried to the brain. The first steps in the excitation are shown in Figure 1, copied from Wald [2]. The 11-cis retinal, also shown in Figure 2, is intimately bound to the opsin. When light is absorbed, the chromophore isomerizes to the all-trans configuration. In the following steps color bleaching is observed, and the chromophore is known to detach from the protein. This results in a conformational transition of the macromolecule, leading to excitation. A more detailed mechanism

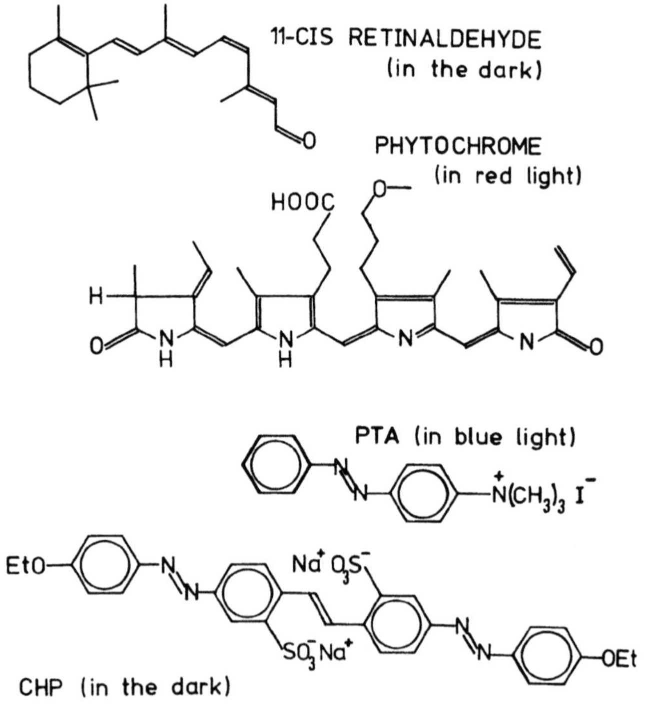

Fig. 2. Schematic representation of the various pigments functioning in photochemical energy transductions.

for the disturbance of the state of polarization of the membrane was recently proposed by Cone [5]. The conformational transition of the opsin might destroy the amphiphilic nature of the protein. The macromolecule is then no longer oriented, and may now be able to undergo rotational diffusion along an axis in the plane of the disc. In this way the molecule could act as a diffusional carrier of ions. In summarizing the molecular events during visual excitation, one could say that the basic feature of the whole process is the action of a chromophore at a water-lipid interface.

1.2. Photoregulation of enzyme inhibition

At Columbia University a group of researchers has been working on the effect of light in enzyme inhibition. To that end they chemically attached azobenzene moieties to well known inhibitors of the enzymes chymotrypsin [6] and acetylcholinesterase [7]. In the case of the latter enzyme the compound they used is called azo-PTA, and is shown as PTA in Figure 2. Bieth *et al.* [7] observe that the cis isomer of this compound is a less effective inhibitor than the trans isomer. Both isomers are stable in the dark and can be interconverted with light of different wavelengths. The difference in activity is demonstrated to be completely reversible. They propose this effect to be a model for such photoresponses in animals, as diurnal changes in biological activity. In a later article from the same laboratory Bartels *et al.* [8] have bound similar azo compounds with quaternary ammonium groups to the acetylcholin-esterase receptor of the electrogenic membrane of the electroplax of the electric eel. The trans form of the compound is a highly potent membrane activator, while the cis form is almost devoid of activity. They note the analogy of this effect with the process of vision. In the case of enzyme inhibition the phenomenon is thought to be due to the mode of binding of the inhibitor [9, 10], while in the case of membrane activation the cis form of the chromophore may not fit at the receptor site [11].

1.3. The photoviscosity effect

Lovrien [12] reported on the influence of light on aqueous solutions of various polymer-dye combinations. A remarkable decrease in viscosity for example was effectuated by light in solutions containing the dis-azo dye Chrysophenine (CHP, Figure 2) and either one of the two polymers poly(vinylpyrrolidone) (PVP) and poly(methacrylic acid) (PMA). A solution containing 0.06 N PMA, 0.0012 N CHP and 0.02 M KCl (pH around 3) showed a lowering of the reduced viscosity of about 50 %.

A very interesting molecular explanation is given by Lovrien. In the dark the dye in its trans configuration, which is quite linear, lines up with the backbone of the macromolecule, as a result of which the polymer coil extends. Upon irradiation to the cis form, the dye tends not to bind so strongly to the polymer. When withdrawing from the polymer backbone, the charges attached to the dye also move away. The polymer will now relax to its unperturbed dimensions.

1.4. The selection of a model system

As was stated by Szent-Györgyi [13]: "Water is the Mater and Matrix of life". This statement should not be interpreted in the sense that everything happens in aqueous solutions. Due to the peculiar properties of water, amphiphilic substances tend to aggregate in aqueous solution. Membrane-like lipid bilayers form spontaneously in water (See e.g. [14]). It is at the interface of such structures that many biological energy transductions take place [15]. This is clearly demonstrated in the molecular mechanism of visual excitation in the vertebrate eye [2]. Our model compounds then should be water soluble and at the same time possess hydrophobic qualities. The simplest artificial polymer combining both features is poly (methacrylic acid). PMA has the additional advantages that it is easy to prepare and has been very extensively studied by many researchers. Once PMA is selected a chromophore is not hard to choose since Lovrien [12] reported interesting interactions between PMA and CHP. A second possible pigment is PTA, used in the group of Erlanger [7]. This dye has a positive charge, contrary to CHP. It has the advantage over CHP that both isomers are stable in the dark.

In Figure 2 the different pigments are listed that act in photoinduced energy transductions. Also included is phytochrome, the photomorphogenetic and phototropic pigment in plants [16]. This dye may act in a way very similar to retinaldehyde in the vertebrate eye [17]. From the different structures shown in Figure 2 one might conclude that the structure of the pigment is not very critical for light-induced energy transductions. The only requirement should be that two states of the chromophore exist that differ in energy content, and that can be interconverted with light. Both states should be relatively stable.

We felt that our investigations should not be confined to dilute aqueous solutions. Such solutions are rare in biological systems. Many processes take place in or on membranes, or in the gel-like cytoplasm. For that reason an extension to gels, preferably in the form of membranes, is very desirable. These gels should exhibit the necessary amphiphilic behavior in water required for a satisfactory model compound. Gels of poly(hydroxyethylmethacrylate) are expected to be excellent candidates, as is discussed in the following section.

A final remark should be made. The model we chose has an attractive aspect, not related to its biological significance. From a physicochemical point of view polymer-ligand interactions are very interesting in themselves. Our model offers the possibility of studying these interactions in a very elegant fashion. One can completely seal the polymer-dye complex in a thermostated glass enclosure. By switching on the light, a well defined property of one component in the system is altered in a way that is known exactly. The effect of many possibly interfering substances and variables is excluded in this way.

2. Materials and Methods

In this section we will briefly discuss some information that will be helpful in the interpretation of the results in the following sections. A more detailed discussion as well as technical information is given by Warren and Prins [18], Galley *et al.* [10], van der Veen and Prins [19, 20], van der Veen *et al.* [21], and van der Veen [22]. All experiments are done at 25 °C, unless stated differently.

2.1. Azo dyes

The two possible forms of azobenzene were first discovered and separated by Hartley [23]. He measured a number of physical characteristics of the two forms, which have been confirmed by later workers. The cis form was found to have a dipole moment of 3.0 D in benzene, as opposed to the trans form that does not have a dipole moment. The lone pairs of electrons of both nitrogen atoms are located on the same side of the azo group, thus making the electron distribution asymmetric. Cis-azobenzene is much more soluble in water and less soluble in petroleum ether, which is a result of the dipole moment. Its melting point is higher by 3 °C. The difference in heat of combustion and heat of melting is about 10 kcal $mole^{-1}$. Contrary to the trans form, which is planar and therefore more stable, the phenyl rings in the cis form are twisted relative to each other. The activation energy of the thermal cis-trans conversion was found to be 23 kcal $mole^{-1}$, the half life period in water at 25°C 600 h.

For our work the change in dipole moment is a very important feature, since it may change the dielectric constant of the medium in which the dye is located. The better water compatibility may play a role in systems in which a balance of hydrophobic and hydrophilic forces exist. Finally it has to be noted that the over-all dimensions of the dye change upon conversion to the cis form. This too may cause changes in binding properties to macromolecules.

2.2. Polymer-ligand interaction

The binding of dye ligands onto macromolecules has been studied extensively because of its relevance to biology. In all enzyme reactions such interactions take place. The subject has been reviewed recently by Steinhardt and Reynolds [24]. Their review includes a treatment of multiple equilibria. The application of this theory to hydrogen ion titrations is also given by Edsall and Wyman [25] and Tanford [26]. The case of dye binding has been treated by Klotz [27].

In dye binding, a change in macromolecular conformation may result from the following factors: (Steinhardt and Reynolds [24]).

(a) electrostatic repulsion between charges on bound ionic ligands;

(b) penetration of a hydrophobic tail into apolar regions of the macromolecule, thus replacing stabilizing segment-segment interactions by segment-ligand interactions;

(c) a ratio of the number of binding sites and their association constants in one conformation to those in a different conformation that favor a transition.

Some of these effects are expected to play a role in systems where light is anticipated to change polymer-dye interactions.

2.3. Poly(methacrylic acid)

Poly(methacrylic acid) (PMA) is an old model compound. Together with poly (acrylic acid) (PAA) it has been studied by many investigators. The interest in both polymers stems from the fact that they are quite simple polyelectrolytes compared with proteins, but still have the polyelectrolyte behavior in common. Thus they served as models for such processes as ion binding and polymer-ligand interactions. A review of the solution properties of both polymers extended to polyelectrolyte solutions in general, has been given by Crescenzi [28].

Contrary to PAA, PMA undergoes a conformational transition upon ionization. In the potentiometric titration curve this transition is observed as an irregularity around α=0.2 (See Figure 3). About the nature of this transition much work has been done by Mandel *et al.* [40] and by Liquori *et al.* [29]. It is generally found that neither the molecular weight nor the tacticity of the polymer is of major influence on the transition.

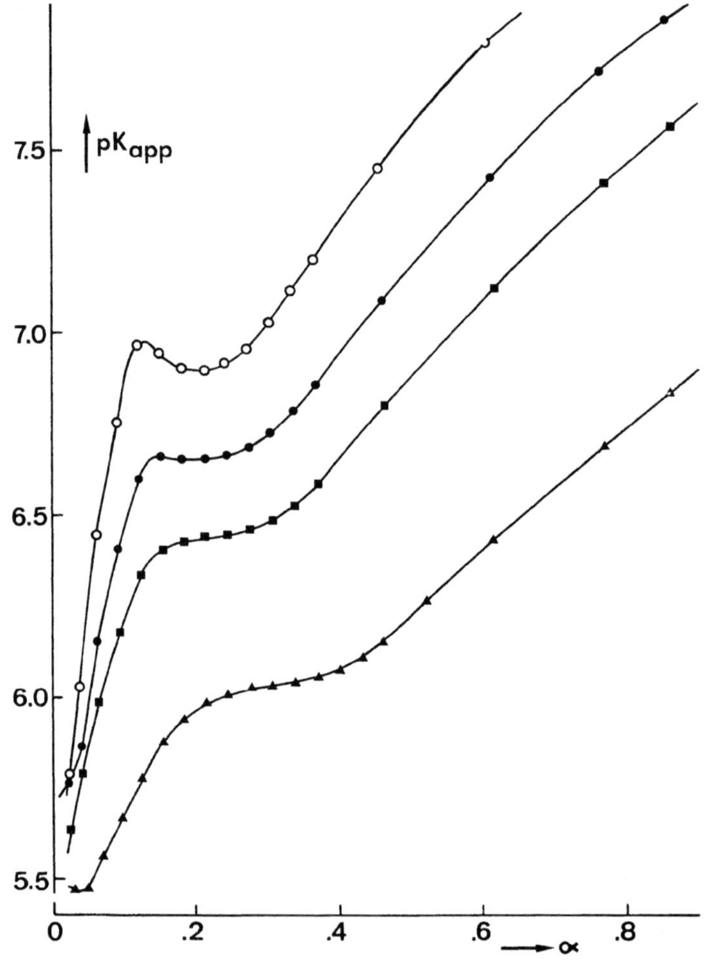

Fig. 3. Titration curves of 0.0033 N PMA in the presence of increasing amounts of KCl; open circles: no salt added, filled circles: 0.000 5 M; squares: 0.00 5 M, triangles: 0.05 M.

The conformational transition can be described briefly as follows. At low ionization the polymer molecule assumes a tight conformation, called the *a*-form by Leyte and Mandel [30], in which the methyl side groups are buried in the interior of the globule as much as possible, while the carboxylgroups tend to be on the surface of the globule, in contact with water. At a degree of ionization around 0.2 the electrostatic repulsion between charged groups on the polymer will overcome the attractive forces that stabilize the compact form, and the molecule will assume a more extended conformation, called *b*-form by the same authors. This change is a true conformational transition, i.e. it shows cooperativity. About the forces that stabilize the compact form no uniformity of opinion exists, although it is generally assumed that hydrophobic forces are of great importance. A discussion about this controversy was recently given by Crescenzi *et al.* [31].

2.4. Poly(hydroxyethylmethacrylate) gels

Poly(2-hydroxyethylmethacrylate) (PHEMA) gels have become of practical interest because of their use as biomedical materials (Wichterle and Lim [39]). They are obtained by copolymerization of 2-hydroxyethylmethacrylate and ethylene glycol dimethacrylate. Upon equilibrating such gels in water, their water content will always be between 40% and 50%, regardless of the crosslink density [32]. Swelling beyond about 45% takes place in mixed solvents of water with ethanol, acetone or ethylene glycol [33, 18]. In these mixtures the swelling exhibits a *S*-shaped curve, indicative of a cooperative transition. Such a transition was strongly supported by results from small-angle light scattering performed by Gouda *et al.* [34]. These authors obtain from their experiments strong evidence for the existence of somewhat ordered microregions in PHEMA gels, resulting from the partially hydrophobic qualities of the polymer. Refojo [33] holds hydrophobic forces responsible for the stabilization of the hydrogel, a view which is supported by Prins and coworkers.

2.5. Potentiometric titration of polyacids

Between the first paper about potentiometric titration by Kern [35] and the most recent one by Meites [36], many papers on this subject have been published. Generally it was attempted in these studies to find a theoretical model which would result in an equation for the titration curve, that could describe the experimental data. Meites [36] who proposed a new approach to the problem, also presented a critical discussion of earlier treatments. A very useful equation for the use in our work was derived by Overbeek [37]. An elegant review of the application of this equation to the titration of polyelectrolytes has recently been given by Nagasawa [38]. For the following reference is made to the last paper.

The titration of a polymeric acid can be represented by the equation:

$$pK = \mathrm{pH} + \log \frac{1-\alpha}{\alpha} - 0.434 \frac{\Delta G_{\mathrm{el}}}{RT}.$$

In this equation the pK is the negative logarithm of the dissociation constant and α

the degree of ionization, varying between 0 and 1. The third term on the right hand side contains the electrostatic free enthalpy. This term represents the extra work that has to be performed to ionize a polyacid. It is additional to the normal work of ionization because the ionizable groups on a polymer are linked together. The electrical free enthalpy can be equated to the product of the unit charge, H^+, and the electrostatic potential at the place of the unionized group (N being Avogadro's number):

$$\Delta G_{el} = -Ne\psi .$$

The electrostatic interaction between the negatively charged groups on a polyacid is of course influenced by the dielectric constant of the medium. The higher the dielectric constant is, the weaker the interaction. The mutual repulsion will also diminish when the groups are shielded by counter-ions. For our purposes it is not useful, if possible at all, to calculate the electrostatic free enthalpy. We will therefore use the titration equation without that term. In so doing the pK will be replaced by an apparent pK, to be called pK_{app}. It is easily seen that this pK_{app} will decrease upon shielding of the ionized groups by counter-ions, as well as upon an increase of the local dielectric constant.

2.6. Dye Irradiation

As can be seen in Figure 4 no wavelength can be selected in the spectrum of CHP where one of the isomers has a low and the other a high absorbance. However, by using light of 400 nm the dye can be converted to the cis form for about 80%, while the trans isomer is formed back in the dark with a half life time of 5 h at room

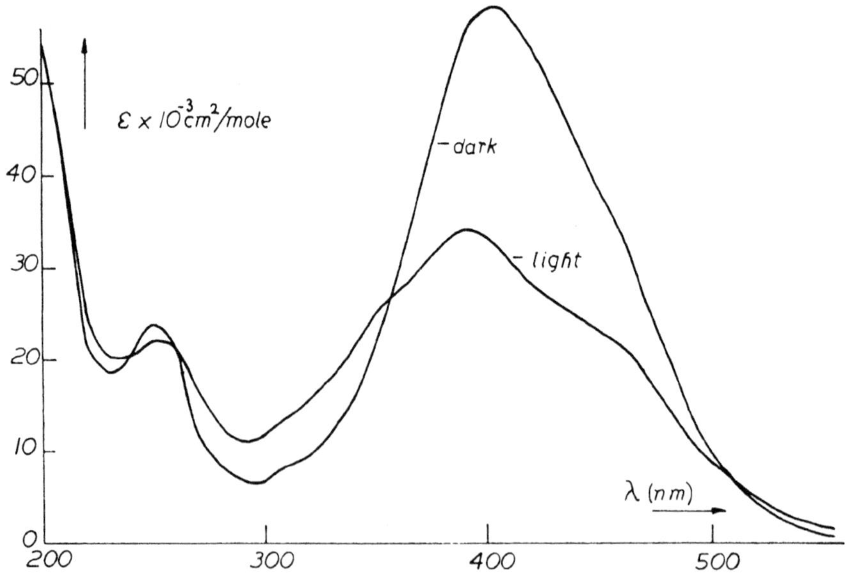

Fig. 4. Spectra of chrysophenine in the dark-adapted and in the photo-stationary state (van der Veen and Prins [20]).

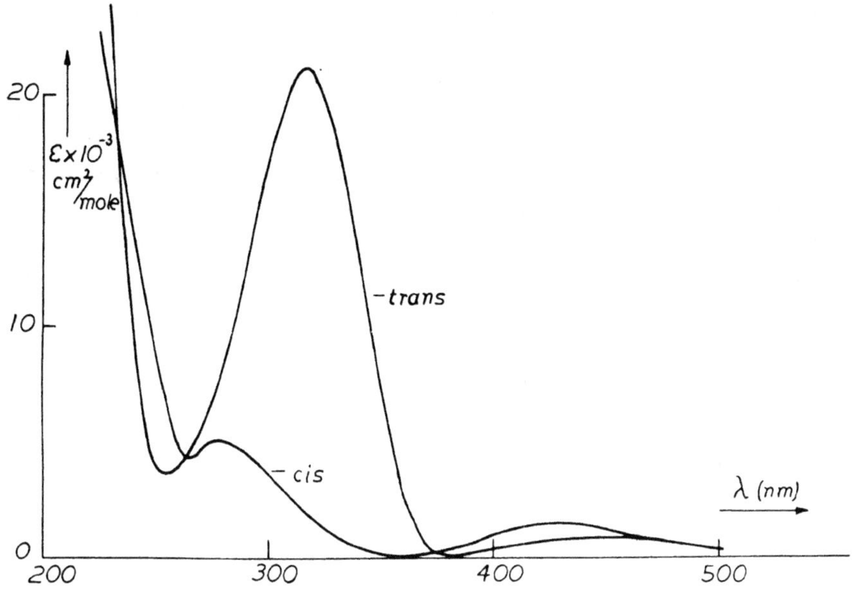

Fig. 5. The spectra of the pure trans form and the pure cis form of the quaternary ammonium azo dye PTA (Chuang [41]).

temperature. The stilbene group does not take part in the photochromism of CHP.

In Figure 5 the spectra of the pure cis and the pure trans isomer of PTA (p-phenylazophenyltrimethylammonium iodide) are shown. Here light of 420 nm can be used to form the trans isomer, while light of 320 nm produces the cis form. In both cases the photostationary state contains about 80% of the desired isomer. Both forms of PTA are stable in the dark.

3. Polymer and Anionic Ligand

In this section we will discuss the influence of light on poly (methacrylic acid) (PMA) in solution, mediated by the negatively charged dis-azo dye chrysophenine (CHP).

3.1. PMA-CHP INTERACTION

For CHP to mediate an effect of light on the conformation of PMA, at least one of its possible isomers should strongly interact with the polymer. This was found to be so, as can be concluded for example from Figure 6. This figure shows the reduced viscosity $\eta_{sp}/c = (\eta - \eta_0)/(\eta_0 c)$ where η and η_0 are the viscosities of the solution and solvent respectively and where c is the normality of PMA. The experiments were performed with PMA at two different concentrations, at pH values around 3, and CHP in its all trans conformation. The viscosity rise at both PMA concentrations takes place in the same concentration range of CHP. The plateau for the highest PMA concentration is at higher viscosity and CHP concentration. The sigmoidal shape of the curves indicates cooperativity of some kind.

Saturated CHP in water at room temperature has a concentration slightly higher than 0.006 M. Figure 6 shows that in the presence of PMA much higher concentrations can be reached. This indicates solubilization of CHP by PMA. Equilibrium dialysis experiments revealed a number of dye molecules bound to PMA upon reaching the plateau of 1 per 10 monomer units. Since 1 CHP molecule carries two charges this confers to the polymer an apparent degree of ionization of 0.2. Figure 3 shows that this is about the charge necessary to bring about unfolding of the polymer globule.

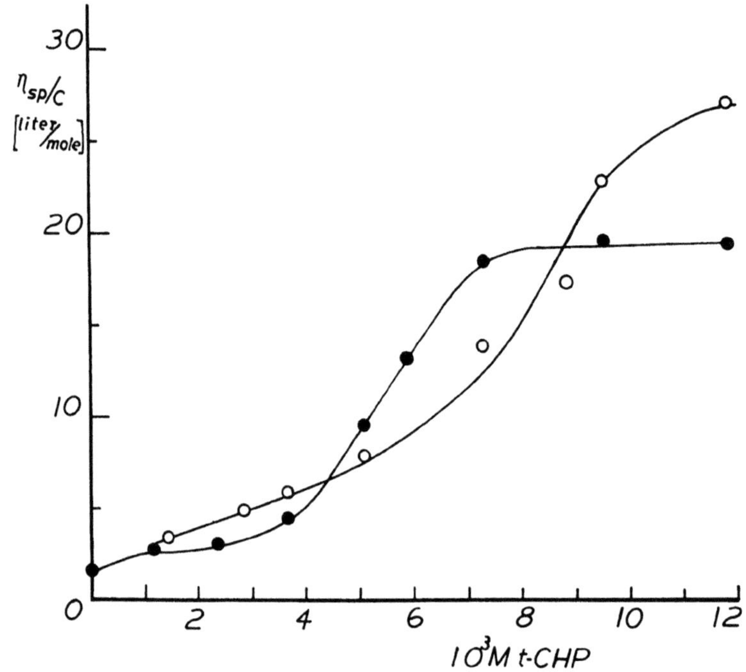

Fig. 6. Reduced viscosity of PMA solutions of two different concentrations, as function of the CHP concentration. Filled circles: 0.06 N PMA, open circles: 0.12 N PMA (van der Veen and Prins [20]).

A PMA-CHP solution in the concentration range where solubilization of CHP occurs, shows an initial decrease in viscosity upon ionization of the polymer [20]. PMA solutions with inorganic salts normally show a continuous increase in viscosity upon ionization, the plot of which is of sigmoidal shape.

The presence of CHP at low concentrations during hydrogen ion titration of PMA has little effect on the shape of the titration curves. A shift to lower values of the apparent pK and to higher values of α is observed. The downshift is more pronounced than with equal normality of potassium chloride. However, around CHP concentrations of 0.001 M an irregularity shows up in the titration curve of the compact form of PMA.

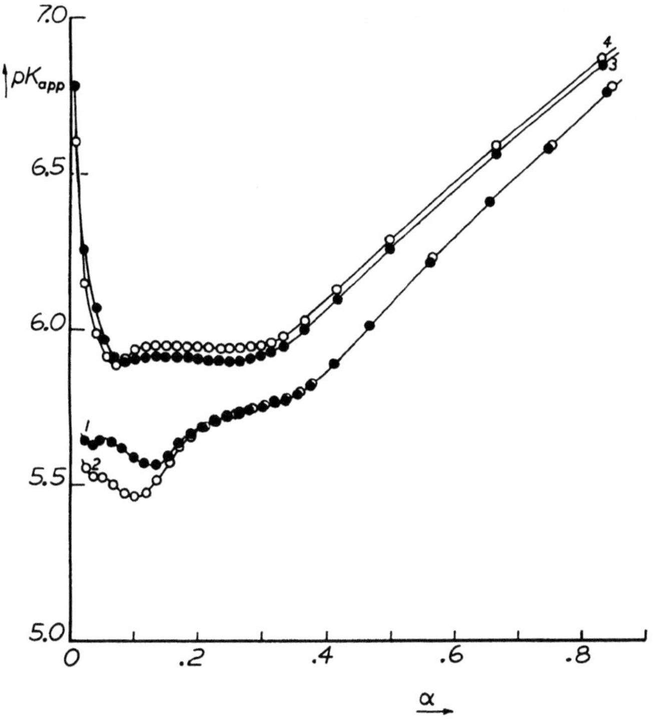

Fig. 7. Titration curves of 0.006 N PMA in the presence of 0.006 M CHP; 1 and 2: undialyzed PMA, 3 and 4: dialyzed PMA; open circles: under irradiation, filled circles: in the dark (van der Veen and Prins [20]. (Fully dialyzed PMA has been used unless specified otherwise.)

It becomes very pronounced when the CHP concentration is further raised, as can be seen in Figure 7. Comparison with Figure 3 shows that at low values of α a higher pK_{app} is measured. In the case of undialyzed PMA, where a low molecular weight fraction of the polymer is present, this pK_{app} increase has the form of a 'shoulder' on the titration curve. Beyond a value of 0.2 the titration curve has a regular shape. Equilibrium dialysis measurements showed that of the strong binding of CHP to PMA at $\alpha=0.01$, little is left when the degree of ionization has increased to $\alpha=0.15$.

3.2. The action of light

Continuous irradiation was applied during titration. Due to the high absorbance of CHP the maximal photostationary state could not always be reached. Best conversion of CHP was obtained in solutions of 0.005 M, where the CHP induced irregularity first showed up. With light the CHP effect at this concentration completely vanishes. At CHP concentrations of 0.006 M still a significant lowering of the pK_{app} at low pH is caused by light, as is demonstrated clearly in Figure 7. This means a transformation of the titration curve, to a more regular shape, as is exhibited in the presence of inert salt. Beyond the CHP-induced irregularity the light and dark curves either coincide, or become parallel to each other with the light curve at higher pK_{app}.

It was also attempted to reproduce the photoviscosity effect reported by Lovrien [12]. To that end a solution was prepared identical to the one used by Lovrien. We observed a drop in viscosity of 5%, which is one tenth of the effect Lovrien observed. In our case the irradiation caused a photoconversion of about 40%. The viscosity reverted to its original value in the dark with a time course typical for the thermal backreaction of CHP.

3.3. Conclusions

From our experiments we cannot completely exclude the possibility of micelle formation of CHP in water. We found however little evidence for such micelles. The viscosity and light absorption of CHP solutions for example, are normal. We conclude therefore that no aggregates are formed in PMA-CHP solution that contain more than one PMA molecule. Our results can be quite well explained by the molecular model proposed by Lovrien [12]. With our results however a more refined mechanism can be formulated. Upon increasing the concentration of chrysophenine in an aqueous solution of PMA, the binding to the macromolecular surface increases. It reaches a critical level around a CHP concentration in solution of 0.001 M. A further increase of the CHP concentration leads to a level of bound dye that forces the polymer globule to open up to some extent. Thus new sites for dye binding become available, etc. Potentiometric titration is a more sensitive technique for recording the start of this process than is viscosity. However either technique demonstrates the cooperative nature of the phenomenon.

Upon binding of CHP to the compact form of PMA, segment-segment interactions are replaced by segment-ligand interactions, resulting in a destabilization of the conformation. Electrostatic repulsion of the charges on the dye is an important force in this mechanism, but probably not the only one. Combination of the results from binding and viscosity experiments indicate the action of other destabilizing interactions. The CHP-induced cooperative effect comes to completion, when a net charge on the macromolecule is reached, where PMA would just start to unfold due to covalently bound charges.

Ionization of PMA leads to the reversal of the process as formulated above. Due to the newly formed negative charges on the macromolecule, the dye is forced out of the polymer domain. Consequently water forces the hydrophobic parts of PMA, that used to be associated with CHP, to fold back on themselves. Hence the process of ionization leads to a decreasing interaction between PMA and CHP. Upon continuing the ionization the polymer unfolds again, due to its own negative charges.

The effect of light on the system is strongest at low degree of ionization of the polymer, which is consistent with the conclusions above. As has been pointed out, light shortens the end to end distance of the dye, and induces a dipole on the azo group. The latter effect results in a higher water solubility, and in a higher local dielectric constant, if the dye remains bound to the macromolecule. However, only the assumption that the dye leaves the polymer domain, as a result of its higher water solubility, does not lead to contradictions with the experimental evidence. This being so, one would expect light to have a similar effect on the system as raising the pH, which also leads to a decreasing interaction between polymer and dye. The drop in viscosity accompanied by a drop in pK_{app} in the beginning of the titration curves is in complete agreement with this interpretation. The decrease of the viscosity is expected for the same reasons as are given for the initial viscosity drop upon ionization. A lowering of the pK is the result of the fact that with the dye, also the negative charges on the dye disappear. These charges opposed the formation of negative charges on the macromolecule as long as they were bound to the polymer. This was expressed in a lower acidity of the complex, which effect is partially undone by light. Since at higher degrees of ionization the interaction between polymer and dye strongly decreases, light should have little effect in that region. In the titration curve this can be seen to be the case. The 'light' and 'dark' curves coincide or are very close. At the lowest CHP concentrations where in the titration curve the irregularity caused by the dye first shows up in the dark, light completely counteracts that effect.

One question still remains to be answered. Although the mechanism proposed by Lovrien seems correct, the photoviscosity effect he observed cannot be explained by that mechanism. The concentration of CHP in solution has to be above 0.001 M to see any effect on PMA at all. A viscosity decrease of 50% would only be possible at much higher CHP concentrations, well above those employed by Lovrien. An almost complete withdrawal of the dye from the polymer domain would then have to be effectuated by light. This is unlikely even at a photo conversion of 80% of the dye, which is about the maximal obtainable. Comparison of our irradiation set-up with Lovrien's suggests that in his case the photoconversion will be lower than the 40% we obtained. The conclusion must be that Lovrien measured a different effect, about the nature of which it is hard to even speculate on the basis of the experimental details given in his paper.

4. Polymer Gel with Azo Ligand

The effect of light on PMA-CHP solutions has interesting extensions to crosslinked waterswollen gels. For example, effects which in solution appear in the viscosity, will in gels result in volume changes; a change in pK_{app} should result in different electrical properties, when a gel is incorporated as a thin membrane in an electrochemical cell. The biological significance of such investigations has been stressed before. In this section we will report on the investigation of a light induced volume change, measured as a change in length. Mainly for practical reasons we used gels of crosslinked poly (hydroxyethylmethacrylate) (PHEMA), rather than of PMA. PHEMA differs from PMA in that the carboxyl groups of the latter are esterified with ethylene glycol. Thus PHEMA is not a polyelectrolyte. In the preceding section it was concluded however, that the attraction between PMA and CHP is based largely on hydrophobic forces. It is strongest in the absence of charges. Therefore PHEMA, in which a delicate balance between hydrophobic and hydrophilic forces is thought to exist, may be expected to react to the binding of CHP also.

4.1. The interaction between PHEMA and CHP

Waterswollen PHEMA gels undergo additional swelling in 0.006 M aqueous CHP. Their water content increases by 20%. At the same time an enrichment of the dye in the gel was observed. The CHP concentration in the gel was found to be 0.009 M. This means that in the gel the number of dye molecules per number of monomer unit is 1 per 400.

Enrichment of dye in the gel, together with twice as large a swelling as in the case of PHEMA, was also observed with PMA gels in CHP solutions of the same concentration.

Refojo and Yasuda [32] reported a similar swelling of PHEMA gels, as a result of immersion in aqueous solutions with even lower concentrations of another organic dye, namely fluorescein. It is quite amazing that relatively little amounts of different organic dyes, not extremely water soluble themselves, have such a pronounced effect on the water content of PHEMA gels. As is discussed in Section 2 PHEMA does not dissolve very well in water either. The reason must be that the dyes, which are amphiphilic in nature, disturb the balance between forces favoring swelling and forces that oppose swelling, which effect is then amplified by the macromolecular matrix.

4.2. The action of light

Details about the preparation of the membranes, as well as about irradiation are reported elsewhere [19]. Thin membranes, $0.1 \times 4 \times 50$ mm in size were swollen in 0.006 M aqueous CHP. After equilibration and inserting markers about 10 mm apart, a gel membrane was suspended in a closed thermostated cell, in air of 100% relative humidity. After equilibration, the membrane was irradiated for 10 min, which caused a length decrease. In the dark the return to the original length was followed with a cathetometer in dim red light. Many cycles were performed in this manner, of which the upper curve of Figure 8 is representative. An identical experiment was performed with a gel not containing dye. These points are shown in the same figure. On some points 90% confidence limits are shown, which are calculated following a procedure in which all points are used. Each point in the figure is an average of two measurements. The differences between each two measurements are used as the set of numbers necessary for the calculation of the confidence limits.

The points for the gel without dye show a greater scatter than those for the gel with CHP, which is possibly related to temperature fluctuations. The dyed gels should be less sensitive to such changes because the presence of the dye limits swelling fluctuations more effectively than the polymer alone.

The lower curve in Figure 8 shows the dark reaction of a CHP solution, as followed in the spectrometer. The time course of the dark recovery is very similar for both curves in Figure 8.

The effect of a temperature increase on the length of PHEMA gels at room tem-

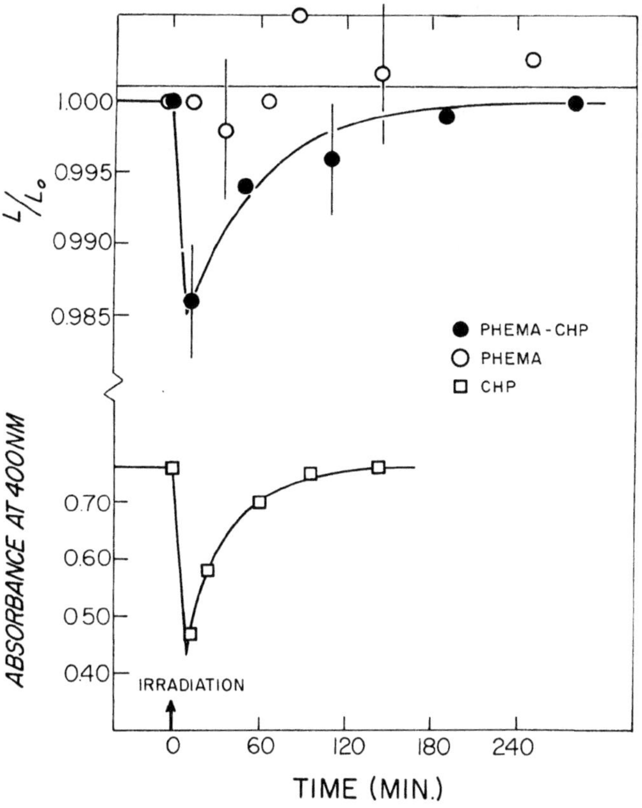

Fig. 8. The effects of 10 min irradiation and dark recovery at 23 °C. Upper curve: relative length changes of PHEMA gels; filled circles: PHEMA-CHP, open circles: PHEMA. Lower curve: changes in absorbance of 0.012 millimolar CHP in water in a 1 cm cell (van der Veen and Prins [19]).

perature would also be a length decrease. Around 50 °C however, PHEMA gels do not change very much with temperature. For that reason the same irradiation experiments have been repeated at 45 °C. As one can see in Figure 9 light has the same effect on the gel length. The same conclusion as above also holds for the dark recovery of the two curves, with of course a different time scale.

A second check was made on the effect of temperature with a copper-constantan thermocouple placed against the back side of the membrane, at the place of irradiation. The 10 min irradiation time gave rise to a temperature increase of 0.1 °C. Independent measurements of the equilibrium swelling of the membranes as a function of the temperature, showed that such a change would cause a 0.2 % shortening at 23 °C, and only 0.01 % at 45 °C. The Figures 8 and 9 show that the contraction due to irradiation is much larger, and also that it is the same at both temperatures.

It is necessary to have an idea about the trans-cis conversion of the dye in the gel. This was obtained in the following manner. Three spectrophotometer cells of one cm

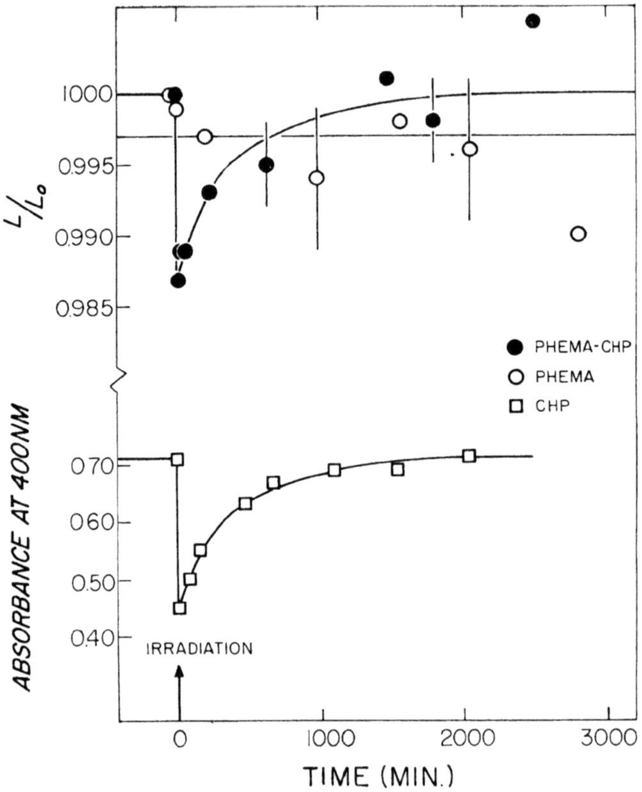

Fig. 9. As Figure 8, except at 45°C and with a different time scale (van der Veen and Prins [19]).

pathlength each, were placed in series in the path of the light, on the same place as the gel, in the same irradiation set-up as used for the gel irradiation. Each cell contained a 0.000015 M aqueous solution of CHP. Upon irradiation with light of 400 nm, the maximal photostationary state in all three cells was reached in less than 5 min. The product of dye concentration and optical path length of the three cells is about half of the same number for the gel membrane. The conclusion seems justified that in an irradiation time of 10 min a sizable photoconversion takes place in the gels.

We conclude from these measurements that the light induced length change we observed, is a true photo-mechanical energy conversion. The average decrease in length of the dyed gels was 1.2% at both temperatures, with a possible error of 0.6% based on the 90% confidence limits. This change in length, corresponding to a 3.6% change in volume, is about one third of the CHP induced swelling of water swollen PHEMA.

4.3. Summary

From the experiments of this section, together with the results from the PMA-CHP investigation in Section 3 a reasonably well-defined model for the photomechanical energy conversion can be postulated. Both PMA and PHEMA swell as a result of binding CHP. In both polymers the dye seems to disturb some pre-existing order. In PMA this is the globular form of the polymer, in PHEMA the micro-mesomorphic organization. Upon irradiation, the interaction between dye and polymer becomes looser. In this way the perturbing effect of the dye on the polymer is partially undone. It is the ability of both polymers to undergo cooperative rearrangements that leads to amplification of the effect of the cis-trans conversion of the dye. From the number of dye molecules per monomer unit of polymer it is quite apparent that the effect is relatively strongest in the PHEMA system. This could very well be explained by the membrane form of PHEMA and the intermolecular order that appears to exist in those membranes. The effects would of course even be enlarged if membranes could be prepared that display a macro anisotropy.

5. Polymer with Cationic Ligand

After investigating the interaction of PMA with a negatively charged dye, it is naturally of interest to undertake similar experiments with a positively charged dye. As has been discussed in Section 1, PTA (p-phenylazophenyltrimethyl ammoniumiodide) is an attractive choice for such an investigation. For details about these experiments one is referred to van der Veen *et al.* [21] and Galley *et al.* [10].

5.1. The interaction between PMA and PTA

PMA does not solubilize PTA. An attempt to dissolve more PTA in a polymer solution than is possible in water, which is 0.001 M, resulted in precipitation of the polymer. Hoguet observed that the viscosity of buffered solutions of PMA drop markedly upon addition of the dye [21].

A piece of crosslinked PMA gel, when immersed in a solution of PTA, shrinked drastically. It showed a decrease in length of 23% and an absorption of the dye from the aqueous surrounding up to a monomer to dye ratio in the gel between 150 and 100. Most of the water is forced out of the gel in this process. DeSorgo observed that the gels become glassy and brittle. A similar observation was made with PHEMA gels [10].

PTA has a marked influence on the titration curve of PMA, as can be seen in Figure 10. The curve through the filled circles represents a titration with trans-PTA. Comparison with Figure 3 shows that a greater pK_{app} change preceeds unfolding, the whole curve is shifted upwards, and the transition region is smoother and not as wide as in the curves of Figure 3. The decrease in width of the transition implies that the free enthalpy difference between the two conformations, or rather the two conformations, or rather the two sets of conformations, has decreased. This quantity is calculated from the surface between the curves of the *a*-form and the extrapolated *b*-form [30]. In our case it dropped from around 130 to 60 as a result of PTA binding in 0.001 M dye solution.

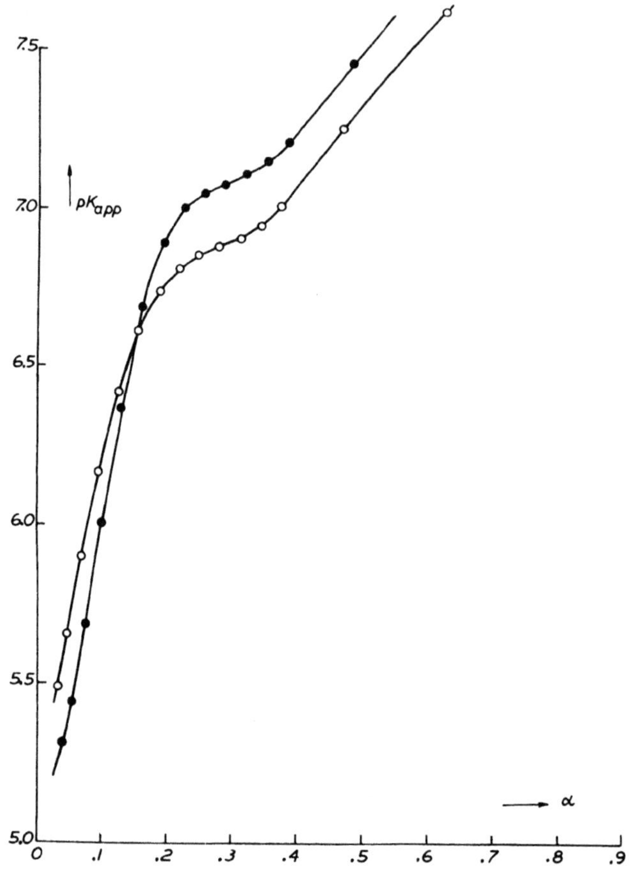

Fig. 10. Titration curves of 0.0033 N PMA in the presence of 0.001 M trans PTA (filled circles) and 0.001 M cis PTA (open circles) (van der Veen, Hoguet and Prins [21]).

5.2. The action of light

In the case of PTA the term 'the action of light' is a little ambiguous, since both conformations can be the result of irradiation. In analogy to CHP, the term should be interpreted as irradiation resulting in the formation of cis PTA.

The conversion of PTA to the cis form has a clearly observable influence on the titration curve of PMA. This influence is the stronger the more PTA is present. For the highest PTA concentration the curve is shown in Figure 10. A downshift in pK_{app} is the most important effect. This means a shift in the direction of a more normal PMA curve, in the presence of neutral salt.

The comparatively large, light-induced pK_{app} drop has interesting possibilities. Even more so than with CHP, light act as a pK_{app} switch. When this switch is used in a buffered system a rise in α will be the result. In the range of the conformational transition the polymer conformation is very sensitive to the degree of ionization. Thus quite large effects may be expected in the viscosity of polymer-dye solutions and the swelling of polymer-dye gels. Both predictions were confirmed by members of the Polymer Research Group at Syracuse University. A viscosity effect was demonstrated by Hoguet [21] and a membrane effect by DeSorgo [41]. A brief discussion will suffice here. A more detailed description is given in the references mentioned.

Conversion of trans PTA to the cis form in solutions buffered with 0.067 M phosphate, containing 0.0017 N PMA and 0.0003 M of the dye, led to a viscosity increase. Since small changes in the pK_{app} are amplified strongest in the range of the PMA conformational transition, one would expect this viscosity effect to be connected in size to that transition. Figure 11 shows both the change in degree of ionization

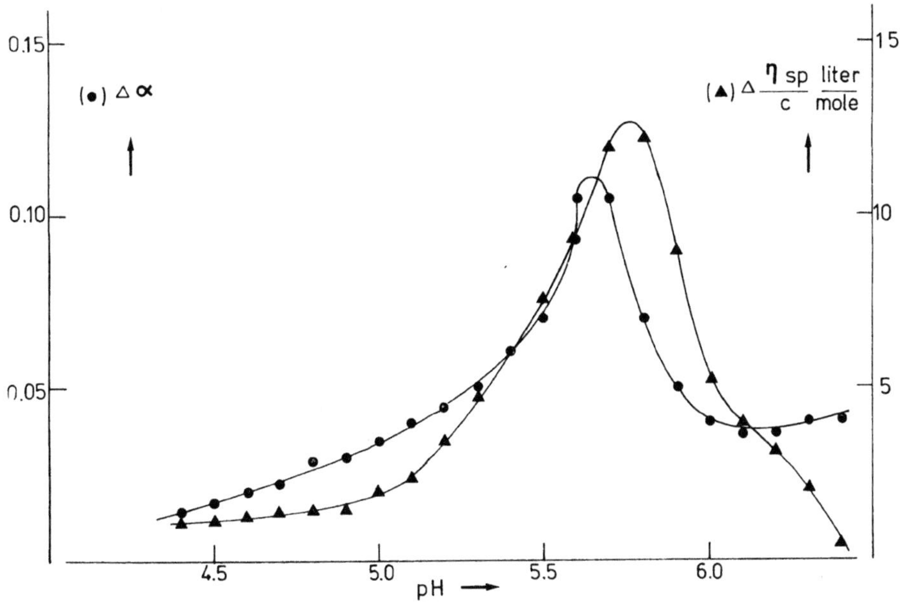

Fig. 11. Photo-induced effects on the degree of ionization and the reduced viscosity as a function of the pH, for solutions with 0.0017 N PMA and 0.0003 M PTA (van der Veen, Hoguet and Prins [21]).

induced by light and the photoviscosity effect as a function of the pH. The difference in degree of ionization was determined from a set of curves like those presented in Figure 10, obtained from titrations in the presence of an equivalent amount of KCl. The viscosity change was obtained by subtracting the reduced viscosity of a buffered solution at a certain pH containing the trans PTA-PMA complex, from the viscosity of an identical solution containing the complex with the cis dye. It is evident that a close relation between the α-change and the viscosity change exist.

A PMA gel, immersed in a buffer, containing 100% trans PTA, shows an elongation of at least 10% upon irradiation with blue light. Only half of this elongation could be reversed under irradiation with blue light. This cycle could be repeated a few times only, because of irreversible side reactions.

5.3. Conclusions

We have not found any indication for supramolecular organization in solutions containing PMA and PTA. The effects found in the PMA-CHP system that could possibly be interpreted that way, have not been found with the PMA-PTA combination. We conclude therefore that each PTA-PMA complex contains a single macromolecule only.

The investigations in this section can be interpreted in the following way. The globular form of PMA is sparingly solvated. When PTA binds on the surface of these coils, water is forced away from the immediate vicinity of the polymer. In solution this can lead to precipitation of the complex, while gels become glassy. Negative charges on the macromolecules will be compensated electrostatically. Initially therefore it will be easier to ionize new carboxyl groups. Due to the replacement of water by the dye, the dielectric constant close to the macromolecular surface has decreased drastically. So electrostatic repulsion of newly ionized groups will be stronger, and will impede further ionization, resulting in a sharp rise in pK_{app} upon ionization.

It is observed that the degree of ionization at which the conformational transition begins, is not influenced very much by the dye. Since the difference in free enthalpy between the *a*- and *b*-form of PMA is lowered significantly, this is surprising at first sight. One would expect that fewer ionized carboxyl groups could bring about the start of the transition. However, upon binding of PTA to PMA in the compact conformation two opposing effects take place. On one hand the globular form is destabilized while at the same time the electrical charges on the polymer are neutralized. Apparently these two effects just cancel each other.

The conversion of trans PTA into the cis form is shown to lower the steepness of the titration curve before the onset of the transition. After this point the cis-PTA-PMA system is more acidic than its trans analogue. Both features can be explained by the induction of a dipole moment on the azogroup. These dipoles, since they will be aligned by the electrostatic field in the polymer domain as far as thermal motion allows, will reduce the strength of this field. Thus the interaction between ionized groups will diminish, resulting in the observed changes in the titration behavior.

6. Evaluation of the Model

The basic principle in the molecular mechanism of biological photoreceptors appears to be the interaction between a photochromic moiety and a macromolecule. The protein-ligand complex as a whole is known to be embedded in a membrane matrix. In a very general way the results of our studies can be summarized by stating that in photochemical energy transductions both features are very important. That is, the interplay of a photochromic ligand with a macromolecule, as well as the fact that this takes place on a water-lipid interface.

The conversion of light into chemical energy, and the subsequent conversion of the latter into mechanical energy, forms the basis of the foregoing machinery. The mechanical effect mentioned has to do with the conformational transition of the macromolecule, and not with the performance of work in these transductions. In photoreceptors, indeed in all sense receptors, we are not dealing with energy storage, but with information transfer. Our main concern therefore is not the efficiency of such machinery. The only requirement is, that the receptors can exist in at least two different states. If the absorption of a single quantum is to lead to perception, the energy difference between these two states should be at least an order of magnitude larger than kT. If that were not the case, thermal energy would cause the receptor to fire too frequently in the absence of a signal, yielding an insufficient signal-to-noise ratio. The study of photochromic macromolecular systems presented in this article was aimed at the elucidation of general molecular mechanisms underlying photoresponses. The investigation was of an exploratory nature, since few studies on model systems of this type have been reported. Three quite different principles were demonstrated to be at the basis of the observed effects.

(1) The reversible photoconversion of a photochromic ligand bound onto a weak polyacid, was shown to lead to a shift in acidity of the polyacid and thus to the regulation of the degree of ionization (Section 5).

(2) The conformation of a macromolecule was reversibly regulated by a photo-induced change of the binding properties of a photochromic ligand (Section 3).

(3) The reversible photoconversion of a photochromic ligand with extensive hydrophobicity led to a shift in the organization of an assembly of amphiphilic macromolecules (Section 4).

In all three cases it is the ability of the macromolecule to undergo a conformational change that led to amplification of three different physico-chemical effects. In case (1) the shielding of electrostatic charges by dipoles was amplified by the sensitivity to changes in the electrostatic potential in the polymer domain of the poly acid in its transition region. In case (2) the decrease in binding of the ligand allowed the polymer to relax to its unperturbed dimensions. The conformational free enthalpy is the driving force in this relaxation. In case (3) the balance of hydrophilic and hydrophobic forces was upset by a change in water compatibility of the photochromic ligand.

For the PHEMA-CHP system it is tempting to calculate the efficiency of the photomechanical cycle. The molarity of the dye in the gel is 0.009. The volume of the gel between the markers is about 4 μl. In the photostationary state 85% of the dye is in the cis form, storing an amount of energy of 10 kcal mole^{-1}. Thus the available energy was 1500 μJ. A weight of 1.5 mg, was lifted 0.1 mm. Therefore the work performed was equal to 0.015 μJ. Hence, the efficiency was 0.01 ‰. It is not surprising that nature does not use this machinery to store energy. In information transfer this part of the energy transduction is not of interest. There the efficiency of an earlier step is of crucial importance, namely that of the transfer of energy from the ligand to the macromolecule. We do have some information about that efficiency also. In the PMA-CHP system 1 dis-azo dye was bound to 10 monomer units. With a conversion of 50% of the azo groups, 1 kcal of energy was available per mole of monomer units, since 10 kcal per mole is stored in each cis isomer. For one monomer to pass from the compact form of PMA to the extended conformation, in the absence of electrical charges, a little over 100 cal per mole monomer units is required. Therefore of the effect we observed with the PMA-CHP complex, the efficiency of energy transfer was 10%.

In biological systems as a result of evolution, higher efficiencies can easily be reached. Moreover, a way in which living systems circumvent the requirement for energy levels functioning in 'switching', viz. to differ by an order of magnitude larger than kT, is by demanding more than one simultaneous trigger for the creation of a nerve impuls. We may therefore conclude that our model fulfills the basic requirements for serving in information transfer.

The sequence of steps in energy transfer is pictured in Figure 12. The first step

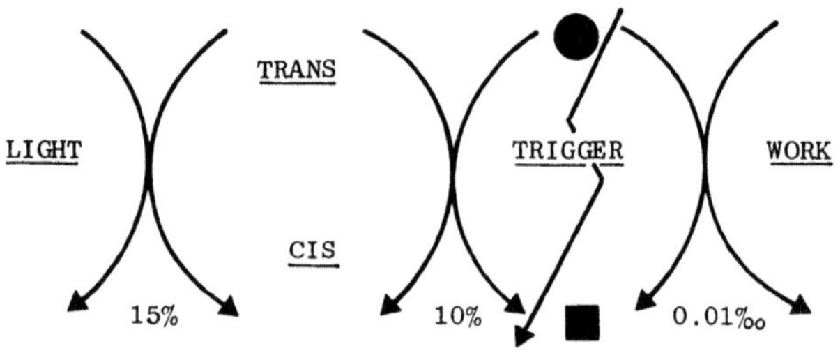

Fig. 12. The different steps of energy transfer in photo-transduction, with the efficiencies calculated from our models.

represents the loss of 60 kcal out of the 70 kcal per einstein transmitted by blue light. Only 10 kcal is stored in the ligand.

The molecular mechanism underlying the effects demonstrated in this article have a wider field of action than in just those effects. An application to for example the photo regulation of membrane permeability is under investigation.

References

1. Morowitz, H. J.: *Energy Flow in Biology*, New York, 1968.
2. Wald, G.: *Nature* **219**, 800 (1968).
3. Heller, J.: in P. D. Boyer (ed.), *The Enzymes* Vol. 6, New York, 1972, p. 573.
4. Green, D. E.: *Science* **174**, 863 (1971).
5. Cone, R. A.: *Nature, New Biology* **236**, 39 (1972).
6. Kaufman, H., Vratsanos, S. M., and Erlanger, B. F.: *Science* **162**, 1487 (1968).
7. Bieth, J., Vratsanos, S. M., Wasserman, N., and Erlanger, B. F.: *Proc. Natl. Acad. Sci. U.S.A.* **64**, 1103 (1969).
8. Bartels, E., Wasserman, N. H., and Erlanger, B. F.: *Proc. Nat. Acad. Sci. U.S.A.* **68**, 1820 (1971).

9. Fournier, M. and Bourdon, J.: *Photochem. Photobiol.* **18**, 103 (1973).
10. Galley, K. T., DeSorgo, M., and Prins, W.: *Biochem. Biophys. Res. Comp.* **50**, 300 (1973).
11. Erlanger, B. F.: Biophysical Society Meeting, Symposium on Photobiology, Toronto, 1972.
12. Lovrien, R.: *Proc. Natl. Acad. Sci. U.S.A.* **57**, 236 (1967).
13. Szent-Györgyi, A.: *Faraday Soc. Disc.* **27**, 111 (1959).
14. Booy, H. L.: in H. R. Kruyt (ed.) *Colloid Science*, Vol. 2, Elsevier, Amsterdam, 1949, p. 701ff.
15. Green, D. E. and Young, J. H.: *Amer. Sci.* **59**, 92 (1971).
16. Kroes, H. H.: *Physiol. Veg.* **8**, 533 (1970).
17. Linschitz, H., Kasche, V., Butler, W. L., and Siegelman, H. W.: *J. Biol. Chem.* **241**, 3395 (1966).
18. Warren, T. C. and Prins, W.: *Macromolecules* **5**, 506 (1972).
19. van der Veen, G. and Prins, W.: *Nature Phys. Sci.* **230**, 70 (1971).
20. van der Veen, G. and Prins, W.: *Photochem. Photobiol.* **19**, 191 (1974).
21. van der Veen, G., Hoguet, R., and Prins, W.: *Photochem. Photobiol.* **19**, 197 (1974).
22. van der Veen, G.: Ph.D. Thesis, University at Groningen, The Netherlands, 1972.
23. Hartley, G. S.: *J. Chem. Soc.* Pt. 1, 633 (1938).
24. Steinhardt, J. and Reynolds, J. A.: *Multiple Equilibria in Proteins*, New York, 1969.
25. Edsall, J. T. and Wyman, J.: *Biophysical Chemistry*, Vol. 1, New York.
26. Tanford, C.: *Physical Chemistry of Macromolecules*, New York, 1961.
27. Klotz, I. M.: in *The Proteins*, New York, 1953, Chapter 8.
28. Crescenzi, V.: *Adv. Polymer Sci.* **5**, 538 (1968).
29. Liquori, A. M., Barone, G., Crescenzi, V., Pispisa, B., Quadriforlio, F., and Vitagliano, V.: *J. Macrom. Chem.* **1**, 291 (1966).
30. Leyte, J. C. and Mandel, M.: *J. Polymer Sci., Part A* **2**, 1879 (1964).
31. Crescenzi, V., Quadrifoglio, F., and Delben, F.: *J. Polymer Sci. Part A-2* **10**, 357 (1972).
32. Refojo, M. F. and Yasuda, H.: *J. Appl. Polymer Sci.* **9**, 2425 (1965).
33. Refojo, M. F.: *J. Polymer Sci., Part A-1* **5**, 3103 (1967).
34. Gouda, J. H., Provodator, K., Warren, T. C., and Prins, W.: *J. Polymer Sci., Part B* **8**, 225 (1970).
35. Kern, W.: *Z. Physik. Chem.* **A181**, 249 (1938).
36. Meites, L.: *J. Polymer Sci., Part A-1* **10**, 779 (1972).
37. Overbeek, J. T. G.: *Bull. Soc. Chim. Belg.* **57**, 752 (1948).
38. Nagasawa, M.: *Pure Appl. Chem.* **26**, 519 (1971).
39. Wichterle, O. and Lim, D.: *Nature* **185**, 117 (1960).
40. Mandel, M., Leyte, J. C. and Stadhouder, M. G. (1967) *J. Phys. Chem.* **71**, 603 (1967).
41. Chuang, J. C., DeSorgo, M., and Prins, W.: *J. Mechanochem. and Cell. Motil.* **2**, 105 (1973).

PHOTOREACTIVE POLYMERS

J. L. R. WILLIAMS

Research Laboratories, Eastman Kodak Company,
Rochester, N.Y. 14650, U.S.A.

1. Introduction

The term 'photoreactive polymer' must be definition limited for our purposes. All polymers respond to sufficiently energetic light. Polymers can dissipate energy by undergoing reactions or by energy transfer to solutes. For our purpose, we are interested in polymers which respond to light in the range of 300 to 700 nm. Furthermore, I am considering only synthetic and specifically excluding natural polymers, though a number of the reactive processes can be similar. We are, therefore, limited to polymers to which photoreactive or photoresponsive groups have been attached chemically or by association.

The majority of photoreactive polymers which have been studied fall into the hydrophobic class of materials. This is for no reason other than that organic soluble polymers have been the polymers of convenience and economic choice. A number of charged hydrophilic polymers have been prepared. In general, they bear the same types of photoresponsive functional groups as do the hydrophobic polymers. Therefore, a discussion of the photoreactive groups in the context of hydrophobic systems can yield useful information relating to the corresponding hydrophobic systems.

I consider it also important to bring into our discussion the subject of photopolymerization, or as it is more correctly described, photoinitiation. Photoinitiation not only lends itself to the preparation of photopolymers from monomers but makes possible photografting of monomers to existing polymers. Thus, we can readily modify solution or other physical properties by means of photografting.

I attempt, herein, to bring together a number of photochemical phenomena and techniques by which the physical and chemical behavior of polymers may be modified. In doing so, I intend to include photochemically reactive groups, optical sensitization, synthesis of typical photoreactive polymers and their charged counterparts.

The response of polymers to light will be divided into a number of groupings based on the end result.

(1) Photochemical Change of Functionality

(2) Photodegradation

(3) Photografting – Photopolymerization

(4) Photoisomerization

(5) Photocrosslinking

(6) Cooperative Effects of Excited States: Macromolecules – Minimolecules

(7) Photochemical Behavior of a Quaternary Photoisomerizable Photocrosslinkable Polymer System and Its Model Compounds

Eric Sélégny (ed.), Polyelectrolytes, 507–528.

2. Discussion

2.1. Photochemical Change of Functionality

Modification of the properties of functional groups by light can lead to changes on physical and/or chemical properties of polymers.

(a) The diazoketone group has been shown to undergo photochemical ring contraction with generation of a carboxylic acid function. The function group thus changes from a non-ionic diazoketone into an ionizable carboxylic acid. Delzenne [1], in his review, cited a group of polymers which were prepared by appending a diazo ketone sulfonic acid chloride to hydroxylic polymers.

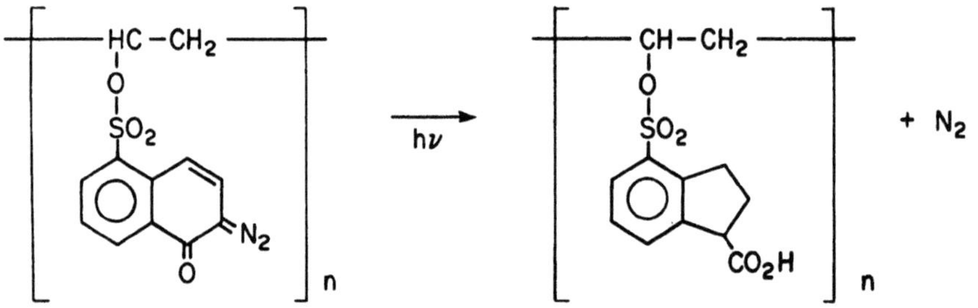

(b) Complex tetraarylborate salts have been shown to decompose with light [2]. The products are generally more soluble in selected solvents than the original salts. When tetraarylborates are complexed by anion exchange with water soluble quaternary polymers, the resulting polymers become hydrophobic. Irradiation of the 'ate' polymer leads to photocrosslinking. This is probably due to interchain interceptions of the reactive species.

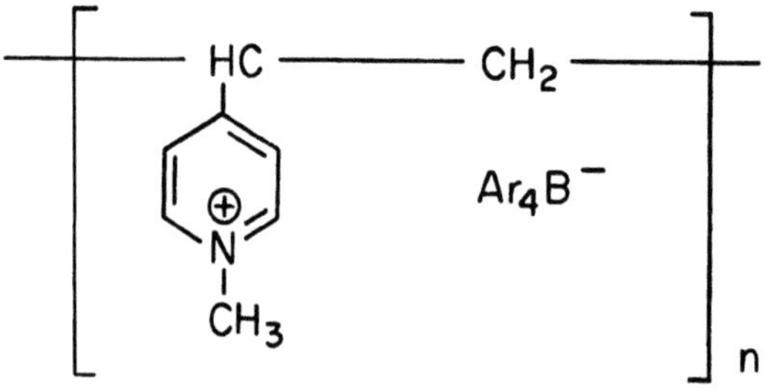

(c) The photoisosolublization of polymeric diazo salts provides a system by which a hydrophilic ionic polymer becomes hydrophobic [3].

(d) The Photochemical Fries Reaction involves migration of an acyl group from a hydroxyl to the aromatic ring of an aryl ester [4, 5, 6].

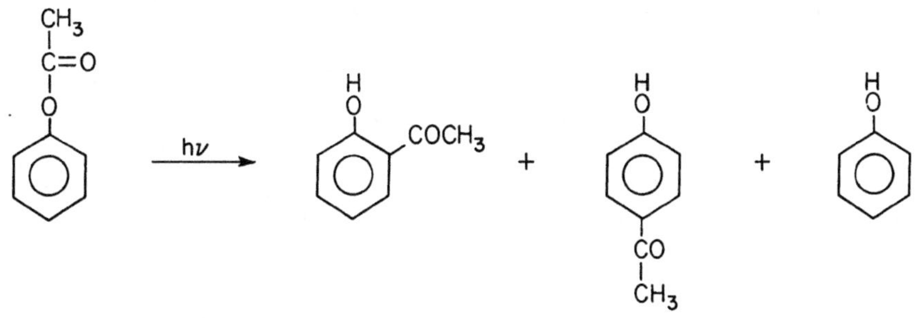

Maerov [7] has reported the rearrangement of the polyester of bisphenol A and isophthalic acid to proceed similarly to the model compound bisphenol A dibenzoate.

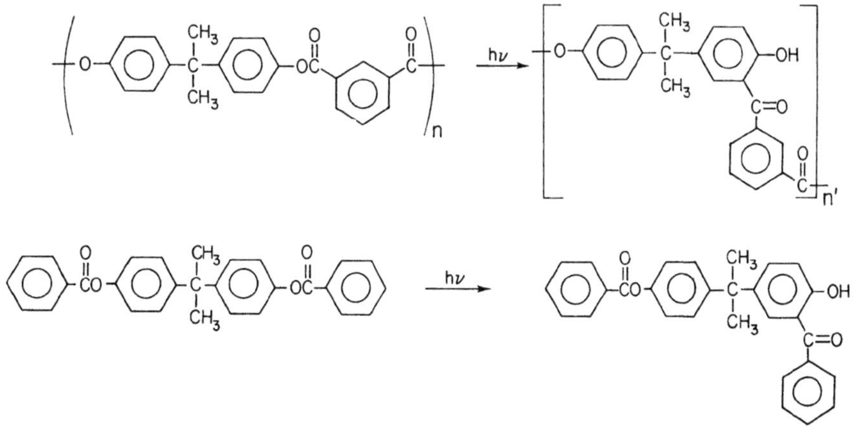

2.2. Photodegradation

A comprehensive survey of photodegradation, and its prevention, has recently appeared [8]. All classes of polymers can be photodegraded. Therefore, one class of polyketones has been chosen for inclusion in this discussion.

David *et al.* [9] have studied the photolysis of poly(phenyl vinyl ketone) in detail. Irradiation in solution or in the solid state reduces the molecular weight and was compared with the photolysis of butyrophenone. The principal mechanism was

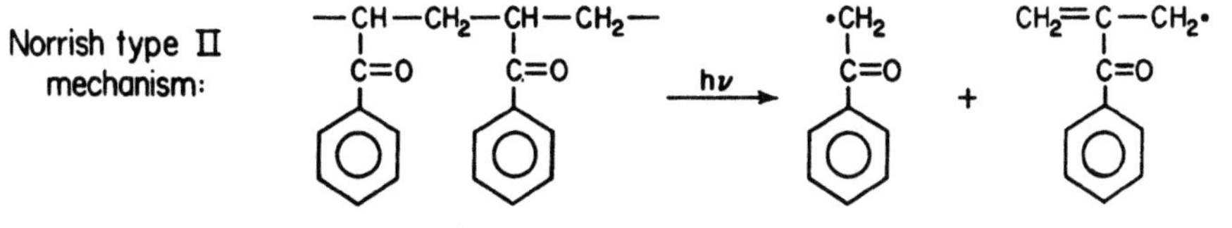

determined to be γ hydrogen abstraction followed by chain scission (Norrish Type II). The Φ of scission in benzene was about 0.3.

In a later paper, the same authors [10] reported that naphthalene inhibited chain scission by a triplet energy transfer mechanism. Golemba and Guillet [11] determined the Φ of chain scission of poly(phenyl vinyl ketone) to be 0.25 of 313 nm. They determined that the excited state lifetime of the carbonyl group on the polymer was of the same order of magnitude as that of the analogous model compound.

2.3. Photografting

The classical example of photografting was carried out by Oster and Shibata [12] who grafted hydrophilic acrylamide units upon natural rubber by means of benzophenone photoinitiation.

Photoinitiation of acrylamide by eosin proceeds through the semiquinone of eosin to yield a growing chain terminated at one end by eosin [13].

Smets *et al.* [13] attached eosin units to poly(vinylamine) through the amino groups. Photopolymerization was initiated at the eosin sites, yielding grafts of poly(vinylamine-eosin-poly(acrylamide) where X is $-CONH_2$.

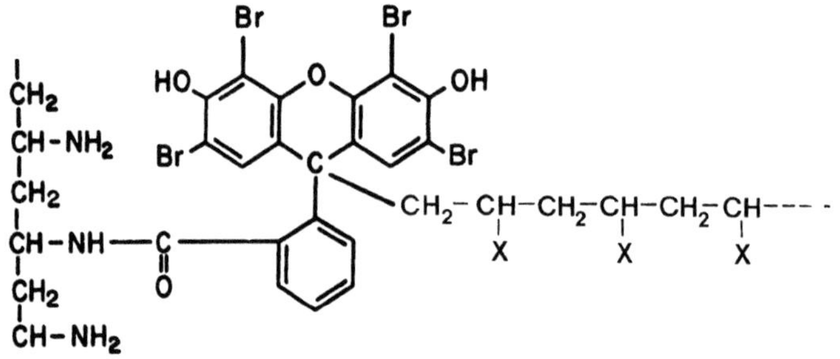

The polymerized acrylamide was completely bound in the graft since no homopolyacrylamide was found. Block polymers have been prepared by linear grafting. Eosin lactone was appended to poly(methyl methacrylate) chains by means of a terminal amino group to provide a polymeric initiator which was then used to photoinitiate polymerization of styrene and acrylamide.

Dhandraj and Guillet [14] reported a new photochemical synthesis of block copolymers by irradiation of poly(tetramethylene sebacate-CO-γ-ketopimelate) in the presence of methyl methacrylate. The chain radicals formed from the cleavage of the γ-ketopimelate unit initiated polymerization of methyl methacrylate leading to blocks the length of which depend on the number of scissions.

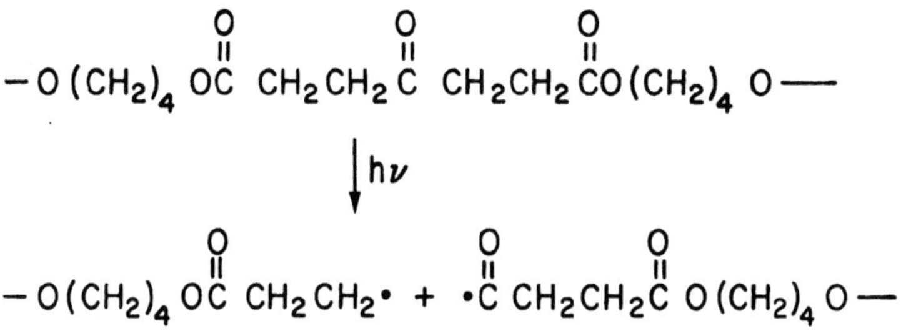

2.4. PHOTOISOMERIZATION

(a) The photoisomerization of a number of quaternary styryl pyridine-type polymers has been studied in our laboratory. The results will be discussed below.

(b) *Loosely Bound and Chemically Bound Groups.* Photochromic macromolecules represent a form of photoisomerizable polymer systems. The photoisomerizable group can be adsorbed or loosely bound with a macromolecule chain in such a way that the dye properties may be modified or in a manner such that the macromolecule then exhibits an effect.

The interaction of a dye with macromolecules which possess a large number of anionic sites in an aqueous solution can result in the formation of new absorption maxima due to 'polymerization' of depolymerization of the dye (or both). The mere occurrence of a large number of anionic groups does not 'a priori' determine the nature of the response [15]. There exist differences between the effects of various peptides and while some cause changes in the spectra of the dyes, others fail to do so. The peptides which show no effect contain bulky aryl groups or have glycine as end groups. On a weight basis, proteins are the most effective as agents which influence the state of aggregation of dyes. The effective sites are the anionic side chains which are disposed along the backbone of the dye. However, the arrangement of these groups is of considerable importance. In the case of nucleotides, the arrangement of the functional groups (guanylic acid, etc.), or the configuration has a greater influence than does the exact nature of the functional groups.

Lovrien and Waddington [16] have prepared polymeric dyes which are, in fact, photochromic macromolecules. These were compared with the analogous free dyes and to bound dyes. The tethered dyes (photoisomerization) photo response and dark relaxation are influenced by the conformation of the polyelectrolyte. Fuoss has shown earlier that the solution viscosity aqueous polyelectrolytes could be altered by the addition of common ions. The same is known for other ionic polymers.

(c) *Photochemical Energy Conversion.* Van der Veen and Prins [17], as described earlier in this conference by Mr Van der Veen, have studied the photochemical energy conversion to work obtained by photoisomerization of a dye adsorbed upon a copolyacrylate membrane.

(d) *Photochromic Spiropyran Polymers.* Chemically bonded or bound spiropyrans attached along a macromolecule chain have been studied by Vanderwijer and Smets [18].

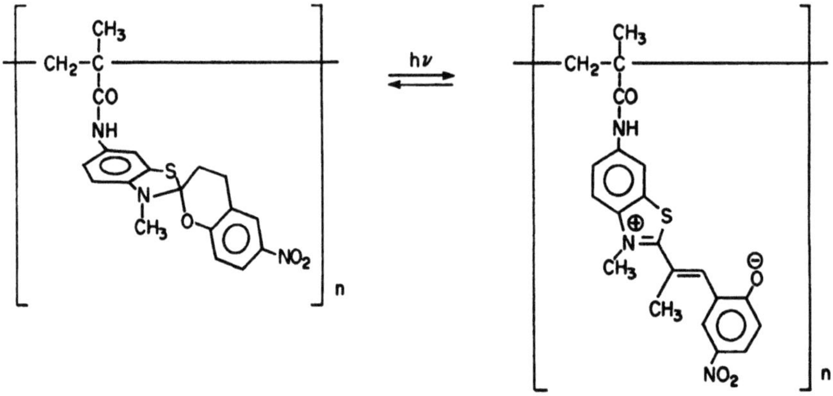

(e) *Chalcone-Type Polymers.* Poly(4′-vinyl-*trans*-benzalacetophenone) was prepared by Unruh [19] by condensation of poly(*p*-vinylacetophenone) with benzaldehyde. In his studies, Unruh compared the photochemical behavior of poly(4′-vinyl-*trans*-benzalacetophenone) with that of the model compound, 4-ethylbenzalacetophenone. Both the polymer and its model showed typical photoisomerization behavior when irradiated in dilute solution. This behavior of the two species was consistent with the results reported for benzalacetophenone. Unruh found, however, that when the *trans* polymer was photoisomerized to the *cis* form, the expected complete return to *trans* form upon addition of acid failed to result. This fatigue or loss was attributed to small amounts of intrachain dimerization or polymerization taking place in the coiled polymer chains. When the polymer was progressively irradiated as a film, a smooth series of curves following through two isosbestic points resulted. A hypsochromic shift of λ_{max} 50 mμ of ultraviolet absorption occurred. These observations indicated a very rapid, clean conversion of *cis* to *trans* isomer. The rapid shift indicated that isomerization preceded polymerization or dimer-type reactions which led to intra- and interchain crosslinking. The film became insoluble in the coating solvent at levels of less than 1% conversion of *trans* to *cis* polymer. Thus photoisomerization and photocrosslinking reactions compete when such films are irradiated.

(f) *Azo Dye-Bearing Polymers.* Fink [20] described a new class of ionic copolyacrylates to which azo groups were appended. The polyanionic, polycationic and polyamphatic molecules all showed photochromism similar to that of model systems with which the polymers were compared.

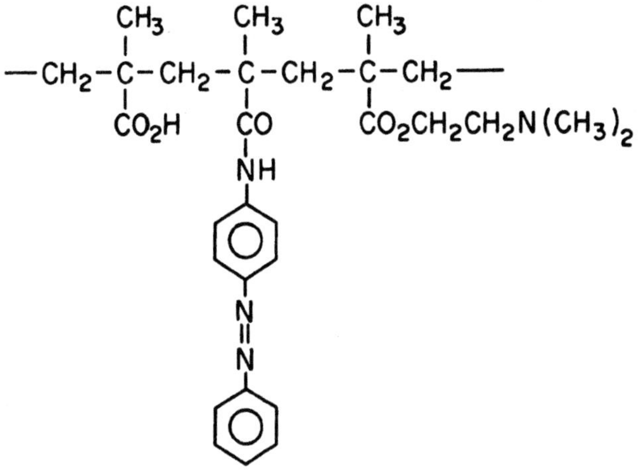

The velocity of the back reaction was determined to be pH dependent.

Paik and Morawetz [21] found that methacrylate or styrene copolymers bearing small numbers of azobenzene or azonaphthalene residues underwent somewhat slower photoisomerization in the glassy state than in solution. The photostationary states were similar in bulk and solution. In analogous polyamides and polyesters, photoisomerization was drastically reduced in bulk.

2.5. Photocrosslinkable polymers

(a) *Structural Variations.* Allen and Van Allan [22] synthesized a polymeric chalcone for use as a photolithographic material from polystyrene and cinnamoyl chloride.

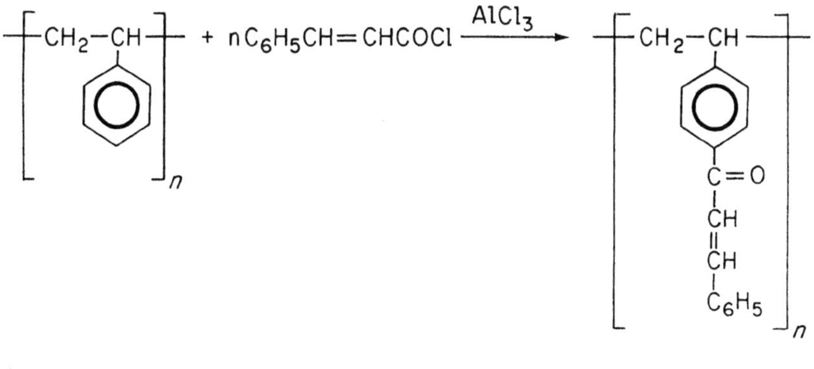

Robertson *et al.* [23] prepared poly(vinyl cinnamate) and found that incorporation of sensitizers such as Michler's ketone into its coatings provided a means of optical sensitization.

Synthesis of a Typical Photoreactive Polymer
Poly(vinylcinnamate)

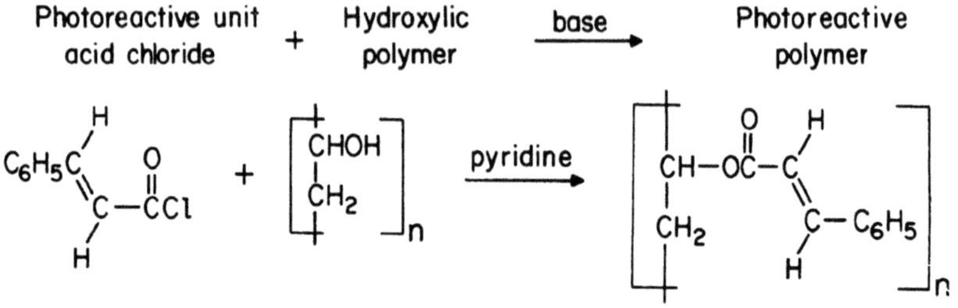

Minsk

Optical sensitization decreased the required exposure time to give imagewise photocrosslinking of poly(vinyl cinnamate).

The basic unit necessary to produce crosslinking in polymers based on poly(vinyl cinnamate) is the $-C{=}C{-}CO-$ unit. This unit can be attached to many polymers in many ways and, depending on its configuration in the polymer chain or when appended to the polymer chain, the light-sensitivity and the physical properties of the polymer can be modified. If a nitro group is placed in the benzene ring of the cinnamoyl residue, the optical response of the polymer can be shifted to longer wavelengths.

Extension of the use of the cinnamoyl unit to photocrosslinkable polymers is found in the numerous chalcone-type polymers prepared by Unruh and Smith [24]. Typical is a group of chalcone polymers which were prepared by reaction of poly(4-vinyl-

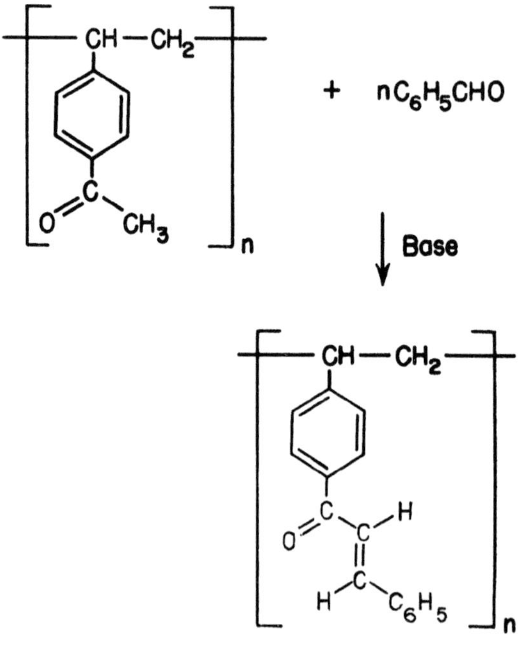

acetophenone) with aromatic aldehydes such as benzaldehyde.

Another quite different group of photocrosslinkable polymers synthesized by Merrill and Unruh [25] was based on the azide group. In this case, an azide group usually attached to an aromatic nucleus is appended to, or incorporated within, a polymer chain. Typical is poly(4-azidostyrene).

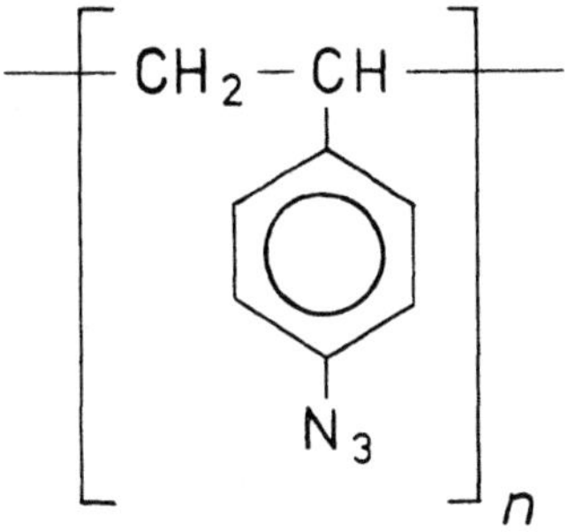

Poly(4-azidostyrene)

The azide polymers have been found to respond quite effectively to sensitization, much in the manner of that found earlier for poly(vinyl cinnamate) [23].

Examples of photosensitive functional groups which have been attached to or incorporated within polymer chains are olefinic, cinnamate, sulfoazide, azido, diazoketone. Units used for attaching or condensing photosensitive groups to polymer chains can be drawn from esters, carbonates, phosphonates, urethanes, ethers, amides and sulfonamides.

(b) *Sensitometry*. A sensitometric method for the evaluation and comparison of light-sensitive polymers was developed by Minsk and co-workers [23]. Coatings of polymers of various compositions or containing different sensitizers were exposed through a step tablet to controlled quantities of light. The unexposed or partially exposed soluble polymer areas were dissolved away by solvents in a developing step. The resulting image was made visible by inking of the residual polymer. By comparison of the minimum amount of light required to produce the weakest appearing step, the comparative speed value of the polymer in question is assessed. Poly(vinyl cinnamate) is taken as a value of 1.0.

Others have used interference microscopic [26] and gravimetric methods [27] to compare the density of the residual polymer on the various steps.

(c) *Spectral Response*. Robertson *et al.* [23] described the use of a wedge spectrograph for the measurement of spectral response. Such a wedge spectrograph consists of a modified Bausch and Lomb monochromator. Comparative spectral response patterns for polymers can be obtained by exposure-development of coatings of various light-sensitive polymers. Wedge spectrograms also provide a very useful and facile method of comparing the effect of the inclusion of other materials, such as sensitizers, in polymer coatings. By this means we can determine whether an addendum produces sensitization or desensitization effects.

(d) *Extension of the Spectral Sensitivity of Photocrosslinkable Polymers.* The range of the spectral sensitivity of light-absorbing photocrosslinkable polymers can be extended in two ways: (i) by modification of the range of light absorption of the crosslinking chromophore in the polymer, or (ii) by inclusion of spectral sensitizers.

(i) *Chromophore modification.* Chromophore modification can be accomplished by extension of the unsaturated chain length. The spectral responses of poly(vinyl acetate cinnamate) and poly(vinyl acetatecinnamylidene acetate) typify this approach.

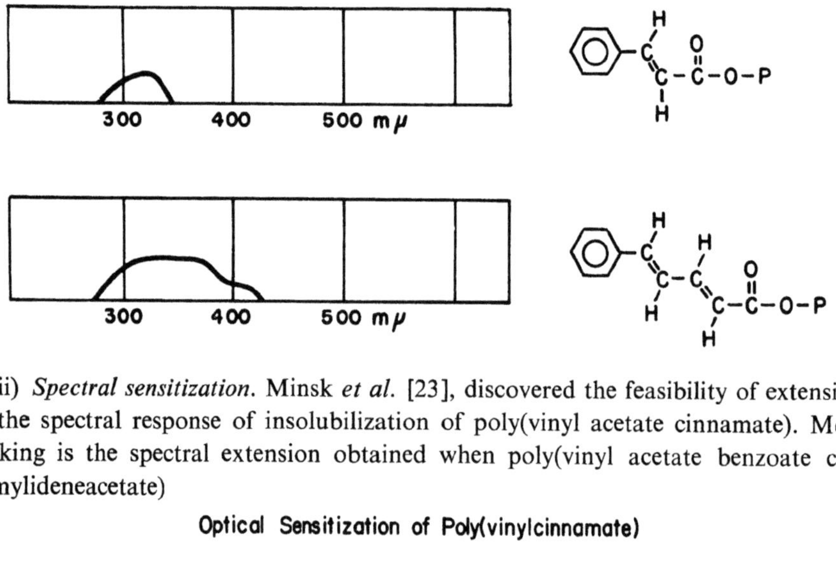

(ii) *Spectral sensitization.* Minsk *et al.* [23], discovered the feasibility of extension of the spectral response of insolubilization of poly(vinyl acetate cinnamate). Most striking is the spectral extension obtained when poly(vinyl acetate benzoate cinnamylideneacetate)

Optical Sensitization of Poly(vinylcinnamate)

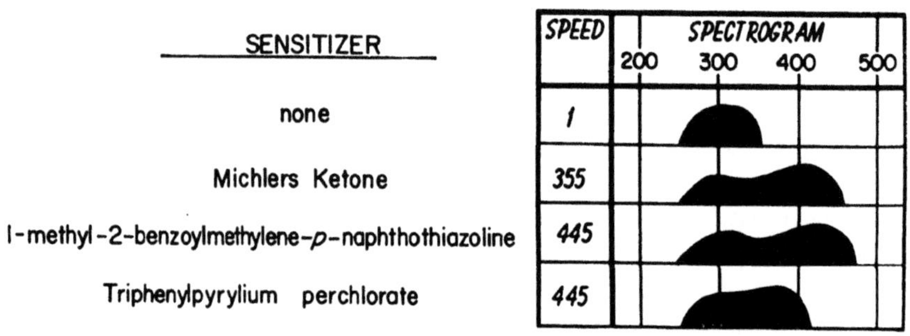

is sensitized with various pyrylium salts.

(e) *Photocrosslinkable Polymers Bearing Ionizable Groups.* A number of photocrosslinkable polymers which bear carboxyl or quaternary amine groups have been prepared. Those bearing carboxyls are generally soluble in dilute alkali or alcohols. A general route to carboxyl bearing polymers is illustrated by use of commercially available copolymers of styrene and maleic anhydride.

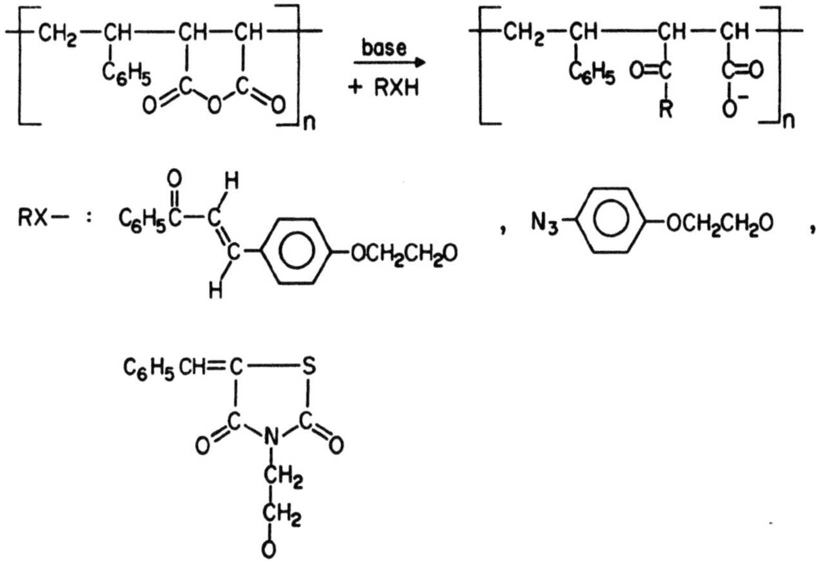

R, the photoreactive group, is attached through the appending group *X* which can be $-NH$ such as in an amide or an oxygen of an ester group. By this means a wide variety of photoreactive groups have been appended to styrene-maleic anhydride copolymers. Merrill and Unruh [28] prepared aromatic azide polymers which are soluble in aqueous bases by reaction of azidophthalic anhydride with poly(vinyl alcohol).

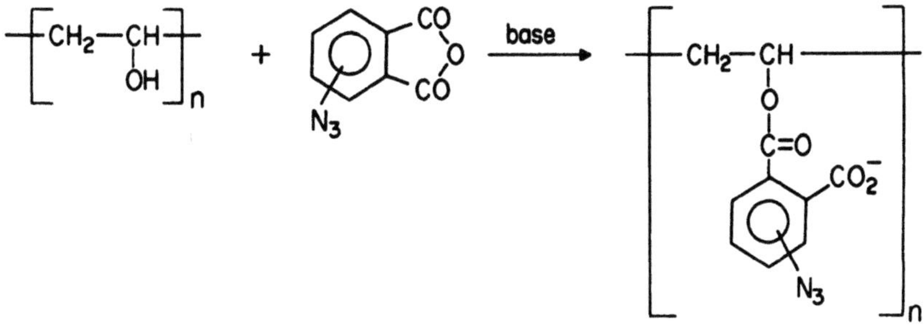

This polymer represents the water soluble equivalent of poly(azidostyrene) and poly(vinyl-*p*-azidobenzoate).

Another approach to the preparation of water soluble polymers has utilized quaternary pyridinium units. Reaction of aromatic aldehydes with poly(5-vinyl-2-methyl pyridinium methosulfate) provide polymers bearing styrylpyridinium units [29].

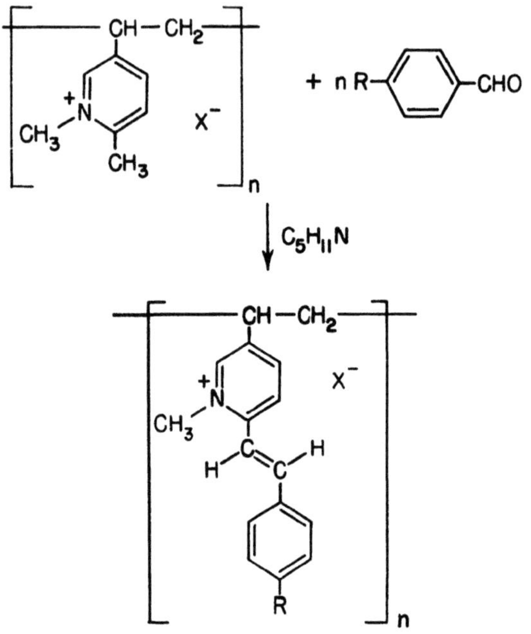

An alternate route to styrylpyridinium polymers [30] uses sulfonate esters of polyvinyl alcohol which are quaternized with alkylpyridines. The quaternary salts are then caused to react with various aldehydes.

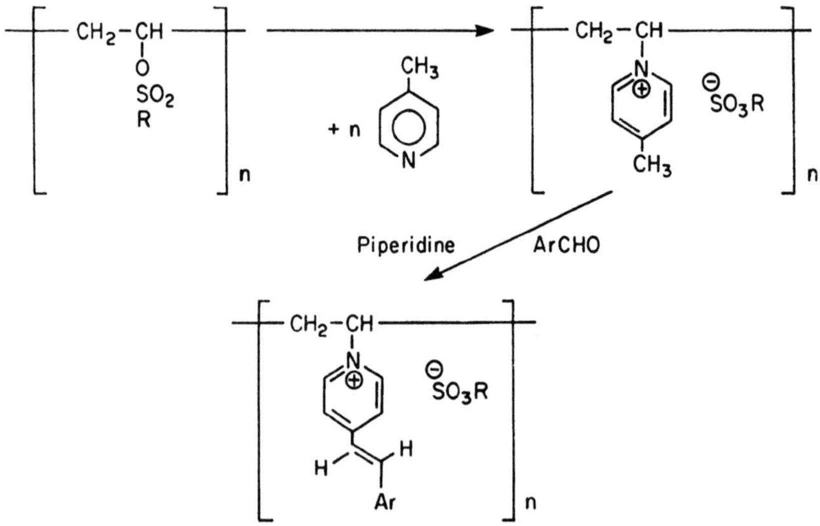

Similar polymers have been prepared by reaction of the salt from 4-picoline and poly(p-chloromethylstyrene) with various aldehydes [29].

2.6. Cooperative effects of polymers and excited states of molecules

In solvents in which a polymer is dissolved, the actual space between polymer molecules which is available for diffusion of another species may be large. The microscopic viscosity in this case will not even approximate the macroscopic viscosity. The available space may be large enough to allow excited receptor species to diffuse at a rapid rate through polymer network as though it were in a pure solvent. Heppel [31] has reported that the translational coefficient of anthracene is unaffected by the presence of polyisobutylene in the solvent. This was determined by the lack of interaction of the polymer at different solution viscosities upon the system of the triplet of anthracene. Even in the gelled polymer, no effect was found. The network is more open than hydrogen bonded viscosity controlled solvent systems, such as glycerol–water mixtures. In such glycerol systems, the microscopic viscosity can be equal to the macroscopic viscosity. However, in dried-down coatings or rigid masses of polymers, the viscosity and network formed can affect the diffusion of molecules as small as oxygen. Buettner [32] has reported this effect for gelatin films containing 2-naphthol.

When a 120 μ thick film of gelatin containing 2-naphthol is flashed repeatedly using a 10-J source, the oxygen quenching is reduced with subsequent flashes.

The film was initially saturated with oxygen by equilibration with the atmosphere and the fast decay after the first flash is due to oxygen quenching of the triplet of 2-naphthol. During the exposures, the dissolved oxygen is consumed at a rate faster than it can be replaced by diffusion from the surface (of the polymer).

At the steady state much less oxygen quenching can occur due to its absence. Further, Buettner found that under a nitrogen atmosphere, the quenching effect could be avoided and that the mean lifetime of the 2-naphthol triplet was 0.47 s at room temperature. At 77 K, McLure found the mean life to be 1.3 s. The rate of decay of the triplet in various polymers in nitrogen and air atmospheres was compared. The triplet of eosin is most stable in polyvinyl alcohol. The curves for polyvinyl alcohol and in gelatin indicates that they are less permeable to oxygen than poly(vinylpyrrolidone) or the copolymer of acrylic acid-ethyl acrylate. This is due to the high degree of hydrogen bonding in the polyvinyl alcohol and gelatin. Anthracene in polystyrene showed no evidence of the triplet until the film had been equilibrated with N_2. Another consequence of the rigid medium was that no self-quenching was observed for any of the above compounds.

Melhuish [33] demonstrated that the decay time of the transient triplet species from benzophenone produced in poly(methyl methacrylate) at 100, 220 and 293 K is the same. This illustrated the utility of polymeric vehicles for flash experiments at various temperatures. Previously, rigid glasses of different chemical composition were required at different temperatures; for example, 77 K-(EPA) and room temperature-borate.

Caution was suggested during the use of polymeric glasses since residual monomer can lead to increased mobilities of phosphorescent species at temperatures above 77 K [34].

Energy transfer from solute to polymer has been demonstrated in a number of systems. As mentioned in the section on photodegradation, inclusion of naphthalene in films of poly(phenyl vinyl ketone) reduces to number of chain scissions by a quenching mechanism [10].

Heskins and Guillet [35] reported a reduction in the number of chain scissions upon irradiation of ethylene-carbon monoxide copolymers when 1,3-cyclooctadiene was present. Complete quenching of the degradative process appeared to be impossible due to the existence of more than one degradative pathway.

Energy transfer from solute-type optical sensitizers appears to remain ill defined. The subject has been dealt with in some detail earlier [36].

Energy transfer from polymeric sensitizers such as poly(vinylbenzophenone) has been undertaken in order to examine whether molecular size influenced the ability to transfer energy to acceptor molecules. To this end we have investigated the ability of poly(vinylbenzophenone) to sensitize the solution photoisomerization of *cis*- and *trans*-stilbene [37].

Poly(vinylbenzophenone) was synthesized by the benzoylation of polystyrene with benzoyl chloride [38]. The number-average molecular weight of the polystyrene used was found to be $68 \pm 7 \times 10^3$ by membrane osmometry. A solution of poly(vinylbenzophenone) in benzene at a concentration of 1.20 g l^{-1} showed a wavelength of λ_{max} at 343 mμ with an absorbance of 0.868. Under similar conditions, 4-methylbenzophenone had a λ_{max} at 343 mμ with $\varepsilon = 169$. Since the latter is an appropriate model for the polymeric chromophore, the polystyrene was approximately 75% benzoylated. We observed no significant difference in sensitizing effect or optical behavior between the two samples. From the phosphorescence spectrum at 77 K of a film cast from benzene, the triplet excitation energy of the polyvinylbenzophenone was determined to be 66 kcal mole^{-1} and that of the model compound, 4-methylbenzophenone, dissolved in a polystyrene film, was also 66 kcal mole^{-1}. Thus, the poly(vinylbenzophenone) triplet possesscs sufficient energy to transfer its excitation to either *cis*- or *trans*-stilbene at the diffusion controlled rate [39].

No change in the quantum efficiency of the benzophenone sensitized solution photoisomerization of stilbene was found when the sensitizer was attached to a polymer chain. The quantum yield for the *trans* to *cis* (poly(vinylbenzophenone)-sensitized photoisomerization of stilbene, $\Phi\ t \rightarrow c$, was 0.42 ± 0.02, 0.44 ± 0.02 at 4.3 and 4.8% conversion. For the *cis* to *trans* poly(vinylbenzophenone)-sensitized reaction, the quantum yield $\Phi\ c \rightarrow t$, was 0.41 ± 0.04 and 0.44 ± 0.02 at 9.3 and 8.7% conversion. When 4-methylbenzophenone was the sensitizer, $\Phi\ t \rightarrow c$ was 0.44 ± 0.02 at 5.3% conversion. Hammond reported [3], Φ, 0.43 ± 0.01 for the latter isomerization.

If efficient energy transfer was possible between the benzophenone units attached to the polymer chain, the effective size of the polymeric sensitizer should be increased by at least a factor of ten relative to that of the monomeric model compound. Efficient energy transfer between nonconjugated but proximate groups has been observed in other systems [40]. According to the theory of the diffusion of molecules in solution, some increase in quantum efficiency should be expected since collision frequency is a function of the sizes of the colliding molecules [41]. Since no increase in quantum efficiency was found, either one or both of two situations must obtain: (1) electronic energy transfer between adjacent benzophenone units does not occur or (2) the increase in effective size of the sensitizer is exactly compensated by its deviation from the spherical molecule requirement of the diffusion theory [41].

In another study [42] polystryrene was naphthoylated to give poly(α-naphthoylstyrene) and poly(β-naphthoylstyrene). The solution photoisomerization of stilbene by the polymer was found to be identical with that resulting from the use of the corresponding ethylnaphylphenyl ketone models. Moser and Cassidy [43] found earlier but reported in an obscure manner that poly(phenyl vinyl ketone) caused the photoisomerization of *cis*-piperylene. The isomerization proceeded both in solution and by use of solid polymer. The remarkable effect of the solid was attributed to rapid diffusion of *cis*-piperylene to the polymer surface.

(7) *Photochemical Behavior of a Quaternary Photocrosslinkable Polymer System and Its Model Compounds*

In his studies of photocrosslinkable polymers, Unruh [19] indicated that the usual chromophoric groups present in such macromolecules are capable of undergoing photoisomerization as well as reactions which lead to photocrosslinking. The distinction between the occurrence of a low-degree-of-polymerization-type mechanism and a 'photodimerization' mechanism of photocrosslinking is of basic importance to understanding such systems. Many model systems which readily dimerize can simultaneously photoisomerize. When bulky groups are present in the *trans*-isomer of a molecule, or more often on the bonds which dimerize, hindrance to the act of dimerization can result. Isomerization to the *cis*-isomer can provide a less hindered route to information of the desired bonds in the dimer structure. As a beginning to investigating the role of photoisomerization in polymer systems we have synthesized and carried out photochemical studies on a number of model compounds. The structural relationship between the model compounds and the macromolecules is easily accomplished. However, the environment of a low molecular weight model compound in dilute solution differs vastly from that of the same model unit when it is attached to or incorporated into a macromolecule. Even in dilute solution a photoreactive group which is attached to a macromolecule is held in close proximity to its nearest neighbor. In films, the units lie even closer. The distance between and the frequency of occurrence of such neighboring groups is determined by the degree of substitution. In charged macromolecules, the distance between groups can be regulated by the degree of coiling of the macromolecule. In coiled macromolecules the situation can be more complex due to the fact that the chains can fold back on themselves. Realizing such limitations do exist, we have begun studies of such systems, using polymers which bear 2-styrylpyridinium groups along their chains.

We have reported [44] the photochemical behavior of a number of 2- and 4-styrylpyridines and their corresponding methiodides which were studied in order to gain insight into the behavior of analogous polymers by means of the model systems.

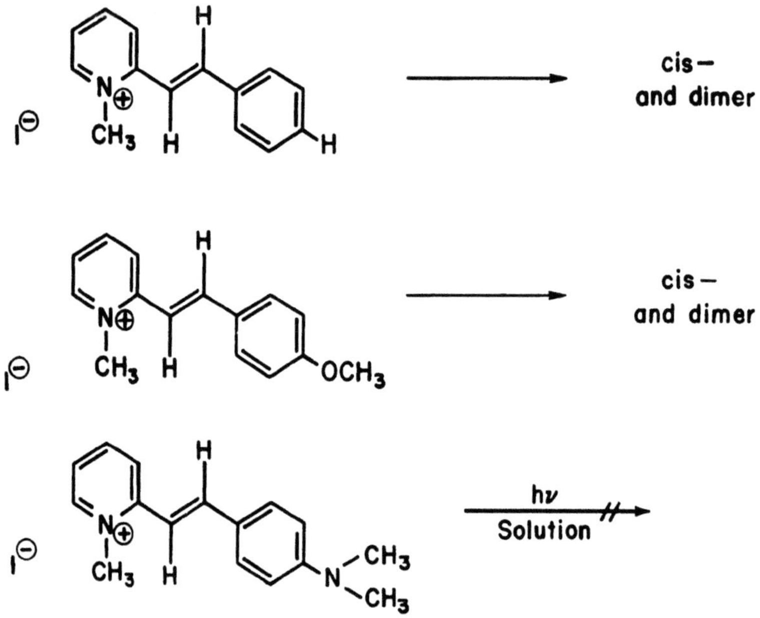

We wish to report the results of a further study which has been extended to three polymeric systems. These systems are based on the polymeric salts obtained by the condensation of the polymer of 2-methyl-5-vinyl-pyridinium methosulfate with benzaldehyde, anisaldehyde and *p*-dimethylaminobenzaldehyde [29] (see Figure 18). Since reaction of the aldehydes with the polymeric salt never approaches 100%, the resulting polymers are, in fact, copolymers. In the present work we have investigated the absorption and emission spectra of the three following *trans*-copolymers, where *R* equals H(**1**), CH_3O(**2**) and $(CH_3)_2N$(**3**).

We have compared the ultraviolet-visible absorption and emission characteristics of the model compounds (**4**, **5**, **6**) with those of the copolymers (**1**, **2**, **3**). The comparisons were conducted in methanol solution and in coating of the polymers upon quartz. In order to compare the model compounds (**4**, **5**, **6**) with the polymers, as coatings, mixtures of the model compounds (**4**, **5**, **6**) in the parent polymer, poly-(5-vinyl-2-methylpyridinium methosulfate), were coated upon quartz plates. The comparative data of the absorption and emission characteristics of the various experimental combinations are summarized in Table I.

The data in Table I indicate that the chromophores in the model compounds are very sensitive to environment. The solvent has a great effect on the wavelengths of maximum absorption and emission both when the chromophores are in the model systems and attached to the polymer chains. If the small bathochromic shift determined in dilute solutions which occurs when the chromophore is attached to a polymer is ignored, it is then possible to estimate the degree of reaction of the corresponding aldehydes with the parent polymer. The degree of substitution determined in this

TABLE I

Spectra of 5-substituted 2-styrylpyridinium model compounds and polymers

Absorption spectra

	Model compound[a]				Polymers			
	MeOH		Coating		MeOH		Coating	
	λ_{max}	A	λ_{max}	A	λ_{max}	A	λ_{max}	A
H	343	0.73	340	1.39	345	0.70	348	1.03
CH_3O	375	0.82	368	1.25	378	1.26	375	1.61
$N(CH_3)_2$	453	0.86	438	0.93	458	0.99	453	0.92

Fluorescent spectra

	λ_{max}			λ_{max}			λ_{max}			λ_{max}		
	Ex.	Em.	In.	Ex.	Em.	In.	Ex.	Em.	In.	Ex.	Em.	In.
H	350	420	79	376	470	179	370	500	51	370	500	357
OCH_3	390	470	48	390	530	348	410	580	580	390	570	2030
$N(CH_3)_2$	465	590	1690	465	600	1340	465	610	55	465	690	381

[a] Model compound coated in poly(2-methyl-5-vinylpyridine methosulfate).

manner is as follows: **1**, 35%; **2**, 32%; and **3**, 30%. It may be pointed out that there occur the following hyposochromic shifts of the wavelengths of λ_{max} of absorption when the model compounds are contained in the matrix of a coating of the parent polymer (**4**, 3 mμ; **5**, 7 mμ; and **6**, 5 mμ).

The absorption spectra of the corresponding polymers **1**, **2**, and **3** show less change when compared in methanol solution to that in film form. The shifts of λ_{max} for both absorption and emission spectra of the hydrogen and methoxyl-substituted model compounds as well as the corresponding polymers are quite consistent. The behavior of the absorption and emission spectra associated with the presence of the 4-dimethylamino group in the model as well as the corresponding polymer is atypical. This is related, we believe, to the failure of the 4-dimethylamino-substituted styrylpyridinium iodide to undergo photoisomerization and dimerization in a manner analogous to the other 4-substituted styrylpyridinium iodides. [45]

We have shown that 4-dimethylamino-2-styrylpyridinium methiodide is photochemically inert. It is therefore surprising that the corresponding polymer, **3**, should be readily crosslinked by light. This is especially true if one assumes a dimerization-type photocrosslinking mechanism to be involved. Wedge spectrograms, as well as the glass factors of polymers **1**, **2** and **3**, were obtained in the usual manner [23]. For purposes of comparison the wedge spectrogram of poly(vinyl cinnamate) is shown together with those of the three polymers **1**, **2** and **3**.

In an earlier report [46], it was shown that 2-styrylpyridine methiodide in water

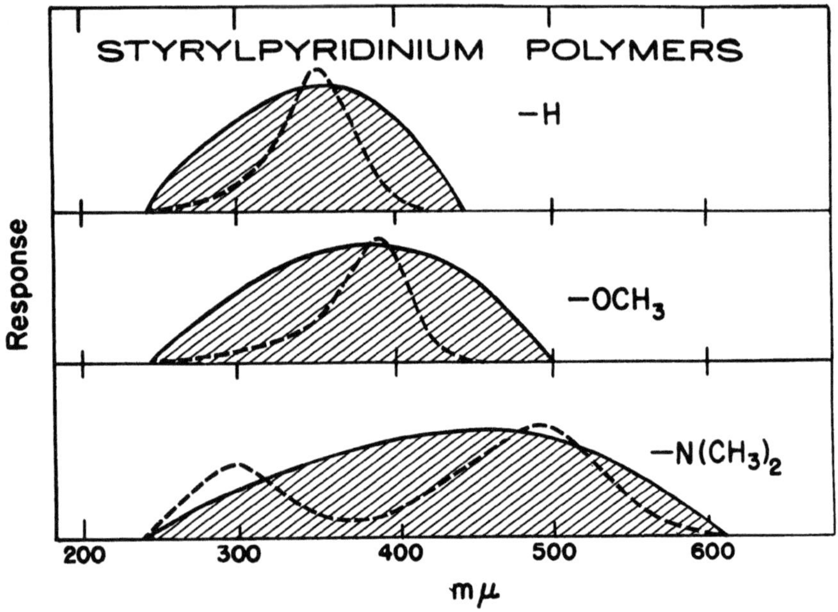

solution could be phototropically cycled without fatigue in the presence of air by choice of the proper wavelengths of the exciting light.

It was of interest to us to find out whether or not the present polymers **1**, **2** and **3** would undergo photoisomerization and phototropic cycling. Since the earlier work had been done using the methiodide salt, we first set out to determine if the anion and the solvent played any part in the process of photoisomerization. The data in Table II illustrate the fact that the nature of the anion plays no role in the reactions when solutions of *trans*-2-styrylpyridinium are converted to the corresponding *cis*-isomers by irradiation. It can also be seen from the data in Table II that the solvent determines the position of the wavelength of λ_{max} but has very little effect on the size of the shifts of λ_{max} of absorption when one compares these values for the *trans*- and *cis*-isomers.

In another study we have found that aqueous solutions of the model compounds **4** and **5** bearing the 5-ethyl group can be phototropically cycled without fatigue. Compound **6** is inert to light under our conditions of study [45]. Thus it appeared possible to us that the occurrence of photoisomerization during photolysis of the corresponding polymers might be detected by cycling experiments. In the model systems the average difference in the wavelength of λ_{max} between pairs of *cis*- and *trans*-isomers of a large group of substituted 2-styrylpyridine methiodides was 15 mμ. However, during phototropic cycling the cycles were never between 100% *cis*- and 100% *trans*-isomer solutions since the shift of λ_{max} was usually about 6 mμ.

We expected a somewhat similar behavior of the model compounds and the cor-

TABLE II

Absorption spectra of quaternary salts of 2-styrylpyridine

Quaternary salt	Solvent	*trans*-isomer		*cis*-isomer		$\Delta\lambda_{max}$
		λ_{max}	A	λ_{max}	A	max mμ
Methosulfate	H_2O	335	0.66	322	0.26	13
	CH_3OH	338	0.7	323	0.26	15
	DMF	340	0.84	325	0.27	15
	Chloroform	343		328		15
Methyl *p*-toluene-sulfonate	H_2O	335	0.74	324	0,29	14
	CH_3OH	340	0.75	325	0.28	15
	DMF	340	0.67	325	0.21	15
	Chloroform	342		328		14
Methyl benzene-sulfonate	H_2O	335	0.73	322	0.29	13
	CH_3OH	340	0.78	325	0.25	15
	DMF	342	0.73	325	0.24	15
	Chloroform	343		328		15
Methiodide	H_2O	333	0.69	320	0.38	13
	CH_3OH	340	0.77	325	0.35	15

responding polymers when they each were studied in dilute solutions (2×10^{-5} molar in water). We erroneously considered only the gross macroscopic concentration and not the fact that the photoisomerizable-photodimerizable groups were held close to each other by the backbone of the polymer chain. The repulsion between the ions along the polymer chain depends on the degree of extension of the macromolecule chain. Addition of a common ion caused as much as tenfold decrease in the value of the inherent viscosity measured in water. These polymers are thus well suited to be used for studies of intra- and intermolecular reactions. Thus, despite the total macroscopic concentration of the polymer in solution, the locale of microscopic concentration cannot be controlled other than by the degree of substitution along the chains. Therefore in even dilute solution the distance to the nearest reactive neighboring group is controlled by the polymer chain substitution. With the exception of the dimethylamino-substituted polymer, **3**, the polymers **1** and **2** photocycled in the presence of oxygen in a manner similar to that of the model compounds. When the methoxy-substituted polymer (**2**) was subjected to the cycling technique, it was possible to only repeat the cycling five times before a 30% loss of total extinction occurred. Polymer **1** was cycled three times before a 50% loss of total extinction occurred. We believe this failure of the polymer to cycle is due to the fact that the local concentration of nearest neighbors along the chain is very high compared to the concentration of nearest neighbors of the model compound in solution. Intrachain reactions result. The corresponding dimethylamino-substituted polymer (**3**), like the model compound is inert to light in solution.

We had shown earlier [45] that despite the fact that the model compounds 4′-

dimethylamino-2-styrylpyridine methiodide (**7a**) and 4′-diethylamino-2-styrylpyridine methiodide (**8a**) were photochemically inert, their hydrochlorides (**7b** and **8b**) could be transformed photochemically into the *cis*-hydrochlorides (**7c** and **8c**). The *cis*-hydrochlorides, **7c** and **8c**, were then treated with sodium bicarbonate to yield the *cis*-isomers of the free methiodides **7d** and **8d**.

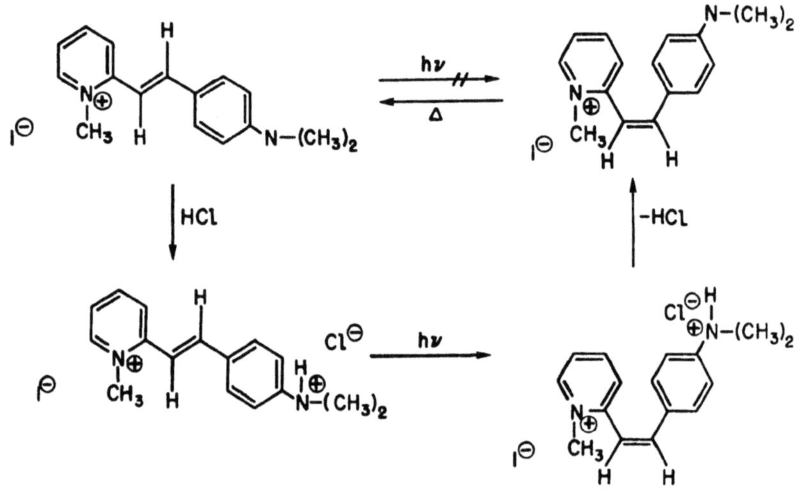

When polymer **3** was treated in a similar manner in acid solution it was transformed into a photolabile species which could be isomerized to the *cis*-isomer.

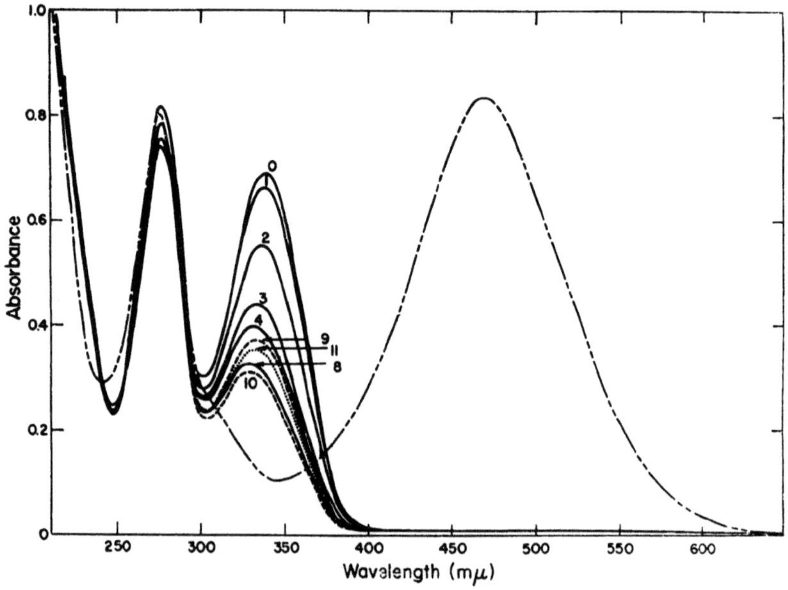

The transformations which were carried out on polymer **3** are shown in the following figure: ----, is polymer **3**; curve 0 is the absorption spectrum of polymer **3** under acid conditions before irradiation. The solid curves, 0–8, therefore, represent a series of cumulative irradiations of the acidic solution using 365 nm light to produce reduction of absorbance to the level shown in curve 8. Curve 9 shows the increase in absorbance resulting from a short irradiation with 254 nm light. Curve 10 shows the reduced absorbance after irradiation with 365 nm light while curve 11 shows the regain of absorbance on irradiation with 254 nm light.

Again, because of the onset of fatigue the reaction could not be cycled. The model compound and the polymer are very weak bases and thus low pH conditions were necessary to completely convert the methosulfate-free-base to the methosulfate-hydrochloride. As coatings, the polymers insolubilize very rapidly with a loss of absorption by the chromophores. They, however, become insoluble before a change of absorbance can be detected. Furthermore, though the methoxyl substituted polymer is more light sensitive (in terms of insolubilization) than is the dimethylamino derivative, it is most surprising that the dimethylamino is more so than the unsubstituted polymer. This is despite the fact that while the unsubstituted and methoxyl substituted model compounds undergo photoisomerization, the dimethylamino model compound does not.

TABLE III

Exposure needed to crosslink polymer **2** versus destruction of chromophore

Exposure (s)	Relative initial intensity	Relative final intensity	Percent decrease	Developed intensity	Percent uncrosslinked
10	1230	1220	0.8	450	63
10	1200	1185	1.0	460	61
20	1250	1210	3.2	680	44
20	1250	1210	3.2	700	42
30	1165	1090	6.4	855	22
30	1155	1110	3.0	885	21
40	1130	1070	5.3	930	13
40	1110	1035	6.8	1020	10
60	1120	1060	9.4	1060	0

Thus we believe that the analogy between model systems and the corresponding polymer systems cannot be carried beyond the dilute solutions. As further illustration of this we have summarized the changes in fluorescent spectrum of coating of polymer **2** which were given cumulative amounts of exposure to light. After each exposure the weight of soluble noncrosslinked material was determined and compared with the intensity of the remaining crosslinked material in Table III.

References

1. Delzenne, G. A.: *European Polymer J. Suppl.*, 55 (1969).
2. Borden, D. G.: SPSE Symposium, October 20, 1971, Washington, D.C.
3. Kosar, J.: *Light Sensitive Systems*, Wiley, New York, 1965.
4. Anderson, J. C. and Reese, C. B.: *Proc. Roy. Soc.*, 217 (1960).
5. Anderson, J. C. and Reese, C. B.: *J. Chem. Soc.*, 1781 (1963).
6. Kobsa, H.: *J. Org. Chem.* **27**, 2293 (1962).
7. Maerov, S. B.: *J. Polymer Sci.* **3**, 487 (1965).
8. Heller, H. J.: *European Polymer J. Suppl.*, 105 (1969).
9. David, C., Demarteau, W., and Geuskins, G.: *Polymer* **8**, 497 (1967).
10. David, C., Demarteau, W., and Geuskins, G.: *European Polymer J.* **6**, 1405 (1970).
11. Golemba, F. J. and Guillet, J. E.: *Macromolecules* **5**, 213 (1972).
12. Oster, G. and Shibota, O.: *J. Polymer Sci.* **26**, 233 (1957).
13. Smets, G., DeWinter, W., and Delzenne, G.: *J. Polymer Sci.* **55**, 767 (1961).
14. Dhandraj, J. and Guillet, J. E.: *J. Polymer Sci.* **C23**, 433 (1968).
15. Kay, R. E., Walwick, E. R., and Gifford, G. K.: *J. Phys. Chem.* **68**, 1896 (1964).
16. Lovrien, R. and Waddington, J. C. B.: *J. Am. Chem. Soc.* **36**, 2315 (1964).
17. Van der Veen, G. and Prins, W.: *Nature* **230**, 70 (1971).
18. Vanderwijer, P. H. and Smets, G.: *J. Polymer Sci.* **22**, 231 (1968).
19. Unruh, C. C.: *J. Appl. Polymer Sci.* **2**, 358 (1959).
20. Fink, H.: *Plaste and Kautschuk* **9**, 645 (1971).
21. Paik, C. S. and Morawetz, H.: *Macromol.* **5**, 171 (1972).
22. Allen, C. F. H. and Van Allan, J. A.: U.S. Patent 2, 566, 302, Sept. 4, 1951.
23. Robertson, E. M., Van Deusen, W. P., and Minsk, L. M.: *J. Appl. Polymer Sci.* **2**, 308 (1959).
24. Unruh, C. C. and Smith, Jr., A. C.: *J. Appl. Polymer Sci.* **3**, 310 (1960).
25. Merrill, S. H. and Unruh, C. C.: *J. Appl. Polymer Sci.* **7**, 273 (1963).
26. Lyalikov, K. S. *et al.*: *Zhur. Nauchn. i Prikl. Fot. i Kinemat.* **10**, 200 (1965).
27. Yashinago, T., Kan, N., and Kikuchi, S.: *J. Chem. Soc. Japan* (Pure Chem. Sec.) **66**, 665 (1963).
28. Unruh, C. C. and Merrill, S. H.: U.S. Patent 3,096,311, July 2, 1963.
29. Leubner, G., Williams, J. L. R., and Unruh, C. C.: U.S. Patent 2, 811, 510, Oct. 29, 1957.
30. Williams, J. L. R. and Borden, D. G.: *Makromolek. Chem.* **713**, 203 (1964).
31. Heppel, G. E.: *Photochem. Photobiol.* **4**, 7 (1965).
32. Buettner, A. V.: *J. Phys. Chem.* **68**, 3253 (1964).
33. Melhuish, W. H.: *Trans. Farad. Soc.* **62**, 3384 (1966).
34. Woods, R. J. and Manville, J. F.: *Can. J. Res.* **49**, 0515 (1971).
35. Heskins, M. and Guillet, J. E.: *Macromolecules* **1**, 97 (1968).
36. Williams, J. L. R.: *Fortschr. Chem. Forsch.* **13**, 227 (1969).
37. Searle, R. *et al.*: *Makromol. Chem.* **107**, 246 (1967).
38. Merrill, S. H. and Unruh, C. C.: U.S. Patent 2, 831, 768, April 22, 1958.
39. Hammond, G. S., Saltiel, J., Lamola, A. A., Turro, N. J., Bradshaw, J. S., Cowan, D. O., Counsell, R. C., Vogt, V., and Dalton, C.: *J. Am. Chem. Soc.* **86**, 3197 (1964)
40. Lamola, A. A., Leermakers, P. A., Byers, G. W., and Hammond G. S.: *J. Am. Chem. Soc.* **87**, 2322 (1965).
41. Debye, P.: *Trans. Electrochem. Soc.* **82**, 265 (1942).
42. Hammond, H. A., Doty, J. C., Laakso, T. M., and Williams, J. L. R.: *Macromolecules* **3**, 711 (1970).
43. Moser, R. E. and Cassidy, H. G.: *Polymer Letters* **2**, 545 (1964).
44. Williams, J. L. R., Carlson, J. M., Reynolds, G. A., and Adel, R. E.: *J. Org. Chem.* **28**, 1317 (1963).
45. Williams, J. L. R., Doty, J. C., Adel, R. E., and Reynolds, G. A.: *Can. J. Res.* **43**, 1345 (1956).
46. Williams, J. L. R.: *J. Am. Chem. Soc.* **84**, 1323 (1962).

INDEX OF SUBJECTS

GPSR Compliance

The European Union's (EU) General Product Safety Regulation (GPSR) is a set of rules that requires consumer products to be safe and our obligations to ensure this.

If you have any concerns about our products, you can contact us on ProductSafety@springernature.com

In case Publisher is established outside the EU, the EU authorized representative is:

Springer Nature Customer Service Center GmbH
Europaplatz 3
69115 Heidelberg, Germany

Batch number: 10815974

Printed by Printforce, the Netherlands